PLASTIC PIPING HANDBOOK

PLASTIC PIPING HANDBOOK

David A. Willoughby, P.O.E.
R. Dodge Woodson
Rick Sutherland

McGRAW-HILL
New York Chicago San Francisco Lisbon London
Madrid Mexico City Milan New Delhi San Juan
Seoul Singapore Sydney Toronto

McGraw-Hill
A Division of The McGraw-Hill Companies

1 2 3 4 5 6 7 8 9 0 DOC/DOC 0 9 8 7 6 5 4 3 2 1

ISBN 0-07-163400-7

The sponsoring editor for this book was Larry S. Hager and the production supervisor was Sherri Souffrance. It was set in Times Roman by Lone Wolf Enterprises, Ltd.

Printed and bound by R. R. Donnelley & Sons Company.

This book is printed on recycled, acid-free paper containing a minimum of 50% recycled, de-inked fiber.

CONTENTS

PREFACE

This is a reference handbook for the engineer, designer, technician, contractor, and utility manager involved in the design, installation, and operation of plastic piping systems. The goal of this handbook is to provide easy-to-access and essential information about the use of plastic pipe, especially in the gas and water industries. There are many manufacturers manuals and documents from organizations such as the Plastic Pipe Institute that provide valuable information concerning the use of plastic pipe. This handbook provides a single text that contains much of the frequently required data and procedures used in the design and installation of plastic pipe systems. This handbook presents the practical aspects of the use of plastic piping systems. It provides the reader with many frequently used equations, charts, and guidelines.

This book is primarily a reference book. It contains many tables that will guide the reader to understand many of the issues about plastic piping systems and sources of additional information. The book also provides example specifications for a natural gas plastic pipe system and horizontal directional drilling with plastic pipe. This book is a ready reference for the field and office.

Thanks to George Fischer Inc. for the use of figures from the *George Fischer Engineering Handbook*.

I hope that readers will find this handbook useful. I invite all readers to suggest any additional data that they would like to see covered in revised versions and anything that they feel will improve this handbook.

David Willoughby, P.O.E.

ABOUT THE AUTHOR

David A. Willoughby, P.O.E.

Mr. Willoughby has 21 years of experience in engineering, technical writing, and management in the natural gas, pipeline, and petroleum facility industry. His experience includes gas transmission and distribution systems, petroleum facilities from conception through design, construction and testing, economic evaluation, and project field supervision. Mr. Willoughby is registered by the Council of Engineering Specialty Boards as a Petroleum Operations Engineer.

Mr. Willoughby is currently a Projects Manager for the consulting engineering firm of Rummel, Klepper & Kahl. Mr. Willoughby works in the Natural Gas, Petroleum, and Pipeline Division out of RK&K's Richmond, Virginia office.

Prior to working for RK&K, Mr. Willoughby served for twenty-four years in the United States Marine Corps, with his last eighteen years of service as a Petroleum Engineer. During this period, Mr. Willoughby held increasingly responsible positions in project management, petroleum operations and engineering, hydraulic design and analysis, pipeline design and operation, and petroleum facility design and operations.

Mr. Willoughby lives in Prince George County, Virginia with his wife, Betsy. They have three daughters, Cheryl, Shannon, and Sarah.

Mr. Willoughby can be reached at his email address; dwilloughby@asme.org.

CHAPTER 1
GENERAL INFORMATION

INTRODUCTION

The purpose of this handbook is to provide accurate and reliable information concerning the application, design, and installation of plastic pipe for water and gas systems.

Thermoplastic piping is the material that has the widest range of applications. Thermoplastic piping includes many materials that have significant differences in characteristics and uses. It is important that the correct thermoplastic material be specified for the various applications. Because of the frequent use of polyethylene (PE) and polyvinyl chloride (PVC) pipe material in the water and gas markets, this handbook will focus primarily on these types of plastic pipe. Other types of plastic pipe and their applications will be introduced to provide the reader with a background in the various possible uses of the material. The design and installation information, however, will deal primarily with PE and PVC pipe.

Each project is different and can have unique conditions. A design or installation necessity for one project might be excessive for another project. The ways the engineer and designer interpret and approach the various conditions are important to achieve an effective and efficient project. The proper design and installation of plastic piping systems require the use of sound engineering judgment and principles. It is the goal of this handbook to provide the information needed by designers, engineers, and installation personnel working in the water and gas fields.

Plastic piping has many applications in today's marketplace and its popularity continues to grow. It is used in a variety of commodities such as acid solutions, chemicals, corrosive gases, corrosive waste, crude oil, drainage, fuel gases, mud, sewage, sludge, slurries, and water. One major reason for the growth in the use of plastic pipe is the cost savings in installation, labor, and equipment as compared to traditional piping materials. Add to this the potential for lower maintenance costs and increased service life and plastic pipe is a very competitive product. The popularity of plastic pipe in the water and natural gas industry has played a significant role in the growth of the industry. The shipment of PE products alone increased by 26 percent from 1996 to 1997 [1].

HISTORY OF PLASTIC PIPE MATERIALS

Plastics have been in use for more than 100 years, and polyethylene, the primary plastic pipe used in the natural gas industry, was invented in the 1930s. Early polyethylenes were low density and were used primarily for cable coatings. World War II provided a catalyst for the development and use of plastic products, largely because of the shortage of other materials. Today's modern polyethylene piping systems began with the discovery of high-density polyethylene in the early 1950s [2].

COMMON APPLICATIONS

Thermoplastics make up the majority of plastic pipe in use today. PVC accounts for the majority of the thermoplastic pipe in use, with PE coming in second. Although thousands of miles of plastic pipe are in service in natural gas and municipal applications, many other uses also exist. Some of the other common uses of plastic piping are:

- Chemical processing
- Food processing
- Power plants
- Sewage treatment
- Water treatment
- Plumbing
- Home fire and lawn sprinkler systems
- Irrigation piping

Detailed information about various piping products and their applications can be obtained from the Plastic Pipe Institute and plastic pipe manufacturers.

In the last 25 to 30 years, plastic piping products have become the predominant piping materials in many markets. As a result of the high demand, the availability and types of plastic piping products in many materials and sizes have increased significantly. This increase provides the piping engineer with many products to choose from when specifying plastic piping products. To select the best product for the desired application, the engineer and designer must have a good knowledge of the plastic piping products available.

DEFINITIONS AND ABBREVIATIONS

adhesive joint: A joint in plastic pipe made by an adhesive substance that forms a continuous bond between the materials without dissolving either of them.
ambient temperature: The prevailing temperature in the surrounding medium usually refers to the temperature of the air surrounding an object.
anchor: A rigid device used to secure the pipe, permitting neither translatory nor rotational displacement of the pipe.
angle of bend: The angle between the radial lines from the beginning and end of the bend to the center.
backfill: The material that is placed around and over the pipe after trench excavation.

primary initial backfill: This part of the backfill supports the pipe against lateral pipe deformation.

secondary initial backfill: This part of the backfill distributes overhead loads and isolates the pipe from any adverse conditions encountered during the placement of the final backfill.

final backfill: The final material inserted in the trench to complete the fill from the initial backfill to the top of the trench.

ball valve: A valve with a ball-shaped disk that has a hole through the center, providing straight-through flow.
blind flange: A flange used to close the end of a pipe.
block valve: A valve used for isolating equipment.
burst pressure: The pressure that can be applied slowly to plastic pipe or component at room temperature for 30 seconds without causing rupture.
burst strength: The internal pressure required to break a pipe or fitting. This pressure will vary with the rate of buildup and the time the pressure is maintained.
butt fusion: A method of joining thermoplastic pipes and components that involves heating the ends of two pieces that are to be joined and quickly pressing them together.
butt joint: A joint between two pipe components in the same plane.
butterfly valve: A valve that gets its name from the wing-like action of the disk.
bypass valve: A valve and loop used to direct the flow in a pipeline around some part of the system.
check valve: A device that allows flow in one direction only in a pipeline.
coefficient of expansion: The increase in unit length, area, or volume for a unit rise in temperature.
compression fitting: A fitting used to join a pipe by pressure or friction.
compression joint: Multi-piece joints with cup-shaped threaded nuts that compress sleeves when tightened so they form a tight joint.
compression strength: The failure crushing load of a pipe or component divided by the number of square inches of resisting area.
control piping: All piping, fittings, and valves used to connect control devices to the piping system components.

creep: Time-dependent strain caused by stress. Creep is a dimensional change with respect to time caused by a load over the elastic deformation.
density: The mass of a substance per unit volume.
depth of fusion: The distance that a fusion extends into the base material.
deterioration: The permanent adverse change in the physical properties of a plastic.
dimension ratio: The diameter of a pipe divided by the wall thickness.
elasticity: The material property that tends to retain or restore the materials original shape after deformation.
elastomer: A material that, under ambient conditions, can be stretched and returns to approximately the original size and shape after the applied stress is released.
elevated temperature testing: Test on plastic pipe above 73°F.
environmental stress cracking: Cracks that develop when the material is subjected to stress in the company of certain chemicals.
expansion joint: A piping component used to absorb thermal movement.
expansion loop: A bend in a pipe run that adds flexibility to the piping system.
flexural strength: The pressure (psi) required to break a piping sample when the pressure is applied at the center and the pipe is supported at both ends.
full port valve: A valve that, when in the fully open position, is equal to an equivalent length of pipe.
gate valve: A valve that opens to the complete cross section of the line. Under most conditions, a gate valve is not used for throttling or control of the flow. It usually is used for complete open or complete shutoff of the fluid flow.
globe valve: A valve used for throttling or control.
haunching: The area from the trench bed to the spring line of the pipe. Provides most of the load bearing for buried piping.
heat joining: The making of a pipe joint in thermoplastic piping by heating the ends of both sections so they fuse when the parts are pressed together.
incomplete fusion: A fusion that is not complete and does not result in complete melting throughout the thickness of the joint.
joint: A connection between two sections of pipe or between a section of pipe and a fitting.
long-term burst: The internal pressure at which a pipe or fitting will fail due to constant internal pressure held for 100,000 hr.
nominal Pipe Size (NPS): A dimensionless designator of pipe size. It indicates standard pipe size when followed by the specific size designation number without an inch symbol (e.g., NPS 2, NPS 10) [3].
non-rigid plastic: A plastic whose modulus of elasticity is not greater than 10,000 psi in accordance with the American Society of Testing and Materials (ASTM) Standard Method of Test for Stiffness in Flexure of Plastics.
pipe alignment guide: A piping restraint that allows the pipe to move freely in the axial direction only [4].
pipe stiffness: A measure of how flexible pipe will be under buried conditions.
pipe supports: Components that transfer the load from the pipe to the support structure or equipment.

plastic: A material that contains an organic substance of high to ultra-high molecular weight, is solid in its finished state, and at some stage of its processing can be shaped by flow.
plastic, semi-rigid: A plastic whose modulus of elasticity is in the range of 10,000-100,000 psi in accordance with the Standard Method of Test for Stiffness in Flexure of Plastics.
plug valve: A valve that consists of a rotating plug in a cylindrical housing with an opening running through the plug.
pressure rating: The maximum pressure that can be inserted in the pipe without causing failure.
reinforced plastic: According to American Society for Testing and Materials, plastics having superior properties as compared to plastics consisting of base resin because of the presence of high-strength filler material embedded in the composition.
relief valve: A safety valve for the automatic release of pressure at a set pressure.
standard dimension ratio (SDR): A series of numbers in which the dimension ratio is constant for all sizes of pipe.
stiffness factor: A property of plastic pipe that indicates the flexibility of the pipe under external loads.
sustained pressure test: A constant internal pressure test for 1,000 hours.
thermoplastic: A plastic that can be softened repeatedly by heating and hardened by cooling. During the soft state, it can be shaped by molding or extrusion.
thermosetting: A plastic that is capable of being changed into an infusible or insoluble product when cured by heat or chemical means.
yield stress: The force required to initiate flow in a plastic.
Young's modulus of elasticity: The ratio of stress in a material under deformation.

ACRONYMS AND ABBREVIATIONS

ASME	American Society of Mechanical Engineers
ANSI	American National Standards Institute
API	American Petroleum Institute
ASCE	American Society of Civil Engineers
ASPOE	American Society of Petroleum Operations Engineers
ASTM	American Society for Testing and Materials
AWWA	American Water Works Association
BBL	Barrel = 42 U.S. gallons
BTU	British thermal unit
CAD	Computer-aided design
FRP	Fiberglass-reinforced plastics
GPM	Gallon per minute
HDPE	High-density polyethylene
LDPE	Low-density polyethylene
MDPE	Medium-density polyethylene

PB	Polybutylene
PE	Polyethylene
PJA	Pipe Jacking Association
PP	Polypropylene
PPFA	Plastic Pipe Fitting Association
PPI	Plastic Pipe Institute
PRI	Plastic and Rubber Institute
PVC	Polyvinyl chloride
VMA	Valve Manufacturers Association

PLASTIC PIPING CODES AND STANDARDS

Codes

Codes establish the minimum requirements for design, fabrication, materials, installation, inspection, and testing for most piping systems. Thermoplastics used for plumbing, sewer, water, gas distribution, and hazardous waste may come under the jurisdiction of a code or regulation. Some of the most frequently used codes for plastic piping products used for water and gas applications are:

BOCA National Mechanical Code

BOCA National Plumbing Code

ASME B31.3 Chemical Plant and Petroleum Refinery Piping

ASME 31.8 Gas Transmission and Distribution Piping Systems

ANSI Z223 National Fuel Gas Code

Code of Federal Regulations (CFR), Title 49, Part 192, Transportation of Natural Gas and other Gas by Pipeline

Code of Federal Regulations (CFR), Title 49, Part 195, Transportation of Liquids by Pipeline

NFPA 54, National Fuel Gas Code

Standards

Standards provide rules that apply to individual piping components and practices. The American Society for Testing and Materials establishes the majority of the standards used in the manufacture of plastic piping products. ASTM develops and publishes voluntary standards concerning the characteristics and performance of materials, products, and services. ASTM standards include test procedures for determining or verifying characteristics such as chemical composition, and measuring performance such as tensile strength. Committees drawn from professional, industrial, and commercial interests develop the standards, many of which are made mandatory by incorporation in applicable codes. Table 1.1 lists the principle ASTM standards that apply to thermoplastic piping products used in water and gas applications.

TABLE 1.1 ASTM Standards for Plastic Piping

	Specifications for
D1600	Abbreviations of terms
F412	Definitions for plastic piping systems
D2749	Symbols for dimensions of plastic pipe fittings
D2581	Polybutylene (PB) plastics molding and extrusion materials
D1228	Polyethylene (PE) plastics molding and extrusion materials
D3350	Polyethylene (PE) plastic pipe and fittings material
D1784	Rigid polyvinyl chloride (PVC) compounds
	Polybutylene (PB) plastic pipe and tubing
F809	Large diameter PB plastic pipe
F845	Plastic insert fittings for PB tubing
D2662	PB plastic pipe SDR
D3000	PB plastic pipe SDR based on outside diameter
D2666	PB plastic tubing
	Polyethylene (PE) plastic pipe, tubing, and fittings
D3261	PE butt heat fusion plastic fittings for PE pipe and tubing
F405	Corrugated PE tubing and fittings
F877	Cross-linked PE (PEX) plastic hot and cold water distribution systems
D2609	Plastic insert fittings for PE plastic pipe
F892	PE corrugated pipe with a smooth interior and fittings
F894	PE large diameter profile wall sewer and drain pipe
D3350	PE plastics pipe and fittings materials
D2239	PE plastic pipe SDR based on inside diameter
F714	PE plastic pipe SDR based on outside diameter
D3035	PE plastic pipe SDR based on controlled outside diameter
D2447	PE plastic pipe, Schedules 40 and 80 based on outside diameter
D2737	PE plastic tubing
D2683	Socket type PE fittings for outside diameter-controlled PE pipe and tubing
F905	Qualification of PE saddle fusion joints
F678	PE gas pressure pipe, tubing, and fittings
D2104	PE plastic pipe Schedule 40
F1055	PE electro-fusion fittings
	Polyvinyl chloride (PVC) plastic pipe, tubing, and fittings
F800	Corrugated PVC tubing and compatible fittings
D3915	PVC and related plastic pipe and fitting compounds

TABLE 1.1 *(continued)* ASTM Standards for Plastic Piping

	Polyvinyl chloride (PVC) plastic pipe, tubing, and fittings
F949	PVC corrugated sewer pipe with a smooth interior and fittings
F679	PVC large diameter plastic gravity sewer pipe and fittings
F794	PVC large diameter ribbed gravity sewer pipe and fittings
D2665	PVC plastic drain, waste, and vent pipe and fittings
D2466	PVC plastic pipe fittings, Schedule 40
D1785	PVC plastic pipe Schedules 40, 80, and 120
D2241	PVC pressure-rated pipe, SDR Series
D2740	PVC plastic tubing
D2729	PVC sewer pipe and fittings
F512	Smooth-wall PVC conduit and fittings for underground installations
D2467	PVC socket-type pipe fittings, Schedule 80
D2464	Threaded PVC plastic pipe fittings, Schedule 80
D2672	PVC plastic pipe, bell end
D3034	PVC plastic sewer pipe and fittings

The American Water Works Association (AWWA) publishes standards for the requirements for pipe and piping components used in water systems. These standards are used for large-diameter piping systems that are not covered by ASME B31, Code for Pressure Piping, or other codes. AWWA standards are incorporated by reference in many codes and by local authorities. Table 1.2 lists the principle AWWA standards that apply to thermoplastic piping products used in water systems.

TABLE 1.2 AWWA Standards for Plastic Piping

C902	PB plastic pipe and tubing for water service
C901	PE plastic pipe and tubing for water service
C906	PE plastic pipe for water distribution and large diameter line pipe
C900	PVC plastic pipe for water distribution
C905	PVC plastic pipe for water distribution

REFERENCES

1. Plastic Pipe Institute (PPI). Annual Statistics for 1997.
2. Chasis, D.A. 1976. *Plastic Piping Systems.* New York: Industrial Press, Inc.
3. ASME. 1989. B31, Code for Pressure Piping, Section B31.8, Gas Transmission and Distribution Piping System. American Society of Mechanical Engineers. New York.
4. Nayyar, M.L. 1992. *Piping Handbook.* New York: McGraw-Hill, Inc.

CHAPTER 2
PLASTIC PIPING CHARACTERISTICS

ADVANTAGES AND LIMITATIONS OF PLASTIC PIPING

Advantages

Plastic piping materials vary greatly in their characteristics and properties. These differences benefit the consumer in two ways:

1. Through proper design, each plastic raw material can be properly utilized and controlled by ASTM Standards.
2. A competitive market exists within the plastic pipe industry because the characteristics and properties of different plastic materials often overlap in piping applications.

Plastic piping materials are designed and selected to satisfy the requirements of the application for which they are to be used. When used in piping applications, plastic materials must withstand decades of stress. Plastic pipe manufacturers test their products for short-term and long-term use. These tests provide the designer with the information required in the selection of a plastic piping material for a particular application.

Thermoplastic piping products are cost-effective solutions to a variety of piping applications and offer many advantages when compared to traditional metal piping materials. Some of these features, which have spurred the widespread acceptance of plastic piping materials for many applications, are:

Corrosion resistance: Plastic piping materials are corrosion resistant and have low flow resistance. Plastic piping systems resist most normal household chemicals and many other substances that might enter a sanitary drainage system. The smooth wall of plastic pipe makes the transport of wastes and water more efficient and effective. Thermoplastic piping materials do not rust or corrode, and resist chemical attack from corrosive soils.

Ease of handling: Plastic piping materials are much lighter than most other piping materials and therefore do not require heavy handling equipment. Cutting, joining, and installing plastic piping is far simpler than the same procedures for other materials. At today's labor rates, the increased productivity is vital to the cost of the overall piping system.

Flexibility and toughness: Most thermoplastic piping materials are flexible, which is an important characteristic for underground applications. The pipe can follow natural contours and transitions around obstacles, which reduces the number of fittings required in most piping applications. Because of their excellent flexibility characteristics, plastic piping materials work well in harsh climate conditions.

Variety of joining methods: Many joining methods are available for plastic pipe. It can be threaded, flanged, cemented, heat-fused, and compression-fitted. The many joining methods make plastic pipe adaptable to most field applications.

Excellent hydraulics: Plastic piping materials provide a smooth pipe wall and have low resistance to flow. They also have a high resistance to scale or build-up.

Lower life cycle cost: Plastic pipe has excellent corrosion resistance and provides a system with a long life. This and other cost benefits make plastic pipe an attractive economic choice.

Long life: The service life of any piping material is important. Millions of plastic piping installations have been in service for more than a quarter of a century and still are functioning well. In most conditions, there is no end of life of a plastic piping system.

Standards: Standards have been developed for many plastic piping materials. Regardless of the manufacturer, these standards make sure that plastic piping products have uniform characteristics.

Easy identification: Plastic piping is marked to aid in identification. Manufacturers mark and test their pipes and fittings according to ASTM Standards. This procedure makes it simple for users to properly identify the many types of plastic pipes and fittings that are available.

Limitations

The primary limitations of thermoplastics come from their relatively low strength and stiffness and their sensitivity to high temperature. Because of these limitations, thermoplastic piping materials have been used mainly in low-pressure applications with low temperature limits. Even with these restrictions, thermoplastic-piping materials meet the design requirements for a wide range of applications.

THERMOPLASTIC PIPING MATERIALS

Principal Materials

Plastics are compounds made up of resins (polymers) and additives. Additives, which are used to obtain specific effects in the plastic material during fabrication or use, expedite processing, heighten certain properties, provide color, and furnish the needed protection during fabrication and use. Some of the key additives used in thermoplastic piping are heat stabilizers, antioxidants, ultraviolet screens, lubricants, pigments, property modifiers, and fillers. Table 2.1 lists some of the main additives used in plastic piping materials and their purpose.

Plastic pipe and components are available in a variety of materials, designs, and diameters. National standards have been established for many different wall constructions, such as double wall, ribbed, and foamed core. The various designs offer

TABLE 2.1 Common Additives in Plastic Piping Material

Additives	Purpose	Benefit
Antioxidants	Inhibit or retard reactions caused by oxygen or peroxides.	Extends the temperature range and service life.
Colorants	Pigments and dyes used to give color to plastic material.	Provides any desired color.
Coupling agents	Improves the bonding characteristics of plastic materials.	Improves the mechanical and electrical properties of the plastic material.
Fibrous reinforcements	Improves the properties of the resin.	Fibers improve the strength to weight ratio.
Fillers and extenders	Improves the physical and electrical properties of resin. Also reduces the cost of higher priced resins.	Plastic materials can be more economically produced without a loss of quality.
Heat stabilizers	Helps prevent the degradation of plastic materials from heat and light.	Helps plastic materials to be stable and retain their physical properties in excessive heat.
Preservatives	Helps prevent degradation of polymers by microorganisms.	Helps prevent fungi and bacteria attack on plastic materials. Makes the plastic material better suited for underground use.
Ultraviolet stabilizers	Helps retard the degradation from sunlight.	Allows plastic material to be used outdoors without any significant changes of the physical properties.

materials with different characteristics, strengths, and stiffness. The Plastic Pipe Institute (PPI) publishes a periodically updated report, PPI TR-5, which includes a listing of North American and international standards for thermoplastic piping. In addition, many plastic piping manufacturers offer product catalogs and manuals that provide excellent information concerning the design and use of their materials.

The principal plastic piping material specifications are issued by the American Society for Testing and Materials (ASTM). Earlier ASTM standards classified plastic materials by type, grade, and class in accordance with three important properties. ASTM used a code that consisted of four digits and a product letter prefix indicating the resin. The four digits stood for:

1st digit	Type of resin
2nd digit	Grade of resin
3rd and 4th digits	Hydrostatic pressure divided by 100.

With the increase in the types and uses for plastic piping materials, the need arose to classify plastic piping materials by more than three properties. To meet this need, a number of ASTM materials standards have gone to a cell classification system. With this system, a property cell number according to the property value defines each of the primary properties. For the designer, this cell classification is a major improvement in specifying piping materials. It is not always sufficient, however, and the manufacturer is still a primary source for information when specifying plastic piping materials.

Thermoplastic piping materials, like many other materials, are affected by weathering, which is a general term used to cover the entire range of outdoor environmental conditions. Thermoplastic piping materials that include appropriate weathering protection have been used in various outdoor applications and have provided many years of service. Plastic piping systems that are intended for continuous outdoor use must have a material composition that provides weather resistance for the specific conditions involved. Most thermoplastic piping has additives, such as ultraviolet absorbers and antitoxins, that prevent the plastic pipe from degrading from weathering.

Available Products

Thermoplastics are the primary plastic piping material in use today. They account for the largest percentage of plastic pipe in use and have the widest range of applications. Polyvinyl chloride (PVC) makes up the majority of the thermoplastic piping market; polyethylene is the second most popular.

Thermoplastics differ significantly in their properties and their suitability for various uses. To properly use thermoplastic piping materials, the engineer and designer must have a good understanding of the different thermoplastic materials and their proper applications.

Thermoplastics are a popular piping material mainly because of their low cost, ease of fabrication (usually by extrusion), and long life. This popularity has increased laboratory and field experience and has helped develop a significant amount of knowledge and technical data. The increased knowledge has resulted in recommendations about the design, installation, use, limitations, and material properties of thermoplastic piping materials. Table 2.2 lists some of the important typical properties and applications of the most popular thermoplastic piping materials.

Polyvinyl Chloride (PVC). This plastic has the broadest range of applications in piping systems and its use has grown more rapidly than that of other plastics. PVC has good chemical resistance to a wide range of corrosive fluids.

The two principal types of PVC used in the manufacture of pipe and fittings are Type I and Type II (ASTM D 1784). Type I, also called unplasticized or rigid PVC, contains a minimum of processing aids and other additives and has maximum tensile and flexural strength, modulus of elasticity, and chemical resistance. It is more brittle, however, and has a maximum service temperature under stress of about 150°F, lower thermal expansion than Type II, and does not support combustion. Type II PVC, which is modified with rubber to render it less rigid and tougher, also is called high-impact, flexible, or non-rigid PVC. It has lower tensile and flexural strength, lower modulus of elasticity, lower heat stability, and less chemical resistance than Type I. With ultraviolet (UV) stabilization, PVC piping material provides

TABLE 2.2 Properties and Applications

Material	Properties	Temperature limit, °F	Joining methods	Application
PVC	Outstanding resistance to most corrosive fluids Offers more strength and rigidity than most other thermoplastic pipe	158	Cementing Threading Heat fusion	Drain, waste, and vent Sewage Potable water Well casings Chemical pro processing
CPVC	Has the same properties as PVC, but can be used at higher temperatures	212	Same as PVC.	Used mainly in high-temperature applications
PE	Offers a relatively low mechanical strength but has good chemical resistance and is flexible at low temperatures	140	Heat fusion Insert fitting	Potable water Irrigation and sprinkler Corrosive chemical transport Gas distribution Electrical conduit
ABS	This pipe is rigid and has high-impact resistance down to −40° F	158	Cementing Threading Mechanical seal devices	Drain, waste, and vent Potable water Sewer Treatment plants
PP	Good high-temperature properties and outstanding chemical resistance	194	Heat fusion Threading	Chemical waste Natural gas Oil field

good long-term service in outdoors applications. The ability of the PVC material to withstand weathering depends on the type of UV stabilization and the amount of UV exposure.

The improvements made through research and the availability of product standards for special uses have increased PVC acceptance by designers, contractors, and building code officials. It is used in drain-waste-vent (DWV) applications, in storm, sanitary, water main, and natural gas distribution, and in industrial and process piping. The fastest growing application in North America is for municipal water and sewer systems. PVC pipe also is used as a conduit for wiring (both electrical and communications). The principle joining techniques for PVC piping is solvent cementing and elastomeric seals.

ASTM has developed a new version of ASTM D 1784 *Standard Specification for Rigid Polyvinyl Chloride and Chlorinated Polyvinyl Chloride Compounds*. This standard classifies PVC materials according to the nature of the polymer and five main properties instead of using the type and grade system. Cell-class limits that describe the polymer and four of the main properties are shown in Table 2.3. Chemical resistance, the fifth main property, is shown in Table 2.4.

Many piping standards still reference the older type and grade designation system. To assist in the conversion, new releases of ASTM D 1784 include a table (see Table 2.5) that cross-references the older with the new cell classifications.

ASTM D 4396 Standard Specification for Rigid Polyvinyl Chloride (PVC) and Related Plastic Compounds for Non-Pressure Piping Products is the PVC specification for non-pressure uses. Table 2.6 lists some physical properties of PVC pipe material.

Chlorinated PVC (CPVC). The basic resin in this plastic is made by post-chlorination of PVC. CPVC has essentially the same properties as Type I PVC material, but it has the added advantage of withstanding temperatures up to 212°F. Although it is suitable for the same piping applications as Type I PVC, the higher cost of CPVC restricts its use to that of conveying hot fluids. CPVC pipe can be used in water distribution lines at up to 100 psi working pressure at 180°F. As a result of the pressure and temperature ratings, CPVC pipe now replaces copper pipe in many areas of Europe and the United States.

Table 2.7 lists some physical properties of CPVC pipe material.

Polyethylene (PE). PE pipe materials are less strong and rigid than PVC materials at ambient temperatures. Because of its flexibility, ductility, and toughness, however, PE pipe materials are the second most widely used. Pipe made from PE has a relatively low mechanical strength but it exhibits good chemical resistance and flexibility and generally is satisfactory for use at temperature below 122°F. The temperature limitation, however, is offset by good flexibility retention down to −67°F. Polyethylene piping plastics are classified into three types based on density: low density (Type I), medium density (Type II) and high density (Type III). The most popular are Types II and III. The mechanical strength and chemical and temperature resistance

TABLE 2.3 Cell Classification Limits for PVC Material ASTM D 1784

Designation order number	Property	Cell limits								
		0	1	2	3	4	5	6	7	8
1	Base resin	Unspecified	Polyvinyl Chloride	Chlorinated polyvinyl chloride	Vinyl copolymer					
2	Minimum Impact strength Ft-lb/in of notch	Unspecified	<0.65	<0.65	1.5	5.0	10.0	15.0		
3	Minimum Tensile strength Psi	Unspecified	<5000	<5000	6000	7000	8000			
4	Minimum Modulus of elasticity in tension Psi	Unspecified	<280,000	<280,000	320,000	360,000	400,000	440,000		
5	Minimum Deflection temperature under load 264 psi	Unspecified	<131	<131	140	158	176	194	212	230

Note: The minimum property value will determine the cell number.

TABLE 2.4 Chemical Resistance ASTM D 1784

	Suffix			
	A	B	C	D
H_2SO_4 (93%), 14 days immersion at 55 +/−2°C				
Change in weight				
Increase, max %	1.0[1]	5.0[1]	25.0	NA
Decrease , max %	0.1[1]	0.1[1]	0.1[1]	NA
Change in flexural yield strength				
Increase, max %	5.0[1]	5.0[1]	5.0	NA
Decrease, max %	5.0[1]	25.0[1]	50.0	NA
H_2SO_4 (80%), 30 days immersion at 60 +/−2°C				
Change in weight				
Increase, max %	NA	NA	5.0	15.0
Decease , max %	NA	NA	5.0	0.1
Change in flexural yield strength				
Increase, max %	NA	NA	15.0	25.0
Decrease, max %	NA	NA	15.0	25.0
ASTM Oil Number 3, 30 days immersion at 23°C				
Change in weight				
Increase, max %	0.5	1.0	1.0	10.0
Decease , max %	0.5	1.0	1.0	0.1

[1] Specimens washed in running water and dried by an air blast or other mechanical means shall show no sweating within 2 hours after removal from the acid bath.

NA = not applicable

TABLE 2.5 Comparison of Older and Newer Designations

Type and grade classification from former specification D 1784-65T	Cell classification class from Tables 2.2 and 2.3
Rigid PVC materials	
Type I, Grade 1	12454-B
Type I, Grade 2	12454-C
Type I, Grade 3	11443-B
Type II, Grade 1	14333-D
Type III, Grade 1	13233
CPVC	
Type IV, Grade 1	23447-B

increases with density, whereas creep diminishes as the density increases. Most pressure PE pipe is made from Type II and Type III materials.

ASTM D 3350 is the primary specification for classifying PE pipe materials. This standard characterizes PE piping materials according to a cell classification system, which sequentially identifies seven physical properties by a matrix with the specified range of cell values for each of the properties. Table 2.8 shows the physical properties specified in ASTM D 3350 and the range for each property.

TABLE 2.6 Physical Properties of PVC Material

ASTM test	Property	Rigid	Flexible
	Physical		
D792	Specific gravity	1.30–1.58	1.20–1.70
D792	Specific volume (in^3/lb)	20.5–19.1	—
D570	Water absorption, 24 hours, ⅛ in. thick (%)	0.04–0.4	0.15–0.75
	Mechanical		
D638	Tensile strength (psi)	6000–8000	1500–3500
D638	Elongation (%)	50–150	200–450
D638	Tensile modulus (10–5 psi)	3.5–10	—
D790	Flexural modulus (10–5 psi)	3–8	—
D256	Impact strength, izod (ft-lb/in. of notch)	0.4–20.0	—
D785	Hardness, Shore	65–85D	50–100A
	Thermal		
C177	Thermal conductivity (10–4 cal-cm/sec-cm–2-°C)	3.5–5.0	3.0–4.0
D696	Coefficient of thermal expansion (10–5in./in.-°F)	1.2–5.6	3.9–13.9
D648	Deflection temperature (°F)		
	At 264 psi	140–170	—
	At 66 psi	135–180	
	Electrical		
D149	Dielectric strength (V/mil) short time, ⅛ in. thick	350–500	300–400
D150	Dielectric constant at 1 kHz	3.0–3.8	4.0–8.0
D150	Dissipation factor at 1 kHz	0.009–0.017	0.07–0.16
D257	Volume resistivity (ohm-cm) at 73°F, 50% RH	>10–16	10–11 to 10–15
D495	Arc resistance(s)	60–80	—

TABLE 2.7 Physical Properties of CPVC Material

Physical Property	ASTM Test Method	
Specific gravity	D 792	1.55
Modulus of elasticity in tension (psi at 73°F)	D 638/D 2105	420,000
Tensile (psi at 73°F)	D 638/D 2105	8400
Flexural Strength (psi)	D 790	15,350
Coefficient of thermal expansion (inch per inch per degree F)	D 696	3.8

An ASTM Material Designation Code, PE 2406 or PE 3408, also identifies thermoplastic PE materials for pressure piping systems. The first two numbers identify ASTM D 3350 cell values for density and slow crack growth resistance. The

TABLE 2.8 Cell Classification Limits for PE Material ASTM D 3350

Property	Test method	0	1	2	3	4	5	6	7
Density, gm/cm3	D 1505	Unspecified	0.910-0.925	0.926–0.940	0.941–0.955	>0.955	—	-	Specify value
Melt index, gm/10 min.	D 1238	Unspecified	>1.0	1.0–0.4	<0.4-0.15	<0.15	†	‡	Specify value
Flexural modulus, 1000 psi	D 790	Unspecified	<20,000	20,000 –<40,000	40,000 –<80,000	80,000 –<110,000	110,000 –<160,000	>160,000	Specify value
Tensile strength,									
1000 psi	D 638	Unspecified	<2200	2200–<2600	2600–<3000	3000–<3500	3500–<4000	>4000	Specify value
Slow crack Growth resistance 1. ESCR									
Test condition									
Test duration	D 1693	Unspecified	A	B	C	C			
Failure, max %			48	24	192	600	—	—	Specify value
			50	50	20	20			
2. PENT (hours)									
Molded plaque,									Specify value
80°C., 2.4 MPa,	F 1473	Unspecified	0.1	1	3	10	30	100	
Notch depth									
Hydrostatic design basis, (psi)	D 2837	NPR *	800	1000	1250	1600			
		A	B	C	D	E			
Color and UV stabilizer	D 3350	Natural	Color	Black with min. 2% carbon black	Natural with UV stabilizer	Color with UV stabilizer			

*NPR = Not Pressure Rated

†Materials with melt index less than cell 4 but which have flow rate <4.0g/10 min when tested according to D 1238, Condition 190/21.6.

‡Material with melt index less than cell 4 but which have flow rate <0.30g/10 min when tested according to D 1238, Condition 310/21.5.

last two numbers identify the materials hydrostatic design stress in psi divided by 100 with tens and units dropped.

Like PVC, PE piping material with ultraviolet (UV) stabilization provides good long-term service in outdoors applications. The ability of the PE material to withstand weathering depends on the type of UV stabilization and the amount of UV exposure.

PE pipe is available in both schedule number and standard dimension (SDR) sizes. Its principal applications are irrigation and sprinkler systems, drainage, chemical transport, gas distribution pipe, and electrical conduit systems. The typical physical properties for PE material are listed in Table 2.9.

Specialty PE Pipes. A relatively new development in PE piping is the introduction of ultrahigh molecular weight (UHMW) PE and cross-linked PE plastic piping materials. The UHMW PE has considerably higher resistance to stress cracking but is more costly than conventional PE piping material. It offers an extra margin of safety when used in sustained pressure conditions in comparison with pipe made from lower molecular weight resin. It is suitable for certain applications in the chemical industry where stress-cracking resistance has been a limiting factor for the conventional PE pipe.

Cross-linked PE piping material, when compared to ordinary PE pipe, displays greater strength, higher stiffness, and improved resistance to abrasion and to most chemicals and solvents at elevated temperatures up to 203°F. Pipe made from cross-linked PE also has high-impact resistance even at sub-zero temperatures. It is used in applications too severe for ordinary PE pipe. The joining technique used is threading.

Acrylonitrile-butadiene-styrene (ABS). ABS plastic is a copolymer made from the three monomers-acrylonitrile (at least 15 percent), butadiene, and styrene. It is a rigid plastic with good impact resistance at temperatures down to -40°F and up to 176°F. ABS is used mainly for drain-waste-ventilation (DWV) pipe and fittings, but it also is used in solvent cement for installing pipe in various applications. The most common applications for ABS pipe material are:

- Slurry lines
- Dewatering lines
- Water lines
- Pump lines.

Like other plastic piping materials, ABS is 70 percent to 90 percent lighter than steel and can be installed without heavy equipment. It offers excellent resistance to most chemicals and has a smooth interior surface that prevents mineral buildup and scaling. Solvent welding or threading can be used to join ADS pipe efficiently. ABS piping material also can be connected to other piping materials with Victaulic couplings or flanges. ABS piping material usually contains carbon

TABLE 2.9 Typical Physical Properties of PE Material

ASTM test	Property	Low density	Medium density	High density	Ultrahigh molecular weight
	Physical				
D792	Specific gravity	0.910–0.925	0.926–0.940	0.941–0.965	0.9258–0.941
D792	Specific volume (in^3/lb)	30.4–29.9	29.9–29.4	29.4–28.7	29.4
D570	Water absorption, 24 hours, ⅛ in. thick (%)	<0.1	<0.1	<0.1	<0.1
	Mechanical				
D638	Tensile strength (psi)	600–2300	1200–3500	3100–5500	4000–6000
D638	Elongation (%)	90–800	50–600	20–1000	200–500
D638	Tensile modulus (10–5 psi)	0.14–0.38	0.25–0.55	0.6–1.8	0.20–1.10
D790	Flexural modulus (10–5 psi)	0.08–0.60	0.60–1.15	1.0–2.0	1.0–1.7
D256	Impact strength, izod (ft-lb/in. of notch)	No break	0.5–16	0.5–20	No break
D785	Hardness, Rockwell R	10	15	65	67
	Thermal				
C177	Thermal conductivity (10–4 cal-cm/sec-cm–2-°C)	8.0	8.0–10.0	11.0–12.4	11.0
D696	Coefficient of thermal expansion (10–5in./in.-°F)	5.6–12.2	7.8–8.9	6.1–7.2	7.8
D648	Deflection temperature (°F)				
	At 264 psi	90–105	105–120	110–130	118
	At 66 psi	100–121	120–165	140–190	170

TABLE 2.9 *(continued)* Typical Physical Properties of PE Material

ASTM test	Property	Low density	Medium density	High density	Ultrahigh molecular weight
	Electrical				
D149	Dielectric strength (V/mil) short time, ⅛ in. thick	460–700	460–500	900 k V/cm	
D150	Dielectric constant at 1 kHz	2.25–2.35	2.25–2.35	2.30–2.35	2.30–2.35
D150	Dissipation factor at 1 kHz	0.0002	0.0002	0.0003	0.0002
D257	Volume resistivity (ohm-cm) at 73°F, 50% RH	10–15	10–15	10–15	10–18
D495	Arc resistance(s)	135–160	200–235	—	—
	Optical				
D542	Refractive index	1.51	1.52	1.54	—
D1003	Transmittance (%)	4–50	4–50	10–50	—

black to provide protection from sunlight. Non-black ABS pipe is not recommended for outdoor use.

Tables 2.10 and 2.11 list some of the properties of ABS piping material.

Polybutylene (PB). Polybutylene piping has practically no creep and has excellent resistance to stress cracking. It is flexible, and in many respects similar to Type III polyethylene, but is stronger. Polybutylene plastic piping is relatively new, and thus far its use has been limited to the conveyance of natural gas and to water distribution systems. Its high temperature grade can resist temperatures of 221-230°F. Table 2.12 lists some important physical properties of PB pipe material.

Polypropylene (PP). Polypropylene (PP) is an economical material that offers a combination of outstanding physical, chemical, mechanical, thermal, and electrical properties not found in other thermoplastics. Compared to low- or high-density PE, PP has a lower impact strength, but superior working temperature and tensile

TABLE 2.10 ABS Physical Properties

	Mechanical		
ASTM test	Property		
D638	Tensile strength at yield	4500 psi	31.0 MPa
D638	Elongation at yield	3.0%	3.0%
D638	Elongation at fail	30%	30%
D838	Modulus of elasticity (in tension)	220,000 psi	1517MPa
D256	Izod impact, notches ½ in. × ½ in. bar, .010 in. notch	7.0 ft/lbs. per inch of notch	0.37J/mn of notch
D690	Thermal expansion (linear)	5.2 × 10.5 in./in./°F	9.4 × 10.5 mm/mm/°C
D792	Specific gravity	1.04	1.04
	Thermal		
D648	Deflection temperature under load ½in. × ½in. bar, injection model	185°F @ 264 psi fiber stress	55°C @ 1.82 MPa fiber stress

TABLE 2.11 Recommended Design Pressures at Elevated Temperatures

Temperature °F	Temperature °C	Percent of rated pressures
73.4	23	100%
90	32	86%
100	38	81%
140	60	60%

TABLE 2.12 Physical Properties of Polybutylene (PB) Material

Physical property	ASTM test method	
Specific gravity	D 792	0.92
Modulus of elasticity in tension (psi at 73°F)	D 638/D 2105	350,000
Tensile (psi at 73°F)	D 638/D 2105	3800
Flexural strength (psi)	D 790	3000+
Coefficient of thermal expansion (inch per inch per degree F)	D 696	7.2

strength. PP is a tough, heat-resistant, semi-rigid material that is ideal for the transfer of hot liquids or gases. Polypropylene-based piping is also the lightest-weight plastic material and generally has better chemical resistance than other plastics. PP is used in some pressure piping applications, but its primary use is in low-pressure lines. Polypropylene plastic pipe is used for chemical (usually acid) waste drainage systems, natural gas and oil-field systems, and water lines. The maximum temperature for non-pressure piping is 194°F. Pipe lengths are joined by heat fusion, threading (i.e., with heavy pipe) and mechanical seal devices. With ultraviolet (UV) stabilization, PP piping material provides good long-term service in outdoors applications. The ability of the PP material to withstand weathering depends on the type of UV stabilization and the amount of UV exposure. See Table 2.13 for properties of PP piping material.

PLASTIC PIPING COMPONENTS

Many plastic piping components are available commercially and the list continues to grow. When considering a plastic piping fitting or valve, manufacturers' catalogs are a valuable source of what is available. Many of the manufactures have Web sites and online catalogs of their equipment. The Plastic Pipe Institute is an excellent source for links to plastic pipe manufacturers and suppliers on the Web and can be found at www.plasticpipe.org.

Thermoplastic fittings usually are injection molded. Molded fittings usually cost less and have higher pressure ratings than fabricated fittings. Most plastic fittings are molded in sizes up to eight inches; most 10 inches and above are fabricated.

Plastic valves fall into the same general categories as metal valves and have the same basic parts, such as stems or shafts, seats, seals, bonnets, hand wheels, and levers. Plastic valves are lighter, usually have better chemical resistance, and have less friction loss through the valve. Plastic valves can be specified to meet the pressure rating of the plastic pipe being used. Valve ends for joining to the pipe are available for socket fusion, threaded, flanged, and spigot ends. Plastic valves also have different types of material for the seats and seals to support the different products being handled by plastic piping systems.

TABLE 2.13 Typical Properties of Polypropylene (PP) Pipe Material

ASTM or UL test	Property	Unmodified resin	Glass reinforced	Impact grade
	Physical			
D792	Specific gravity	0.905	1.05–1.24	0.89–0.91
D792	Specific volume (in^3/lb)	30.8–30.4	24.5	30.8–30.5
D570	Water absorption, 24 hours, ⅛in. thick (%)	0.01–0.03	0.01–0.05	0.01–0.03
	Mechanical			
D638	Tensile strength (psi)	5000	6000–14,500	2800–4400
D638	Elongation (%)	10–20	2.0–3.6	350–500
D638	Tensile modulus (10–5 psi)	1.6	4.5–9.0	1.0–1.7
D790	Flexural modulus (10–5 psi)	1.7–2.5	3.8–8.5	1.2–1.8
D256	Impact strength, izod (ft-lb./in. of notch)	0.5–2.2	1.0–5.0	1.0–15
D785	Hardness, Rockwell R	80–110	110	50-85
	Thermal			
C177	Thermal conductivity (10–4 cal-cm/sec-cm–2-°C)	2.8	—	3.0–4.0
D696	Coefficient of thermal expansion (10–5in./in.-°F)	3.2–5.7	1.6–2.9	3.3–4.7
D648	Deflection temperature (°F)			
	At 264 psi	125–140	230–300	120–135
	At 66 psi	200–250	310	160–210
UL94	Flammability rating	HB	HB	HB
	Electrical			
D149	Dielectric strength (V/mil) short time, ⅛ in. thick	500–660	475	500–650
D150	Dielectric constant at 1 kHz	2.2–2.6	2.36	2.3
D150	Dissipation factor at 1kHz	0.0005–0.0018	0.0017	0.0003
D257	Volume resistivity (ohm-cm) at 73°F, 50%RH	10–17	2 × 10–16	10–15
D495	Arc resistance(s)	160	100	—

REFERENCES

1. Chasis, D.A. 1988. Plastic Piping Systems. New York: Industrial Press, Inc.
2. American Society of Mechanical Engineers. 1989. ASME B31, Code for Pressure Piping, Section B31.8, Gas Transmission and Distribution Piping Systems. New York.
3. Nayyar, M.L. 1992. Piping Handbook. New York: McGraw-Hill, Inc.
4. Blaga, A. 1981. Use of Plastics as Piping Materials. Division of Building Research, National Research Council of Canada. Ottawa (CBD 219).
5. Plastic Pipe Institute, 1999. Weathering of Thermoplastic Piping Systems, TR-18/99.

CHAPTER 3
FLUID FLOW

GENERAL

The main purpose of piping systems is to transport fluids from one location to another. Numerous standard fluid flow equations are used to calculate the flow and pressure drop of pipe and fittings. Each equation is unique and might have limitations associated with its use. This chapter will describe many of the various equations used for fluid flow calculations, with a focus on plastic piping systems. It is important that the engineer uses an equation that is appropriate for the flow condition being analyzed.

A fluid is any liquid or gas that cannot sustain its shape when subjected to a tangential or shearing force when at rest. This continuous and irrecoverable change of position of one part of the material relative to another part when under shear stress constitutes flow, a characteristic property of fluids. Liquids and gases are classified together as fluids because, over a wide range of situations, they have identical equations of motion and exhibit the same flow phenomena. Liquids change their volume slightly with significant variations in pressure, while gases tend to expand and completely fill any container. With gases, a change in pressure is accompanied by a change in volume.

LIQUID FLOW

The following variables and nomenclature are used throughout the liquid flow section:

C_w = Hazen-Williams Coefficient

D = Inside pipe diameter, ft

d = Inside pipe diameter, in.
f = Friction factor, dimensionless
g = Gravitational acceleration, ft/sec^2
g_c = Gravitational constant, 32.174 ft/sec^2
h_p = Head gain, ft
h_L = Head loss, ft
h_f = Friction head loss, ft
h_m = Head loss due to minor loss valve or fitting, ft
h_w = H_2O pressure, in.
K = Resistance coefficient for valve or fitting
k = Internal pipe wall roughness, ft
L_f = Pipe length, ft
P = Pressure, lb/in^2 (psia)
p = Pressure, lb/ft^2, psf
ΔP = Change in pressure, psia
P_1 = Inlet or upstream pressure, psia
P_2 = Outlet or downstream pressure, psia
Q = Flow rate, gallons/min
Q_h = Volumetric flow rate, ft3/hr (cfh)
Re = Reynolds number, dimensionless
Sg = Specific gravity, dimensionless
v = Velocity, ft/sec
W = Weight, lb
z = Elevation, ft
Δz = Change in elevation
ν = Kinematic viscosity, ft^2/sec
μ = Absolute (dynamic) viscosity, lbm/ft-sec
ρ = Density of fluid, lb/ft^3
γ = Specific weight, lb/ft^3

Table 3.1 lists some general formulas used for liquid hydraulics.
Table 3.2 lists some conversion factors used in liquid hydraulics.

The Energy Principle

Although there is no such thing as a truly incompressible fluid, this term is used for liquids. The first law of thermodynamics states that for any given system, the change in energy is equal to the difference between the heat transferred to the sys-

TABLE 3.1 General Formulas Used for Liquid Hydraulics

Formulas	Symbols
$A = \frac{\pi D^2}{4}$	A = Cross-sectional area of pipe, ft^2
$H = \frac{2.31P}{Sg}$	D = Inside diameter of pipe, ft
$H = \frac{P}{0.433Sg}$	H = Pressure measured in ft of head
$P = \frac{HSg}{2.31}$	P = Pressure measured in lb/in^2
$Sg = \frac{W\,(\text{liquid})}{W\,(\text{water})}$	Q = Flow rate in ft^3/sec
$V = \frac{Q}{A}$	Sg = Specific gravity
	V = Velocity in ft/sec
	W = Specific weight
Volume of Pipeline Fill	
$r = \frac{ID}{2}$	r = Pipe radius
$G = \pi r^2(L \times 12)0.004329$	ID = Inside pipe diameter
$V = \frac{G}{L}$ or $V = \frac{B}{L}$	L = Length in ft
$B = \frac{ID^2 5.13L}{5280}$	V = Pipeline fill per ft
	G = Pipeline fill per length in gal
	B = Pipeline fill per length in barrels

TABLE 3.2 Conversion Factors

1 ft^3 = 7.48 gallons	ft^3/sec = 642 BPH
1 barrel = 42 gallons	ft^3/sec = 449 GPM
1 gallon = 231 in^3	1 GPM = 1.43 BPH
1 ft^3 = 1728 in^3	

tem and the work done by the system on its surroundings during a given time interval. This energy represents the total energy of the system. In piping applications, energy often is converted into units of energy per unit weight resulting in units of length. Engineers use these length equivalents to get a better feel for the resulting behavior of the system. In pipeline hydraulics, we express the state of the system in terms of "head" or feet of head. The energy at any point in a piping system often is identified as:

Pressure head $= p/\gamma$

Elevation head $= z$

Velocity head $= v^2/2g$

These quantities can be used to express the head loss or head gain between two locations using the energy equation.

The Energy Equation

In addition to pressure head, elevation head, and velocity head, head also can be added to the system (usually by a pump) and head can be removed from the system due to friction or other disturbances within the system. These changes in head are referred to as head gains and head losses. By balancing the energy between two points in the system, we can obtain the energy equation (Bernoulli's Equation):

$$\frac{p_1}{\gamma} + z_1 + \frac{v_1^2}{2g} + h_p = \frac{p_2}{\gamma} + z_2 + \frac{v_2^2}{2g} + h_L \tag{3.1}$$

The basic approach to all piping systems is to write the Bernoulli Equation between two points, connected by a streamline, where the conditions are known. The total head at point 1 must match with the total head at point 2, adjusted for any increases in head because of pumps, losses because of pipe friction, and so-called "minor losses" because of entries, exits, fittings, etc. The parts of the energy equation can be combined to express two useful quantities, the hydraulic grade and the energy grade.

Hydraulic and Energy Grades

The hydraulic grade line (HGL) and the energy grade line (EGL) are two useful engineering tools in the hydraulic design of a system that is in a dynamic state. The hydraulic grade is the sum of the pressure head and the elevation head. This represents the height that a water column would raise in a piezometer. When plotted in a profile, this is referred to as the hydraulic grade line or HGL (see Figure 3.1).

The energy grade is the sum of the hydraulic grade and the velocity head and represents the height that a column of water would raise in a pitot tube. When plotted in a profile, this is referred to as the energy grade line, or EGL (see Figure 3.1).

Pipe Sizing

Fluid flow is a basic component of sizing a piping system. The fluid flow design determines the minimum acceptable pipe diameter required for transferring the fluid efficiently. The main factors in determining the minimum acceptable pipe diameter are the design flow rates and pressures losses. The design flow rates are based on system demands that usually are established in the design phase of a project. Before the determination of the minimum inside diameter can be made, service conditions must be reviewed to determine operational requirements, such as the recommended fluid velocity, and liquid characteristics, such as viscosity, temperature, and solids density.

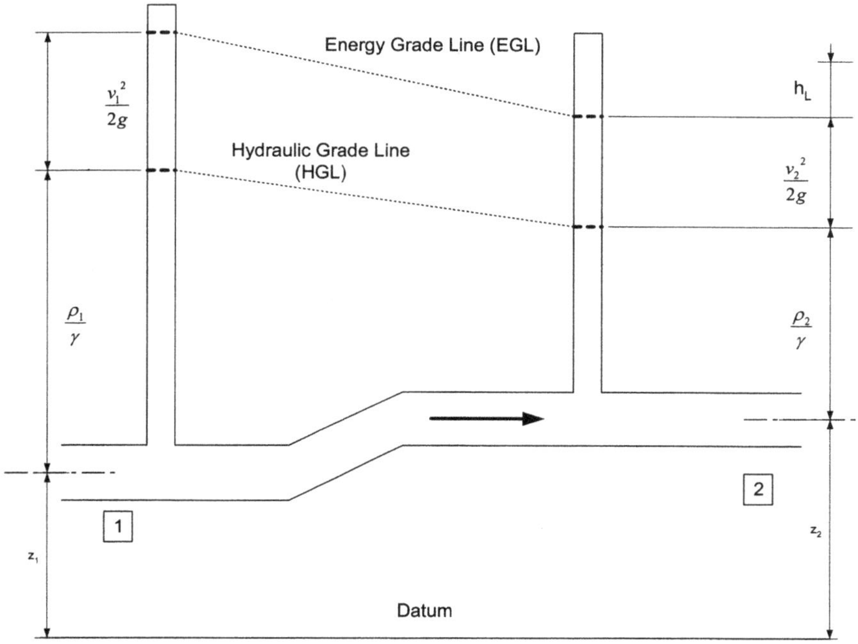

FIGURE 3.1 Energy grade line.

For normal liquid service applications, the acceptable fluid velocity in pipes is around 7 ft/sec ± 3 ft/sec. The maximum velocity at piping discharge points usually is limited to 7 ft/sec. These velocity ranges are considered reasonable design targets for normal applications. Other limiting factors, such as pressure transient conditions, however, can overrule. In addition, some applications can allow greater velocities based on general industry practices, such as boiler feed water and petroleum liquids.

Pressure losses throughout a piping system should be designed to provide an optimum balance between the installed cost of a piping system and operating cost of the system. The primary factors that will affect the cost and system operating performances are the inside pipe diameter (and the resulting fluid velocity), materials of construction, and pipe routing.

Energy Losses in Pipes

When a fluid is transported inside a pipe, the pipe's inside diameter determines the allowable flow rate. Several factors might cause the energy loss (hL) in a piping system, with the main cause friction between the fluid and the pipe wall. Liquids in the pipe resist flowing because of viscous shear stresses within the fluid and friction along the pipe walls. This friction is present throughout the length of the pipe. As a result, the energy grade line (EGL) and the hydraulic grade line (HGL) drop linearly in the direction of flow. Flow resistance in pipe results in a pressure drop, or loss of head, in the piping system.

Localized areas of increased turbulence and disruption of the streamlines are secondary causes of energy loss. These disruptions usually are caused by valves, meters, or fittings and are referred to as minor losses. When considered against the friction losses within a piping system, the minor losses often are considered negligible and sometimes are not considered in an analysis. While the term minor loss often is applicable for large piping systems, it might not always be the case. In piping systems that have numerous valves and fittings relative to the total length of pipe, the minor losses can have a significant impact on the energy or head losses.

Pressure Flow of Liquids

Many equations approximate the friction losses that can be expected with the flow of liquid through a pressure pipe. The two most frequently used equations in plastic piping systems are:

Darcy-Weisbach Equation

Hazen-Williams Equation

The Darcy-Weisbach Equation applies to a wide range of fluids, while the Hazen-Williams Equation is based on empirical data and is used primarily in water modeling applications. Each of these methods calculates friction losses as a function of the velocity of the fluid and some measure of the pipe's resistance to flow (pipe wall roughness). Typical pipe roughness values for these methods are shown in Table 3.3. These values can vary depending on the product manufacturer, workmanship, age, and many other factors.

Darcy-Weisbach Equation. Friction losses in a piping system are a complex function of the system geometry, the fluid properties, and the flow rate in the system. By observation, the head loss is roughly proportional to the square of the flow rate in most engineering flows (fully developed, turbulent pipe flow). This observation leads us to the Darcy-Weisbach Equation for head loss from friction:

$$h_f = f \frac{L_f}{D} \frac{v^2}{2g_c} \tag{3.2}$$

The Darcy-Weisbach Equation is a generally accepted method for calculating friction losses from liquids flowing in full pipes. It recognizes the dependence on pipe diameter, pipe wall roughness, liquid viscosity, and flow velocity. Darcy-Weisbach is a general equation that applies equally well at any flow rate and any incompressible fluid.

Depending upon the Reynolds number, the friction factor is a function of the relative wall roughness of the pipe, the velocity of the fluid, and the kinematic

TABLE 3.3 Pipe Roughness Values

Material	Hazen-Williams *Cw*	Darcy-Weisbach roughness height k (feet)
Asbestos cement	140	0.000005
Brass	135	0.000005
Brick	100	0.002
Cast iron; new	130	0.00085
Concrete		
Steel forms	140	0.006
Wooden forms	120	0.002
Copper	135	0.000005
Corrugated metal	—	0.15
Galvanized iron	120	0.0005
Glass	140	0.000005
Plastic	150	0.000005
Steel; new unlined	145	0.00015
Wood stave	120	0.0006

viscosity of the fluid. Liquid flow in pipes can be laminar or turbulent, or it can be in a transition between the two. For laminar flow (Reynolds number below 2000), the head loss is proportional to the velocity rather than the velocity squared and the pipe wall roughness has no effect. The friction factor calculation is:

$$f = \frac{64}{R_E} \tag{3.3}$$

Laminar flow can be characterized as consisting of a series of thin shells that are sliding over one another. The velocity of the fluid is the greatest at the center and the velocity at the pipe wall is zero.

In the turbulent flow region, it is not possible to obtain an analytical solution for the friction factor as we do for laminar flow. Most of the data available for evaluating the friction factor in turbulent flow have been derived from experiments. For turbulent flow (Reynolds number above 4000), the friction factor is dependent upon the pipe wall roughness as well as the Reynolds number. For turbulent flow, Colebrook (1939) found an implicit correlation for the friction factor in round pipes. This correlation converges well in a few iterations.

$$\frac{1}{\sqrt{f}} = -2\log\left[\frac{k}{3.7d} + \frac{2.51}{R_E\sqrt{f}}\right] \tag{3.4}$$

$$R_E = \frac{vD}{vy} \tag{3.5}$$

or

$$R_E = \frac{3162Q}{dk} \quad (3.6)$$

The familiar Moody Diagram is a log-log plot of the Colebrook correlation on an axis of the friction factor and the Reynolds number, combined with the $f = 64/\text{Re}$ result for laminar flow.

For turbulent flow, appropriate values for the friction factor can be determined using the Swamme and Jain Equation, which provides values within 1 percent of the Colebrook Equation over most of the useful ranges:

$$f = \frac{1.325}{\left[In\left(\frac{k}{3.7d} + \frac{5.74}{R_E^{0.9}}\right)\right]^2} \quad (3.7)$$

Hazen-Williams Equation The Hazen-Williams Equation is used primarily in the design and analysis of pressure pipe for water distribution systems. This equation was developed experimentally with water and, under most conditions, should not be used for other fluids. The Hazen-Williams formula for water at 60°F, however, can be applied to liquids that have the same kinematic viscosity as water. This

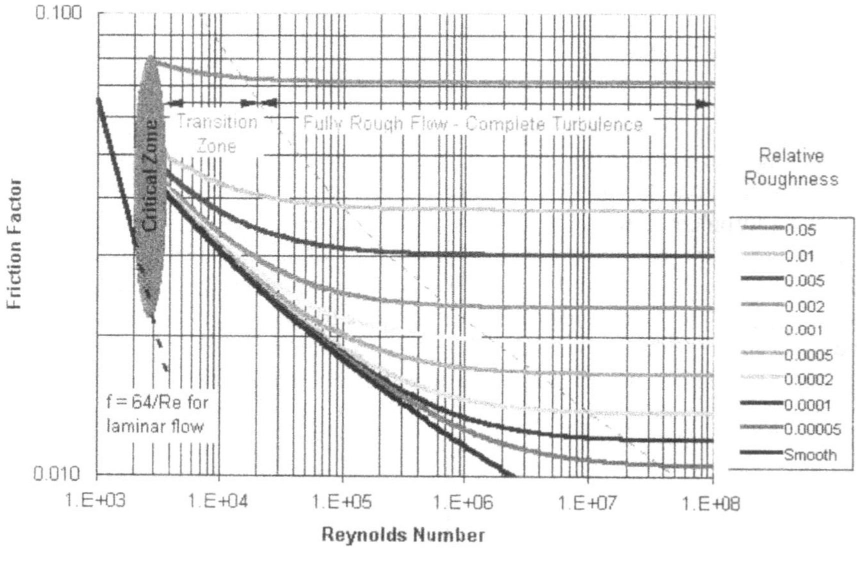

FIGURE 3.2 Reynolds number.

equation includes a roughness factor *Cw*, which is constant over a wide range of turbulent flows and an empirical constant.

$$h_f = \frac{3.02L_f}{D^{1.16}}\left(\frac{v}{Cw}\right)^{1.85} \tag{3.8}$$

For a simpler solution to fluid flow in plastic pipe, consider this version of the Hazen-Williams formula:

$$\Delta P_{100} = \frac{452Q^{1.85}}{Cw^{1.85}d^{4.86}} \tag{3.8}$$

where ΔP = Friction pressure loss, psi, per 100 feet of pipe.

The coefficient *Cw* is essentially a friction factor. Table 3.1 lists *Cw* values for various types of pipe.

The designer must use proper judgment to select pipe sizes that best meet the project conditions. The following considerations may be helpful:

- At a given flow rate, a larger diameter pipe will have a lower velocity and less pressure drop.
- At a given flow rate, a smaller diameter pipe will have higher velocity and increased pressure drop.
- The frictional head loss is less in larger diameter pipes than smaller pipe flowing at same velocity.

Minor Losses. Fluids flowing through a valve or fitting will have a friction head loss. Minor losses in pipes at these areas are caused by increased turbulence, which causes a drop in the energy and hydraulic grades at that point in the pipe system. The magnitude of the energy losses primarily depends on the shape of the fitting. The head or energy loss can be expressed by using the applicable resistance coefficient for the valve or fitting. The Darcy-Weisbach Equation then becomes:

$$h_m = K\frac{v^2}{2g_c} \tag{3.10}$$

$$K = f\frac{L_f}{D} \tag{3.11}$$

Equation 3.10 can be rearranged to express the fitting head loss as feet of straight pipe having the same head loss as the fitting.

$$L_f = \frac{KD}{f} \tag{3.12}$$

To calculate head losses in piping systems with both pipe friction and minor losses use:

$$h_f = \left(f\frac{L_f}{D} + \sum K\right)\frac{v^2}{2g_c} \tag{3.13}$$

Typical K values for the fitting loss coefficients are in Table 3.4.

Table 3.5 lists the estimated pressure drop for thermoplastic lined fittings and valves.

Water Hammer/Pressure Surge

Flowing liquid has momentum and inertia. When flow is stopped suddenly, the mass inertia of the flowing stream is converted into a shock wave. Consequently, a high static head exists on the pressure side of the pipeline. Quick surge pressures are shock waves known as water hammer. Water hammer, or hydraulic transients, is caused by opening and closing (full or partial) valves, starting and stopping pumps, changing pump or turbine speed, reservoir wave action, and entrapped air. The pressure wave from water hammer races back and forth in the pipe, getting progressively weaker with each "hammer." Maximum surge pressure results when the time required to change a flow velocity a given amount is equal to or less than:

$$t \leq \frac{2L_f}{S} \tag{3.13}$$

TABLE 3.4 Fitting Loss Coefficients

Fitting	Description	K Value
Pipe entrance	Sharp edged	0.5
	Inward projected pipe	1.0
	Rounded	0.05
Pipe exit	All	1.0
Bends	90° standard elbow	0.9
	45° standard elbow	0.5
Tee	Standard, flow through run	0.6
	Standard, flow through branch	1.8
Valves	Globe, fully open	10
	Angle, fully open	4.4
	Gate, fully open	0.2
	Gate, ½ open	5.6
	Ball, fully open	4.5
	Butterfly, fully open	0.6
	Swing check, fully open	2.5

Notes: Hydraulic Institute, Pipe Friction Manual, 3rd Ed., Crane Company, Technical Paper 410.

TABLE 3.5 Estimated Pressure Loss for Thermoplastic Lined Fittings and Valves

Size Inch	Standard 90° elbow	Tee through run	Tee through branch	Plug valve	Diaphragm valve	Vertical check valve	Horizontal check valve
1	1.8	1.2	4.5	2.0	7.0	6.0	16
1½	3.5	2.3	7.5	4.2	10	6.0	23
2	4.5	3.0	10	5.5	16	10	45
2½	5.5	4.0	12	NA	22	11	50
3	7.0	4.1	15	NA	33	12	58
4	10	6.0	20	NA	68	20	65
6	15	10	32	NA	85	31	150
8	19	14	42	NA	150	77	200
10	25	19	53	NA	NA	NA	NA

Notes: Data is for water expressed as equal length of straight pipe in feet.
NA = Part is not available from source.

Source: *"Plastic Lined Piping Products Engineering Manual," page 48.*

where L_f = is the length of the pipeline, feet
S = is the speed of the pressure wave, feet per seconds
t = is the time, seconds.

S is determined by the following:

$$S = \sqrt{\frac{(144E)K}{\left(\frac{w}{g}\right)\left(144E + \frac{KD}{t}\right)}} \tag{3.14}$$

where K = Bulk modulus of the liquid, psi (300,000 psi for water)
E = Modulus of elasticity of the pipe material, psi
w = Unit weight of fluid, lb/ft^3.

The excess pressure caused by the water hammer can be calculated by:

$$P_s = \frac{wSv_c}{144g} \tag{3.15}$$

where P_s = Change in pressure, psi
vc = Change in velocity, ft/sec, occurring within critical time.

Performing a water hammer analysis of a piping system is a complex task. Factors to be considered include pumping characteristics, fluid velocity, elevation changes, valve closing times, and piping geometry. Equation 3.15 calculates the maximum surge pressure for the given velocity change. Keeping the time to stop the flow at more than t (Equation 3.13) can minimize pressure changes. The greatest effects on the velocity of the liquid occur during the final stage of valve closure. A general guideline for gate valves with linear closure characteristics is to maintain a valve closure time of 10 times t. This should keep the pressure surge at about 10 percent to 20 percent of the surge developed by the t closure time.

Plastic piping materials have different characteristics and handle the effects of pressure surges differently. The designer should consult with the plastic pipe manufacturer concerning their products ability to handle pressure surges. For example, polyethylene (PE) pipe can handle short-term pressure surges above the design pressure rating of the pipe because of its short-term strength and flexibility. When under similar conditions, surge pressures in PE pipe are significantly less than surges seen in rigid pipe. For the same liquid and velocity change, surge pressures in PE pipe are about 50 percent less than PVC pipe. The fatigue endurance of the plastic piping material must be taken into account if the piping system has frequent or continuous pressure surges. A piping system encountering repeated stress could have a long-term strength loss. If the piping system will see frequent cyclical surge pressure, the total system pressure (including surge pressure) should not exceed the design pressure rating of the material.

COMPRESSIBLE GAS FLOW

Compressible flow implies that variations exist in the density of a fluid. The variations are caused by pressure and temperature changes from one point to another. The rate of change is important in the analysis of compressible flow and is connected closely with the velocity of sound. When dealing with compressible fluids, when the density change is gradual and not more than a few percent, the flow can be treated as incompressible by using an average density. If the change in pressure divided by the initial pressure is greater than 0.05, however, the effects of compressibility must be considered. In plastic piping systems, compressible flow is encountered most often in gases, such as natural gas. The following section provides many of the frequently used formulas in the design of plastic piping for natural gas applications. The following variables will apply to each equation described in the following pages:

Cw = Hazen-Williams Coefficient

d = Inside pipe diameter, inches

E = Pipeline efficiency factor

e = 2.71828, natural logarithm base

F = Transmission factor, dimensionless

f = Friction factor, dimensionless

G = Specific gravity of gas, dimensionless

H = Elevation, ft

ΔH = Change in elevation

h = H_2O pressure, in.
k = Internal pipe wall roughness, in.
L_f = Pipe length, ft
L_m = Pipe length, miles
P = Pressure, lb/in^2 (psia)
ΔP = Change in pressure, psia
P_1 = Inlet or upstream pressure, psia
P_2 = Outlet or downstream pressure, psia
P_{atm} = Average atmospheric pressure, psia
P_{avg} = Average pressure along the pipeline segment, psia
P_b = Base pressure, psia
Q_h = Volumetric flow rate, ft^3/hr (cfh)
Q_d = Volumetric flow rate, ft^3/day (cfd)
Re = Reynolds Number, dimensionless
ΔT = Change in temperature
T_{avg} = Average gas flowing temperature, Rankine
T_b = Base temperature, Rankine
T_1 = Initial temperature of the gas, Rankine
T_2 = Temperature of the gas under the second conditions, Rankine
V_1 = Volume of the gas in original condition, ft^3
V_2 = Volume of the gas in second set of conditions, ft^3
v = Gas velocity, ft/sec
W = Weight of the gas, lb
Z = Gas compressibility factor, dimensionless
ν = Kinematic viscosity, ft^3/sec
μ = Absolute (dynamic) viscosity, lbm/foot-second
ρ = Density of fluid, lb/ft^3

Table 3.6 lists some general formulas used for gas hydraulics:

Gas Laws

Boyle's Law. If the temperature of the gas remains constant, the volume of a quantity of gas will vary inversely as the absolute pressure. This is expressed mathematically by Boyle's Law as:

$$\frac{V_1}{V_2} = \frac{P_2}{P_1}$$

When using Boyle's Law, we are usually interested in the volume at the second set of conditions. For this purpose, the equation often is rewritten as:

$$V_2 = V_1 \frac{P_1}{P_2}$$

EXAMPLE 1

A quantity of gas at 70 psia has a volume of 1000 cubic feet. If the gas is compressed to 150 psia, what volume would it occupy? The barometric pressure is 14.7 psia and the temperature remains constant.

$$V_2 = 1000 \frac{70 + 14.7}{150 + 14.7} = 514.3 \text{ ft}^3$$

Charles' Law. Charles' Law states that the volume occupied by a fixed amount of gas is directly proportional to its absolute temperature, if the pressure remains constant. This empirical relation was formulated by the French physicist J.A. Charles about 1787 and reaffirmed later by Joseph Gay-Lussac. Charles' Law is expressed as:

$$\frac{V_1}{V_2} = \frac{T_1}{T_2}$$

Like the example above, we usually are interested in the volume at a second set of temperature conditions, so this equation often is rewritten as:

$$V_2 = V_1 \frac{T_1}{T_2}$$

Charles' Law also states that if the volume of a quantity of gas does not change, the absolute pressure will vary directly as the absolute temperature. This is expressed as:

$$\frac{P_1}{P_2} = \frac{T_1}{T_2}$$

If we are interested in the pressure at a second temperature condition, the equation can be expressed as:

$$P_2 = P_1 \frac{T_2}{T_1}$$

EXAMPLE 2

A gas has a volume of 500 cubic feet when the temperature is 45°F and the pressure is 20 pounds per square inch gauge (psig). If the temperature is changed to 90°F and the pressure stays the same, what will be the gas volume?

$$V_2 = 500 \frac{90 + 460}{45 + 460} = 544.6 \text{ ft}^3$$

What would the pressure be for the gas above if the volume remains constant and the temperature changes from 45°F to 80°F? Atmospheric pressure is 14.7 psia.

$$P_2 = (20 + 14.7)\frac{90 + 460}{45 + 460} = 37.8 \text{ psia}$$

Boyle's and Charles' laws can be combined and expressed as:

$$\frac{P_1V_1}{T_1} = \frac{P_2V_2}{T_2}$$

We can substitute known values in the above equation and solve for any one unknown.

Avogadro's Law. Avogadro's Law states that equal volumes of gases at the same temperature and pressure contain an equal number of molecules. From this law, we see that the weight of a volume of gas is a function of the weight of the molecules. In addition, at a certain volume, the gases weigh in pounds the numerical value of its molecular weight. This is known as the mol-volume. The mol-volume is 378.9 cubic feet for gases at 60°F and 14.73 psia. Table 3.7 lists the molecular weights for some of the compounds often associated with natural gas. The molecular weight for methane is 16.043. From the mol-volume explanation above it can be determined that 378.9 cubic feet of methane at 60°F and 14.73 psia weighs 16.043 pounds.

Ideal Gas Law. The Ideal Gas Law is the basic law for gas equations. It is used in many arrangements but often is written as:

$$PV = nRT$$

where P = Pressure of the gas
V = Volume of the gas
n = Number of pound-mols of the gas
R = Universal gas constant which varies depending on the pressure, volume, and temperature of the gas.

The number of pound-mols is equal to the weight of the gas divided by the molecular weight of the gas. Therefore, we write the Ideal Gas Law as:

$$PV = 10.722\frac{W}{M}T$$

where P = Pressure of the gas
V = Volume of the gas
W = Weight of the gas, pounds
M = Molecular weight of the gas
T = Temperature of the gas, Rankine.

The constant 10.722 is from the generally used value for the universal gas constant of 1544 when the pressure is in lb/ft^2 absolute.

TABLE 3.6 General Formulas for Gas

Formulas	Symbols
Gas velocity: $V = \frac{748Q_{Mcfh}}{d^2P_2}$ Alternate method with gas temperature as a factor: $V = \frac{748Q_{Mcfh}}{d^2P_2}$	V = Velocity in feet per second D = Inside pipe diameter inches Q_{Mcfh} = Flow rate in *Mcfh* Q_{cfh} = Flow rate in *cfh* P_2 = Downstream pressure in psia T = Gas temperature in degrees Rankine
Pipeline pack for gas pipelines: $v = d^2P_1.378$ For $P_{avg} < 100$; $v = \frac{d^2P_1.378}{z}$ For $P_{avg} > 100$; $P_{avg} = \frac{2}{3}\left(P_1 + P_2 - \frac{P_1P_2}{P_1 + P_2}\right)$; $z = \frac{1}{\left[1 + \frac{P_{avg}344400(10)^{(1.785Sg)}}{T^{3.825}}\right]^2}$	V = Pipeline volume in scf per 1000 feet D = Inside pipe diameter P_1 = Upstream pressure in psia P_2 = Downstream pressure in psia P_{avg} = Average pressure in the pipeline Z = Gas compressibility factor
Pipeline blowdown time: $BT_m = .0588\frac{P_1^{1/3}Sg^{1/2}d^2LF_c}{d_{blowdown}}$	BT_m = Blowdown time in minutes D = Inside pipe diameter $D_{blowdown}$ = Inside diameter of blowdown pipe inches P_1 = Upstream pressure in psia P_2 = Downstream pressure in psia Sg = Gas specific gravity L = Length of pipe section in miles F_c = Blowdown valve choke factor: Ideal nozzle: 1.0 Through gate: 1.6 Regular gate: 1.8 Regular lube plug: 2.0 Venturi lube plug: 3.2

The above equation also can be written in many forms to find an unknown. An often-used equation to find the weight of a quantity of gas is:

$$W = 0.0933\frac{MVP}{T}$$

EXAMPLE 3

Find the weight of a gas in a 2000-cubic foot tank. The gas pressure is 200 psig at 90°F. The molecular weight of the gas is 16.535 and the barometric pressure

TABLE 3.7 Molecular Weights

Compound	Molecular weight
Methane	16.043
Ethane	30.070
Propane	44.097
Butane	58.124
Pentane	72.151
Hexane	86.178
Heptane	100.205
Carbon dioxide	44.011
Nitrogen	28.016
Oxygen	32.00
Air	28.967

is 14.7 psia. This problem can be written as:

$$W = \frac{.0933 \times 16.535 \times 2{,}000 \times (200 + 14.7)}{90 + 460} = 1204 \text{ pounds}$$

Gas Pipeline Hydraulics

Several equations are used in gas pipeline hydraulics and selecting the proper equation is as important as doing the calculations correctly. Selecting the correct equation requires an understanding of the equations and parameters for the system being designed. Because of the assumptions made in each equation, a slight difference in the calculation can result when the equations are compared. The following section provides some of the frequently used equations and when they should be used.

Colebrook-White Equation. The Colebrook-White Equation is recommended for those who are not familiar with pipeline flow equations because it produces the greatest consistency of accuracy over a wide range of variables.

$$Q_h = 234.8\frac{T_b}{P_b}\left(\frac{1}{f}\right)^{.5}\left(\frac{\Delta P}{GT_{avg}L_eZ}\right)^{.5} D^{2.5}E$$

where $\Delta P = P_1^2 - P_2^2$

and

$$Re \leq 2000, \left(\frac{1}{f}\right)^{.5} = .25(Re)^{.5}$$

$$Re \geq 2000, \left(\frac{1}{f}\right)^{.5} = -4\log_{10}(Re)^{.5}\left[\frac{k}{3.7D} + \frac{1.255}{Re}\left(\frac{1}{f}\right)^{.5}\right]$$

Panhandle A Equation. This equation is best used for pipelines with Reynolds numbers in the range of 5×10^6 to 11×10^6. The average pipeline efficiency factor used in this equation is 92 percent. This number is based on actual empirical experience with the metered gas flow rates corrected to standard conditions. For larger diameter pipelines, the pipeline efficiency factor can be as high as 98 percent. With this equation, pipeline efficiency factors should be reduced for smaller pipe diameters. The Panhandle A Equation provides a reasonable approximation for partially turbulent flow. For fully turbulent flow, this equation does not produce accurate results. In the fully turbulent flow region, the Panhandle B Equation is recommended.

$$Q_d = 435.87\left(\frac{T_b}{P_b}\right)^{1.0778} D^{2.6182} E \left(\frac{\dfrac{\Delta P - (.0375 G \Delta H P_{avg}^2)}{T_{avg} Z}}{G^{.8539} L_m T_{avg} Z}\right)^{.5394}$$

where $\Delta P = P_1^2 - P_2^2$

and $P_{avg} = \dfrac{2}{3}\left[P_1 + P_2 - \dfrac{(P_1 P_2)}{P_1 + P_2}\right]$

$E = 1$ for new straight pipes without fittings or pipe diameter changes.

$E = 0.95$ for good pipe typically during the first 1-2 years of operation.

$E = 0.92$ for average operating conditions.

$E = 0.85$ for poor operating conditions.

Panhandle B Equation. This equation is best used for pipelines with Reynolds numbers in the range of 4×10^6 to 40×10^6. The Panhandle B Equation is used in the design of large diameter, high-pressure, long pipelines. The average pipeline efficiency factors are the same as the Panhandle A Equation. The Panhandle A Equation provides a reasonable approximation for turbulent flow conditions.

$$Q_d = 737\left(\frac{T_b}{P_b}\right)^{1.020} D^{2.53} E \left(\frac{\dfrac{\Delta P - (.0375 G \Delta H P_{avg}^2)}{T_{avg} Z}}{G^{.961} L_m T_{avg} Z}\right)^{.51}$$

where $\Delta P = P_1^2 - P_2^2$

and $P_{avg} = \dfrac{2}{3}\left[P_1 + P_2 - \dfrac{(P_1 P_2)}{P_1 + P_2}\right]$

Weymouth Equation. This is one of the older equations, but it still is used widely for natural gas distribution and gathering systems. The Weymouth Equation originally was developed from data taken from low- to medium-pressure pipelines. The results obtained are conservative when the equation is used for higher-pressure pipelines. The Weymouth Equation typically is used for pipe diameters less than 6 inches in pressure ranges greater than 1.5 psig and less than 300 psig. This equation is not recommended for gas transmission through long pipelines.

$$Q_d = 433.5\left(\frac{T_b}{P_b}\right)\left[\frac{P_1^2 - P_2^2}{GL_m T_{avg} Z}\right]^{.5} D^{2.667}E$$

where $P_{avg} = \frac{2}{3}\left[P_1 + P_2 - \frac{(P_1 P_2)}{P_1 + P_2}\right]$

IGT-Improved Equation. This equation is used widely for natural gas distribution systems. When used for higher-pressure pipelines, the results obtained are conservative. The IGT Equation typically is used for pressure ranges between 1.5 and 100 psig. This equation is not recommended for gas transmission through long pipelines.

$$Q_h = 664\frac{T_b}{P_b}\frac{1}{\mu^{.111}}\left(\frac{\Delta P}{G^{.8}T_{avg}L_f}\right)^{.556} D^{2.667}E$$

Mueller-High-Pressure Equation. This equation is used in natural gas distribution systems with pressures above 1 psig.

$$Q_h = 2826\left(\frac{\Delta P}{G^{.739}L_f}\right)^{.575} D^{2.275}E$$

where $\Delta P = P_1^2 - P_2^2$

Mueller-Low-Pressure Equation. This equation is used in natural gas distribution systems with pressures below 1 psig.

$$Q_h = 735.4\frac{T_b P_{atm}^{.575}}{P_b \mu^{.15} G^{.425}}\left(\frac{\Delta h}{T_{avg}L_f}\right)^{.575} D^{2.725}E$$

where $\Delta h = h_1 - h_2$

Spitzglass Equation. This equation is best used for pipe diameters of 10 inches or less with pressures between 1.5 and 50 psig.

$$Q_h = 2209\frac{T_b}{P_b}\left(\frac{1}{f}\right)^{.5}\left(\frac{P_1^2 - P_2^2}{GT_{avg}L_f}\right)^{.5} D^{2.5}E$$

where $$\frac{1}{f} = \left(\frac{1}{1 + \frac{3.6}{D} + 0.03D}\right)$$

Another popular version of the Spitzglass Equation often is used for natural gas distribution systems operating at 1 psig or less:

$$Q_h = 3550 \sqrt{\frac{\Delta h_w d^5}{GL_f + \left(\frac{3.6}{D} + 0.03D\right)}}$$

where Q_h = Flow rate, cubic feet per hour at 14.7 psia and 60°F
h = Static pressure head, inches of water.

REFERENCES

1. "Charles's law," Encyclopædia Britannica Online. *http://subscribe.eb.com* [Accessed February 9, 2000].
2. "Fluid," Encyclopædia Britannica Online. [Accessed February 12, 2000].
3. ASME B31, Code for Pressure Piping, Section B31.8, Gas Transmission and Distribution Piping Systems, American Society of Mechanical Engineers, New York, 1989 ed.
4. Nayyar, M.L. 1992. *Piping Handbook.* New York: McGraw-Hill, Inc.
5. Chevron, *Plexco/Spirolite Engineering Manual.* 1998. Vol 2, 2d Edition.
6. Irving Granet, 1996. *Fluid Mechanics.* New Jersey: Prentice Hall.

CHAPTER 4
GENERAL DESIGN PROCEDURES

INTRODUCTION

This chapter covers the general design concepts that apply to various thermoplastic-piping systems. Regardless of the type of material being used, the effective use of thermoplastic piping systems depends on a thorough knowledge of pipe system design and the characteristics of the plastic material being used. When designing a thermoplastic piping system careful attention should be given to the unique properties of plastics and their effects on the piping system design, installation, and operation. In the majority of thermoplastic piping systems the main properties of concern are:

- Pressure limitations
- Temperature limitations
- Piping expansion and contraction

Thermoplastic piping systems are composed of various additives to a base resin or composition. Thermoplastics are characterized by their ability to be softened and reshaped repeatedly by the application of heat. Because of the slightly different compositions, the properties of plastic piping materials may vary between different manufacturers. As a result, designs and specifications need to address specific material requirements on a type and grade basis, which may need to be confirmed with the manufacturer.

DESIGN METHODOLOGY

The Design Analyses

All piping system designs are based on designing a piping system that will effectively and efficiently transport the required product. The design criteria for all piping applications are based on experience and applicable codes, standards, environmental requirements, and other parameters that may effect or constrain the work. An effective piping system design must include a methodical, step-by-step approach that gives proper consideration to the purpose of the piping system, how the system will be used, the conditions that the piping system will encounter, and the type of materials to be used.

Engineering calculations performed during the design analyses document the piping system design. Combined with the piping system criteria, the calculations define the process flow rates, system pressure and temperature, pipe strength requirements, pipe wall thickness, pipe stresses, and pipe support requirements. To be effective, design calculations should be clear, concise, and complete. The design calculations should document assumptions made, design data, and the sources of the data. All references (such as manuals, handbooks, and catalog cuts), alternate designs considered, and planned operating procedures should be included in the calculation records. Computer-aided design programs are helpful and time saving, but they should not be a substitute for the designer's understanding of the design process.

Another main task during the design is the development of an accurate piping system description. The piping system description should provide the function and major features of the system. The piping system description should contain the design bases, operating modes, control concepts, and both system and component performance ratings. Figure 4.1 lists the typical contents of a piping system description. The piping system description should provide enough information to develop:

- Process flow diagrams (PFDs)
- Piping and instrumentation diagrams P&IDs)
- And to obtain any required permits or approvals

Based on the design calculations and the piping system description, the last step of the design analyses is the development of the specifications for the piping system. Piping system specifications define the materials, fabrication, installation and construction, and service performance requirements of the materials to be used. There are two methods for identifying piping system specifications. One method identi-

Piping System Description
1. Function
2. Bases of Design Environmental Safety Performance Requirements Codes and Standards
3. Description General Overview System Operation Major Components

FIGURE 4.1 Piping system description.

fies the specific materials or installation or construction practices to be followed, while the other is based on performance criteria. Performance specifications (as they are frequently called) list the required performance or capabilities but do not specify how the performance shall be achieved. The primary reason for performance-based specifications is to allow more vendors and manufacturers to compete for providing their products and services at the best cost or benefit to the client. There are some government contracts that require the use of performance-based specifications.

DRAWINGS

Plastic piping system drawings can take many forms depending on the application. For natural gas and water distribution systems they are primarily utility type drawings. For water, gas, and oil pipelines they are often in an alignment sheet format. However, regardless of the type of drawing they usually provide common information. Plastic piping system drawings normally include piping layout drawings, fabrication or detail drawings, instrumentation drawings, and pipe support drawings.

Process Flow Diagram (PFD)

PFD's are the schematic illustrations of the piping system description. PFD's show the relationship between the major piping system components. PFD's do not show pipe ratings or designations, minor piping systems, for example, sample lines or valve bypass lines; instrumentation or other minor equipment, isolation valves, vents, drains, etc. Figure 4.2 lists the typical items found on a PFD.

Piping and Instrumentation Diagram (P&ID)

P&ID's schematically illustrate the functional relationship of the piping, instrumentation, and system equipment components. P&ID's show all of the piping, including the intended physical sequence of branches, reducers, and valves, etc.;

Process Flow Diagrams
1. Major Equipment Symbols, Names, Identification
2. System Piping
3. Control Valves and Other Valves that Affect Operations
4. Piping System Interconnections
5. Piping System Ratings and Operational Variables Maximum, average, minimum flow Maximum, average, minimum pressure Maximum, average, minimum temperature
6. Type of Product Being Handled

FIGURE 4.2 Process flow diagrams.

equipment; instrumentation and control. Figure 4.3 lists the typical items contained on a P&ID, and Figure 4.4 depicts a small and simplified P&ID.

DESIGN BASES

The bases of design are the physical and material parameters that are considered in the detailed design of a piping system to ensure a reasonable life cycle. The main considerations in the design of a piping system are loading and service condition and environmental factors. The basis of the design must be developed in order to perform the design calculations and prepare the drawings. Pre-design surveys are often required in the design of a new piping system and are a necessity for the renovation or expansion of existing systems. A site visit provides an overview of the project and the conditions at the location. Site visits are useful in obtaining the design requirements from the client and acquiring an overall sense of the project.

SERVICE CONDITIONS

Thermoplastic piping systems must be designed to provide reliable service under the service conditions it will encounter during the life of the system. The thermoplastic piping material must accommodate all combinations of loading situations that may occur such as pressure changes, temperature changes, thermal expansion and contraction, and other forces that may be present in the piping system. These conditions are referred to as the service conditions of the piping system. Service conditions are used to set design stress limits and may be defined or specified by applicable codes, or are determined based on the piping system description, site survey, and other design bases.

A main factor in the successful design of any thermoplastic piping system is determining and using the correct codes and standards. These are reviewed based on the project descriptions to determine and verify applicability. Codes and standards for the various plastic piping systems provide the required design criteria. These criteria are rules and regulations to follow when designing a plastic piping

Piping and Instrumentation Diagrams
1. Mechanical Equipment
2. All Valves
3. Instrumentation and designations
4. All Piping Sizes and Identification
5. All Miscellaneous Appurtenances including Vents, Drains, Special Fittings, Reducers, etc.
6. Direction of Flow
7. Class Change
8. Interconnections
9. Control Inputs and Outputs

FIGURE 4.3 Table of piping and instrumentation diagrams.

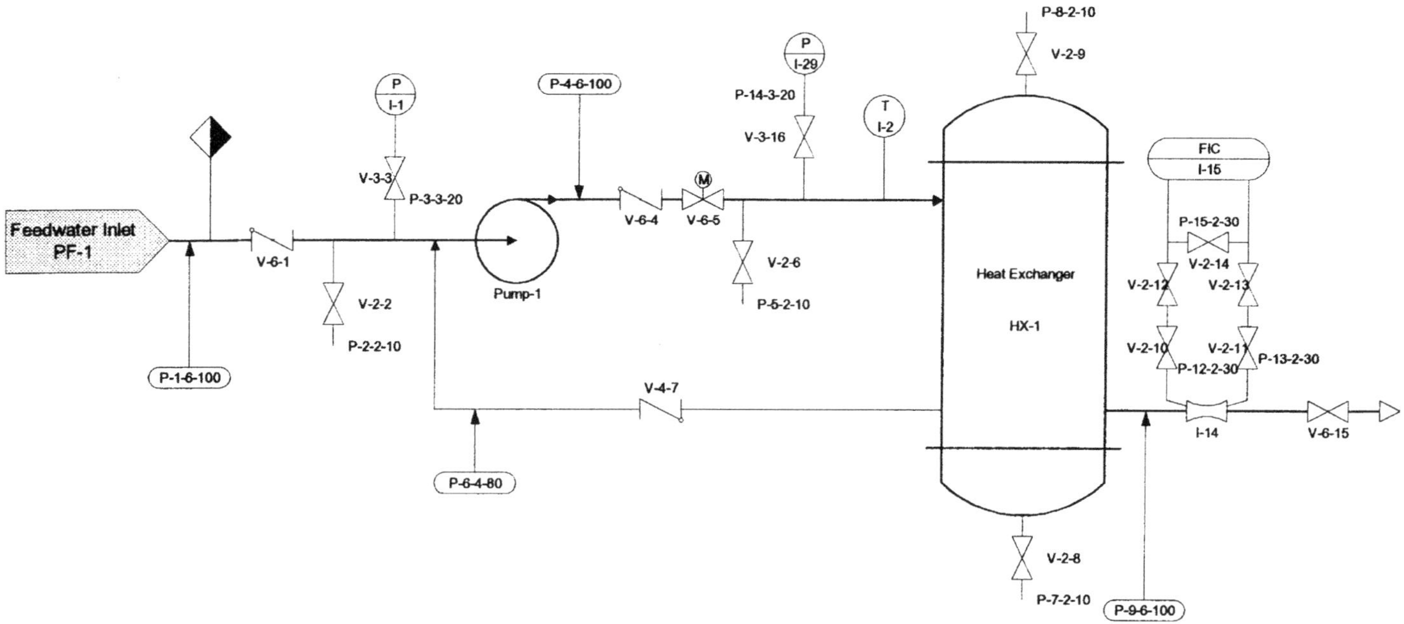

(A)

FIGURE 4.4 Piping and instrumentation diagrams.

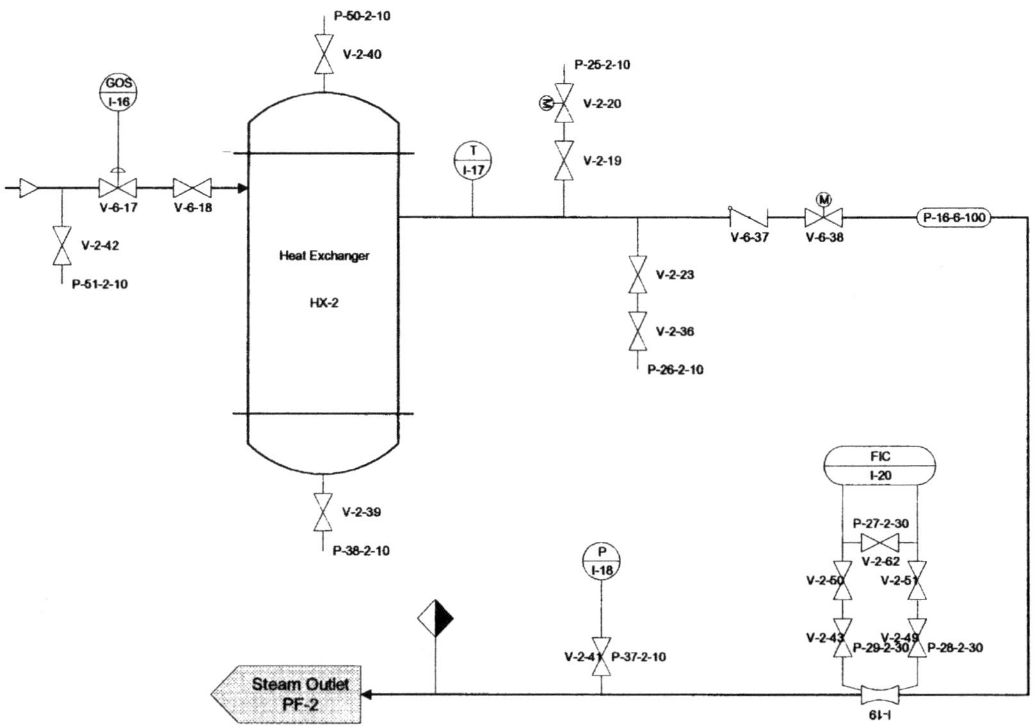

FIGURE 4.4 *(continued)* Piping and instrumentation diagrams.

system. The following list is a sample of some of the parameters that are covered by design criteria found in piping codes:

- Allowable stresses and stress limits
- Allowable loads and load limits
- Materials
- Minimum pipe wall thickness or SDR
- Maximum deflection
- Seismic loads
- Thermal expansion

Standards provide the required design criteria and rules for components such as fittings, valves, and meters. The purpose of standards is to specify rules for the manufacturer of the components. Standards apply to both dimensions and performance of system components and are prescribed when specifying a plastic piping system.

LOADING CONDITIONS

The stresses on a piping system define the service conditions of the piping system and are a function of the loads on that system. The sources of these loads are internal pressure, piping system dead weight, differential expansion due to temperature changes, wind loads, and snow or ice loads. Loads on a piping system are classified as sustained or occasional loads.

Sustained Loads

Sustained loads are those loads that do not vary considerably over time and are constantly acting on the system. Examples of sustained loads are the pressures, both internal and external, acting on the system and the weight of the system. The weight of the system includes both that of the piping material and the operating fluid.

The sustained maximum system operating pressure is the basis for the design pressure. The design temperature is the liquid temperature at the design pressure. The minimum wall thickness of the pipe and the piping components pressure rating is determined by the design temperature and pressure. Although the design pressure is not to be exceeded during normal, steady state operations, short-term system pressure excursions in excess of the design pressures occur. These excursions are acceptable if the pressure increases over time and the time durations are within code-defined limits.

Piping codes provide design guidance and limits for design pressure excursions. If a code does not have an over-pressure allowance, transient conditions are accounted for within the system design pressure. A reasonable approach to overpressure conditions for applications without a specific design code is:

1. For transient pressure conditions that exceed the design pressure by 10 percent or less and act for less than 10 percent of the total operating time, neglect the transient and do not increase the design pressure
2. For transients whose magnitude or duration is greater than 10 percent of the design pressure or operating time, increase the design pressure to encompass the range of the transient

Dead weight is the dead load of a piping system or the weight of the pipe and system components. Dead weight generally does not include the weight of the system fluid. The weight of the fluid is normally considered an occasional load by code. For buried piping, dead weight is not a factor. However, a sustained load that is analyzed is the load from the earth above the buried piping. Because of the different potential for deformation, the effects of an earth load on flexible piping and rigid piping are analyzed differently.

Occasional Loads

Occasional loads are those loads that act on the system on an intermittent basis. Examples of occasional loads are those placed on the system from the hydrostatic leak test, seismic loads, and other dynamic loads. Dynamic loads are those from forces acting on the system, such as forces caused by water hammer, and the energy released by a pressure relief device. Another type of occasional load is caused by the expansion of the piping system material. An example of an expansion load is the thermal expansion of pipe against a restraint due to a change in temperature.

Wind load is a transient, live load (or dynamic load) applied to piping systems exposed to the effects of the wind. Obviously the effects of wind loading can be neglected for buried or indoor installation. Wind load can cause other loads, such as vibratory loads, due to reaction from a deflection caused by the wind.

Snow and ice loads are live loads acting on a piping system. For most heavy snow climates, a minimum snow load of 25 pounds per square foot (psf) is used in the design. In some cases, local climate and topography dictate a larger load. This is determined from ANSI A58.1, local codes, or by research and analysis of other data. Snow loads can be ignored for locations where the maximum snow is insignificant. Ice buildup may result from the environment, or from operating conditions.

The snow loads determined using ANSI A58.1 methods assume horizontal or sloping flat surfaces rather than rounded pipe. Assuming that snow lying on a pipe will take the approximate shape of an equilateral triangle with the base equal to the pipe diameter, the snow load is calculated with the following formula:

$$W_S = -nD_oS_L$$

where W_S = design snow load acting on the piping, lb/ft
D_o = pipe (and insulation) outside diameter, in.
S_L = snow load, lb/ft^2
n = conversion factor, 0.083 ft/in

Ice loading information does not exist in data bases like snow loading. Unless local or regional data suggests otherwise, a reasonable assumption of 2 to 3 inches maximum ice accumulation is used to calculate an ice loading:

$$W_I = n_3 S_I t_I (D_o t_I)$$

where W_I = design ice load, lbs/ft
S_I = specific weight of ice, 56.1 lbs/ft^3
t_I = thickness of ice, in.
D_o = pipe (and insulation) outside diameter, in.
n^3 = conversion factor, 6.9×10 ft^2/in^2.

Seismic loads induced by earthquake activity are live (dynamic) loads. These loads are transient in nature. Appropriate codes are consulted for specifying piping systems that may be influenced by seismic loads. Seismic zones for most geographical locations can be found in American Water Works Association (AWWA) D110, AWWA D103, or CEGS 13080, Seismic Protection for Mechanical Electrical Equipment. ASME B31.3 (Chemical Plant and Petroleum Refinery Piping) requires that the piping is designed for earthquake induced horizontal forces using the methods of ASCE 7 or the Uniform Building Code.

Forces resulting from thermal expansion and contraction can cause loads applied to a piping system. A load is applied to a piping system at restraints or anchors that prevent movement of the piping system. Within the pipe material, rapid changes in temperature can also cause loads on the piping system resulting in stresses in the pipe walls. Finally, loads can be introduced in the system by combining materials with different coefficients of expansion.

Movements exterior to a piping system can cause loads to be transmitted to the system. These loads can be transferred through anchors and supports. An example is the settlement of the supporting structure. The settling movement transfers transient, live loads to the piping system.

Live loads can result from the effects of vehicular traffic and are referred to as wheel loads. Because aboveground piping is isolated from vehicle traffic, these live loads are only addressed during the design of buried piping. In general, wheel loads are insignificant when compared to sustained loads on pressure piping except when buried at "shallow" depths. The term shallow is defined based upon both site-specific conditions and the piping material. "However, as a rule, live loads diminish rapidly for laying depths greater than about four feet for highways and ten feet for railroads." Wheel loads are calculated using information in AASHTO H20 and guidance for specific materials such as AWWA C900 (PVC), and AWWA C950 (FRP).

PIPING LAYOUT

The bases of design establish the factors that must be included in liquid process piping design. The preparation of the piping layout requires a practical understanding of complete piping systems, including material selections, joining methods, equipment connections, and service applications. The standards and codes previously introduced establish criteria for design and construction but do not address the physical routing of piping.

COMPUTER AIDED DRAFTING AND DESIGN

Computer based design tools, such as computer aided draft and design (CADD) software, can provide powerful and effective means to develop piping layouts. Much of the commercially available software can improve productivity and may also assist in quality assurance, particularly with interference analyses. Some CADD software has the ability to generate 3-dimensional drawings or 2-dimensional drawings, bills of material, and databases.

Piping Layout Design

System P&IDs, specifications, and equipment locations or layout drawings that are sufficiently developed to show equipment locations and dimensions, nozzle locations, and pressure ratings are needed to develop the piping layout. A completely dimensioned pipe routing from one point of connection to another with all appurtenances and branches as shown on the P&ID should be prepared.

Pipe flexibility is required to help control stress in liquid piping systems. Stress analysis may be performed using specialized software. Considerations that must be accounted for in routing piping systems in order to minimize stress include:

- Avoiding the use of a straight pipe run between two equipment connections or fixed anchor points
- Locate fixed anchors near the center of pipe runs so thermal expansion can occur in two directions
- Provide enough flexibility in branch connections for header shifts and expansions

In addition, the piping layout should utilize the surrounding structure for support where possible. Horizontal and parallel pipe runs at different elevations are spaced for branch connections and also for independent pipe supports.

Interferences with other piping systems, structural work, electrical conduit and cable tray runs, heating, ventilation and air conditioning equipment, and other process equipment not associated with the piping system of concern must be avoided.

Insulation thickness must be accounted for in pipe clearances. To avoid interferences, composite drawings of the facility are typically used. This is greatly aided by the use of CADD software. However, as mentioned previously in this chapter, communications between engineering disciplines must be maintained as facilities and systems are typically designed concurrently though designs may be in different stages of completion.

Piping connections to pumps affect both pump operating efficiency and pump life expectancy. To reduce the effects, the design follows the pump manufacturer's installation requirements and the Hydraulic Institute Standards, 14th Edition. The project engineer should be consulted when unique piping arrangements are required.

Miscellaneous routing considerations are:

- Providing access for future component maintenance
- Pipe tracing access
- Hydrostatic test fill and drain ports
- Air vents for testing and startup operations

System operability, maintenance, safety, and accessibility are all considerations that are addressed in the design.

PRELIMINARY DESIGN DATA

Thermoplastic pipe material offers the advantages of flexibility, resiliency, and toughness. They differ from rigid plastic and steel pipe in some ways. Thermoplastic pipe is strong, extremely tough, and very durable. It can stand-alone or be considered as a portion of the environment. Thermoplastic piping installations act as a "system" within the environment and can gather additional strength from its surroundings and are responsive to changes in its physical environment.

There are many items to consider in the proper selection, design, and use of thermoplastic piping systems. Proper system design should give consideration to the following design criteria:

PIPELINE LIFE REQUIREMENTS

A determination should be made of the estimated "life" of the installed thermoplastic pipe system. When designing a city sewer system a life expectancy of 50 years is normal. A temporary, gravity-flow, mine slurry line may be in service only five years before operations are moved. A specialized chemical process plant may be obsolete or renovated in 15 to 20 years due to technology changes. The design parameters will vary depending on the intended use and desired life expectancy.

FLOW REQUIREMENTS

Thermoplastic piping material provides a smooth pipe wall. Because it is smoother than many common pipe materials such as steel or concrete, it will transport more products in comparable pipe sizes. When compared to steel or concrete, smaller diameter thermoplastic pipe can carry an equivalent volumetric flow rate at the same pressure. This makes it ideal for relining pipes while maintaining identical flow capabilities.

Compatibility

There are many fluid property and chemical handling problems that can be solved with thermoplastic pipe. Because of its inherent chemical composition and structure, thermoplastic pipe does not react with most products being transported. There are only a very few strong chemicals which affect it. When considering chemical compatibility it is helpful to keep the following three factors in mind.

1. The chemical resistance of thermoplastic pipe is related to the chemical itself, the operating temperature and the concentration of the chemical
2. Strong oxidizing agents such as nitric acid, sulfuric acid, chlorine gas and liquid bromine are most aggressive and deserve special consideration
3. Permeation of the pipe wall is negligible for most products. However, aromatic hydrocarbon permeation rates should be reviewed

Specific Gravity. A close approximation should be made of the fluid's density or specific gravity for later use in flow calculations and/or installation calculations.

Viscosity. Some measure of the fluid's viscosity should be made or approximated over the system's operating temperature range. Flow and pressure calculations are related to this property. For example, if oil is the medium being pumped during summer and winter at constant pressure, the winter output (flow) decreases as its viscosity increases (i.e. gets colder).

Operating Pressures. When setting the design limits for a thermoplastic pipe system the engineer needs to consider the interdependence of the operating pressure, the operating temperature, the safety factor, and the expected life. Pipelines often do not operate at a stable pressure and the engineer needs to estimate the operating pressure range. The design should consider the highest long term operating pressure with recognition given to the additional safety factor gained at lower operating pressures.

Surge Pressures (Water Hammer). Special consideration should be given to pressurized systems with valves or shut-offs. The time of operation of a valve may convert the mass inertia of the flowing fluid into a high static head on the pressure side of the pipeline. Because of the flexibility, resilience and toughness of thermoplastic pipe it can absorb significant surge pressures.

Solids Content. Thermoplastic pipe is often used to convey slurry mixtures in processing pipelines and waste conveying pipelines. The design in these applications should consider the slurry solids content, its particle structure, abrasiveness, size distribution, and net specific gravity.

Fluid Temperature Range. The temperature of the fluid being conveyed has an effect on the service capability of thermoplastic pipe. Thermoplastic piping loses stiffness and tensile strength as temperature increases. As temperature rises, the normal operating pressure of the pipe should be derated or a heavier wall pipe should be specified to hold the same pressure at higher temperatures.

As the temperature decreases, the pipe gains strength. The pipe may be designed to hold rated pressure at 73.4°F with recognition of a greater safety factor at lower temperatures. Allowances for thermal expansion and contraction should be engineered into any installation based upon the fluid or environmental temperature.

Environmental Temperature Range. The design of thermoplastic pipe systems must take into consideration the environmental conditions. Piping system in the hot climates with high daily temperatures obviously must differ from installations in cold regions below the frost line. The system engineer needs to consider both the fluid temperature and environmental temperature when determining which SDR to specify, as well as in compensating for thermal expansion and contraction.

INSTALLATION CONSIDERATIONS

When designing a thermoplastic piping system, the designer should categorize each installation. Each installation requires selection of the proper SDR to support the external earth or traffic loads imposed on the pipe. Simultaneously, earth compaction factors should be specified for burial installations to ensure that the true earth and traffic loads do not exceed the system design limits. Some important factors to consider for each installation are:

Loads on Supported Pipelines. Supports must be spaced frequent enough to prevent the pipeline from any sag caused by the weight of the pipe and contents. The proper spacing of supports also allow for control or restraint of thermal expansion.

Loads on Exposed Pipelines. Chapter 5 will fully cover aboveground piping installation. Pipelines installed overland are exposed to numerous hazards. Changes in temperature causes the pipe to contract or expand as all materials do. The pipe movement caused by this thermal movement should be controlled by such means as snaking, anchoring or shallow trenching. This will prevent abrasion due to movement as well as possible kinking. When the temperature increases, the pressure rating of the pipe is decreased. When the temperature decreases, the safety factor on pipe pressure is increased.

Loads on Buried Pipelines. When designing buried thermoplastic piping system the design must consider the earth load, loads imposed by settling, loads imposed by variable water tables, etc.

Live Traffic Loads. Traffic operating over a buried pipeline (or even near a pipeline) causes the earth to move under its weight. This ever-so-slight movement is a dynamic load transfer from the vehicle to the ground. The heavier the vehicle, the greater the load transfers. To distribute and reduce the load on the pipe, it can be buried deeper (within limits) and/or located farther from traffic. The stress on the pipe may also be reduced by increasing the soil compaction (density) or eliminated by conducting it through a casing pipe. The system designer should review the various traffic weight classes, soil compaction factors and the associated stress presented in the design section.

When designing a thermoplastic piping system it is often justified to select a pipe size, wall thickness or SDR other than that determined through an engineering analysis. Selecting a thicker wall is often specified in slurry applications to lengthen the service life and maintain the pipe pressure rating as it wears. Upgrading may also be performed as a safeguard against unknowns such as variable operating conditions, system abuse, suspicious soil conditions, etc. The designer may select the next thicker wall and higher pressure rating to reduce hoop stress and increase the functional factor of safety. If a piping installation offers a high risk of damage and serious economic consequences or the public safety is involved, the engineers best judgment may be to incorporate additional safety factors in the pipeline design.

DESIGN CONSIDERATIONS FOR PLASTIC PIPING SYSTEMS

Thermoplastic piping systems, commonly referred to as plastic piping systems, are composed of various additives to a base resin or composition. Thermoplastics are characterized by their ability to be softened and reshaped repeatedly by the application of heat. Figure 4.5 lists the chemical names and abbreviations for a number of thermoplastic piping materials. Because of the slightly different formulations, properties of plastic piping materials (for example, polyvinyl chloride/PVC) may vary from manufacturer to manufacturer. Therefore, designs and specifications need to address specific material requirements on a type or grade basis, which may have to be investigated and confirmed with manufacturers.

Abbreviation	Chemical Name
ABS	Acrylonitrile-Butadiene-Styrene
CPVC	Chlorinated Poly(Vinyl Chloride)
ECTFE	Ethylene-Chlorotrifluoroethylene
ETFE	Ethylene-Tetrafluoroethylene
FEP	Perfluoro(Ethylene-Propylene) Copolymer
PE	Polyethylene
PFA	Perfluoro(Alkoxyalkane) Copolymer
PP	Polypropylene
PTFE	Polytetrafluoroethylene
PVC	Poly(Vinyl Chloride)
PVDC	Poly(Vinylidene Chloride)
PVDF	Poly(Vinylidene Fluoride)

FIGURE 4.5 Abbreviations for thermoplastic materials.

Corrosion

Unlike metallic piping, thermoplastic materials do not display corrosion rates. That is, the corrosion of thermoplastic materials is dependent totally on the material's chemical resistance rather than an oxide layer, so the material is either completely resistant to a chemical or it deteriorates. This deterioration may be either rapid or slow. Plastic piping system corrosion is indicated by material softening, discoloration, charring, embrittlement, stress cracking (also referred to as crazing), blistering, swelling, dissolving, and other effects. Corrosion of plastics occurs by the following mechanisms:

- Absorption
- Solvation
- Chemical reactions such as oxidation (affects chemical bonds), hydrolysis (affects ester linkages), radiation, dehydration, alkylation, reduction, and halogenation (chlorination)
- Thermal degradation which may result in either depolymerization or plasticization
- Environmental-stress cracking (ESC) which is essentially the same as stress-corrosion cracking in metals
- UV degradation
- Combinations of the above mechanisms

If reinforcing is used as part of the piping system, the reinforcement is also a material that is resistant to the fluid being transported. Material selection and compatibility review should consider the type and concentration of chemicals in the liquid, liquid temperature, duration of contact, total stress of the piping system, and the contact surface quality of the piping system.

Operating Pressures and Temperatures

The determination of maximum steady state design pressure and temperature is similar to that described for metallic piping systems. However, a key issue that must be addressed relative to plastic piping systems is the impact of both minimum and maximum temperature limits of the materials of construction.

Sizing

One of the basic principles of designing and specifying thermoplastic piping systems for liquid process piping pressure applications is that the short and long term strength of thermoplastic pipe decreases as the temperature of the pipe material increases.

Thermoplastic pipe is pressure rated by using the International Standards Organization (ISO) rating equation using the Hydrostatic Design Basis (HDB) as contained in ASTM standards and Design Factors (DF's). The use of DF's is based on the specific material being used and specific application requirements such as temperature and pressure surges. The following is the basic equation for internal hydraulic pressure rating of thermoplastic piping:

$$P_R = 2(HDS)\left(\frac{t}{D_m}\right)$$

where P_R = pipe pressure rating, psi
t = minimum wall thickness, in.
D_m = mean diameter, in.
$HDS = (HDB)(DF)$

It should not be assumed that thermoplastic fittings labeled with a pipe schedule designation would have the same pressure rating as pipe of the same designation. A good example of this is contained in ASTM D 2466 and D 2467, which specify pressure ratings for PVC schedule 40 and 80 fittings. These ratings are significantly lower than the rating for PVC pipe of the same designation. For thermoplastic pipe fittings that do not have published pressure ratings information similar to ASTM standards, the fitting manufacturer shall be consulted for fitting pressure-rating recommendations.

Joining

Common methods for the joining of thermoplastic pipe for liquid process waste treatment and storage systems are contained in Figure 4.6. In selecting a joining method for liquid process piping systems, the advantages and disadvantages of each method are evaluated and the manner by which the joining is accomplished for each liquid service is specified. Recommended procedures and specification for these joining methods are found in codes, standards and manufacturer procedures for joining thermoplastic pipe. Figure 4.7 lists applicable references for joining thermoplastic pipe.

Thermal Expansion

When designing a piping system where thermal expansion of the piping is restrained at supports, anchors, equipment nozzles, and penetrations, large thermal stresses and loads must be analyzed and accounted for within the design. The system PFDs and P&IDs are analyzed to determine the thermal conditions or modes to which the piping system will be subjected during operation. Based on this analysis, the

Joining Method	ABS	PVC	CPVC	PE	PP	PVDF
Solvent Cementing	X	X	X			
Heat Fusion				X	X	X
Threading*	X	X	X	X	X	X
Flanged Connectors**	X	X	X	X	X	X
Grooved Joints***	X	X	X	X	X	X
Mechanical Compression****	X	X	X	X	X	X
Elastomeric Seal	X	X	X	X	X	X
Flaring				X		

FIGURE 4.6 Thermoplastic joining methods.

Reference	Key Aspects of Reference
ASTM D 2657	Recommended practice for heat fusion.
ASTM D 2855	Standard practice for solvent cementing PVC pipe and fittings.
ASTM D 3139	Elastomeric gasketed connections for pressure applications.
ASTM F 1290	Recommended practice for electrofusion.

FIGURE 4.7 Thermoplastic joining methods.

design and material specification requirements from an applicable standard or design reference are followed in the design.

A basic approach to assess the need for additional thermal stress analysis for piping systems includes identifying operating conditions that will expose the piping to the most severe thermal loading conditions. Once these conditions have been established, a free or unrestrained thermal analysis of the piping can be performed to establish location, sizing, and arrangement of expansion loops, or expansion joints (generally, bellows or slip types).

If the application requires the use of a bellow or piston joint, the manufacturer of the joint shall be consulted to determine design and installation requirements. When expansion loops are used, the effects of bending on the fittings used to install the expansion loop are considered. Installation of the loop should be performed in consultation with the fitting manufacturer to ensure that specified fittings are capable of withstanding the anticipated loading conditions, constant and cyclic, at the design temperatures of the system. Terminal loadings on equipment determined from this analysis can then be used to assess the equipment capabilities for withstanding the loading from the piping system. It should also be noted that this termination analysis at equipment and anchor terminations should consider the movement and stress impacts of the "cold" condition.

No rigid or restraining supports or connections should be made within the developed length of an expansion loop, offset, or bend. Concentrated loads such as valves should not be installed in the developed length. Piping support guides should restrict lateral movement and should direct axial movement into the compensating configurations. Calculated support guide spacing distances for offsets and bend should not exceed recommended hanging support spacing for the maximum temperature. If that occurs, distance between anchors will have to be decreased until the support guide spacing distance equals or is less than the recommended support spacing. Use of the rule of thumb method or calculated method is not recommended for threaded Schedule 80 connections. Properly cemented socket cement joints should be utilized.

Expansion loops, offsets and bends should be installed as near as possible at the mid point between anchors.

Values for expansion joints, offsets, bends and branches can be obtained by calculating the developed length from the following equation:

$$L = n_1 \frac{3ED_o\sqrt{e}^{1\backslash 2}}{S}$$

where L = developed length, ft.

n_1 = conversion factor, $1/12$ ft/in.

E = tensile modulus of elasticity, psi

D_o = pipe outer diameter, in.

e = elongation due to temperature rise, in.

S = maximum allowable stress, psi

In determining the elongation due to temperature rise information from the manufacturer on the material to be used should be consulted. For example, the coefficient of expansion is 3.4×10^{-5} in/in/F) for Type IV Grade CPVC and 2.9×10^{-5} in/in/F) for Type I Grade I PVC. Other sources of information on the thermal expansion coefficients are available from the plastic pipe manufacturers.

PVC and CPVC pipe does not have the rigidity of metal pipe and can flex during expansion, especially with smaller diameters. If expansion joints are used, axial guides should be installed to ensure straight entrance into the expansion joint, especially when maximum movement of the joint is anticipated. Leakage at the seals can occur if the pipe is cocked. Independent anchoring of the joint is also recommended for positive movement of expansion joints.

PIPING SUPPORT AND BURIAL

Support for thermoplastic pipe follows the same basic principles as metallic piping. Spacing of supports is crucial for plastic pipe. Plastic pie will deflect under load more than metallic pipe. Excessive deflection will lead to structural failure. Therefore, spacing for plastic is closer than for metallic pipe. Valves, meters, and fittings should be supported independently in plastic pipe systems, as in metallic systems.

In addition, plastic pipe systems are not located near sources of excessive heat. The nature of thermoplastic pipe is that it is capable of being repeatedly softened by increasing temperature, and hardened by decreasing temperature. If the pipe is exposed to higher than design value ambient temperatures, the integrity of the system could be compromised.

Contact with supports should be such that the plastic pipe material is not damaged or excessively stressed. Point contact or sharp surfaces are avoided as they may impose excessive stress on the pipe or otherwise damage it.

Support hangers are designed to minimize stress concentrations in plastic pipe systems. Spacing of supports should be such that clusters of fittings or concentrated loads are adequately supported. Valves, meters, and other miscellaneous fittings should be supported exclusive of pipe sections.

Supports for plastic pipe and various valves, meters, and fittings, should allow for axial movement caused by thermal expansion and contraction. In addition, external stresses should not be transferred to the pipe system through the support members. Supports should allow for axial movement, but not lateral movement. When a pipeline changes direction, such as through a 90° elbow, the plastic pipe should be rigidly anchored near the elbow.

Plastic pipe systems should be isolated from sources of vibration such as pumps and motors. Vibrations can negatively influence the integrity of the piping system, particularly at joints. Support spacing for several types of plastic pipe are found in Figures 4.8, 4.9, and 4.10. Spacing is dependent upon the temperature of the fluid being carried by the pipe.

Nominal Pipe Size, inch	Maximum Support Spacing, ft at 60°F	Maximum Support Spacing, ft at 80°F	Maximum Support Spacing, ft at 100°F	Maximum Support Spacing, ft at 120°F	Maximum Support Spacing, ft at 140°F*
1	6.0	5.5	5.0	3.5	3.0
1.5	6.5	6.0	5.5	3.5	3.5
2	7.0	6.5	6.0	4.0	3.5
3	8.0	7.5	7.0	4.5	4.0
4	9.0	8.5	7.5	5.0	4.5
6	10.0	9.5	9.0	6.0	5.0
8	11.0	10.5	9.5	6.5	5.5
10	12.0	11.0	10.0	7.0	6.0
12	13.0	12.0	10.5	7.5	6.5
14	13.5	13.0	11.0	8.0	7.0

Note: The above spacing values are based on test data developed by the manufacturer for the specific product and continuous spans. The piping is insulated and is full of liquid that has a specific gravity of 1.0.
* The use of continuous supports or a change of material (e.g., to CPVC) is recommended at 140°F.

FIGURE 4.8 Support spacing for schedule 40 PVC pipe. Maximum support spacing in feet at various temperatures.

Nominal Pipe Size, inch	Maximum Support Spacing, ft at 68°F	Maximum Support Spacing, ft at 104°F	Maximum Support Spacing, ft at 140°F	Maximum Support Spacing, ft at 176°F
1	3.5	3.0	3.0	2.5
1.5	4.0	3.0	3.0	3.0
2	4.5	4.0	3.0	3.0
3	5.5	4.0	4.0	3.5
4	6.0	5.0	4.0	4.0
5	7.0	6.0	5.0	4.5

Note: The above spacing values are based on test data developed by the manufacturer for the specific product and continuous spans. The piping is insulated and is full of liquid that has a specific gravity of 1.0.

FIGURE 4.9 Support spacing for schedule 80 PVDF pipe. Maximum support spacing in feet at various temperatures. *(Source: Asahi/America, Piping Systems Product Bulletin P-97/A)*

The determining factor to consider in designing buried thermoplastic piping is the maximum allowable deflection in the pipe. The deflection is a function of the bedding conditions and the load on the pipe. The procedure for determining deflection is as follows:

$$\%\text{deflection} = \frac{100Y}{D_o}$$

where: Y = calculated deflection
D_o = outer pipe diameter, in.

$$Y = \frac{K_x d_e w}{0.149(PS).061(E)}$$

Nominal Pipe Size, inch	Maximum Support Spacing, ft at 73°F	Maximum Support Spacing, ft at 100°F	Maximum Support Spacing, ft at 120°F	Maximu m Support Spacing, ft at 140°F	Maximum Support Spacing, ft at 160°F	Maximum Support Spacing, ft at 180°F
1	6.0	6.0	5.5	5.0	3.5	3.0
1.5	7.0	6.5	6.0	5.5	3.5	3.0
2	7.0	7.0	6.5	6.0	4.0	3.5
3	8.0	8.0	7.5	7.0	4.5	4.0
4	8.5	8.5	8.5	7.5	5.0	4.5
6	10.0	9.5	9.0	8.0	5.5	5.0
8	11.0	10.5	10.0	9.0	6.0	5.5
10	11.5	11.0	10.5	9.5	6.5	6.0
12	12.5	12.0	11.5	10.5	7.5	6.5

Note: The above spacing values are based on test data developed by the manufacturer for the specific product and continuous spans. The piping is insulated and is full of liquid that has a specific gravity of 1.0.

FIGURE 4.10 Support spacing for schedule 80 CPVC pipe. Maximum support spacing in feet at various temperatures.

where Y = calculated deflection
K_x = bedding factor (see Figure 4.11)
d_e = deflection lag factor (see Figure 4.12)
w = weight per length of overburden, lb/in.
PS = pipe stiffness, psi
E = soil modulus, psi (see Figure 4.13)

$$w = \frac{HD_o\gamma}{144}\pi D_o$$

where w = weight per length of overburden, lb/in.
H = height of cover, ft.
D_o = outer pipe diameter, in.
γ = density of soil lb/ft^3
π = soil overburden pressure, psi

$$PS = \frac{EI_a}{.149R^3}$$

where PS = pipe stiffness, psi
E = modulus of elasticity of pipe, psi
I_a = area moment of inertia per unit length of pipe, in4/in
R = mean radii of pipe, psi

$$R = \frac{D_o t}{2}$$

where R = mean radii of pipe, psi
D_o = outer pipe diameter, in.
t = average wall thickness, in

$$I_a = \frac{t^3}{12}$$

where I_a = area moment of inertia per unit length of pipe, in^4/in
t = average wall thickness, in.

Proper excavation, placement, and backfill of buried plastic pipe is crucial to the structural integrity of the system. It is also the riskiest operation, as a leak in the system may not be detected before contamination has occurred. A proper bed, or trench, for the pipe is the initial step in the process. In cold weather areas, underground pipelines should be placed no less than one foot below the frost line. The trench bottom should be relatively flat, and smooth, with no sharp rocks that could damage the pipe material. The pipe should be bedded with a uniformly graded material that will protect the pipe during backfill. Typical installations use an American Association of State Highway Transportation Officials (AASHTO) #8 aggregate, or pea-gravel for six inches below and above the pipe. These materials can be dumped in the trench at approximately 90-95 percent Proctor without mechanical compaction. The remainder of the trench should be backfilled with earth, or other material appropriate for surface construction, and compacted according to the design specifications.

Type of Installation	K_x
Shaped bottom with tamped backfill material placed at the sides of the pipe, 95% Proctor density or greater	0.083
Compacted coarse-grained bedding and backfill material placed at the side of the pipe, 70-100% relative density	0.083
Shaped bottom, moderately compacted backfill material placed at the sides of the pipe, 85-95% Proctor density	0.103
Coarse-grained bedding, lightly compacted backfill material placed at the sides of the pipe, 40-70% relative density	0.103
Flat bottom, loose material placed at the sides of the pipe (not recommended); <35% Proctor density, <40% relative density	0.110

FIGURE 4.11 Bedding factor, K_x.

Installation Condition	d_c
Burial depth <5 ft. with moderate to high degree of compaction (85% or greater Proctor, ASTM D 698 or 50% or greater relative density ASTM D-2049)	2.0
Burial depth <5 ft. with dumped or slight degree of compaction (Proctor >85%, relative density >40%)	1.5
Burial depth >5 ft. with moderate to high degree of compaction	1.5
Burial depth >5 ft. with dumped or slight degree of compaction	1.25

FIGURE 4.12 Deflection lag factor, d_c.

Soil Type and Pipe Bedding Material	Dumped	Slight <85% Proctor >40% relative density	Moderate 85-95% Proctor 40-70% relative density	High >90% Proctor >70% relative density
Fine-grained soils (LL>50) with medium to high plasticity CH, MH, CH-MH	No data available – consult a soil engineer or use E = 0	No data available – consult a soil engineer or use E = 0	No data available – consult a soil engineer or use E = 0	No data available – consult a soil engineer or use E = 0
Fine-grained soils (LL<50) with medium to no plasticity CL, ML, ML-CL, with 25% coarse-grained particles	50	200	400	1000
Fine-grained soils (LL<50) with no plasticity CL, ML, ML-CL, with >25% coarse-grained particles	100	400	1000	2000
Coarse-grained soils with fines GM, GC, SM, SC contains >12% fines	100	400	1000	2000
Coarse-grained soils with little or no fines GW, SW, GP, SP contains <12% fines (or any borderline soil beginning with GM-GC or GC-SC)	200	1000	2000	3000
Crushed rock	1000	3000	3000	3000

Notes: LL = liquid limit

FIGURE 4.13 Values of E^o Modulus of soil reaction for various soils. *E* for degree of compaction of bedding. lbs/ft^2. *(Source: AWWA C900, Table A.4)*

GENERAL DESIGN CONSIDERATIONS

The design of a thermoplastic piping system is a straightforward process. It is a repetitive procedure that requires attention to detail and the exercise of some judgment. Each design can be different and operating conditions are diverse. Regardless of design criteria, the design process is relatively simple. Basically, it involves the selection of a pipe; sized to transport the required flow and the selection of a pipe wall thickness adequate handle the system pressure safely. The system pressure and temperature determine the pipe's wall thickness. Flow velocity and pressure drop determine the pipes inside diameter and the system flow rate. The pipe selection process is a repetitive procedure which matches and balances pipe I.D. and wall thickness in order to optimize pressure and flow capabilities at a reasonable cost.

The following items should be evaluated when designing a specific thermoplastic piping system.

- Consider service life requirements
- Consider the pipe inside diameter required to meet flow requirements

- Consider the pipe and wall thickness required to meet pressure requirements
- Work with the pipe size and wall thickness until the flow and pressure in the pipe selected are acceptable for both
- Consider the external "Earth Load and Live Loads"
- Adjust pipe wall thickness as required for external loads
- Consider rerating the pipe based upon the environmental or operating temperature
- Review the final pipe size and wall thickness to meet flow, pressure and external load requirements at a given temperature and system life expectancy

SYSTEM FLOW REQUIREMENTS

Thermoplastic pipe has an exceptionally smooth pipe wall. This results in its excellent flow capacity. Thermoplastic pipes have less drag, less tendency for flow turbulence, no corrosion, and are less susceptible to deposits or bacterial growth. Because of its excellent flow properties, a smaller diameter pipe can often be specified to carry a given volume when compared with steel, cast iron or concrete.

Pressurized Full Flow

Many equations are available to show the relationship between fluid flow and pressure drop in a given pipeline. The equations typically involve a friction factor that is dependent on the pipe materials. Darcy-Weisbach and Hazen-Williams are commonly used equations.

Initial Flow Estimates. If the inside diameter of a particular pipe size is known, the flow rate, in gallon per minute, can be calculated by assuming a nominal velocity.

$$Q = 2.449VD^2$$

where Q = Gallons per minute
V = Velocity (ft./sec.)
D = $I.\ D.$ (inches)

By knowing the required gallon per minute and assuming a nominal flow velocity we can calculate the pipe diameter or velocity.

$$D = .639\sqrt{\frac{Q}{V}}$$

$$V = .408\left(\frac{Q}{D^2}\right)$$

By using these three formulas we can establish a required range for the pipe diameter, flow rate, and flow velocity.

The designer will need to exercise judgment in the selection of pipe sizes to best meet the design parameters and goals. The following items can prove helpful:

- For a given flow rate, a larger diameter pipe will have a lower velocity and a lower pressure drop
- For a given flow rate, a smaller diameter pipe will have higher velocity and a higher pressure drop
- Higher velocity in larger diameter pipes produced less frictional head loss when compared to the same velocity in a smaller pipe
- For laminar flow, the loss in pressure through a pipe is inversely proportional to the fourth power of the pipe inside diameter
- For turbulent flow, the loss in pressure through a pipe is directly proportional to the square of the flow rate and inversely proportional to the fifth power of the inside diameter

PRESSURE LOSS IN FITTINGS

Valves, fittings, and other piping system components in the thermoplastic piping system result in additional pressure loss within the piping system. The designer should consider the additional pressure losses. There are several methods for calculating the pressure loss caused by piping system components that are available in many fluid flow and piping handbooks. Listed below are various common thermoplastic piping system components and the associated pressure loss through the fitting expressed as an equivalent length of straight pipe in terms of diameters. The inside diameter (in feet) multiplied by the equivalent length diameters gives the equivalent length (in feet) of pipe. This equivalent length of pipe is added to the total footage of the piping system when calculating the total system pressure drop.

These equivalent lengths should be considered an approximation suitable for most installations.

Fabricated Fitting	Equiv. Length
Running Tee	20 D
Branch Tee	50 D
90(Fabricated Elbow	30 D
60(Fabricated Elbow	25 D
45(Fabricated Elbow	18 D
45(Fabricated Wye	60 D
Conventional Globe Valve (Full Open)	350 D
Conventional Angle Valve (Full Open)	180 D
Conventional Wedge Gate Valve (Full Open)	15 D
Butterfly Valve (Full Open)	40 D

Once the proper pipe sizes have been chosen based upon the flow requirements, the following items should be considered:

- The total system pressure drop should not exceed the pressure rating of the thermoplastic pipe selected
- The pump pressure should exceed the system pressure drop, but should not exceed the pressure rating of the pipe selected

GRAVITY FLOW

Gravity Flow systems are common in industrial and municipal waste and sewer lines as well as water and slurry pipelines. Gravity flow systems may operate under full flow or partially full conditions. Because of the superior wall smoothness and excellent flow characteristics of thermoplastic pipe, they are an excellent choice for gravity flow piping systems.

The designer can specify smaller pipe diameters to support the required flow resulting in reduced system costs. In addition, the maintenance costs for thermoplastic piping systems are less than metal systems. The reduced operating costs and reliability of thermoplastic piping systems often result in improved service.

Full Flow. There are three key items required to properly select and size thermoplastic piping for a full flow gravity system:

1. GPM flow-rate requirements
2. The slope of the pipeline
3. Selection of an appropriate pipe inside diameter

For a full flow condition, the gallon per minute flow rate can be calculated using the Manning equation as follows:

$$Q = 98.3ARh^{2/3}S^{1/2}$$

where Q = Flow in gpm
Rh = Hydraulic radius (ID)(4)(inches)
S = Slope (ft./foot)
A = Cross sectional area of pipe I.D. in sq. inches
V = Velocity (ft./sec)
ID = Inside diameter in inches

(Note: Above formula includes $\eta = .009$)

The velocity can be calculated by:

$$V = 31.5R_h^{2/3}S^{1/2} = \frac{.320Q}{A}$$

The inside diameter by:

$$ID = \sqrt[2.67]{\frac{.03279Q}{S^{1/2}}}$$

And the slope by:

$$S = \frac{.001075Q^2}{ID^{5.34}}$$

Partial Flow. Gravity pipelines have a higher liquid flow capacity when flowing at 85-95 percent full than when 100 percent full. This is caused by the effect of reduced friction due to the liquid's contact with less pipe wall surface. The following illustrates the changes in velocity and flow capacity when compared to full flow.

Flow Capacity of Partially Full Pipes

% Full	Velocity (% of full)	Flow Capacity (% of full)
100	100	100
95	111	106.3
90	115	107.3
80	116	98
70	114	84
60	108	67
50	100	50
40	88	33
30	72	19
25	65	14
20	56	9
10	36	3

For gravity partial flow pipelines, the GPM flow rate can be calculated through the use of the Manning equation as follows:

$$Q = 98.3AR_h^{2/3}S^{1/2}$$

where Q = Flow in gpm
A = Pipeline cross-sectional flow area in square inches
R_h = Hydraulic Radius in inches
R_h = Flow area in sq. inches divided by wetted perimeter in inches
S = Slope or gradient (ft/ft)
V = Velocity in ft/sec

$$V = .320\left(\frac{Q}{A}\right)$$

It is normal practice to consider a partially full gravity flow pipeline as a full flow pipeline of a smaller, "equivalent" diameter. The "equivalent" diameter matches all the hydraulic characteristics of the larger, partial flow gravity pipeline. The velocity, GPM flow rate and slope are identical in each case. The equivalent diameter is four times the hydraulic radius. The hydraulic radius for partial flow gravity pipelines is defined as the ratio of the cross-sectional flow area divided by the wetted perimeter.

SLIPLINING FLOW CAPACITY

Thermoplastic piping, most often PE pipe, is frequently used for slip lining of sewers and waste handling systems. Determining the proper pipe size for sliplining is done by determining the maximum size of liner that can be inserted into the existing line and the flow required through the new liner. Manning's formula can be used to determine the flow of sewage. With the Manning Formula, a relationship of pipe diameters is established so that the liner size may be calculated that is required to restore the sewer to its original capacity. A good rule of thumb in sizing the slipline pipe is to allow 10 percent of the diameter as a clearance gap between the larger pipe and the liner.

Slurry Critical Flow

Thermoplastic piping has demonstrated its superior abrasion resistance when compared to conventional materials and field tests have demonstrated that thermoplastic piping outlast steel piping as much as 3 to 1. This section discusses some of the major design topics to be considered when designing a slurry pipeline. The designer is encouraged to pursue an in-depth study of each of these topics as they apply to a particular situation.

Slurry is a two-phase mixture of solid particles and a fluid. The two phases do not chemically react and can be separated by mechanical means. There are two types of slurry systems:

1. Non-settling slurries
2. Settling slurries

Non-settling slurries have the hydraulic flow characteristics of a viscous fluid. These type of slurry systems are designed according to standard procedures with allowances for the higher viscosity. The majority of slurry systems are the settling type. In settling systems the solids will tend to settle out of the carrier fluid. When the flow velocity is reduced in these systems, the fluid flow goes through settling phases. The tendency for settling is often countered by increasing the flow velocity.

Flow Phases. The flow velocity of a slurry affects the mode of the flow. When the flow velocity is high and then is gradually slowed, the slurry passes through four flow modes:

- Homogeneous Flow: This term describes a system in which the solids are uniformly distributed throughout the liquid. This is the most desirable of all flow modes because the particles do not contact the wall as frequently, thus reducing abrasion
- Heterogeneous Flow: The solids tend to flow nearer the bottom of the pipe but do not actually slide on the pipe bottom. This is the most economical flow mode and is typically used for sand sized solids
- Saltation Flow: In this mode, solid particles tend to bounce along the bottom of the pipe. This flow is particularly aggressive in its abrasion of pipe.
- Sliding Bed Flow: This mode of flow is generally unsatisfactory. Solids slide and roll on the pipe bottom. Excessive corrosion in the pipe bottom occurs quickly. Blockages can occur frequently

Increasing the flow velocity will often avoid sliding bed and saltation flow. However, the operational cost, such as high power requirements, could increase significantly. The critical velocity of a slurry is the velocity range that the solid particles tend to drop out of suspension and settle to the bottom of the pipe. The critical velocity is determined by the particle size and shape, size distribution, concentration, particle density and carrier fluid density. Some particle solids form a viscous fluid with the liquid carrier. This is a homogeneous mode. For these materials when the flow velocity makes a transition from turbulent flow to laminar flow, the viscous, homogeneous fluid makes a transition from a smooth mixture to a separated mixture. When turbulence stops and laminar flow develops, the homogeneous mode of flow ends and the saltation or sliding bed mode begin. Turbulent flow is essential to keeping these type solids in suspension. When designing a piping system for this type of slurry consider:

- As the slurry viscosity increases, the flow velocity must be increased to prevent settling
- As the solids concentration increases, the flow velocity must be increased to prevent settling
- As the particle size decreases, the flow velocity must be increased to prevent settling

Slurries with high concentrations of fine size particles can be more abrasive than slurries with larger size particles. The basic reason is that the particle/wall contact is greater and more frequent with fine slurry.

GAS FLOW

The flow capacity for thermoplastic piping transporting a gaseous product may be found through the use of the Mueller relationship for pressure drop in a plastic pipe as noted below.

$$Q = \frac{2826}{G^{0.425}} \sqrt[0.575]{\frac{P_1^2 - P_2^2}{L}}\, D^{2.725}$$

where Q = Gas flow rate in standard cubic feet per hour, SCFH
G = Specific gravity (Air = 1.0)
P_1 = Inlet pressure, psia
P_2 = Outlet pressure, psia
L = Pipeline length, feet
D = Inside diameter, inches

LIFE EXPECTANCY

Thermoplastic piping systems offer long, trouble free service. Piping manufacturers are continuously testing their products in accordance with ASTM procedures to ensure the long-term strength and performance of their products. Standard samples of pipe are held at constant pressure and temperature according to specifications. Manufacturers use high pressures and temperatures so that some samples are forced to fail. They then develop statistical predictions based on the test data as to the service life and long term strength of the pipe based upon the number of forced failures and the time it took them to fail. This data is plotted and a stress-life curve graph is developed. The stress-life curve provides a relationship between the expected life of the pipe and the internal stress at a given working pressure and temperature. A basis recommendation for all HDPE pipe systems is to expect a 50-year life from the system. The hydrostatic design stress according to the industry accepted formula as defined in ASTM D-2837.

$$P = \frac{2St}{(D - t)}$$

where S = Hydrostatic design stress
P = Working pressure
D = Average outside diameter
t = Minimum wall thickness

SYSTEM PRESSURE

Thermoplastic piping systems are designed for one of three types of service:

1. Pressurized flow
2. Non-pressurized flow
3. Vacuum flow

When designing a pressurized pipe system, the pipe selected must hold the internal pressure safely and continuously. In a non-pressurized system such as a gravity flow sewer, pipe selection depends on other factors. Vacuum piping systems must use a pipe that resists collapse. The design engineer will use different design criteria and calculations for each type of installation.

POSITIVE PRESSURE PIPELINES

Many manufacturers of thermoplastic piping material use the standard dimension ratio (SDR) method of rating pressure pipe. SDR is the ratio of the pipe O.D. to the minimum thickness of the wall of the pipe. It can be expressed mathematically as:

$$\mathrm{SDR} = \frac{D}{t}$$

where SDR = Standard Dimension Ratio
D = Pipe outside diameter in inches
t = Pipe minimum wall thickness in inches

For a given SDR the ratio of the O.D. to the minimum wall thickness remains constant. An SDR 11 means the O.D. of the pipe is eleven times the thickness of the wall. This remains true regardless of diameter. For example, a 14-inch O.D. pipe with a wall (t) of 1.273-inch is an SDR 11 pipe. An 8-inch O.D. pipe with a wall (t) of .785-inch is also an SDR 11 pipe. Common SDR ratios are SDR 9.3, SDR 11, SDR 13.5, SDR 15.5, SDR 17, SDR 19, SDR 21, SDR 26 and SDR 32.5. For high SDR ratios, the pipe wall is thin in comparison to the pipe O.D. For low SDR ratios, the wall is thick in comparison to the pipe O.D. Given two pipes of the same O.D., the pipe with the thicker wall will be stronger than the one with the thinner wall. Pipes with a high SDR rating have low-pressure ratings and pipe with low SDR ratings have high-pressure ratings because of the relative wall thickness.

The pressure rating of thermoplastic pipe is mathematically calculated from the SDR and the allowable hoop-stress. The allowable hoop-stress is commonly known as the long-term hydrostatic design stress. This is the stress level that can exist in the pipe wall continuously with a high degree of confidence that the pipe will operate under pressure for at least 50 years with safety. The American Society for Testing and Materials (ASTM) and the Plastics Pipe Institute (PPI) has adopted the following formula relating SDR and hydrostatic design stress as the standard for the industry.

$$P = \frac{2St}{(D - t)}$$

where P = Pressure rating (psi)
D = Pipe OD (inches)
t = minimum wall thickness (inches)
S = Hydrostatic Design Stress

$$\mathrm{SDR} = \frac{D}{t}$$

From the formula you can see that all pipes of the same SDR (regardless of diameter) will have the same pressure rating for a given design stress.

Water Hammer/Pressure Surge. The effects and calculations for water hammer or pressure surge were covered in detail in chapter 3. The following section contains some key design considerations for handling the effects within thermoplastic piping systems.

Since all moving objects have mass and velocity, any flowing liquid has momentum and inertia. When flow is suddenly stopped, the mass inertia of the flowing stream is converted into a shock wave or high static head on the pressure side of the pipeline. Some of the more common causes of hydraulic transients are

1. The opening and closing (full or partial) of valves
2. Starting and stopping of pumps
3. Changes in turbine speed
4. Changes in reservoir elevation
5. Reservoir wave action
6. Liquid column separation
7. Entrapped air

Thermoplastic piping materials are well suited to handle occasional surge pressures. Some PE manufactures provide pipe that can withstand occasional surge pressures up to 2.5 times the rated pressure capability of the pipe without a cumulative effect. This is due to the long-term modulus of the material being only a fraction of the short-term modulus.

In general, good system design will eliminate quick opening/closing valves on anything but very short lines. The design engineer should use judgment with regard to the addition of surge pressures to operating pressures when selecting pipe SDRs. The following rules of thumb may be of help:

- Occasional shock pressures can be accommodated within the design safety factor. Due to the short time duration of the surge pressure, occasional shock wave surge pressures to 2.5 times the SDR pressure rating at 73.4°F are usually allowable

- If surge pressure or water hammer is expected in a system, keep the flow velocity on the low side of the velocity range

If surge pressure or water hammer is expected, maximize the time required to shut off a valve or reduce flow. A shutoff cycle 6–10 times the time period 2L/S is suggested to minimize surge pressures by gradually slowing the fluid flow. If constant and repetitive surge pressures are present, the excess pressure should be added to the nominal operating pressure when selecting the pipe SDR.

Cyclic Overpressure

There are many causes to short-term cyclic overpressure. Activities such as a stuck relief valve, a plugged discharge line or repetitive freeze/thaw cycles can cause extended rises in pressure. In steel, concrete, PVC or fiberglass pipe, the effect of the increased pressure can be devastating. With many of the thermoplastic piping materials the pipe wall is able to stretch (strain) with the freezing water and the pipe will return to its original condition after the frozen water has thawed. Although some residual strain may be evident, the physical properties of the pipe resin are not adversely affected and the performance of the pipe at normal operating conditions is not affected. Due to the innate elastic characteristics of many thermoplastic piping materials, they are capable of withstanding these types of cyclic loadings without damage to the pipe's performance. The effects of extended and repeated over pressure can be tolerated within specific limits. Many thermoplastic pipes have an inherent ability to recover from the strain of overpressures. If the recovery period at a normal level of stress is equal to or greater than the duration of the over pressure, the pipe can be subjected to the stresses for short periods without affecting their long term strength, endurance, and performance.

The basic limitation on short-term overpressure cycles is to stay within the elastic limits of the pipe material. If the system pressure exceeds 2.5 times the rated pressure of the pipe for any length of time, permanent strain or deformation of the pipe occurs. As a result, the expected service life of the pipe can be dramatically reduced. When overpressure cycling is expected as a regular condition of operation, the highest pressure anticipated the majority of the operating times should be considered as the operating pressure and it should be treated as though it would persist continuously for the design life of the system.

Longitudinal Stress from Internal Pressure

When a fully restrained pipeline such as a buried or anchored pipeline is pressurized, longitudinal stresses develop in the pipe wall. The longitudinal stress is calculated as follows:

$$S_L = \frac{\mu P(D - t)}{2t}$$

where S_L = Longitudinal tensile stress, psi
μ = Poisson's ratio
P = Internal operating pressure, psi
D = Pipe outside diameter, inches
t = Pipe wall thickness, inches

Thermal Expansion and Contraction

Thermal expansion and contraction are key items of concern in the design of a thermoplastic piping system. Design parameters should be developed and incorporated into the installation specifications. Thermoplastic piping materials have a higher coefficient of expansion than some other common pipeline materials, however, the forces generated by thermal stresses are much lower because the modulus of elasticity is lower and it is capable of stress relaxation.

There are many methods available to the designer to control expansion or contraction. One method is to install the pipeline when it is within 10 to 15 degrees Farenheit of its operating temperature. Other methods of controlling expansion/contraction are pertinent to certain types of installations and are briefly discussed.

Supported Pipelines

A common practice is to install the pipe in a warm condition in a straight line while it is in an expanded state. As the pipeline cools it develops a tensile stress and the pipeline remains straight between supports. As the pipe warms to its installation temperature due to seasonal change or operating conditions, it returns to its installation condition and straightness. In this manner, sag between supports is minimized.

Overland Pipes

Controlling the expansion and contraction of overland surface lines is difficult because the uneven soil friction between the pipe and the ground does not allow distributed lateral deflections to occur uniformly. In the worse case, all deflection may occur in one area where friction is low and the pipe may kink. This condition will most likely occur in empty lines or where large, sudden operating temperature changes occur. If overland pipelines are installed in a snaked pattern, thermal expansion/contraction can be controlled through control of lateral deflection. During pipeline warming, the "S" configuration becomes slightly greater. As the pipe cools, the pipeline becomes straighter. Surface lines that are continuously operated full of fluid normally experience small, slow temperature variations and are easy to control. The weight of the fluid also increases friction and reduces deflection. However, it may necessary to anchor the line at intervals to direct and limit the deflection in any one segment of the pipeline.

Buried Pipelines

Buried pipeline installations offer a significant degree of restraint due to soil friction. This is further controlled because the pipe usually lies in a slight "S" curve in the trench as it is installed. Because the temperature of the soil is fairly constant, temperature changes that do occur take place over a yearly season. Due to the enormous heat sink capability of the earth, the magnitude of any temperature change is reduced and the time required to effect that temperature change extended.

A buried process pipeline operating at a specific temperature may develop some initial thermal stress upon start-up. This is usually restrained by soil friction and dissipated with time by stress relaxation. As the pipeline continues operation, it tends to bring the soil envelope surrounding the pipeline into equilibrium with the operating temperature. If a minor temperature change does occur, its effect is further minimized by the massive thermal inertia within the pipe wall and in the soil surrounding the pipeline.

TRANSITION CONNECTIONS

The stress and the corresponding force developed by temperature change in a restrained pipeline are independent of the length and the burial conditions of the pipe. If pipe movement at the end sections cannot be tolerated, the pipe must be anchored mechanically to resist the thermal forces. A normal design practice is to use concrete collars to transfer the thermal force into the soil enveloping the pipe. Adequate frictional resistance must also be provided to transfer the force from the pipe into the concrete collar. If the pipe is not anchored at the ends to resist movement, portions may expand or contract as the temperature changes. This change in length will extend into the burial trench to a point at which the frictional resistance of the backfill is equal to the thermal force. The normal design practice for considering the movements or forces are to; isolate the end connection, by means of an anchor or collar, from the effects of thermal movement of the rest of the pipeline. The following example may be helpful.

EXAMPLE

Assume a 4 ινψη diameter SDR 15.5 process pipeline is buried five feet deep in dense sandy soil with a high water table. The ground temperature is 60°F. Under intermittent operating conditions, it must carry 40°F water 1000 ft. to a remote part of the plant in a straight path. Calculate the following:

- The temperature change
- The theoretical strain

- The theoretical length change
- The instantaneous tensile stress in the pipe wall
- The tensile force
- Design a collar to isolate the terminal connection from the effects of thermal contraction

Pipe:		4-inch SDR 15.5
Pipe Wall Cross-sectional Area		3.83 square inches
Linear Coefficient of Thermal Expansion		1.2×10^{-4} in./in./μF
Instantaneous Modulus of Elasticity	E	180,000 psi at 73.4 μF
Temperature Change	μT	20 μF
Soil Coefficient of Friction	μ	0.10
Length of Run		L 1000 feet
Soil Density	μ	130 pounds per cubic foot
Depth of Burial	h	5 feet

Calculate:

Thermal Strain:

$\epsilon = \alpha \Delta T$

$\mu = \mu 1.2 \times 10^{-4}$ in./in./ (F)(20 (F) = .0024 inch per inch

Theoretical-Instantaneous Unrestrained Contraction:

$\Delta L = L\epsilon$

$$\Delta L = 1000 \text{ ft} \left(\frac{12 \text{ inch}}{\text{ft}}\right)\left(\frac{.0024 \text{ inch}}{\text{inch}}\right) = 28.8 \text{ inches}$$

Note: Since the soil restrains the pipe, it will not change length but will instead develop tensile stress due to contraction.

Theoretical Tensile Stress: $E = \dfrac{\sigma}{\epsilon}$

$\sigma = E\epsilon = (180{,}000 \text{ psi})(.0024) = 432 \text{ psi}$

Actual Tensile Stress: (per ASTM D2513)

$$\sigma = \frac{432}{2} = 216 \text{ psi}$$

Tensile stress

Actual Tensile Force: $F = \sigma \xi A$

$F = (216 \text{ psi})(3.83 \text{ sq. in.}) = 827.3$ lbs. (tensile)

Soil Frictional Resistance: $f = \mu N$

where Soil Pressure $= h = \dfrac{(130 \text{ lbs})(5 \text{ ft})}{\text{cubic ft}} = 650$ psf $= 4.5$ psi

Normal Pressure $= N =$ Normal force due to soil pressure on circumference of pipe ring one inch wide

$N = ((D) \times (1'' \text{ ring}) \times (\text{soil pressure})$
$= ((4.5'') \times (1'') \times (4.5 \text{ psi})$
$= (14.14 \text{ sq. in.})(4.5 \text{ psi}) = 63.63$ lbs. per in. of pipe

Frictional Resistance $f = (N = (0.10)(63.63 \text{ lbs./in.})$
$= 6.363$ lb. per inch of pipe due to soil friction

Note: Beyond 122.8 inches (10.2 ft.) the soil friction will overcome the tensile stress force developed by thermal contraction of the pipeline. This is calculated by dividing the tensile force in the pipe by the frictional resistance of the soil (i.e.: 827.3 lbs. (6.363 lbs./in. = 130 inches).

Theoretical Movement of Unrestrained Ends:

$\Delta L = (130 \text{ inch}).0024 \text{ in/in} = .312$ inches

DESIGN OF COLLAR

By pouring a square concrete collar around the pipe and branch saddles into undisturbed soil, the tensile force of 827.3 lbs. is removed from the pipe connection and is evenly distributed into the soil. Assume a collar 12 inches square and 6 inches wide is used.

Area of Collar $= A = (12'' \times 12'') -$ (Cross Sectional Area of Pipe)
$= (144 - 16)$ sq. in. $= 128$ sq. in. surface area

Compressive Stress on soil due to load transfer by collar face:

$$S = F\,(A = 827.3 \text{ lbs. } (128 \text{ sq. in.} = 6.5 \text{ psi}$$

DESIGN CONSIDERATIONS FOR VARIOUS THERMOPLASTIC PIPE MATERIALS

Polyvinyl Chloride (PVC)

Polyvinyl chloride (PVC) is the most widely used thermoplastic piping system. PVC is stronger and more rigid than the other thermoplastic materials. When specifying PVC thermoplastic piping systems particular attention must be paid to the high coefficient of expansion-contraction for these materials in addition to effects of temperature extremes on pressure rating, viscoelasticity, tensile creep, ductility, and brittleness.

PVC pipe is available in sizes ranging from ¼ to 16 inch, in Schedules 40 and 80. PVC piping shall conform to ASTM D 2464 for Schedule 80 threaded type; ASTM D 2466 for Schedule 40 socket type; or ASTM D 2467 for schedule 80 socket type.

Maximum allowable pressure ratings decrease with increasing diameter size. To maintain pressure ratings at standard temperatures, PVC is also available in Standard Dimension Ratio (SDR). SDR changes the dimensions of the piping in order to maintain the maximum allowable pressure rating.

For piping larger than 4 inches diameter, threaded fittings should not be used. Instead socket welded or flanged fittings should be specified. If a threaded PVC piping system is used, two choices are available, either use all Schedule 80 piping and fittings, or use Schedule 40 pipe and Schedule 80 threaded fittings. Schedule 40 pipes will not be threaded. Schedule 80 pipes would be specified typically for larger diameter pipes, elevated temperatures, or longer support span spacing. The system is selected based upon the application and design calculations.

The ranking of PVC piping systems from highest to lowest maximum operating pressure is as follows: Schedule 80 pipe socket-welded; Schedule 40 pipe with Schedule 80 fittings, socket-welded; and Schedule 80 pipe threaded. Schedule 40 pipe provides equal pressure rating to threaded Schedule 80, making Schedule 80 threaded uneconomical. In addition, the maximum allowable working pressure of PVC valves is lower than a Schedule 80 threaded piping system.

Acrylonitrile-Butadiene-Styrene (ABS)

Acrylonitrile-Butadiene-Styrene (ABS) is a thermoplastic material made with virgin ABS compounds. Pipe is available in both solid wall and cellular core wall, which can be used interchangeably. Pipe and fittings are available in 1½ inch through 12 inch. The pipe can be installed above or below grade.

ASTM D 2282 specifies requirements for solid wall ABS pipe. ASTM D 2661 specifies requirements for solid wall pipe for drain, waste, and vent pipe and fittings with a cellular core. Solid wall ABS fittings conform to ASTM D 2661. ASTM D 3311 specifies the drainage pattern for fittings.

ABS compounds have many different formulations that vary by manufacturer. The properties of the different formulations also vary extensively. ABS shall be specified very carefully and thoroughly because the acceptable use of one compound does not mean that all ABS piping systems are acceptable. Similarly, ABS compositions that are designed for air or gas handling may not be acceptable for liquids handling.

Pigments are added to the ABS to make pipe and fittings resistant to ultraviolet (UV) radiation degradation. Pipe and fittings specified for buried installations may be exposed to sunlight during construction, however, and prolonged exposure is not advised.

ABS pipe and fittings are combustible materials, however they may be installed in noncombustible buildings. Most building codes have determined that ABS must be protected at penetrations of walls, floors, ceilings, and fire resistance rated assemblies. The method of protecting the pipe penetration is using a through-penetration protection assembly that has been tested and rated in accordance with ASTM E 814. The important rating is the "F" rating for the through penetration protection assembly. The "F" rating must be a minimum of the hourly rating of the fire resistance rated assembly that the ABS plastic pipe penetrates. Local code interpretations related to through penetrations are verified with the jurisdiction having authority.

Chlorinated Polyvinyl Chloride (CPVC)

Chlorinated polyvinyl chloride (CPVC) is more highly chlorinated than PVC. CPVC is commonly used for chemical or corrosive services and hot water above 140°F and up to 210°F. CPVC is commercially available in sizes of ¼ inch to 12 inch for Schedule 40 and Schedule 80. Exposed CPVC piping should not be pneumatically tested, at any pressure, due to the possibility of personal injury from fragments in the event of pipe failure.

ASTM specifications for CPVC include: ASTM F 437 for Schedule 80 threaded type; ASTM F 439 for Schedule 80 socket type; and ASTM F 438 for Schedule 40 socket type. However, Schedule 40 socket may be difficult to procure.

Polyethylene (PE)

Polyethylene (PE) piping material properties vary as a result of manufacturing processes. Figure 4.14 lists the common types of PE, although an ultra high molecular weight type also exists. PE should be protected from ultraviolet radiation by the addition of carbon black as a stabilizer; other types of stabilizers do not protect adequately. PE piping systems are available in sizes ranging from ½ inch to 30 inch. Like PVC, PE piping is available in SDR dimensions to maintain maximum allowable pressure ratings.

Type	Standard	Specific Gravity
Low Density (LDPE)	ASTM D 3350, Type I	0.91 TO 0.925
Medium Density (MDPE)	ASTM D 3350 Type II	0.926 to 0.940
High Density (HDPE)	ASTM D 3350 Type III and ASTM D 1248 Type IV	0.941 to 0.959

FIGURE 4.14 Polyethelene designations.

CHAPTER 5
ABOVE GROUND PIPE DESIGN

The installation of plastic pipe above ground is similar to the installation of other types of pipe. There are, however, differences, and these differences must be taken into account when planning and installing a system of plastic pipe to deliver material from one point to another. Not all designers and installers have an extensive understanding of the requirements for working with plastic pipe. In the scheme of things, plastic pipe is fairly new to the piping industry. Yes, the pipe has been available for many years, but it's use was slow to catch on in some regions.

When we talk of plastic pipe it might seem as though we are referring to one type of pipe. In one sense, we are, but there are many types of plastic pipe, and the various types of materials can require different installation guidelines. Whenever you are working with plastic pipe it is wise to obtain specifications and instructions directly from the manufacturer of the product. There is no substitute for the authority of recommendations provided by a product manufacturer.

There are critics of plastic pipe. Some people feel that the material is not up to the rigors which can be withstood by other types of materials.

In fact, manufacturers have insisted on exacting quality-control procedures to produce products that are dependable for their intended uses. Table 5.1 shows the application ranges of plastic pipe. In addition to the manufacturers, many testing agencies have been involved in the evolution of plastic piping. For example, when problems were experienced with polybutylene (PB) piping having faulty joints and unexpected ruptures, the development of cross-linked polyethylene (PEX) came to the industry. Professional associations, contractors, testing agencies, manufacturers, and consumers all have an eye on the use of plastic pipe.

As good as plastic pipe is, it is only as good as its installation will allow for. If the product is not installed properly, problems are likely to occur. This means that individuals responsible for the installation of plastic pipe must be trained to work with the type of material they are installing. Designers must take into account individual characteristics for various types of plastic pipe when they create working drawings and specifications. Application ranges of plastic pipe are shown in Table 5.1.

If you or your crews are not familiar with plastic piping, you should seek competent training and instruction. Sources for such services and information can include manufacturers, distributors, experienced contractors, books, and training seminars. You should learn all you can about the materials you will be installing before attempting an installation. Cutting corners can come back to haunt you quickly in the form of costly repairs, a damaged reputation, insurance claims for losses or damage, or worse.

Many contractors have learned that a little in-house training can go a long way in reducing on-the-job problems. While the training of crews is an overhead expense, it can be some of the best money a company will spend. Sending crews to a training seminar is good business, but how much will they learn from a few handouts and the spoken word of experts? Potentially, they will learn a great deal, but there is no substitute for hands-on training. Investing the time and money to train crews before putting them in the field may be one of the best investments a contractor can make.

TABLE 5.1 Application Range of Plastic Pipe

Application Range of Plastic Pipe

	Transported Medium	Recommended Material	Conditionally Suitable
Laid above ground	Solvents	PVDF	PP, PE
	Acids	PVDF, PVC	PP, PE
	High Purity Acids	PVDF	
	Alkalis	PVC, PP, PE	
	Hot Medium	PVDF, PP	
	Swimming Pool Water	PVC	PE
	Rain Water	PVC	
	Waste Water	PP, PE	
	High Purity Water	PVDF	PP, PVC
	Potable Water	PVC, PE	
Buried in ground	Potable Water	PVC, PE	
	Waste Water	PVC, PE	
	Gas	PVC, PE	
	Drainage	PVC	
	Cable Protection	PVC, PE	

(Courtesy George Fischer Engineering Handbook)

A number of professions require continuing education in order to remain licensed for the profession. This is true of the trades in some states, but many states don't require ongoing education for licensed trades. In any case, it is the responsibility of the contractor to guarantee that crews sent to do a job are competent for the task at hand. If something goes wrong in the field, who is going to be blamed? The individuals performing the work will be the first target, but the company they work for will also feel the heat. The odds of having your crews and company dragged into controversy and potential lawsuits can be turned in your favor with adequate training programs.

When I speak of adequate training programs, I am not suggesting that you put your people through a rigorous ordeal that lasts for months. Depending upon the type of work you are doing, a single day may be adequate for the training process. A few days of training will almost always be adequate for experienced tradespeople who simply need to be brought up to speed with a particular type of material. Don't overlook the training opportunity. More and more contractors are seeing the value of in-house training, and you should give it serious consideration.

RECEIVING MATERIALS

Receiving materials on a job site is such a common practice that workers often take it for granted. This can be a mistake. How the materials are handled can affect the entire installation. A successful installation starts with design and ends with testing and putting the system into operation. This process includes the unloading of materials.

When materials are not handled properly, they can be damaged. The damage may go unnoticed until a system is installed and tested. This has happened more than once. For example, there was a crew installing a PVC pipeline in cold weather. The installation went well, until the test was performed on the system. It was then that long cracks in the PVC pipe were discovered.

After investigating the incident, It was discovered that some of the pipe had been handled roughly during the installation. Workers had been dropping the pipe, instead of laying it down. The rocky ground provided a hard surface for the pipe to land on. Cold PVC and hard rocks are not a good combination. The cracks were thin enough to avoid visual detection, but they leaked when the test was applied. This meant cutting out sections of the pipeline and replacing the damaged sections. Not only was this time consuming and expensive, it was embarrassing. If the workers had handled the pipe properly, the entire problem would have been avoided.

Instructing your workers in proper pipe handling is essential to your success. Know and understand the materials you are working with. Handle them accordingy. Once the materials are on the job site, you will need to provide proper storage conditions. Thermal expansion and contraction of plastic pipe is illustrated in

Figure 5.1. Solvent welded pressure ratings vs. service temperature for CPVC and PVC pipe is shown in Figure 5.2.

Storing plastic pipe does not require sophisticated facilities. Plastic offers excellent resistance to weathering and can usually be stored outside. How the pipe is stored depends on the type of plastic you are working with and the conditions that the pipe will be exposed to. For example, you could place PVC pipe on a rack and leave it exposed to a hot sun for several days without any damage. But, don't try this with ABS pipe, unless the pipe is fully supported. By nature, ABS pipe sags when laid in direct, hot sunlight. On the other hand, ABS can take cold weather very well without becoming brittle. PVC pipe cannot. You must know the characteristics and needs of the specific materials you are working with.

Linear Expansion and Contraction

Coefficient of Thermal Linear Expansion

PVC $= 2.8 \times 10^{-5}$ in/in/°F
CPVC $= 3.4 \times 10^{-5}$ in/in/°F

To Calculate:
ΔL = Change in pipe length due to thermal changes
L = Straight runs of pipe with no changes in direction.
Y = Coefficient of thermal expansion (see above).
ΔT = maximum change in temperature between installation and operation (T MAX. - T. MIN.)

$\Delta L = Y \times L \times \Delta T$

Example:
• A system has 350 feet (4,200") of straight run (L) with no direction change.
• Pipe material is CPVC. Coefficient (Y) is 3.4×10^{-5} (0.000034").
• Pipe is installed at an ambient temperature of 60°F. Maximum anticipated operating temperature is 140°F. The difference (ΔT) is 80°F.

$\Delta L = 0.000034 \times 4200 \times 80$

$\Delta L = 11.4"$ of linear expansion in 350 ft. in pipe.

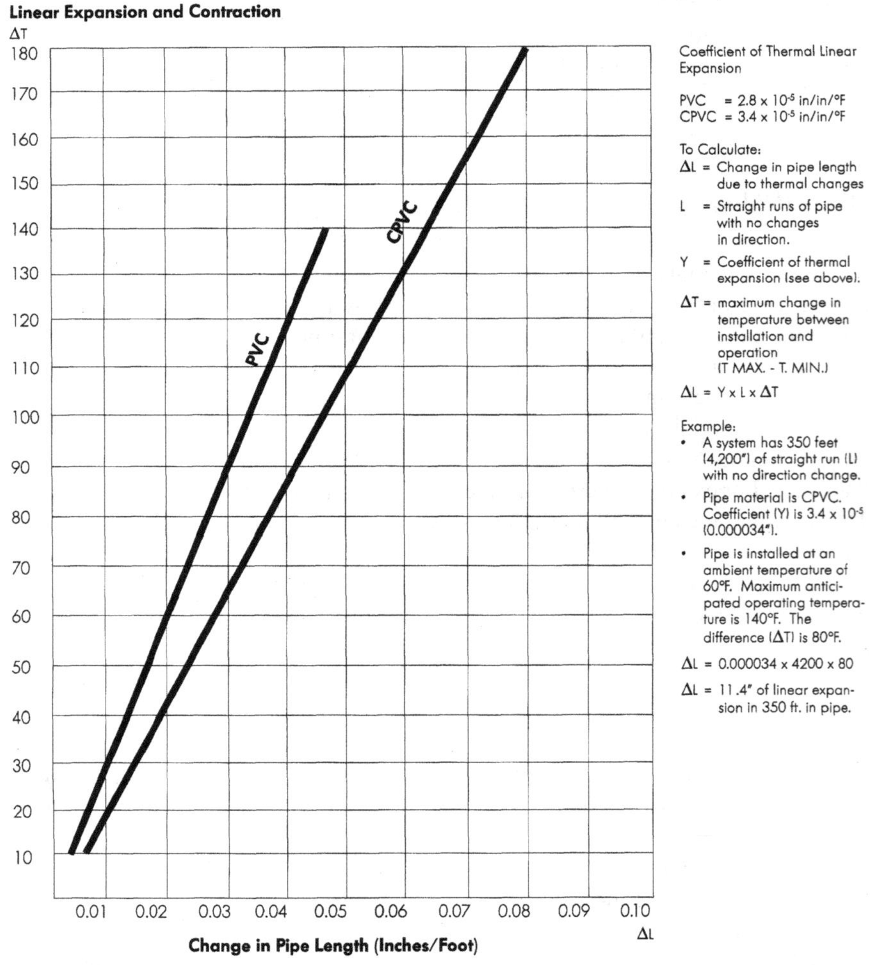

FIGURE 5.1 Thermal expansion and contraction. *(Courtesy George Fischer Engineering Handbook)*

Solvent-Welded Pressure Rating vs. Service Temperature — CPVC and PVC

				P														
				73°F		90°F	100°F	110°F	120°F		130°F		140°F		160°F	180°F	200°F	210°F
	D			PVC	CPVC	PVC	PVC	PVC	PVC	CPVC	PVC	CPVC	PVC	CPVC	CPVC	CPVC	CPVC	CPVC
Nom.	Outside	t	DR=	f=1	f=1	f=0.75	f=0.62	f=0.50	f=0.40	f=0.65	f=0.30	f=0.57	f=0.22	f=0.50	f=0.40	f=0.25	f=0.20	f=0.16
Size	Dia.	Wall		S=2000	S=2000	S=1500	S=1240	S=1000	S=800	S=1300	S=600	S=1135	S=440	S=1000	S=800	S=500	S=400	S=320
1/2	.840	.147	5.714	848	848	636	526	424	339	552	255	484	187	424	339	212	170	136
3/4	1.050	.154	6.818	688	688	516	426	344	275	447	206	392	151	344	275	172	138	110
1	1.315	.179	7.346	630	630	473	390	315	252	410	189	359	139	315	252	158	126	101
1 1/4	1.660	.191	8.691	520	520	390	322	260	208	338	156	296	114	260	208	130	104	83
1 1/2	1.900	.200	9.500	471	471	353	292	235	188	306	141	268	104	235	188	118	94	75
2	2.375	.218	10.894	404	404	303	251	202	162	263	121	230	89	202	162	101	81	65
2 1/2	2.875	.276	10.417	425	425	319	263	212	170	276	127	242	93	212	170	106	85	68
3	3.500	.300	11.667	375	375	281	233	188	150	244	113	214	83	188	150	94	75	60
4	4.500	.337	13.353	324	324	243	201	162	130	210	97	185	71	162	130	81	65	52
6	6.625	.432	15.336	279	279	209	173	140	112	181	84	159	61	140	112	70	56	45
8	8.625	.500	17.250	246	246	185	153	123	98	160	74	140	54	123	98	62	49	39

P - Pressure rating of pipe at service temperatures (psi)
S - Hydrostatic design stress (psi)
D - Outside diameter of pipe (inches)

1) Figures for pressure rating at 73°F are rounded off from actual calculated values. Pressure ratings for other temperatures are calculated from 73°F values.
2) Pressure rating values are for PVC (12454-B) and CPVC (23447-B) pipe and for most sizes are calculated from the experimentally determined long-term strength of PVC1 and CPVC extrusion compounds. Because molding compounds may differ in long-term strength and elevated temperature properties from pipe compounds, piping systems consisting of extruded pipe and molded fittings may have lower pressure ratings than those shown here, particularly at the higher temperatures. Caution should be exercised when designing PVC systems operating above 100°F and CPVC systems operating above 180°F.
3) The pressure ratings given are for solvent-cemented systems. When adding valves, flanges or other components, the system must be derated to the rating of the lowest component. (Pressure ratings: molded or cut threads are rated at 50% of solvent-cemented systems; flanges and unions are 150 psi; for valves, see manufacturer's recommendation.)

FIGURE 5.2 Solvent-welded pressure rating vs service temperature for CPVC and PVC pipe. *(Courtesy George Fischer Engineering Handbook)*

When plastic pipe and fittings are to be stored for an extended time, they should be kept under a light tarp or in a ventilated, covered area. How often have you seen these materials stored in the large trailers retired from use with 18-wheelers? These storage trailers keep the materials dry, but excessive heat can build up in the container, and heat can be a problem for some plastics. This is why a well-ventilated, covered area is more desirable for storage.

Sunlight can degrade plastics with the ultraviolet rays associated with the rays of the sun. Plastic products should be kept clean, dry, free of ice, and ready for instant installation. The sun is not the only enemy of plastic pipe. You must protect the pipe from extraordinary deflection. If pipe is stacked too high on itself, the pipe diameter can be affected to an unacceptable level. A rule-of-thumb for stacking plastic pipe is to avoid pipe stacks that are more than 3 feet in height. It is common to double-stack bundled pipe.

Belled pipe should be stored so that alternate rows of bells are inverted. This minimizes the loading on the bell. This procedure is especially important if belled pipe is to be stored for an extended time. As with any pipe rack for plastic pipe, racks for belled pipe should be smooth and free of any burrs or sharp edges that might compromise the integrity of the plastic pipe.

GENERAL ABOVE-GROUND RECOMMENDATIONS

There are general above-ground recommendations for the installation of plastic pipe. But, never make assumptions based on average installations.

Confirm manufacturer's recommendations for all of your installation projects. The anchoring, support spacing, and hanger designs used with plastic pipe can differ from the procedures used with other types of pipe. Another special consideration with plastic pipe is the potential risk of damage from different types of impact. Let's go over some of these basics.

Support Spacing

The rules for support spacing when working with plastic pipe are different than those used for metal pipe. Both the tensile and compression strengths of plastic pipe are lower than those of metal pipe. Recommended spacing is illustrated in Table 5.2.

This means that additional support is needed for plastic pipe. Additionally, the tensile strength of thermoplastic pipe decreases when the pipe gets hot. Therefore, more support is needed. Conditions can exist when temperature ratings are so high that thermoplastic pipe will require continuous support.

Thermoset requirements for support are not as extensive as those used with thermoplastic pipe. In fact, thermoset installations are much more in line with the requirements set forth for metal pipes. Check your local code requirements for specific spacing data on supports.

TABLE 5.2 Recommended Support Spacing (in feet)

Recommended Support Spacing* (In Feet)

Nom. Pipe Size (In.)	PVC Pipe										CPVC Pipe					
	Schedule 40					Schedule 80					Schedule 80					
	Temp. °F					Temp °F					Temp. °F					
	60	80	100	120	140	60	80	100	120	140	60	80	100	120	140	180
1/2	4 1/2	4 1/2	4	2 1/2	2 1/2	5	4 1/2	4 1/2	3	2 1/2	5 1/2	5 1/2	5	4 1/2	4 1/2	2 1/2
3/4	5	4 1/2	4	2 1/2	2 1/2	5 1/2	5	4 1/2	3	2 1/2	5 1/2	5 1/2	5 1/2	5	4 1/2	2 1/2
1	5 1/2	5	4 1/2	3	2 1/2	6	5 1/2	5	3 1/2	3	6	6	6	5 1/2	5	3
1 1/4	5 1/2	5 1/2	5	3	3	6	6	5 1/2	3 1/2	3	6 1/2	6 1/2	6	6	5 1/2	3
1 1/2	6	5 1/2	5	3 1/2	3	6 1/2	6	5 1/2	3 1/2	3 1/2	7	7	6 1/2	6	5 1/2	3 1/2
2	6	5 1/2	5	3 1/2	7	7	6 1/2	6	4	3 1/2	7	7	7	6 1/2	6	3 1/2
2 1/2	7	6 1/2	6	4	3 1/2	7 1/2	7 1/2	6 1/2	4 1/2	4	8	7 1/2	7 1/2	7 1/2	6 1/2	4
3	7	7	6	4	3 1/2	8	7 1/2	7	4 1/2	4	8	8	8	7 1/2	7	4
4	7 1/2	7	6 1/2	4 1/2	4	9	8 1/2	7 1/2	5	4 1/2	9	9	9	8 1/2	7 1/2	4 1/2
6	8 1/2	8	7 1/2	5	4 1/2	10	9 1/2	9	6	5	10	10 1/2	9 1/2	9	8	5
8	9	8 1/2	8	5	4 1/2	11	10 1/2	9 1/2	6 1/2	5 1/2	11	11	10 1/2	10	9	5 1/2
10	10	9	8 1/2	5 1/2	5	12	11	10	7	6	11 1/2	11 1/2	11	10 1/2	9 1/2	6
12	11 1/2	10 1/2	9 1/2	6 1/2	5 1/2	12	11	10	7	6	12 1/2	12 1/2	12 1/2	11	10 1/2	6 1/2
14	12	11	10	7	6	13 1/2	13	11	8	7						
16	12 1/2	11 1/2	10 1/2	7 1/2	6 1/2	14	13 1/2	11 1/2	8 1/2	7 1/2						

Note: This data is based on information supplied by the raw material manufacturers. It should be used as a general recommendation only and not as a guarantee of performance or longevity.

*Chart based on spacing for continuous spans and for uninsulated lines conveying fluids of specific gravity up to 1.00.

(Courtesy George Fischer Engineering Handbook)

Hangers

The types of hangers used with plastic pipe can be critical to the success of the pipe's function. Using the wrong type of hanger can cause stress on a pipe that will shorten the useful life of the conduit. Overall, most designers choose hangers that have a large bearing area to disperse the load of the pipe over the largest area feasible. A variety of hangers are illustrated in Figures 5.3, 5.4, 5.5, 5.6, and 5.7.

Hangers that are manufactured for metal pipe can often be modified and used with plastic pipe. Horizontal pipe is often hung with either a sling clamp or a clevis hanger.

Shoe supports can be used when conditions are favorable for them. Remember to choose hangers that offer the largest area of support that is practical.

It is not unusual to find a sleeve of sheet metal installed between the pipe and its hanger. Why is this done? The sheet metal spreads the load over a larger area of the pipe to reduce stress on the pipe.

When U-bolt hangers and roller hangers are used, the plastic pipe should be fitted with a protective sleeve. Medium-gage sheet metal is often used to fabricate these sleeves. Another type of sleeve is a section of plastic pipe that has been cut in half to fit over the pipe being secured. A rule-of-thumb for plastic pipe installations where excessive temperature calls for continuous support is to use a smooth structural angle or channel.

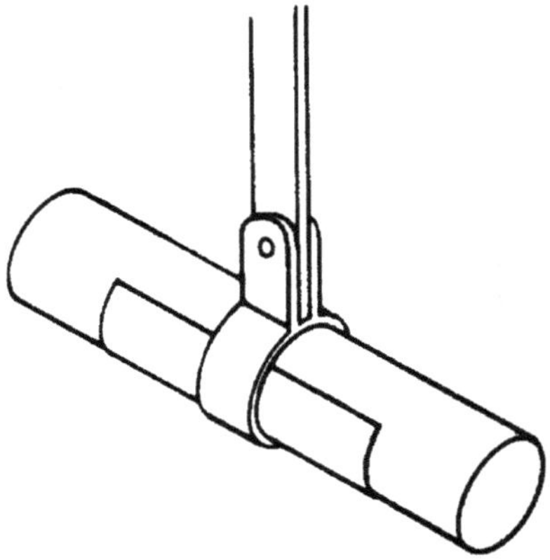

FIGURE 5.3 Band hanger with protective sleeve. *(Courtesy George Fischer Engineering Handbook)*

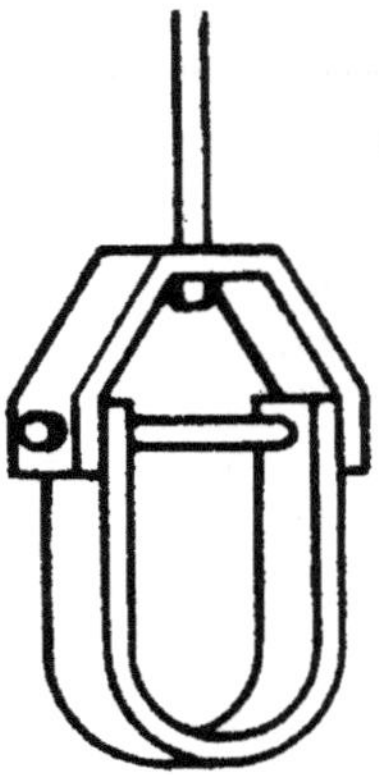

FIGURE 5.4 Clevis hanger. *(Courtesy George Fischer Engineering Handbook)*

Roller hangers are recommended when a pipe might move axially.

Thermal expansion is possible due to fluid or environmental temperature variations. This may cause movement that is best handled by roller hangers. In such cases, the pipe should be fitted with a protective sleeve.

Plastic pipe that rubs against a steel support can be damaged to a point that the useful life of the pipe is reduced. Any abrasive surface can be destructive to plastic pipe. Wood is sometimes used to protect plastic pipe from abrasive surfaces, but wood can deteriorate. A thermoplastic pad, such as PVC or polyethylene

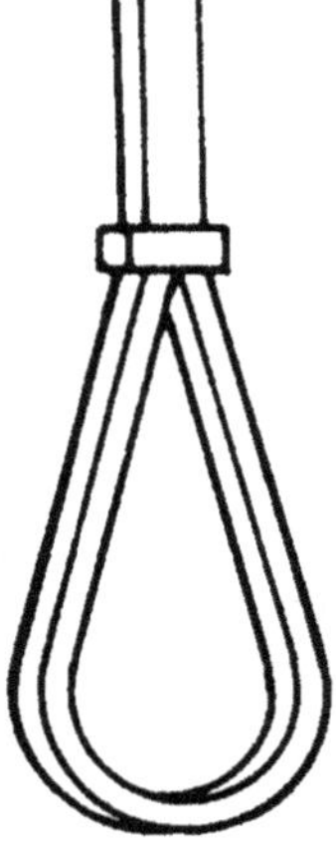

FIGURE 5.5 Adjustable solid ring swivel type. *(Courtesy George Fischer Engineering Handbook)*

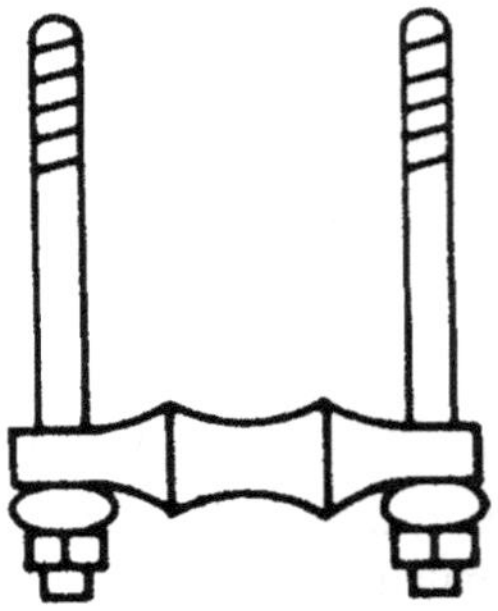

FIGURE 5.6 Single pipe roll. *(Courtesy George Fischer Engineering Handbook)*

makes a better protective surface. Pipe roll plates as illustrated in Figure 5.8 are also effective supports.

Your local code will dictate where hangers must be installed, but remember to install them as close as possible to all 90-degree bends.

Vertical installations must be supported in compliance with code regulations. The support intervals are not as frequent as those used for horizontal piping, but they are just as necessary. The base of all stacks must be supported. From there, the vertical intervals vary, so check your local code requirements.

It is preferable to avoid heavy weight loads on the base of vertical runs. The weight load can be controlled with riser clamp or double-bolt pipe clamps, illustrated in Figures 5.9 and 5.10.

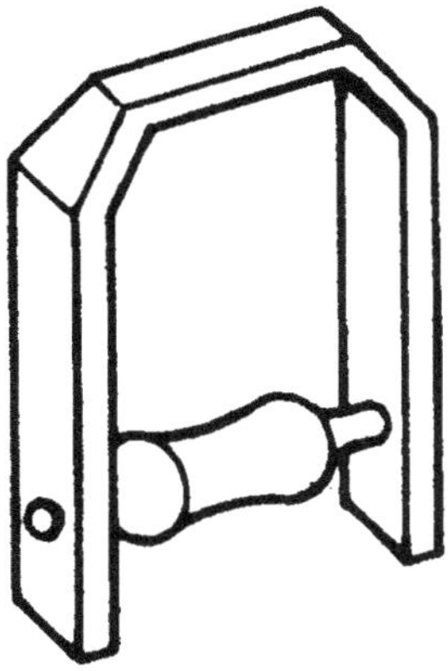

FIGURE 5.7 Roller hanger. *(Courtesy George Fischer Engineering Handbook)*

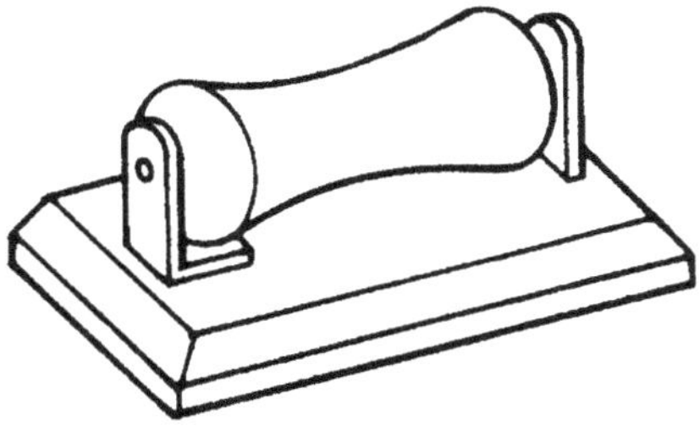

FIGURE 5.8 Pipe roll and plate. *(Courtesy George Fischer Engineering Handbook)*

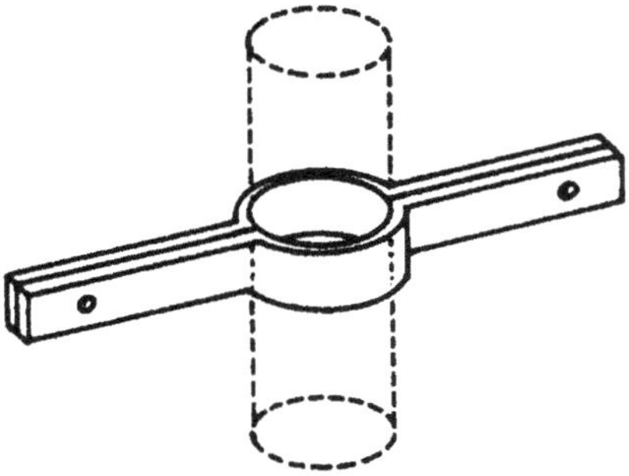

FIGURE 5.9 Riser clamp. *(Courtesy George Fischer Engineering Handbook)*

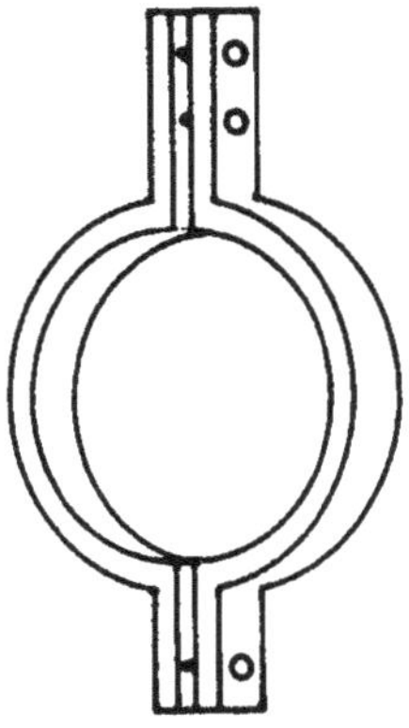

FIGURE 5.10 Double-bolt clamp. *(Courtesy George Fischer Engineering Handbook)*

When using these devices, you must avoid tightening the supports to a point where they will compress the wall of the pipe being secured. It's common to install these supports directly beneath couplings. In this way, the shoulder of the coupling rests on the support for maximum results. A trick of the trade when you need a support in a location where there is no fitting is to cut the hub of a fitting from the fitting and bond it to the vertical pipe. The shoulder of the hub can then rest on your support. Placing your supports under the shoulders of fittings makes it easier to maintain support without compressing the pipe too much. Continuous support arrangements are illustrated in Figure 5.11.

Valves are a common part of a piping installation. When the valves are larger than 2 inches in diameter, they should be supported. Unsupported valves can put stress on the joints between the valve and the pipe ends. In horizontal installations, it's a good idea to support the pipe on both sides of a valve, near the point of connection. Support arrangements are illustrated in Figure 5.12

Pipe movement must be controlled. This is most often done with the use of anchors and guides. These devices can direct the motion of a pipe within a defined range. Once an anchor is installed, there is no axial or transverse movement of the pipe. When axial movement is allowable, but transverse movement is not, guides are installed. Achorage methods are illustrated in Figures 5.13 and 5.14.

The use of anchors and guides should be designed in a way to function without point loading the pipe. You can expect to find anchors and guides whenever expansion joints are used. Long runs or piping are logical places to install anchors and guides. Directional changes in piping also call for anchors. When 90-degree bends are installed, anchors should be installed as close to the offsets as possible. These are illustrated in Figure 15.5.

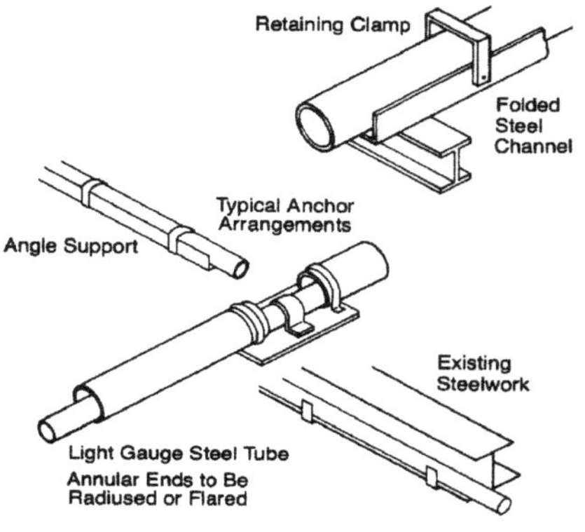

FIGURE 5.11 Continuous support arrangements. *(Courtesy George Fischer Engineering Handbook)*

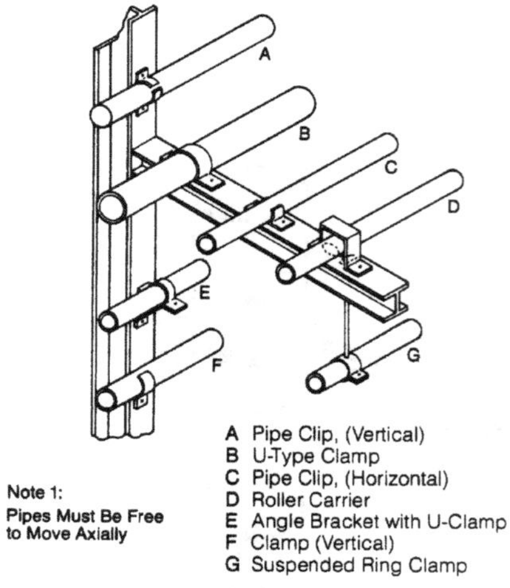

FIGURE 5.12 Typical support arrangements. *(Courtesy George Fischer Engineering Handbook)*

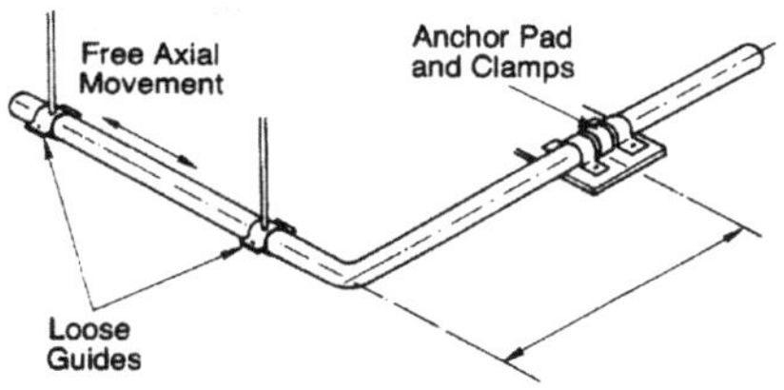

FIGURE 5.13 Continuous support arrangements. *(Courtesy George Fischer Engineering Handbook)*

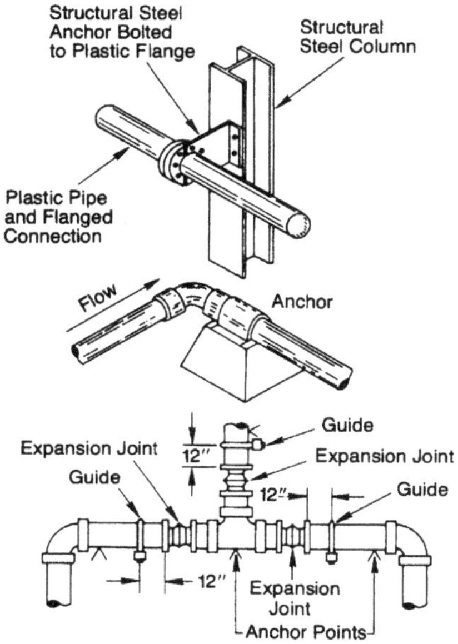

FIGURE 5.14 Anchoring methods. *(Courtesy George Fischer Engineering Handbook)*

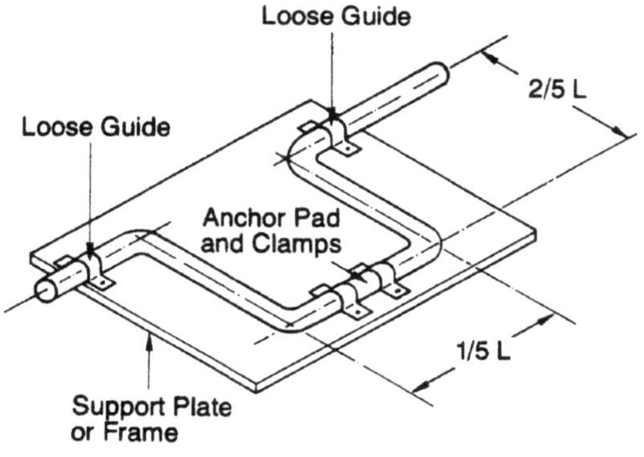

FIGURE 5.15 Anchoring changes in direction. *(Courtesy George Fischer Engineering Handbook)*

POLYETHYLENE PIPE

Polyethylene pipe (PE) is just one type of pipe that you may consider using in above-ground applications. The toughness, flexibility, light weight, and joint integrity of polyethylene pipe make it an excellent choice for above-ground installations. What can PE pipe be used for? It has seen use for a number of applications. Gas and oil transportation are among its uses. PE pipe has been used for temporary water lines, many types of bypass lines, dredge lines, fines disposal, and mine tailings, among other things. Slurry transport in many industries is another point of use for PE pipe. Not only is PE pipe versatile, it is economical.

Design criteria which may influence the use of PE pipe can include, temperature, ultraviolet radiation, any potential impact or loading, and chemical exposure. Most resources agree that PE pipe can be used in temperature extremes ranging from a low of –75F to a high of 150F. This is quite a range. However, the wide range of temperatures can call for specific engineering design to accommodate the extreme temperatures. Individual pipe manufacturers can provide detailed instructions on what to look out for when using PE pipe in extreme temperatures.

The pressure capability of PE pipe is established by the Hydrostatic Stress Board of the Plastics Pipe Institute. A hydrostatic design stress (HDS) is recommended by the board. Pressure capability is based on the long-term hydrostatic strength (LTHS) of the polymer used in the manufacturing of the pipe. The strength of a pipe is classified into one of a series of hydrostatic design bases (HDB) in accordance with ASTM D2937. For example, the HDB of a PE3408 piping material is 1600 pounds per square inch (PSI) at 73.4F. This provides a HDS of 800 psi at the same temperature.

If the temperature of PE pipe is exposed to rises or decreases, the pressure capability of the pipe varies. When the temperature exceeds 73.4F. the ratio of outside diameter to wall thickness decreases the LTHS. However, if the temperature is lowered, the LTHS rises. The HDS also rises as the temperature drops below 73.4F. A rise in temperature will decrease the modulus of elasticity. Embrittlement of the pipe material is a concern when extremely low temperatures are experienced. It is true, however, that many types of PE materials have been tested at extremely low temperatures without showing any signs of embrittlement.

Expansion

Expansion and contraction are always considerations when working with plastic pipe. These same factors are in play with metal and concrete pipe. PE pipe can expand up to 10 times more than metal or concrete. On the surface, this can appear to be a strike against PE pipe. However, while PE pipe will expand much more than metal or concrete, or even vitrified clay pipe, the modulus of elasticity for the plastic pipe is much lower than it is for the other types of pipe. CPVC expansion loop specifications are shown in Table 5.3. Offsets and changes in direction are shown in Table 5.4.

What does all of this mean? PE pipe is likely to move more when exposed to temperature extremes, but the stress on the plastic pipe will be substantially less, and this is a good thing. The bottom line is that PE pipe is a good choice for above-ground installations and it is an economically-wise choice.

TABLE 5.3 CPVC Expansion Loops

CPVC		**Length of Run (feet)**									
		10	**20**	**30**	**40**	**50**	**60**	**70**	**80**	**90**	**100**
Pipe Size (in.)	**O.D. of Pipe (in.)**	**Minimum Deflected Pipe Length (DPL) (inches)**									
1/2	**0.840**	15	21	26	30	33	36	39	42	44	47
3/4	**1.050**	17	23	29	33	37	40	44	47	50	52
1	**1.315**	18	26	32	37	41	45	49	52	55	58
1 1/4	**1.660**	21	29	36	42	46	51	55	59	62	66
1 1/2	**1.900**	22	31	39	44	50	54	59	63	67	70
2	**2.375**	25	35	43	50	56	61	66	70	75	79
3	**3.500**	30	43	52	60	67	71	80	85	91	95
4	**4.500**	34	4	59	68	77	84	91	97	103	108
6	**6.625**	42	59	72	83	93	102	110	117	125	131
8	**8.625**	47	67	82	95	106	116	125	134	142	150
10	**10.750**	53	75	92	106	118	130	140	150	159	167
12	**12.750**	58	81	100	115	129	141	152	163	173	182

(Courtesy George Fischer Engineering Handbook)

TABLE 5.4 CPVC Offsets and Change of Direction

CPVC		**Length of Run (feet)**									
		10	**20**	**30**	**40**	**50**	**60**	**70**	**80**	**90**	**100**
Pipe Size (in.)	**O.D. of Pipe (in.)**	**Minimum Deflected Pipe Length (DPL) (inches)**									
1/2	**0.840**	21	30	36	42	47	51	55	59	63	66
3/4	**1.050**	23	33	40	47	22	57	62	66	70	74
1	**1.315**	26	37	45	52	58	61	69	74	78	83
1 1/4	**1.660**	29	42	51	59	66	72	78	86	88	93
1 1/2	**1.900**	31	44	54	63	70	77	83	89	94	99
2	**2.375**	35	50	61	70	79	86	93	99	105	111
3	**3.500**	43	60	74	85	95	105	113	121	128	135
4	**4.500**	48	68	84	97	108	119	128	137	145	153
6	**6.625**	59	53	102	117	131	144	155	166	176	186
8	**8.625**	67	95	116	134	150	164	177	189	201	212
10	**10.750**	75	106	130	150	167	183	198	212	224	237
12	**12.750**	81	115	141	163	182	200	216	230	244	258

(Courtesy George Fischer Engineering Handbook)

Ultraviolet Effects

Ultraviolet effects on PE pipe can degrade the material. This is easy enough to overcome. The secret is to use a PE pipe that contains a minimum of 2 percent carbon black. If you must use non-black pipe, check with the manufacturer for recommendations in regards to the effects of ultraviolet rays.

Durability

PE pipe is extremely durable and is often used to convey chemical-based products. The pipe will not rust or corrode. Neither chemical, electrolytic, or galvanic action will cause PE pipe to rot or pit. Some strong oxidizing agents, such as sulphric or nitric acids, can cause problems for PE piping. Hydrocarbons, such as fuel oils and diesel fuels pose threats to PE pipe.

Extended exposure to strong oxidizing agents can lead to crack formation or crazing on the pipe surface. PE pipe that is exposed to hydrocarbons for extended periods of time can reduce the pressure capability of the pipe. You might notice swelling in a PE pipe that has had extended exposure to hydrocarbons. Reduced tensile strength is another side effect of long-term exposure to hydrocarbons. In time, the hydrocarbons may eat through the pipe and result in leaching of the material being conveyed through the pipe. Before installing PE pipe that will have long-term exposure to hydrocarbons or strong oxidizers you should consult the pipe manufacturer for recommendations.

EXTERNAL DAMAGE

External damage is a potential risk for any type of exposed piping. Pipe must be protected from be flattened out, gouging, and deflecting. When pipe is installed in a high-traffic area or any other location where the risk of damage is higher than normal, extra protection must be provided for the pipe. Some situations will allow the pipe to be protected by the construction of a berm. In other cases, the pipe must be encased in a protective covering.

In the event PE pipe is damaged, it may have to be replaced. If the pipe is gouged in excess of 10 percent of its minimum wall thickness, the pipe should be replaced. When PE pipe has been flattened or deflected, watch for stress-whitening, crazing, cracking, and any other visible signs of damage. If any of these conditions appear, replace the section of pipe that is affected.

INSTALLING PE PIPE ON GRADE

There are two ways to install PE pipe above ground. You can install it by laying it on top of the ground, or you can install it on some type of designed support.

Laying the pipe directly on the ground is known as installing it on grade. The pipe might be laid in an unrestrained method. This means draping the pipe over the ground and leaving it alone. Temperature changes will result in expansion and contraction, but this will not be a concern in an unrestrained system. However, if there is concern that the pipe may move too much, it can be anchored to the ground to control the range of motion.

PE pipe that is laid unrestrained on grade is usually laid in a snaking fashion. This allows slack in the pipe to account for contraction. The surface on which the pipe is laid must be clear of any obstacles that might damage the pipe. When there are temperature changes, the pipe will move. Any movement of the pipe over sharp or abrasive surfaces is likely to result in pipe damage.

Pipe installations have a beginning and an end. Both are usually fixed connections. It is not wise to make fixed connections at the end of unrestrained pipe runs. You should stabilize the pipe run when it approaches a fixed connection. The distance from the fixed location to stabilize should be a minimum of 3 pipe diameters away from the rigid connection. By doing this, the stress-concentrating effect of lateral pipe movement at the fixed connection is controlled. A consolidated earthen berm makes an effective stabilizer.

ANCHORED PIPE RUNS

Anchored pipe runs are desirable when too much pipe wandering will create problems. There are many methods of retention available They include:

- Earthen berms
- Pylons
- Augered anchors
- Concrete cradles
- Thrust blocks

If you elect to use earthen berms to restrain an above-ground installation, you have two options. The pipe can be covered continuously with a light layer of earth. Or, you can use earth at specific intervals to hold the pipe in place. A continuous layer of earth is not only very effective in restraining pipe movement, it also provides a level of protection for the pipe from temperature fluctuations. When the temperature is more stable, so is the pipe. Securing pipe with earth at selected intervals is less expensive that covering the pipe with a full coat of earth. When this method is used, the earth should encase the pipe for a distance equal to one to 3 pipe diameters.

Pipe that is anchored at specific intervals will deflect laterally when temperatures fluctuates. What is an acceptable spacing pattern for interval supports? It is generally thought that the spacing of support intervals is based on economics. On occasions when lateral deflection must be seriously controlled, the distance between supports must be reduced. The greater the distance between supports, the greater the lateral movement. Of course, you must also pay attention to the maximum allowable lateral movement allowable for the type of pipe being secured.

INSTALLING PE PIPE ABOVE GRADE

When installing PE pipe above grade, you must still consider lateral movement. In addition to this, you must take into consideration beam deflection and the support or anchor configuration. The spacing of supports for above-grade, suspended or supported pipe depends upon simple-beam or continuous-beam analysis. Spacing of supports is based on limiting bending stress.

If excessive temperature ranges are expected, PE pipe should be supported continuously. Any support that simply cradles a pipe, rather than gripping it, should be at least one-half to one pipe diameter in length and should support at least 120 degrees of the pipe diameter. Of course, no support should subject the plastic pipe to sharp or abrasive surfaces.

When supports are used for PE pipe, they should have enough strength to restrain the pipe from lateral or longitudinal deflection under anticipated service conditions. Supports intended to offer free movement of pipe during expansion must provide a guide without restraint in the direction of movement. A support designed to grip a pipe firmly must offer either a flexible mount or have adequate strength to stand up under anticipated stresses. PVC expansion loop specifications are shown in Table 5.6. PVC offsets and changes of direction are illustrated in Figure 5.6.

Any heavy fitting, valve, or flanges must be fully supported and restrained for a distance of at least one full pipe diameter. Such fittings, valves, and flanges are considered to be rigid structures. Any rigid structure in a flexible pipe system should be fully isolated from bending stresses associate with beam sag or thermal deflection. Some of the typical pipe hanger types include: pipe stirrup supports, clam shell supports, and suspended I-beam or channel continuous supports.

FOLLOWING INSTRUCTIONS

The key to a successful pipe installation lies in following instructions and knowing how to work the specific material being installed. Manufacturers are very willing to provide designers and contractors with a wealth of information about plastic products.

Many professional organizations offer extensive research data that proves helpful to both designers and installers. The data exists and is accessible. Anyone with an intent to make a safe and successful pipe installation can find plenty of information to work with.

Contractors and installers can usually rely on drawings and specifications provided by designers. When designers run into questionable areas, they can turn to

TABLE 5.5 PVC Expansion loops

PVC Expansion Loops

PVC		Length of Run (feet)									
		10	20	30	40	50	60	70	80	90	100
Pipe Size (in.)	O.D. of Pipe (in.)	Minimum Deflected Pipe Length (DPL) (inches)									
1/2	0.840	11	15	19	22	24	27	29	31	32	34
3/4	1.050	12	17	21	24	27	30	32	34	36	38
1	1.315	14	19	23	27	30	33	36	38	41	43
1 1/4	1.660	15	22	26	30	34	37	40	43	46	48
1 1/2	1.900	16	23	28	33	36	40	43	46	49	51
2	2.375	18	26	32	36	41	45	48	51	55	58
3	3.500	22	31	38	44	49	54	58	62	66	70
4	4.500	25	35	43	50	56	61	66	71	75	79
6	6.625	30	43	53	61	68	74	80	86	91	96
8	8.625	35	49	60	69	78	85	92	98	104	110
10	10.750	39	55	67	77	87	95	102	110	116	122

(Courtesy George Fisher Engineering Handbook)

TABLE 5.6 PVC Offsets and Change of Directions

PVC Offsets and Change of Directions

PVC		Length of Run (feet)									
		10	20	30	40	50	60	70	80	90	100
Pipe Size (in.)	O.D. of Pipe (in.)	Minimum Deflected Pipe Length (DPL) (inches)									
1/2	0.840	15	22	27	31	34	37	41	43	46	48
3/4	1.050	17	24	30	34	38	42	45	48	51	54
1	1.315	19	27	33	38	43	47	51	54	57	61
1 1/4	1.660	22	30	37	43	48	53	57	61	65	68
1 1/2	1.900	23	33	40	46	51	56	61	65	69	73
2	2.375	26	36	45	51	58	63	68	73	77	81
3	3.500	31	44	54	62	70	77	83	88	94	99
4	4.500	35	50	61	71	79	87	94	100	106	112
6	6.625	43	61	74	86	96	105	114	122	129	136
8	8.625	49	69	85	98	110	120	130	139	147	155
10	10.750	55	77	95	110	122	134	145	155	164	173
12	12.750	60	84	103	119	133	146	158	169	179	189

(Courtesy George Fisher Engineering Handbook)

plastic manufacturers. There is no excuse for failure when there is so much material available to ensure success.

This chapter has provided solid, basic information about the installation of above-ground piping systems. There are, of course, numerous options in the design and construction of above-ground pipelines.

The Internet is an invaluable source of information. Web sites include consultants, manufacturers, contractors, professional organizations, professional bulletin boards, newsgroups, and more. On the local level, you can talk with suppliers and distributors. Code officials are often helpful in resolving questions about an installation. My experiences have shown that code officers are very receptive to legitimate questions about the code. In many cases, local designers and contractors will share information with other professionals.

In general, contractors and installers should be able to turn to the designers of systems to be installed. Designers can talk with manufacturers to obtain product information, formulas, and other detailed information. Don't be afraid to ask questions. It is far better to seek competent answers than it is to make costly mistakes. A little research can go a long way in the development of a successful piping system.

CHAPTER 6
BURIED PIPE DESIGN

Plastic pipe is an ideal material for underground use. Corrosion can be a problem with metal pipe, but plastic pipe is not subject to corrosion. Plastic pipe is light in weight and cost-effective. Throw in the fact that plastic pipe is easy to join together and you've got the best choice for buried pipe installations. For everything from water services for individual homes to sewers to major pipelines, plastic pipe is the answer. And, there are plenty of types of plastic pipe available for underground use.

If there is a complaint against using plastic pipe below grade, it is that the pipe is hard to identify once it is buried. This can cause trouble when excavation is done around the pipe after installation. It is said that most pipeline damage occurs as a result of accidental contact when excavating. With this being the only major drawback to installing plastic pipe underground, you can see why plastic installations are so popular. The problems associated with locating buried plastic pipe can be overcome by installing conductive wire next to the pipe in a trench. Then electronic instruments can pick up a signal from the wire to identify the location of the pipeline.

There are two basics methods for installing plastic pipe in trenches. Pipe with a large diameter is usually joined above ground and then lowered into a narrow trench. Smaller pipe, usually pipe with a diameter up to 8 inches, is often joined in a trench. This requires a much wider trench and it means that installers must be in the trench. When this is the case, appropriate safety procedures must be employed to protect the workers from trench failure and cave-ins. Before we get into deep details, let's discuss the basics of an underground installation.

PRELIMINARY WORK

There is preliminary work required before pipe can be installed below grade. Material must arrive on site. Once the material has been unloaded and stored properly, the site is ready for work. Sometimes a trench is dug prior to pipe being put in the vicinity. Since open trenches pose some threat, most contractors prefer to have pipe, fittings, and other installation supplies close at hand, so that the pipe installation can proceed quickly and allow the trench to be closed as soon as possible.

If you are stocking a job where the trench has not yet been dug, you should position the pipe and other materials far enough away from the trench path to ensure that the materials will not be damaged by the trenching process. When laying out the pipe in the trench area, this is know as stringing, string it so that the socket ends are pointing in the direction that the work will be progressing.

When stringing pipe along an open trench, place the pipe as close to the trench as is reasonably possible. This will expedite work once the joining process begins. Some regions suffer from vandalism. It is sad but true that job sites do come under attack from vandals and thieves. If you are stringing material that will be left unattended, you should not string more material than what you must in order to keep crews busy while more materials are brought into place.

DIGGING

The digging of trenches must be accomplished before pipe can be installed below ground. Trench depths vary widely. The type of piping application can also affect the trenching process. For example, a pipeline that is going to be pressurized, such as a water main, does not require a trench that has a set grade on its bed. In contrast, a trench for a sewer will require a steady grade to allow for a gravity flow through the pipe. The width of a trench is also a factor as illustrated in Figure 6.1.

As stated earlier, pipe with a diameter of 8 inches or less is sometimes installed by workers who are inside the trench. Naturally, this requires a much wider trench. When a backhoe, crane, or other lifting equipment will be used to lower pre-joined plastic pipe into a trench, the width of the trench can be much narrower. It is not uncommon for this type of trench to be no wider than 3 times the diameter of the pipe being installed.

Safety is a top concern with any trenching operation. No corners should be cut when it comes to protecting workers and the public in regards to open trenches. This may mean installing highly visible barriers to avoid accidents around a trench. It is common to put proper shoring equipment in place prior to allowing workers to enter a deep trench. The purpose of this chapter is not to explore and expound on all safety requirements for trenching, but it is a major factor in the development of an underground piping system.

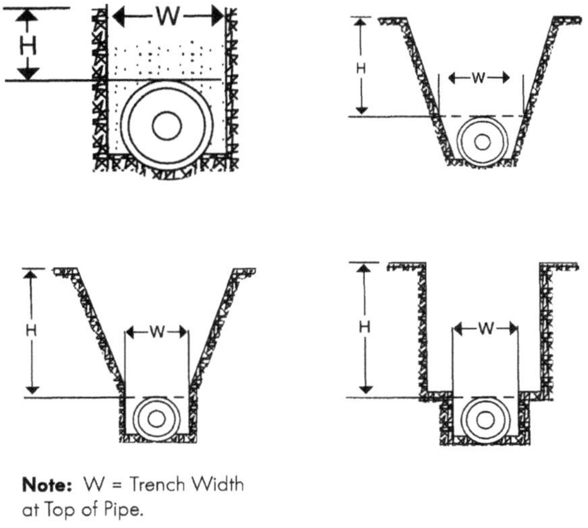

FIGURE 6.1 Trench widths for PVC pipe. *(Courtesy George Fischer Engineering Handbook)*

THE BED

The bed of a trench must be prepared properly to accept the installation of plastic pipe. A first concern is to make sure that the trench bed is continuous, fairly smooth, and free of rocks and other objects that might damage pipe or fittings placed in the trench. In the case of a gravity-type pipeline, the trench bed must be dug to an appropriate grade factor. Having a solid trench bed is another major consideration. It is not acceptable to have a bed that may sink in some sections after an installation is backfilled. There are times when natural obstacles are encountered that are not practical to remove. Examples of this could include ledge or bedrock and large boulders. Blasting these obstacles out of the trench may not be necessary. However, plastic pipe should never be installed in a manner in which the pipe makes contact with rocks or other abrasive materials.

To overcome bedrock and boulders that do not have to be removed, it is possible to install a softer surface over the rock. Sand is a good solution, but earth can also be used. It is not, however, acceptable to simply dump sand or dirt in on top of the rock and spread it out. The padding material must be compacted. A rule-of-thumb is to install 4 to 6 inches of padding material over the rock. Once the padding material, the sand or earth, is compacted properly, it will protect the pipe from direct contact with objects that may damage the pipe. It may be necessary to install the padding in layers and to compact each layer as it is installed. For example, you may find that the best job can be done by installing a 3-inch layer of sand and compacting it prior to installing an additional 3 inches of sand that will then also be compacted.

Trenches that are to be dug on a specific grade must be checked periodically as the trench is dug to ensure that the grade is being maintained. During the digging process it is common for digging errors to occur and for obstacles to require removal. This can compromise the steady grade of the trench. However, there are ways to overcome these problems.

Assume that you are having a trench dug for a sewer that requires a steady downstream grade of one-eighth of an inch per foot. The equipment operator is very experienced and is doing a good job. But then a large rock is encountered that needs to be removed. When the rock is removed it will leave a hole in the trench bed. It is not acceptable to install pipe over the hole. If this were done, the pipe would not have continuous support. So, how do you fix the hole and maintain the grade of the trench? If necessary, you excavate the hole to make it large enough to work with. Then you begin filling it with sand or dirt. Fill the hole in layers and compact each layer prior to placing a new layer. The compacted fill will come up to the bed level of the trench and you will no longer have a problem.

PLACING PIPE

Placing pipe in a trench may be done in a number of ways. The method of placement depends on the type of trench being used.

When a wide trench is used, with workers in the trench, pipe is often passed from workers above the trench to workers in the trench. Pipe should never be rolled or tossed into a trench. Occasionally, pipe is lowered into this type of workplace with equipment, such as a backhoe. In all cases, the pipe should be handled carefully and laid in the trench gently.

Narrow trenches call for pipe to be assembled prior to installation in the trench. Due to this fact and the larger size of pipe normally installed in this manner, equipment is used to lower the pipe into the trench. Circumstances of a specific job can dictate the type of equipment that will be used. Backhoes and cranes are both used to lower assembled pipe into trenches. Telescoping lifting rigs are also used to lower pipe into trenches.

When equipment is used to lower assembled pipe sections, precautions must be taken to protect the pipe and the integrity of pipe joints. Standard procedure usually involves the use of rope or slings. The slings or rope should be positioned in a way to support the pipe sections adequately. At no time should pipe be rolled into a trench. It is important that the pipe not be twisted during the installation process. Improper handling can result in pipe damage and leaks.

THERMAL CONTRACTION

Thermal contraction is a consideration when installing plastic pipe in a trench. We talked about snaking pipe for above-ground installations in Chapter 5. Snaking is also used with underground pipe installations. Offsetting the pipe by snaking it is done to allow for thermal contraction. By increasing the length of the pipe by snaking it, you are providing additional pipe length to compensate for contraction. An exception to this rule is when you are working with pipe that is joined with O-rings. The O-ring connection allows for contraction within the O-ring. Pipe that is fused or solvent welded is the type of pipe that requires snaking. Figure 6.2 illustrates snaking pipe in a trench. Snaking length versus offset to compensate for thermal contraction is illustrated in Table 6.1.

AVOIDING BENDING AND STRESS

Avoiding bending and stress in a pipe system is a factor in the installation of plastic pipe. Unlike steel pipe, plastic pipe is not intended to support heavier weights, such as valves, anchors, and so forth. Plastic pipe is designed to support soil loads and internal pressures up to a specified hydrostatic pressure rating. When accessories are placed in the pipeline, such as valve boxes, the accessories must be supported to prevent additional bending and stress on the plastic pipe.

THRUST BLOCKS

Concrete thrust blocks are used to anchor plastic pipelines. It is not acceptable for plastic pipe to be in direct contact with concrete or other abrasive materials. When a concrete thrust block is used, the pipe must be protected from the concrete. Wrapping the pipe with rubber, or some other suitable sleeve, to prevent direct contact with the concrete is necessary.

Axial movement may be a cause for pipe restriction. When this is the case, it is common to apply split collars around the outside diameter of the pipe with solvent-welded joints to protect the pipe from contact with concrete. It is recommended that the solvent-welded joints between the collars and the pipe exterior be allowed to dry for at least 48 hours prior to pouring concrete in the area.

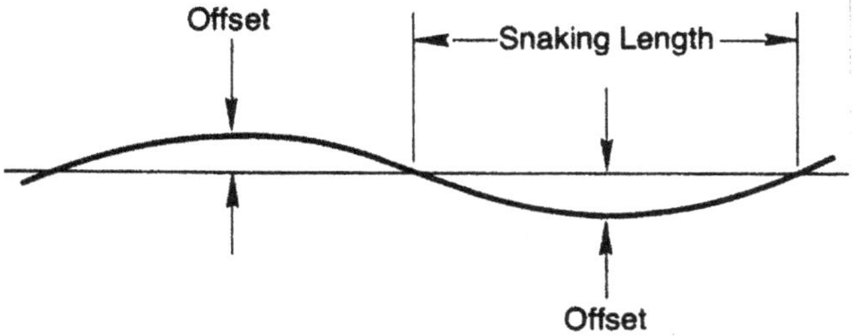

FIGURE 6.2 Snaking pipe in a trench. *(Courtesy George Fischer Engineering Handbook)*

TABLE 6.1 Snaking Length vs. Offset (in inches) to Compensate for Thermal Contraction

Snaking Length, (ft.)	Maximum Temperature Variation (°F) Between Time of Cementing and Final Backfilling									
	10°	20°	30°	40°	50°	60°	70°	80°	90°	100°
	Loop Offset, (in.)									
20	2.5	3.5	4.5	5.20	5.75	6.25	6.75	7.25	7.75	8.00
50	6.5	9.0	11.0	12.75	14.25	15.50	17.00	18.00	19.25	20.25
100	13.0	18.0	22.0	26.00	29.00	31.50	35.00	37.00	40.00	42.00

(Courtesy George Fischer Engineering Handbook)

VERTICAL PIPE SECTIONS

Vertical pipe sections, also known as risers, are sometimes installed in underground piping systems. In the case of sewers, risers will come to the surface at grade level and will be fitted with a cleanout fitting and plug. This allows the pipe to be rodded out if an obstruction occurs in the pipe and blows sewage flow. Risers should not be installed to support above-ground metal valves or other heavy objects. The stress on the exposed pipe may prove to be more than the pipe can withstand.

When vertical pipe sections rise above grade they are exposed to the risk of impact and damage. Protection of such piping is in order. Such protection could come in the form of an enclosed box or area around the pipe. Another method of protection might be the installation of a metal pipe sleeve over the plastic pipe. If this is done, the metal pipe must have a smooth interior and it must be properly supported to withstand impact. Another consideration in exposed risers is the risk of excessive heat, such as sunlight, degrading the pipe material to a dangerous point.

PLOWING

Plowing pipe into a trench is a cost-effective means of installation when practical. If a job lends itself to having the pipe plowed into a trench, the time required to make an installation can be reduced greatly. Plastic pipe that is subject to hot temperatures should not be plowed. High temperature weakens the pipe and opens the door to the risk of damage. Before pipe is plowed into a trench, the pipe should be tested for leaks. This is done with a low-pressure test to determine that all joints are sealed satisfactorily to maintain the pipe's full hydrostatic pressure rating.

Plowing is not to be confused with pulling. Assembled pipe should never be pulled into a trench. A tractor is used to plow pipe into trenches. The tractor is fitted with a plowing chute that the pipe is fed through. This chute is located on the rear of a plow blade. A plow chute must be sized appropriately for the size of the pipe being installed. When this method is available and suitable, it is well worth considering.

BACKFILLING

The backfilling of a trench is not as simple as taking a bulldozer and pushing mounds of earth in on top of the pipe. In fact, the backfilling process is a critical part of a pipeline installation. Done improperly, backfilling can damage buried pipe in many ways. Sharp objects, such as rocks, can cut a pipe. Heavy loads of dirt dumped on some types of pipe can crimp or collapse the pipe. This is especially true when the pipe is subjected to hot temperatures and become softer than normal. Once a pipe is installed in a trench and is ready for backfilling, it is a good idea to put some pressure in the pipeline. A pressure of just 25 psi can help in keeping a pipe from collapsing during the backfilling process.

Backfill material should be checked before it is used. If the fill material contains rocks, pieces of concrete, or other items that might damage the pipe, don't use it. Backfill material should be free of debris and suitable for satisfactory compaction. The first layer of backfill material shouldn't be more than about 6 inches in depth. Get this layer in the trench and compact it to protect the pipe as more backfilling is done. In many cases, the trench will continue to be covered with similar layers of backfill that will be compacted before new layers are added. Large loads of backfill should not be pushed into a trench until the pipe is fully protected with an adequate depth of compacted fill to accept the weight of the larger loads without damaging the pipe.

GENERAL DESIGN PROCEDURE

One of the first steps in designing a buried pipeline is determining dead loads and surcharge loads. Other factors that come into play include prism loads, soil arching, Marston loads, soil creep, distributed loads, and more. Let's look at these design factors individually.

Dead Load

A dead load is the load that is applied to a pipeline at all times. This includes the weight of soil on top of the pipe. In addition to the weight of soil, the weight of any other permanent load over a pipe is considered a dead load. An example of this could be the pavement of a highway that passes over a pipeline.

Since polyethylene pipe (PE) is such a common plastic pipe, let's talk about dead loads and PE pipe. The overburden load applied to the pipe crown is usually considered to be equal to the weight of the soil column that projects above the pipe. A soil column of this type is often called a prismatic element and is involved in the use of a prism load.

Prism Load

A convention used to calculate the earth pressure on a pipe when estimating vertical defection is the prism load. The true load transmitted to a pipe from soil mass is subject to the stiffness of the soil and the pipe. The prism load may be deceiving when working with flexible plastic pipe. In reality, the load applied to such a pipe can be considerably less than the prism load might show. This is because the shear resistance transfers part of the soil load directly above the pipe into the trench sidewalls and the embedment. This process is called arching. For an accurate assessment, designers frequently use both the prism load and the Marston method to determine the proper design for buried pipe.

The simple way to determine vertical earth load on a horizontal pipe in a mass of soil is when the soil has uniform stiffness and weight throughout. This assumes that there are no large voids or buried structures in the area of the pipe. Under these conditions, the vertical earth pressure acting on a horizontal pipe at a depth is equal to the prism load per unit area.

$$\text{Prism Load, } P_E = wH$$

where PE = vertical soil pressure, lb/ft^2
w = unit weight of soil, lb/ft^3
H = soil height above pipe crown, ft

Arching

As already discussed, PE pipe rarely shares the same stiffness as the soil encasing it. This throws off the results of calculations made using the prism load. The load could be more or less than the results of the calculations would indicate. In the case of PE pipe, and most flexible plastic pipe, the soil above the pipe disperses its load away from the pipe and into the soil beside the pipe. This is arching. Think of arching as the difference between the applied load and the prism load.

When there is a reduction in vertical load, you have arching. If the vertical load is more than the prism load, you have reverse arching. The downward movement of backfilled soil is what causes arching. Pipe deflection can initiate the arching process. Any compression of deeper layers of backfill or any settlement in a trench bed can be responsible for arching. In the case of plastic pipe, vertical deflection of the pipe crown is usually what starts the arching process.

Arching is generally permanent. It occurs in most stable applications. The arching is maintained by soil shear stresses. When large vibrating machines operate over a pipeline the arching may not be permanent. This is also true of situations where there is light cover over the buried pipe. Soft and unstable backfill can also prevent permanent arching. Any pipeline placed under roadways may not experience permanent arching.

Marston Load

The Marston load is generally used along with the prism load when working with plastic pipe installations and design. A more realistic value can usually be obtained for plastic pipe when the Marston load is used. This method dates back to 1930 and has proved itself within the industry. A review of ASCE Manual No. 60 will reveal more on the method.

$$\text{Marston Load, } P_M = C_D w B_D$$

where terms are previously defined, and

e = natural log base number, 2.71828

K = Rankine earth pressure $\tan^2\left(45 - \frac{\phi}{2}\right)$

ϕ = Internal friction angle, degrees
H = Soil cover height, feet
u' = Soil cover height, feet

Typical Ku' values are:

Saturated Clay	0.110
Ordinary Clay	0.130
Saturated Top Spoil	0.150
Sand and Gravel	0.165

Loads applied to pipes in embankments are generally higher than they would be on pipes in trenches. Actual load depends on the relative stiffness between the embankment soil and the pipe. The prism load is most often used when pipe is to be placed in an embankment. By using the prism method, the vertical pressure on flexible pipe in an embankment can be calculated.

Soil Creep

When backfill material consists of cohesionless soil and analytical methods are not available for precise calculations, designers don't normally factor in soil creep. Plastic pipe tends to creep faster than cohesionless soil.Clayey soil can creep much more than cohesionless soil. This is especially true if the clay is saturated. Arching can be high when clayey soil is used as a backfill material. The soil creep moves more soil towards the buried pipe. A conservative design approach is called for under these conditions. This means that a low friction angle is used when working with the Marston equation. It's common for a factor of a 11-percent angle to be assigned to ordinary clay and an angle of 8-percent to be used for saturated clay.

Surcharge Loads

Surcharge loads are often temporary loads, but they can be permanent loads. Any load created by a structure or a vehicle can be considered a surcharge load. In the case of vehicular loads, they are called live loads. There are many types of potential surcharge loads. A footing or foundation for a building is considered a surcharge load. Point loads, as they are called, can be from the tires from vehicles. Any of these loads can be distributed through soil in a way that reduces pressure with an increase in depth of horizontal distance from the surcharged area.

Common design practice is to equate the load on a buried pipe from a surcharge load with the downward pressure acting at the plane of the pipe crown. When the surcharge load is determined, the total load acting on the pipe is the sum of the earth load and the surcharge load.

Wall Compressive Strength

Compressive thrust can occur in the wall of a non-pressurized pipe that is confined in a dense embedment. This happens when the pipe is subjected to a radially directed soil pressure. When there is compressive stress within the pipe wall it can create internal pressure. Physical properties of PVC and CPVC are shown in Table 6.2.

Radial soil pressure that is resulting in stress is no usually uniform. Interestingly enough, it is generally assumed that the radial soil pressure is uniform and equal to the vertical soil pressure at the crown of a pipe. It's very possible for buried pressure pipe to have an internal pressure greater than the radial external pressure applied by the soil. Therefore, wall compressive stress is rarely factored in when working with pressurized pipe. Thermodynamic properties of PVC and CPVC pipe are shown in Table 6.3.

Shallow Cover

When shallow cover is provided over a plastic pipe, there are design factors that must be taken into consideration. For example, will floatation of the pipe, due to shallow cover, become a problem? Will there be upward buckling due to flooding? How likely is it that a high groundwater table will float the pipe? Will the pipe be adversely affected by live loads?

How much cover is enough? This depends on the type of pipe being installed and the job conditions that the pipe will be subjected to. A rule-of-thumb cover depth is enough cover material to equal at least the height of the pipe diameter, or 18 inches, whichever is greater. It is generally accepted that there should never be less than 12 inches of cover over a pipe.

GROUNDWATER

Groundwater can cause buried pipe to float. This is especially true when shallow cover is applied and when the cover material is light in weight. Floatation is much more likely when working with plastic pipe. Flooding and high water tables can float plastic pipe when the water produces a force greater than the downward force of the soil prism above the pipe. Pipe weight and the weight of its contents are also a factor.

Flooding can cause some types of soil to lose cohesiveness. This, too, can result in upward movement of a pipe. Any long-term ground saturation can cause a

TABLE 6.3 Thermodynamics of PVC and CPVC Thermoplastic Materials

Properties	Unit	PVC	CPVC	Remarks	ASTM Test
Coefficient of Thermal Linear Expansion per °F	in/in/°F	2.8×10^{-5}	3.4×10^{-5}		D-696
Thermal Conductivity	BTU/hr/ft^2/°F/in	1.3	0.95	Average Specific Heat of 0-100°C	C-177
Specific Heat	CAL/g/°C	0.20-0.28		Ratio of Thermal Capacity to that of Water at 15°C	
Maximum Operating Temperature	°F	140	210	Pressure Rating is Directly Related to Temperature	
Heat Distortion Temperature @ 264 PSI	°F	158	217	Thermal Vibration and Softening Occurs	D-648
Decomposition Point	°F	400+	400+	Scorching by Carbonization and Dehydrochloration	

(Courtesy George Fischer Engineering Handbook)

TABLE 6.2 Physical Properties of Rigid PVC and CPVC Thermoplastic Materials

The following table lists typical physical properties of PVC and CPVC thermoplastic materials. Variations may exist depending on specific compounds and product.

Mechanical

Properties	Unit	PVC	CPVC	Remarks	ASTM Test
Specific Gravity	g/cm^3	1.40 ± .02	1.55 ± .02		D-792
Tensile Strength @ 73°F	PSI	7,200	8,000	Same in Circumferential Direction	D-638
Modules of Elasticity Tensile @ 73°F	PSI	430,000	360,000	Ratio of Stress on Bent Sample at Failure	D-638
Compressive Strength @ 73°F	PSI	9,500	10,100		D-695
Flexural Strength @ 73°F	PSI	13,000	15,100	Tensile Stress on Bent Sample at Failure	D-790
Izod Impact @ 73°F	Ft-Lbs/In of Notch	1.0	1.5	Impact Resistance of a Notched Sample to a Sharp Blow	D-256
Relative Hardness @ 73°F	Durometer "D" Rockwell "R"	80 ± 3 110-120	— 119	Equivalent to Aluminum —	D-2240 D-785

(Courtesy George Fischer Engineering Handbook)

reduction in soil support for a pipe. These conditions can result in pipe buckling from external hydrostatic pressure.

Pipelines that are installed over groundwater are less prone to buckling, due to design considerations prior to the installation. When pipelines run full of liquid at all times they are less likely to float and buckle. Most designers agree that floatation is unlikely when a pipe is buried in common saturated soil with a cover depth equal to at least one-and-one-half times the pipe diameter.

MANHOLES

Manholes installed in conjunction with pipelines are at more risk to floatation than the piping is. Since manholes are attached to vertical risers and don't have cover, they are much more vulnerable to floatation. A solution to this problem can be as simple as the installation of manhole anti-floatation anchors. These anchors are made with reinforced concrete. The concrete slabs are placed over manhole stub-outs. Once the concrete is in place, its weight offsets the risk of floatation.

DESIGNING FOR A WATER ENVIRONMENT

Pipe systems installed in a water environment require design consideration for external hydraulic pressure, submergence weighting, and floatation. Any river, lake or stream crossing is considered a water environment. Wetlands and marshes are also considered water environments.

The flattening of pipes carrying gasses and pipes that are carrying partial loads of liquids are a concern when the internal pressure of a pipe is less than the static external hydraulic load. Flattening is usually not a concern for outfall and intake lines. When pipe ends are open the pressure is balanced. Water and wastewater pipelines that pass under water are protected due to the static head in the full pipe.

UNCONSTRAINED BUCKLING

Unconstrained buckling of pipe walls can occur when excessive external pressure is encountered.

This causes a flattening of a pipe. Stiffness is the factor involved in the maximum external load capacity. Material strength might seem like the proper factor, but stiffness is the true variable. Pipes will flatten when the bending moment, due to the load, exceeds the resisting moment due to elastic stresses in the pipes. To determine the critical external pressure above which a round pipe will flatten you can use the Love's equation.

TABLE 6.4 Friction Loss in Equivalent Feet of Pipe for Schedule 80 Thermoplastic Fittings

Feet of Pipe — Schedule 80 Thermoplastics Fittings

Nominal Pipe Size, In.	3/8	1/2	3/4	1	1 1/4	1 1/2	2	2 1/2	3	3 1/2	4	6	8
Tee, Side Outlet	3	4	5	6	7	8	12	15	16	20	22	32	38
90° Ell	1 1/2	1 1/2	2	2 3/4	4	4	6	8	8	10	12	18	22
45° Ell	3/4	3/4	1	1 3/8	1 3/4	2	2 1/2	3	4	4 1/2	5	8	10
Insert Coupling	—	1/2	3/4	1	1 1/4	1 1/2	2	3	3	—	4	6 1/4	—
Male-Female Adapters	—	1	1 1/2	2	2 3/4	3 1/2	4 1/2	—	6 1/2	—	9	14	—

(Courtesy George Fischer Engineering Handbook)

TABLE 6.5 Friction Loss in Schedule 40 Pipe

Carrying Capacity and Friction Loss — Schedule 40 Thermoplastics Pipe

Independent variables: Gallons per minute and nominal pipe size O.D. (Min. I.D.)
Dependent variables: Velocity, friction head and pressure drop per 100 feet of pipe, interior smooth.

GALLONS PER MINUTE	VELOCITY FEET PER SCOND	FRICTION HEAD FEET	FRICTION LOSS POUNDS PER SQUARE INCH	VELOCITY FEET PER SCOND	FRICTION HEAD FEET	FRICTION LOSS POUNDS PER SQUARE INCH	VELOCITY FEET PER SCOND	FRICTION HEAD FEET	FRICTION LOSS POUNDS PER SQUARE INCH	VELOCITY FEET PER SCOND	FRICTION HEAD FEET	FRICTION LOSS POUNDS PER SQUARE INCH	VELOCITY FEET PER SCOND	FRICTION HEAD FEET	FRICTION LOSS POUNDS PER SQUARE INCH	VELOCITY FEET PER SCOND	FRICTION HEAD FEET	FRICTION LOSS POUNDS PER SQUARE INCH	VELOCITY FEET PER SCOND	FRICTION HEAD FEET	FRICTION LOSS POUNDS PER SQUARE INCH
		1/2 in.			**3/4 in.**			**1 in.**			**1 1/4 in.**			**1 1/2 in.**			**2 in.**			**3 in.**	
1	1.13	2.08	0.90	0.63	0.51	0.22															
2	2.26	4.16	1.80	1.26	1.02	0.44	0.77	0.55	0.24	0.44	0.14	0.06	0.33	0.07	0.03						
5	5.64	23.44	10.15	3.16	5.73	2.48	1.93	1.72	0.75	1.11	0.44	0.19	0.81	0.22	0.09	0.49	0.066	0.029	0.30	0.015	0.007
7	7.90	43.06	18.64	4.43	10.52	4.56	2.72	3.17	1.37	1.55	0.81	0.35	1.13	0.38	0.17	0.69	0.11	0.048	0.49	0.021	0.009
10	11.28	82.02	35.51	6.32	20.04	8.68	3.86	6.02	2.61	2.21	1.55	0.67	1.62	0.72	0.31	0.98	0.21	0.091	0.68	0.03	0.013
15		**4 in.**		9.48	42.46	18.39	5.79	12.77	5.53	3.31	3.28	1.42	2.42	1.53	0.66	1.46	0.45	0.19	1.03	0.07	0.030
20	0.51	0.03	0.013	12.65	72.34	31.32	7.72	21.75	9.42	4.42	5.59	2.42	3.23	2.61	1.13	1.95	0.76	0.33	1.37	0.11	0.048
25	0.64	0.04	0.017		**5 in.**		9.65	32.88	14.22	5.52	8.45	3.66	4.04	3.95	1.71	2.44	1.15	0.50	1.71	0.17	0.074
30	0.77	0.06	0.026	0.49	0.02	0.009	11.58	46.08	19.95	6.63	11.85	5.13	4.85	5.53	2.39	2.93	1.62	0.70	2.05	0.23	0.10
35	0.89	0.08	0.035	0.57	0.03	0.013				7.73	15.76	6.82	5.66	7.36	3.19	3.41	2.15	0.93	2.39	0.31	0.13
40	1.02	0.11	0.048	0.65	0.03	0.013				8.84	20.18	8.74	6.47	9.43	4.08	3.90	2.75	1.19	2.73	0.40	0.17
45	1.15	0.13	0.056	0.73	0.04	0.017		**6 in.**		9.94	25.10	10.87	7.27	11.73	5.08	4.39	3.43	1.49	3.08	0.50	0.22
50	1.28	0.16	0.069	0.81	0.05	0.022	0.56	0.02	0.009	11.05	30.51	13.21	8.08	14.25	6.17	4.88	4.16	1.80	3.42	0.60	0.26
60	1.53	0.22	0.095	0.97	0.07	0.030	0.67	0.03	0.013				9.70	19.98	8.65	5.85	5.84	2.53	4.10	0.85	0.37
70	1.79	0.30	0.13	1.14	0.10	0.043	0.79	0.04	0.017							6.83	7.76	3.36	4.79	1.13	0.49

(Courtesy George Fischer Engineering Handbook)

TABLE 6.5 *(continued)* Friction Loss in Schedule 40 Pipe

GALLONS PER MINUTE	VELOCITY FEET PER SCOND	FRICTION HEAD FEET	FRICTION LOSS POUNDS PER SQUARE INCH	VELOCITY FEET PER SCOND	FRICTION HEAD FEET	FRICTION LOSS POUNDS PER SQUARE INCH	VELOCITY FEET PER SCOND	FRICTION HEAD FEET	FRICTION LOSS POUNDS PER SQUARE INCH	VELOCITY FEET PER SCOND	FRICTION HEAD FEET	FRICTION LOSS POUNDS PER SQUARE INCH	VELOCITY FEET PER SCOND	FRICTION HEAD FEET	FRICTION LOSS POUNDS PER SQUARE INCH	VELOCITY FEET PER SCOND	FRICTION HEAD FEET	FRICTION LOSS POUNDS PER SQUARE INCH	VELOCITY FEET PER SCOND	FRICTION HEAD FEET	FRICTION LOSS POUNDS PER SQUARE INCH
75	1.92	0.34	0.15	1.22	0.11	0.048	0.84	0.05	0.022							7.32	8.82	3.82	5.13	1.28	0.55
80	2.05	0.38	0.16	1.30	0.13	0.056	0.90	0.05	0.022		**8 in.**					7.80	9.94	4.30	5.47	1.44	0.62
90	2.30	0.47	0.20	1.46	0.16	0.069	1.01	0.06	0.026							8.78	12.37	5.36	6.15	1.80	0.78
100	2.56	0.58	0.25	1.62	0.19	0.082	1.12	0.08	0.035	0.65	0.03	0.012				9.75	15.03	6.51	6.84	2.18	0.94
125	3.20	0.88	0.38	2.03	0.29	0.125	1.41	0.12	0.052	0.81	0.035	0.015		**10 in.**					8.55	3.31	1.43
150	3.84	1.22	0.53	2.44	0.40	0.17	1.69	0.16	0.069	0.97	0.04	0.017							10.26	4.63	2.00
175	4.48	1.63	0.71	2.84	0.54	0.235	1.97	0.22	0.096	1.14	0.055	0.024								6.16	2.67
200	5.11	2.08	0.90	3.25	0.69	0.30	2.25	0.28	0.12	1.30	0.07	0.030	0.82	0.027	0.012		**12 in.**			7.88	3.41
250	6.40	3.15	1.36	4.06	1.05	0.45	2.81	0.43	0.19	1.63	0.11	0.048	1.03	0.035	0.015					11.93	5.17
300	7.67	4.41	1.91	4.87	1.46	0.63	3.37	0.60	0.26	1.94	0.16	0.069	1.23	0.05	0.022						
350	8.95	5.87	2.55	5.69	1.95	0.85	3.94	0.79	0.34	2.27	0.21	0.091	1.44	0.065	0.028	1.01	0.027	0.012			
400	10.23	7.52	3.26	6.50	2.49	1.08	4.49	1.01	0.44	2.59	0.27	0.12	1.64	0.09	0.039	1.16	0.04	0.017			
450				7.31	3.09	1.34	5.06	1.26	0.55	2.92	0.33	0.14	1.85	0.11	0.048	1.30	0.05	0.022			
500				8.12	3.76	1.63	5.62	1.53	0.66	3.24	0.40	0.17	2.05	0.13	0.056	1.45	0.06	0.026			
750							8.43	3.25	1.41	4.86	0.85	0.37	3.08	0.28	0.12	2.17	0.12	0.052			
1000							11.24	5.54	2.40	6.48	1.45	0.63	4.11	0.48	0.21	2.89	0.20	0.087			
1250										8.11	2.20	0.95	5.14	0.73	0.32	3.62	0.31	0.13			
1500										9.72	3.07	1.33	6.16	1.01	0.44	4.34	0.43	0.19			
2000													8.21	1.72	0.74	5.78	0.73	0.32			
2500													10.27	2.61	1.13	7.23	1.11	0.49			

(Courtesy George Fischer Engineering Handbook)

$$P_{CR} = \frac{2E}{1 - \mu^2}\left(\frac{1}{DR - 1}\right)^3$$

where P_{CR} = critical flattening pressure, lb/in^2
E = elastic modulus, lb/in^2
μ = Poisson's Ratio
= 0.45 for polyethylene (long term)
DR = pipe dimension ratio

Pipe that is submerged in a body of water displaces its volume of the water. If the pipe and its contents are heavy enough, it will sink. Otherwise, it will float. In order to keep lightweight pipe submerged, the use of added weight is needed.

Most submergence weights are made with reinforced concrete. Since concrete forms can be made in various shapes, there is quite a bit of flexibility in the design and implementations of the weights.

Concrete weights are normally formed in two, or more, sections that clamp around a pipe over an elastomeric padding material. It is important that there be adequate clearance between weight sections to avoid the weights sliding along the pipe. Most weights are formed with flat bottoms and are bottom heavy. The reason for this design is to reduce rolling when cross-current conditions exist. Any fasteners used to secure weights to pipes must be approved for the intended use and must be resistant to the specific water environment.

DIFFERENT TYPES OF PIPE

Installation requirements for buried pipe vary with different types of pipe. For example, the installation methods for PE pipe can be very different than the requirements for PVC pipe. Installation details are covered in Chapter 7.

CHAPTER 7
PIPE HANDLING AND CONSTRUCTION

Conditions for pipe handling and construction vary depending upon the type of pipe being installed. For example, PVC pipe is often used for both drainage and water supply. The pipe design and installation for these two types of piping applications are different. Principles and fundamentals can be similar, but specific requirements often vary. In view of this, we will talk about different types of pipe individually. Long term behaviors of PVC, CPVC, PP, and HDPE are illustrated in Figures 7.1, 7.2, 7.3, 7.4, and 7.5.

PVC PIPE

Sewers are often made of PVC pipe. The diameter of the pipe can range from small to large. How the PVC pipe for sewers is handled and installed is essential to a successful installation. Some characteristics of PVC sewer pipe make it subject to damage that might not affect another type of pipe. For example, PVC pipe used in sewers can become brittle in cold weather. Dropping cold PVC pipe on a hard surface can crack the pipe. While the crack may not be very visible, it will cause a leak. Water can create extreme problems when joining the pipe with fittings. Any water in the joint can create a void that will leak. Mud is also a problem when joining PVC pipe and fittings.

A good installation begins in the field with the handling and storage of pipe and fittings. Any incoming pipe should be checked for visual defects upon delivery. Unloading the pipe, whether by hand or machine, should be done in a manner that will not damage the pipe. Sliding cold PVC off a truck and allowing it to drop onto a concrete floor is not the right way to unload a truck. The pipe should

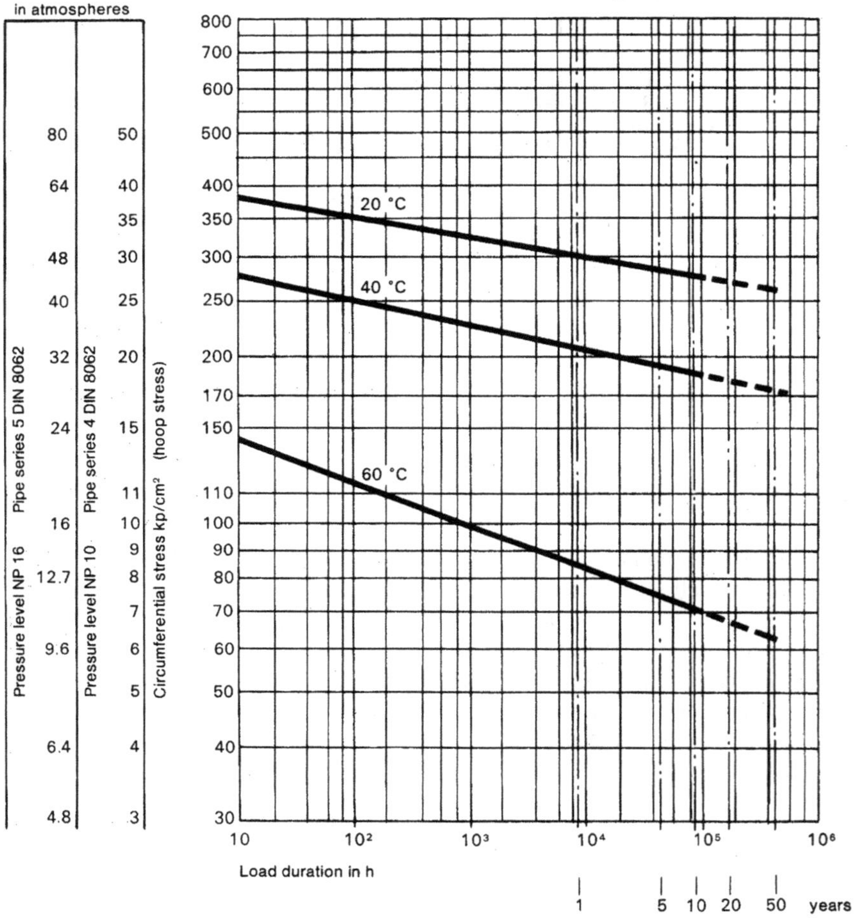

FIGURE 7.1 Long-term behavior of PVC. *(Courtesy George Fischer Engineering Handbook)*

never be dropped. While PVC pipe is resilient, it can crack, even in warm weather, if it is mishandled.

It is best to store PVC pipe in a protected location. Keeping the pipe dry is extremely advantageous. If the pipe is not contained by bindings or containers, it should be blocked to avoid a rollout collapse of the pipe. Never stack the pipe so high as to create a safety risk.

When a bell-type pipe is used, the bell should be laid out by the trench with the bell ends pointing in the direction that work will progress. If the ground surface is wet or muddy, the pipe should be laid out on a waterproof groundcover, such as a tarp. Keep the pipe and fittings dry and clean.

All trenches should be created with safety in mind. Deep trenches should be fitted with appropriate protection for workers. It is also important to keep the trench

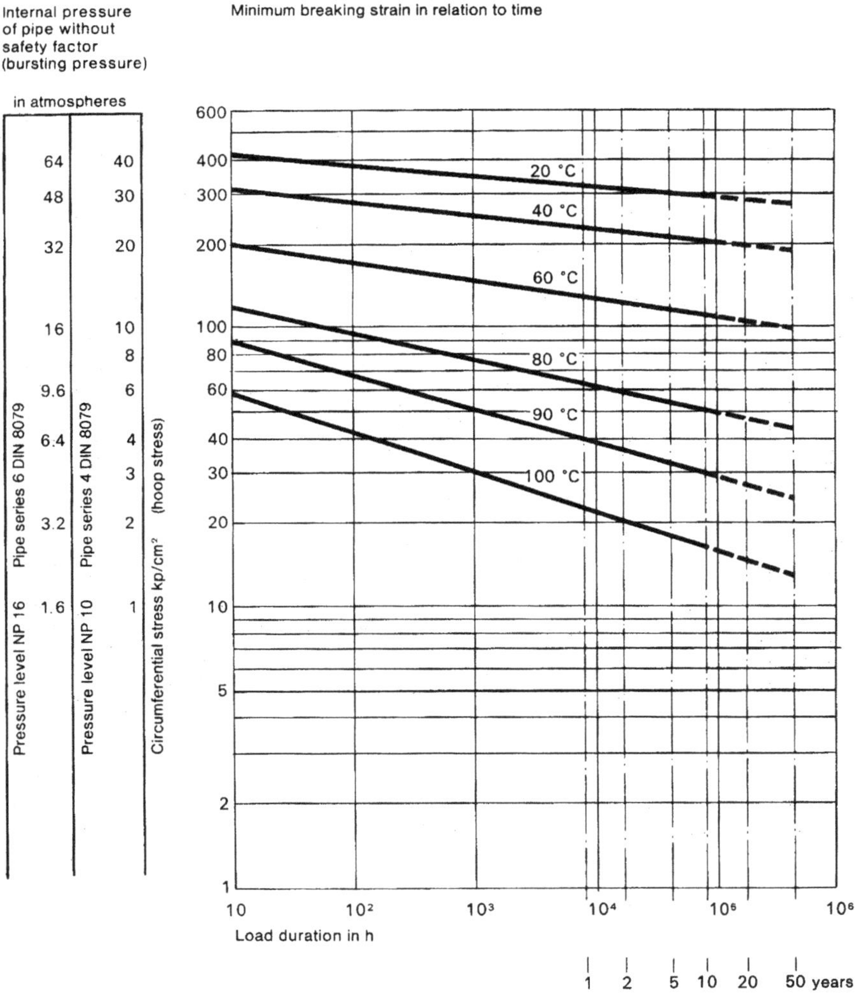

1 atmosphere = 14.7 psi
1 kp/cm² = 14.22 psi

FIGURE 7.2 Long-term behavior of CPVC. *(Courtesy George Fischer Engineering Handbook)*

bed as dry and firm as possible. Once installed, PVC pipe can float if a trench is subjected to a collection of water. This problem can be solved with backfilling. As a rule-of-thumb, cover the pipe with backfill material so that the backfill height is at least 1.5 times the size of the pipe diameter. As an example, a 4-inch pipe would require at least 6 inches of backfill to prevent floatation from flooding.

PVC is fairly easy to cut. The methods used to cut the pipe vary with installers and pipe size. Pipe with a small diameter is sometimes cut with roller-type cut-

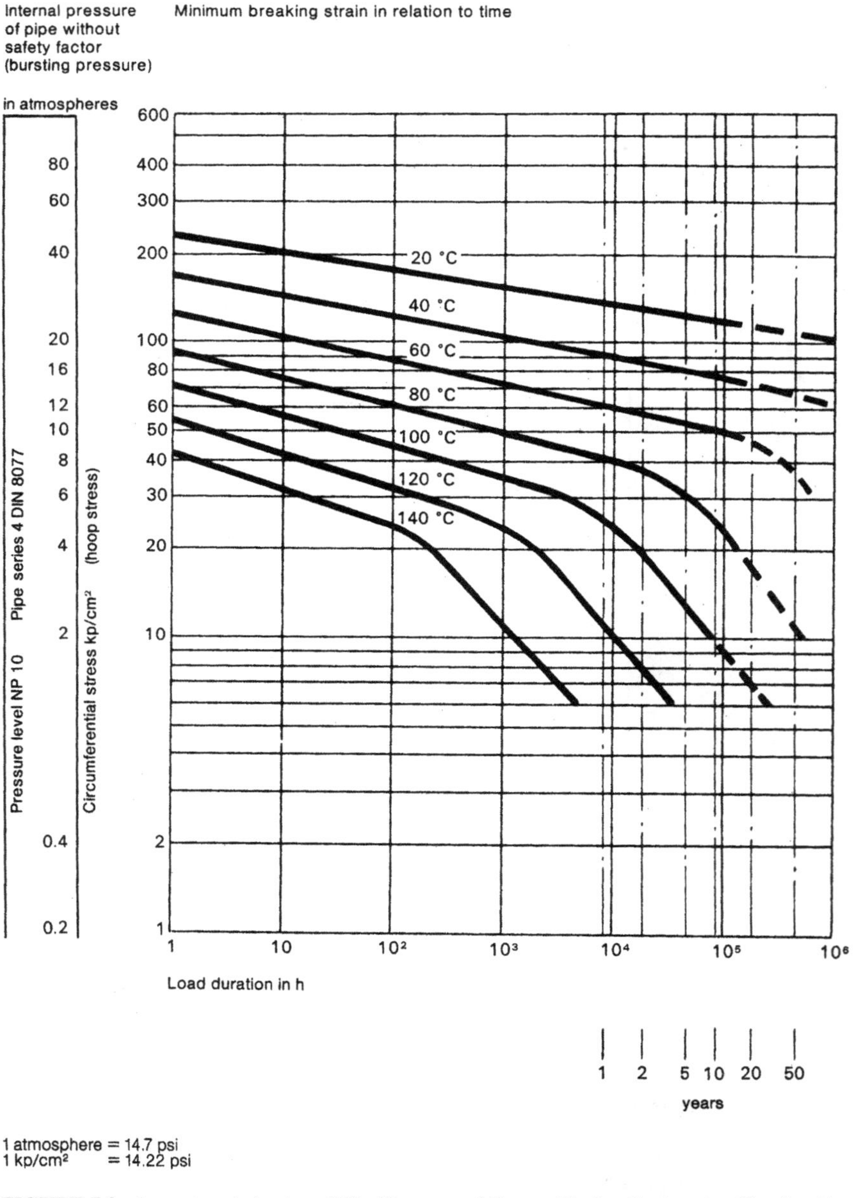

FIGURE 7.3 Long-term behavior of PP. *(Courtesy of George Fischer Engineering Handbook)*

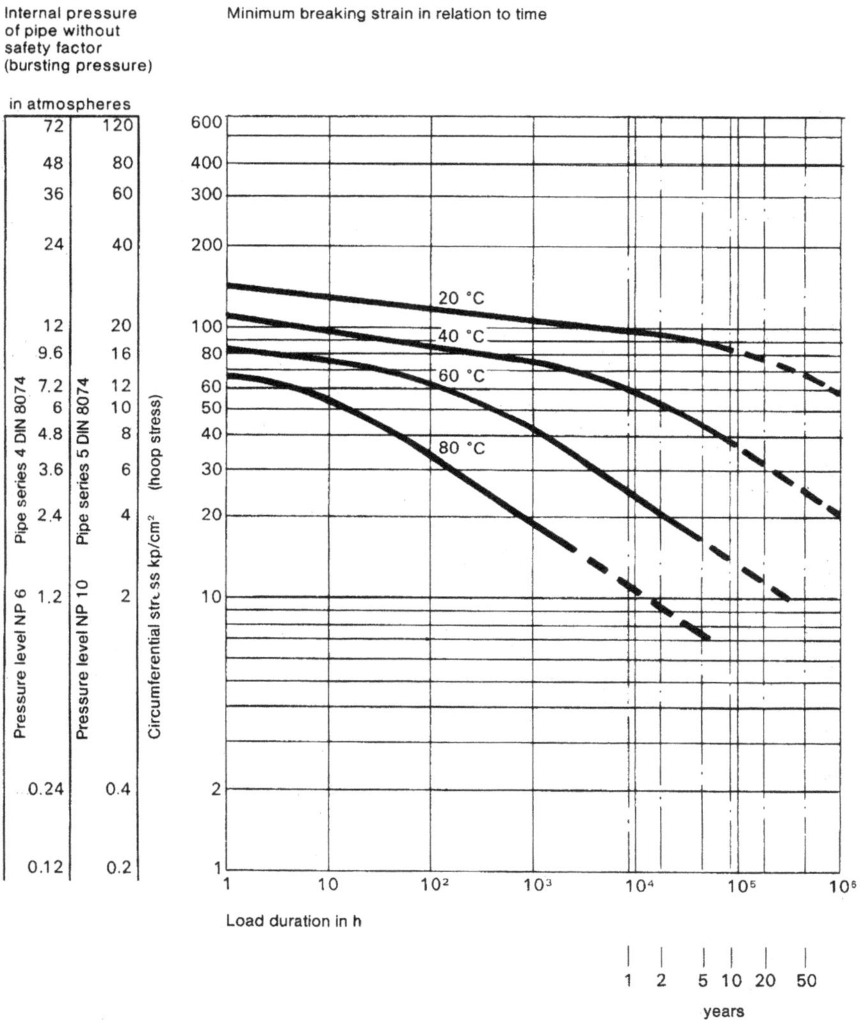

1 atmosphere = 14.7 psi
1 kp/cm² = 14.22 psi

FIGURE 7.4 Long-term behavior of HDPE. *(Courtesy of George Fischer Engineering Handbook)*

ters, similar to the type used to cut copper or steel pipe. Handsaws, chop saws, and other types of electric saws can be used. The key is to cut the pipe end squarely. There will be burrs on the end of the pipe. These burrs should be removed and the pipe end should be beveled. This can be done with a beveling tool, a wood rasp, or even a power sander. Eye protection should be worn to protect workers cutting pipe or in the area where pipe is being cut. Working conditions for PVC fittings are illustrated in Figure 7.6.

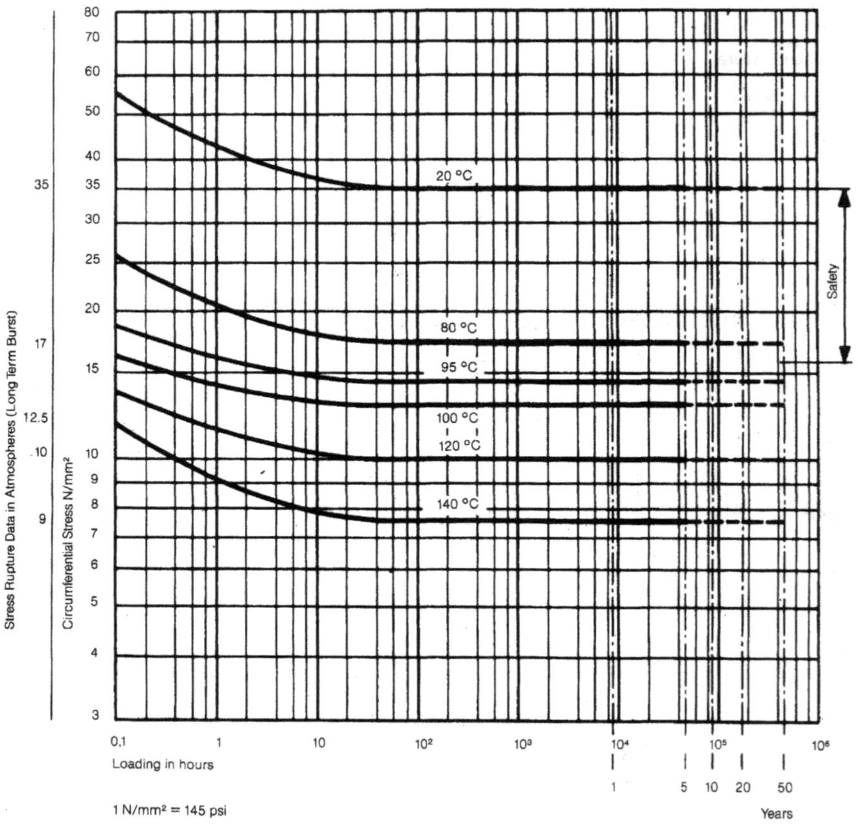

FIGURE 7.5 Long-term behavior of SYGEF. *(Courtesy of George Fischer Engineering Handbook.)*

SOLVENT-WELDED JOINTS

Solvent-welded joints are common when working with small-diameter sewers. The joining process for this type of pipe and fittings is simple. The surface of the pipe ends and the inner fitting hubs should be dry, clean, and free of any debris. A primer is first applied to the pipe ends and the fitting hubs. Then a solvent-weld material, normally called glue by workers in the field, is applied to the pipe ends and fitting hubs. The pipe end is inserted into the fitting and turned to assure full coverage of the bonding material. While these joints are fairly secure at this point, they should not be exposed to water, impact, or extreme movement. It can take several hours, depending upon climatic conditions, for a solvent-weld joint to dry to a satisfactory condition. Testing for leaks should not be done immediately. Wait until you are sure the joints have had time to cure properly before testing. Thermal application range of sealing materials is illustrated in Figure 7.7.

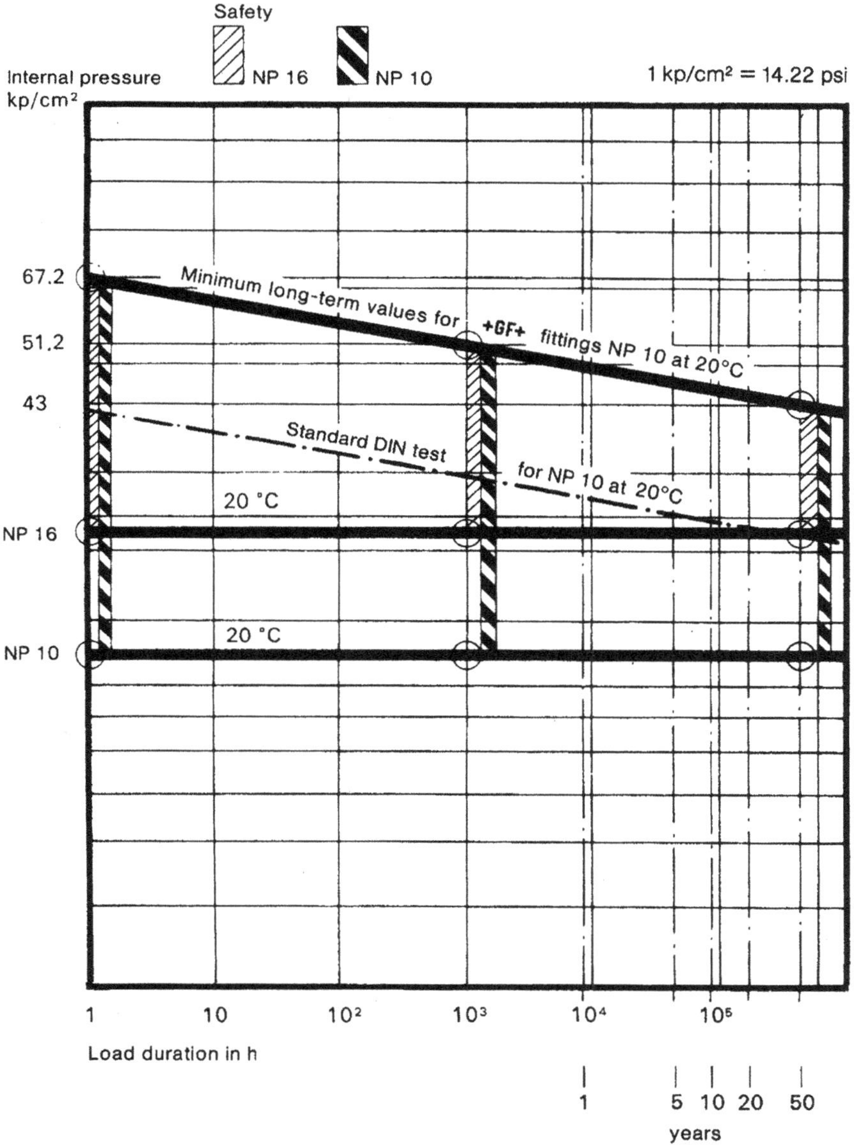

FIGURE 7.6 Working conditions for PVC fittings. *(Courtesy George Fischer Engineering Handbook)*

Refers to safe media, e.g. water/under static load

long-term service   short-term peak

NR Natural rubber

NBR Nitrile (butadiene) rubber

EPDM Ethylene propylene rubber

CR Chloroprene rubber (e.g. Neoprene®)

CSM Chlorine sulphynol polyethylene (e.g. Hypalon®)

IIR Isobutene isoprene (butyl) rubber

FPM Flourine rubber (e.g. Viton®)

PTFE Polytetraflouroethylene (e.g. Teflon®)

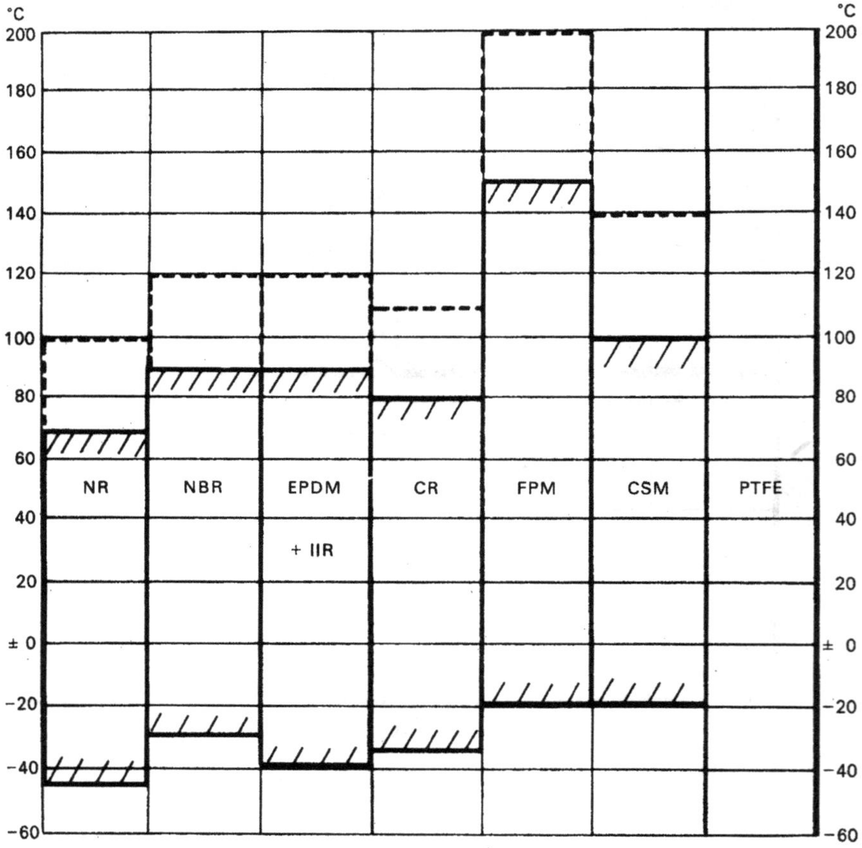

®Du Pont's registered trade name

FIGURE 7.7 Thermal application range of sealing materials. *(Courtesy George Fischer Engineering Handbook)*

GASKET JOINTS

Some types of PVC pipe are connected with gasket joints. This requires a different type of installation procedure. Sometimes the gaskets are supplied separately from the pipe. In other cases, the gaskets are provided as a part of the pipe. When installing gaskets that were not shipped as a part of the pipe being installed, be sure that the gaskets are clean and dry before using them. Both the gasket groove and the spigot should be clean and dry prior to connection. Pipe that is shipped with gaskets in place will require installers to merely make sure that the gaskets are clean. The gaskets should not be removed for cleaning.

A manufacturer recommended lubricant is needed to make a connection with gasket-type pipe. The lubricant is applied to the bevel of the spigot end of pipe. This lubricant is applied from the end of the pipe to a point about halfway to the insertion line. You should not place any lubricant in the bell end of the pipe.

The next step is to insert the spigot end of the pipe into the bell end. This can be a tricky task to perfect in some cases. Large-diameter pipe requires the use of various types of mechanical devices to create a joint. Smaller pipe can be joined by hand. A good joint is made when the insertion line of the spigot end of a pipe is lined up with the edge of the bell that the spigot is being inserted into.

If there seems to be a problem with making a connection, you should check the gasket material to see if it is in good condition. Any gasket that appears to be defective or damaged should be replaced. When the gasket seems okay, seat the gasket in the bell properly and attempt the joining process again.

When mechanical assistance is required to create a joint, there are a few additional considerations to keep in mind. Installers must be sure that the spigot end of a pipe is not inserted too far into the bell end of a pipe. There is some risk of existing joints being damaged if too much force is used to join pipe sections. Many problems can arise if excessive force is used to create a joint. Gaskets can roll up and cause a poor connection. The pipe bell might split. A pipeline might not pass an air test if excessive force is used to create joints.

One common means of mechanical assistance is very simple. All it involves is the use of a wooden block and a lever, that is also frequently made of wood. A typical 2-×-4 wall stud can be used to build such a device. A short piece of the lumber is attached to the lever. Once a section of pipe is aligned for insertion, the block of wood on the lever is placed against the pipe end with the lever at a backward angle. The lever is then pushed forward. This motion provides the power needed to insert a pipe end into a bell.

TRENCH INSTALLATION

A trench installation with PVC pipe calls for several levels of fill material in a trench. The first is the bedding. Bedding is usually less than 6 inches thick and is used to provide continuous support of the pipe. The next layer of fill is the haunching that comes from the top of the bedding to a point about half way up the pipe diameter. Then there is initial backfill and finally there is the final backfill.

Bedding is crucial since it supports the underside of a pipe. Haunching is also important. It is a major factor in controlling pipe performance and deflection. This also provides pipe support. Compaction is needed to assure proper pipe placement and support. Initial backfill material must be free of rocks and other objects that might damage a pipe when the backfill is placed in a trench. Generally, initial backfill should be at least 6 inches in depth. This layer of backfill serves as protection for the pipe when other backfill material is installed. Final backfill should be free of large rocks, frozen ground, rubble and other items that might cause voids or sinking in the backfill layer over a period of time.

POLYETHYLENE PIPE

Polyethylene (PE) pipe is flexible. Several types of pipe can be considered flexible and all of them need to be protected from deflection. When installed below ground, flexible pipe support is a major requirement of trench bedding and backfilling. Above-ground installations require adequate support in the form of hangers or clamps. As with any piping material, PE pipe should be handled and stored in a way that will not damage the pipe. PE pipe should never be placed in contact with sharp objects.

Deflection is an increase or decrease in pipe diameter. Kinking of the pipe and crushing of the pipe must be avoided. Where there is an increase in diameter the condition is called a rise. Squeezing the pipe causes a decrease in diameter. When buried, the embedment materials surrounding PE pipe must be planned carefully.

Buried PE pipe depends on a trench bed to maintain adequate support. The in-situ soil, also known as the native soil, of the trench is not as important as the backfill material used under, around, and over the pipe. A designer must factor in all soil conditions when specifying embedment materials. The categories of embedment material for PE pipe is the same as those discussed for PVC pipe. Some considerations for backfill material include how the material will stand up to the final installation, the ease of placement and compaction, and the cost and availability of the material.

Class l and class ll soils are considered good materials for use in trenches with PE pipe. Both classes of material are granular and provide excellent pipe support. A class l material is a manufactured aggregate and is most often crushed stone. The class ll material is more of a sandy material. The two types of embedment materials can be combined and they allow for good drainage. It is important to keep the backfill material within acceptable ranges for the size of the pipe being supported. For example, a 4-inch pipe should have an embedment material surrounding it that is no more than one-half an inch in size. Pipe with a diameter of 18 inches or more should be surrounded by fill material that is not more than 1.5 inches in size.

The use of cement-stabilized sand is sometimes called for as an embedment material. This is a mixture of cement and sand. Normally, the cement content is about 3 to 5 percent of the sand volume. The mixture is placed with compaction rather than in a pouring fashion. Moisture is added to the mix before placement and compaction is done. When allowed to cure overnight, a cement-stabilized sand mixture installed before backfilling can reduce pipe deflection.

Class lll material and class lVa material do not provide the stiffness that a class l or class ll material can. This is due to increased clay content. Very little pipe support is offered by either class lVb and class V materials. It's not unusual for pipe to float when installed with such materials.

Compaction is a factor in most buried-pipe installations. The compaction is sometimes done with hand-held, manual tampers. Gasoline-powered tamping machines are often used. Other forms of power compactors can be used. In all cases, the compactors must be employed in a way that does not have a negative impact on the pipe being installed. Above-ground installation instructions were given in Chapter 5. Therefore, we will now move into the discussion of joining PE pipe.

PIPE-JOINING PROCEDURES

Pipe-joining procedures for PE pipe are different than the joining methods used for PVC pipe. This type of joining is done with either heat fusion or mechanical fittings. Plastic pipe may be joined to other types of material with the use of compression fittings, flanges, or other qualified types of transition fittings. The type of joint used is up to designers and installers, but it can be subject to the application. Let's start with heat fusion.

Heat fusion offers 3 types of joints. They are butt joints, saddle joints, and socket fusion. There are two spinoff joints for producing socket- and saddle-heat-fusion joints. The concept of a heat-fusion joint is simple. You heat two surfaces to a specific temperature and them fuse them together with an application of sufficient force. Done correctly, the two materials mix together and form a joint. This type of joint can be as strong as the pipe being joined. When a fused joint cools to near ambient temperature the joint can be handled and worked with.

There are basically two ways to create fused joints. The first type uses heating tools to heat the materials to be joined. A second type uses electric current to make an electrofusion joint. This is done for socket joints and saddle-type joints.

Butt fusion is one of the most common ways for connecting sections of large-diameter PE pipe. There are only 6 steps in this process. A butt-fused joint is inexpensive, permanent, and does not restrict flow in the pipe. Making butt-fusion joints in the field is not difficult when the proper equipment is available to work with. The joining equipment is available for pipe with a diameter of up to 72 inches. What are the six steps required? First, an installer must securely fasten the components that are to be fused together. Then they have to face the pipe ends. The pipe profile must be aligned properly. Melting the pipe interfaces is next. Both profiles are joined together and then held under pressure. Sounds simple, but there is more to know about this process.

Securing the pipe requires that the pipe will not move during the joining process. The facing process must put the pipe ends in a parallel mating surface. A rotating planer block design is normally installed in equipment that is used for facing. There has to be a perfectly square face that is perpendicular to the pipe centerline on each pipe end. It is unacceptable for any detectable gap between pipe ends to exist.

Alignment of the pipe ends is essential. This is usually done by adjusting the joining equipment. Then the ends of the pipe sections are heated. The temperature is determined by the manufacturer's recommendations. During the heating process, the pipe must maintain interface pressure and the time for making the joint must be monitored and meet the manufacturer's recommendations. A molten bubble or bead will be visible when this procedure in done correctly. The joining machine being used should monitor temperature ratings. Due to local conditions, the thermometer ratings can be fooled. To compensate for this risk, a pyrometer should be used periodically to keep track of temperature requirements. If any molten plastic residue is present on the heater face, it must be cleaned to ensure good fusion joints.

After the heating application reaches the proper temperature, the heating device is removed and the pipe ends are put together to form the fusion joint. Consult manufacturer recommendations for interface pressure and the bead size of molten material required to make a solid joint.

Once a molten joint is formed it must be held in place so that it can cool and create a satisfactory joint. Consult manufacturer recommendations for cooling times. If joint beads need to be removed, there is equipment available for such a process.

Installing a saddle with sidewall fusion is not uncommon. This process requires the use of a saddle-fusion machine. The process can be done without a machine, but it is generally not recommended. Since this process is generally discouraged, it will not be discussed here.

There are eight steps involved in making a saddle-fusion joint. The first step is to clean the pipe where a joint will be made. Once this is done, install a saddle-adapter heater of the proper size. Make sure that the saddle is in direct and tight contact with the pipe. However, do no overtighten. Bring the heater up to a temperature recommended by the manufacturer. The next step is to install the saddle-fusion machine. Again, be careful not to tighten the device to a point of collapsing the pipe.

You are now ready to clean the joint area again. It is highly recommended that you rough up the joint surfaces with utility cloth. The cloth should have a grit rating of 50 or 60. Don't use sandpaper or other material that may leave grit or other unwanted deposits on the joint surfaces.

Check to be certain that the proper saddle-fitting holding inserts are in the fusion machine. Place the fitting on the pipe and place the fitting into the insert. Slight downward force is then applied on the fitting. Make sure that there is a good fit between the pipe and fitting. Double check the fit between the fitting and pipe.

Heater temperature should be monitored periodically. Follow manufacturer recommendations when heating the pipe and fitting. Once the materials to be fused are up to temperature, remove the heater and quickly inspect the melt pattern on the pipe and fitting. Then proceed to join the pipe and fitting with proper fusion force. Maintain a stable fit on the joint until it has cooled to ambient temperature.

There are only 5 steps in socket fusion. The proper equipment for the job must be selected. Pipe ends are squared and prepared and then the portions to be fused are heated. The parts are joined and then allowed to cool to ambient temperature.

Electrofusion

Electrofusion is another way to join plastic pipe with a fused joint. The main difference between electrofusion and regular fusion joining is the process of heating the materials. An electrofusion joint is made by heating internally rather than externally. This can be done with either a wire coil at the interface of the joint or by the use of a conductive polymer. Electrical current creates the heat needed to fuse a joint.

When fused joints are not wanted, you can turn to a wide variety of mechanical connections to join PE pipe and fittings. Mechanical compression fittings are commonly used on PE pipe that has a diameter of 2 inches, or less. A bolt-type mechanical coupling can be used to join the ends of PE pipe. This same device can be used to join PE pipe with steel pipe. Comparisons of the strength characteristics of metal and plastics are illustrated in Figure 7.8. Stab-type mechanical fittings are another option. It is safe to say that there is no shortage of options when it comes to joining PE pipe and fittings.

CPVC PLASTIC PIPE

CPVC plastic pipe is approved for use with both hot and cold water. Small sizes of CPVC pipe, up to 2 inches in diameter, are joined with solvent cement. Larger CPVC pipe is schedule 80 pipe and is threaded to iron pipe size outside diameters. The joining process for solvent-weld connections is very similar to the process used to join PVC pipe.

Several types of transition fittings are available for connecting CPVC pipe to other types of materials. There are unions, compression fittings, and metal fittings with CPVC socket connections on one end. One of the most commonly used transition fittings is a CPVC male or female adapter that is screwed into or onto the different type of material. There are, however, two potential problems when using straight CPVC threaded adapters. If the adapter is subjected to a wide temperature range a drip leak may occur. It is also possible for some thread sealants to chemically attack the composition of the CPVC. The simple solution is to use metal threaded fittings that provide a socket for a solvent-welded joint on one end.

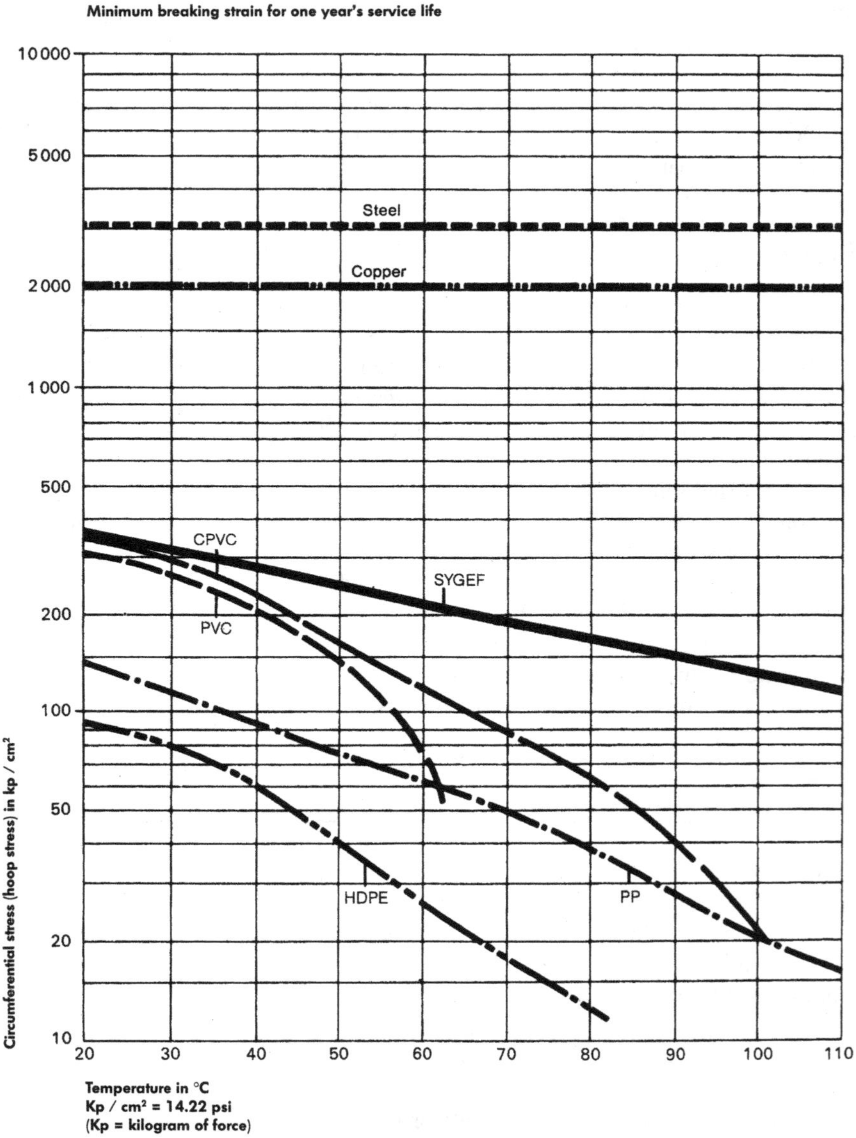

FIGURE 7.8 Comparison of the strength characteristics of metal and plastics from 20 degrees C. to 110 degrees C. *(Courtesy George Fischer Engineering Handbook)*

OTHER TYPES

There are other types of plastic pipe in use for various purposes. PEX pipe has become very popular for water distribution and for in-floor radiant heating. This pipe, or tubing as it may be called, comes in rolls and is easy to cut and install. ABS pipe has seen extensive use as a drainage and vent pipe, but PVC pipe has overshadowed it in recent years. Corrugated pipe is used in nonpressure applications, such as sewers, culverts, and subdrainage systems.

Regardless of the type of pipe you are working with, you should refer to manufacturer's recommendations and follow them. There are many general principles that can be applied to installation methods, but there is no substitute for specific instructions provided by manufacturers and appropriate building, gas, and plumbing codes.

CHAPTER 8
HORIZONTAL DIRECTIONAL DRILLING

Horizontal Directional Drilling (HDD) is probably the fastest growing technology in the trenchless industry. HDD has gained widespread acceptance in the construction industry over the past decade. Horizontal directional drilling in North America has grown from 12 operational units in 1984 to more than 2000 units operating in 1995. The reason for the popularity is that HDD represents a significant improvement over the traditional open-cut and cover methods for installing pipelines beneath obstacles.

Equipment and installation techniques used in the HDD process are an outgrowth of the technologies from the oil field and water well industries. There is a wide range of directional boring units in use today, from mini drilling rigs, which are used for small pipes and conduits to maxi rigs, which are capable of installing large diameter pipelines. The maximum length of a HDD is determined by many parameters including rig size, soil conditions and pipe diameter. Installations as long as 6000 feet have been successfully completed. Current HDD equipment can operate in a wide range of soil conditions, from extremely soft soils to full-face rock formations with unconfined compressive strengths of 40,000 psi.

ADVANTAGES OF HDD

Traditional open trenching methods for installing pipe can be expensive, particularly in congested urban areas. Construction involves digging around existing utilities to get to the required depth, which slows down the operation. Also, not only must the trench be backfilled, often sidewalks, pavement, brick paving, sod, or other surfaces must be replaced. In addition, open cut operations often cause interruption of traffic and disruption of near by commercial activities. Excavation requirements in HDD are minimal. As a result, in crowded urban areas, HDD is increasingly viewed as the preferred technology. It minimizes the negative impact on residents and businesses, and eliminates the need for the removal and restoration of expensive landscaping. In open areas, HDD provides an efficient method for crossing obstacles such as rivers, highways, rail tracks, or airfield runways. The HDD method also eliminates the cost and time associated with installing de-watering facilities for operations carried out below the ground water table level.

Applications

The market for horizontal directional drilling is experiencing a continuous growth worldwide. The installation of pipe and utility conduits in urban areas and across rivers and highways is the mainstay of the horizontal directional drilling industry. It is common practice to use HDD for the installation of new networks of power, natural gas, and telecommunications.

Municipal applications are perhaps the most promising future market for horizontal directional drilling. Recent advancements in equipment and tracking systems make the use of HDD cost efficient for projects that involve larger diameter products and stricter placement tolerances, as it is the case in many municipal applications. As an increasing number of municipal engineers become aware of the technology and its advantages, this market is expected to grow rapidly over the next five years. The oil, gas, and petrol-chemical industries are another important market for the directional drilling industry.

HDD PROCESS

Installation of a pipe by HDD is usually accomplished in two stages. The first stage involves directionally drilling a small diameter pilot hole along a designed directional path. The second stage consists of enlarging (reaming) the pilot hole to a diameter that will support the pipeline and pulling the pipeline back into the enlarged hole. A HDD drill rig is used to drill and ream the pilot hole and pull the pipeline back through the hole. HDD drill rigs provide torque, thrust, and pullback to the drill string. The drill drive assembly resides on a carriage that travels under hydraulic power along the frame of the drill rig. The thrust mechanism for the carriage can be a cable, chain, screw, or a rack and pinion system. Table 8.1 lists the three general categories of drilling rigs used in the industry.

Mini Rigs are mounted on a trailer, truck, or a self-propelled track vehicle. These systems are designed for drilling in relatively soft semi-consolidated formations and are used primarily for installation of utility conduits and small diameter pipelines in congested urban areas. They are not suitable for drilling gravel, cobble, or other formations where borehole stability is difficult to maintain.

Medium drilling rigs are used to install larger conduits and pipelines, normally up to 12 inches in diameter, with drill lengths ranging up to 1900 feet. They are particularly suitable for the installation of municipal pipelines, as they are sufficiently

TABLE 8.1 Typical Characteristics of HDD Rigs

	Mini Rigs	Midi Rigs	Maxi Rigs
Thrust/Pullback	<20,000 lbs.	20,000–80,000 lbs.	>80,000 lbs.
Maximum Torque	<2000 ft.lbs.	2000–20,000 ft.lbs.	>20,000 ft. lbs.
Drilling Speed	>130 RPM	130–200 RPM	<200 RPM
Carriage Speed	>100 ft/min.	90–100 ft./min.	<90 ft./min.
Carriage Drive	Cable or Chain	Chain or Rack & Pinion	Rack & Pinion
Drill Pipe Length	5-10 ft.	10–30 ft.	30–40 ft.
Drilling Distance	<700 ft.	700–2000 ft.	>2000 ft.
Power Source	<150 HP	150–250 HP	>250 HP

compact to be used in urban areas, while at the same time have the capacity of installing large diameter products beneath highways, subdivisions, and rivers. Bores can be installed in unconsolidated to consolidated sediments.

Maxi rigs typically involve a large operation with multiple trailer-mount support equipment and substantial mobilization and demobilization periods. High operating costs make their use somewhat prohibitive in the utility installation market, and they are employed primarily in the pipeline industry. These large units may be used in the installation of large diameter pipes (24-48 inches) and or exceptionally long bores.

In addition to the drilling rig, a variety of support equipment may be required. Depending on the HDD project, a drilling fluid or mud cleaning and recirculation unit, drill pipe trailer, water truck, and pump and hoses may be required. An excavator is needed to dig the entry, exit, and recirculation pits. In urban or environmentally sensitive areas a vacuum truck may be required to handle the fluid in the return pits or inadvertent returns.

Bore Installation

The bore is launched from the surface and the pilot bore proceeds downward at an angle until the necessary depth is reached. A small diameter drill string penetrates the ground at a prescribed entry point and the design entry angle, normally between 8-12 degrees. At a prescribed depth or point the drill pipe is bent to follow the proposed drill path and the designed bending radius. Then the path of the bore is gradually brought to the horizontal, followed by another bend before the bore head is steered to the designated exit point where it is brought to the surface. Choosing the proper drill pipe is a key element in the HDD process. The outer diameter and the wall thickness of the drill pipe have limitations that influence the bend radius of the bore. Larger diameter drill pipe cannot bend in short distances and cannot be used on short bores. Smaller drill pipes are more flexible and suited for short bores in the right soil conditions.

During the drilling process the bore path is traced by interpretation of electronic signals sent by a monitoring device, located near the head of the drilling string. At any stage along the drilling path the operator receives information regarding the position, depth, and orientation of the drilling tool, allowing him to navigate the drill head to its target. After the pilot string breaks the surface at the exit location, the bit is removed from the drill string and replaced with a back-reamer.

The pilot hole is then back-reamed, enlarging the hole to the desired diameter while simultaneously pulling back the line product behind the reamer.

This is typically referred to as a "continuous" borehole. In some situations with small diameter product pipe or conduit, the pipe can be pulled straight into the pilot hole after the drill is completed. However, in most HDD operations the borehole has to be reamed to enlarge the hole to accommodate pulling in the product pipe. Generally the borehole is reamed to 1.5 times the outside diameter of the product pipe. The purpose of this is to provide an annular void between the product pipe and the drillhole for the drilling fluids and spoils and for the bending radius of the product pipe.

Sometimes it is necessary to ream the borehole without pulling back the pipe. After the drillhole is reamed the product pipe or conduit is pulled back through the reamed hole filled with the drilling fluids. It is best to fabricate the product pipe on the exit side in one section so it can be tested and pulled in one continuous pullback. The drill pipe is connected to the product pipe or conduit using a pullhead or swivel. The swivel is used to prevent rotational torque from spinning the product pipe. A reamer is placed between the pullhead and the drill string to keep the drillhole open.

Drilling and Steering

Drilling curved and horizontal boreholes requires specialized drilling equipment. This equipment is contained in a bottom hole assembly (BHA) that consists of a drilling tool, a bent sub-assembly, and a steering/tracking tool. Pilot hole directional control is achieved by using a non-rotating drill string with an asymmetrical leading edge. The asymmetry of the leading edge results in a steering bias. When a change of direction is required, the drill string is rotated so that the direction of the bias is the same as the desired change of direction. The drill string may also be continuously rotated when directional control is not required. Normally, the leading edge will have an angular offset created by a bent sub or bent motor housing. The most common types of down-hole drilling/steering tools used in the HDD industry are compaction tools and down-hole mud motors.

Compaction heads consists of a wedge shaped drilling bit, which is used for cutting and displacing the soil as well as for steering. To bore a straight hole the drill string is rotated and pushed simultaneously. When a correction in direction is required, rotation stops and the drilling head is preferentially oriented in the borehole. Then the drill rig pushes the entire drill string forward. As the slant on the face of the wedge is pushed against the soil, the entire assembly is deflected in the desired direction. After the steering correction is completed, rotation is resumed until another correction is needed. Compaction type drilling tools are most often used in mini and midi size drill rigs to drill through soft to medium consolidated soils, as well as loose and dense sands. When gravel or hard clay is encountered, compaction heads tend to wear rapidly. They are not suitable for drilling in rock formations.

When drilling with compaction heads, steering difficulties are often encountered when trying to drill in very soft soils. This is caused when the resistance to the deflector plate is not sufficient to offset the tendency of the drill string to drop vertically under its own weight. To solve this problem use a larger deflector plate. Steering can be improved by increasing the flexibility at the head of the drill string. A common method is to add a length of smaller diameter more flexible drill rod behind the drill bit.

Mud (down hole) motors are used in ground conditions ranging from hard soil to rock. Mud motors convert hydraulic energy from the drilling mud being pumped from the surface to mechanical energy at the drill bit. This allows for the bit to rotate without drill string rotation. Positive displacement motors are typically used in HDD operations. These motors generate torque and rotation at the drill bit from the flow output of the mud pump. Directional control is obtained by a small bend in the drill string just behind the cutting head. As with the compaction heads, once the correction is made, the complete drill string is rotated to continue boring straight in the new direction. This method costs more than compaction heads and is less common in the utility installation industry.

The advantage of mud motors is that cutting of the formation is done by the mud motor, reducing the drill string rotation requirements, thus making it possible to drill long boreholes to substantial depths. The main disadvantage to mud motors is that they are more expensive in comparison to compaction heads and require hundreds of gallons of drilling fluids per minute.

Tracking

In HDD applications tracking is the ability to locate the position, depth, and orientation of the drilling head during the drilling process. The ability to accurately track the drill is essential to the completion of a successful bore. The drill path is tracked by taking periodic readings of the inclination and azimuth of the leading edge of the drill string. Readings are recorded with a probe that is inserted in the drill collar as close as possible to the drill bit. The three most common type of tracking tools are:

1. Electronic beacon systems (walkover)
2. Combination magnetometer-accelerometer systems
3. Inertial navigation systems

A "walkover system" consists of a transmitter, receiver, and a remote monitor. A battery-powered transmitter is located in the bottom hole assembly near the front of the drill string and emits a continuous magnetic signal. The receiver is a portable, hand held unit, which measures the strength of the signal sent by the transmitter. This information is used to determine the drill heads position, depth, and orientation. The remote monitor is a display unit installed at the drilling rig in front of the operator. It receives and displays the information provided by the receiver. This information is used to navigate the drilling head below the surface. The data is recorded to provide the as-built profile of the bore path.

When access to a location directly above the borehole alignment is not possible, or when the depth of the bore exceeds 100 feet, other types of navigation systems should be used. Two systems commonly employed are the magnetometer-accelerometer system and the inertial navigation system. The magnetometer-accelerometer system uses three magnetometers to measure the position (azimuth) of the tool in the earth's magnetic field and three accelerometers to measure the position (inclination) of the tool in the earth's gravitational field. The steering tool sends information via a wire line to a computer at the surface where the azimuth, inclination, and tool face orientation are calculated. As far as operating depth and distance from the drilling rig, this steering tool does not impose any limitation on the rig's operating range. Disadvantages of this system include susceptibility to magnetic inferences from buried metal objects and power lines. Some magnetic-accelerometer systems use a secondary survey system to account for local magnetic influences on the downhole probe. The secondary survey system induces a known magnetic field at the ground surface through a copper wire surface grid. A computer program connected to both the surface magnetic field and the steering tool compares the magnetic field measured by the steering tool and the theoretical magnetic field induced by the system, and compensate for local magnetic interference.

The inertial navigation system uses a system of three gyroscopes and three accelerometers to measure the azimuth and the inclination of the steering tool, respectively. The gyroscopes are aligned to true north at the ground surface before the survey is made. Any deviation from true north during the survey is detected by the gyroscopes and relayed to the surface where the azimuth, inclination, and drilling tool orientation are calculated by a computer. Because of the cost and sensitivity of these systems they are used mainly for calibration purposes.

Drilling Fluids

Drilling fluids are commonly called drilling mud or slurry. Drilling mud is mixed on the surface and pumped down the drill string. The mud comes out at the drill bit and is either left in the annulus of the borehole or circulated back to the surface. Drilling mud is a mixture of water, premium bentonite, and if needed, small amounts of polymer. Bentonite is a non-hazardous material.

Drilling fluids have many uses or functions. The main purposes of HDD drilling fluids are:

- To establish and maintain the borehole integrity
- To transport drill cuttings to the surface by suspending and carrying them in the fluid stream that flows in the annulus between the wellbore and the drill rod.

- To clean the build-up on the drill bits or reamer cutters by directing high-velocity fluid streams at the cutters. This also cools the bits and electronic equipment
- To reduce the friction between the drill string and the borehole wall aided by the lubricating properties of the drilling fluid
- To stabilize the borehole, especially in unconsolidated soils, by building a low permeability filter or mud cake lining and exerting a positive hydrostatic pressure against the borehole wall preventing collapse as well as preventing formation fluids from flowing into the borehole or drilling fluids from exiting the borehole into the formation (lost of circulation)
- To provide hydraulic power to downhole mud motors if used

A drilling fluid is composed of a carrier fluid and solids (clay or polymer). The carrier fluid carries the solids down the borehole where they block off the pore spaces on the borehole wall. The blockage is referred to as a filter or mud cake. The ideal mud cake will form quickly during construction of the wellbore and prevent intrusion of drilling fluid into the formation. At times additives such as detergents are added to the drilling fluids to counteract some of the formation characteristics such as swelling and stickiness.

Drilling fluids that are not properly contained on the surface can cause problems. A drilling plan should include the procedures for handling the drilling fluids as they return to the surface. Pre-dug pits and trenches or a vacuum truck should be a part of the bore planning. In addition, a drilling fluid disposal plan is a requirement for the HDD project. After all the federal, state, and local regulations are met, spreading the used bentonite slurry on pastures and fields, or pipeline rights of way with the land owner's permission can benefit the contractor and the land owner.

HDD CONSIDERATIONS

There are several factors to take into account when considering the feasibility of HDD applications.

Feasibility

The design and engineering of a HDD project is affected by actual site conditions including soil formations, terrain, existing utilities, and equipment set-up restraints. The final bore may differ from the original design because of the limitations of downhole tooling and the actual drilling conditions encountered. There are many factors to consider when deciding if HDD is the best installation method for a pipeline or utility conduit. Economics or cost is always a primary factor. In many instances HDD provides the best economical choice and in others it does not. However, determining if the potential HDD application is technically feasible is the key factor. Due to the capabilities of today's HDD tools and drill pipe there are limitations on the length of a bore and the diameter of the product pipe or conduit. The equipment in use today involves thrusting pipe from the surface to drill a pilot hole. There are limitations on the amount of thrust that can be applied to the drill pipe. In addition, control of the drill path diminishes over long lengths.

Subsurface Soil Material

While many factors have an impact on the feasibility of a HDD, the feasibility is primarily limited by the subsurface conditions. Subsurface conditions consisting of large grain content (gravel and cobbles) and excessive rock strength and hardness can prevent a successful HDD application. Coarse-grained soil materials are a serious limitation on the feasibility of a HDD. It is hard for the drilling fluid to fluidize the coarse soil material. Boulders and cluster of cobbles can remain in the drill path and obstruct the drill bit, reamer, or pipeline. Exceptionally strong and hard rock may hamper all phases of a HDD operation. There have been successful HDD operations in rock with unconfined compressive strengths exceeding 12,000 psi and Mohs Scale of Hardness factors above 7. However, these conditions are usually difficult to penetrate, especially at depths. When pushing against hard rock the drill string tends to deflect rather than penetrate. Table 8.2 provides some general guidelines for the feasibility of HDD based on earth material type and gravel percent by weight. Practical experience and engineering judgment must be used applying the guidelines shown in Table 8.2.

Design Factors

Typical pipe products installed by the HDD method include steel, High Density Polyethylene (HDPE), Polyethylene (PE), and Polyvinyl Chloride (PVC) conduits, as well as direct buried cables. During the HDD installation, the pipe product will experience a combination of tensile, bending, and compressive stresses. The magnitude of these stresses is a function of the approach angle, bending radius, product diameter, length of the borehole, and the soil properties at the site.

By properly selecting the radius of curvature and type of product, the design engineer can ensure that these stresses do not exceed the product pipe capacity during the installation. Ideally, the design should call for a minimum number of joints. If joints are necessary, flush joints (butt fusion) are preferable to glued or threaded joints that tend to increase the drag on the product in the borehole. Other considerations include minimum cover, minimum separation from existing utilities, tolerances for deviation in the vertical, and horizontal profiles and maximum true depth.

TABLE 8.2 HDD Feasibility

Earth Material	Gravel % by Weight	HDD Feasibility
Very soft to hard strength, possibly slickensided clay	NA	Good to Excellent. Penetration of strong clay surrounded by looser soils may result in the bit skipping at the interface. Bit steering may be difficult when passing through soft soil layers.
Very loose to very dense sand with or without gravel traces.	0 to 30	Good to Excellent. Gravel may cause steering problems.
Very loose to very dense gravelly sand	30 to 50	Marginal. In these conditions drilling fluid characteristics are critical to success. Bit steering may be inaccurate.
Very loose to very dense sandy gravel	50 to 85	Questionable. Horizontal penetration for any appreciable distance will be extremely difficult. Bit steering will be inaccurate.
Very loose to very dense gravel	85 to 100	Unacceptable. With current technology horizontal penetration is almost impossible. This type of material must be avoided or penetrated at a steep angle.
Rock	NA	Excellent to Unacceptable. Softer or weathered materials offer good HDD characteristics. Penetrating solid rock after passing through soil may be difficult due to the bit's tendency to skip on the lower hard surface. Rock in the rounded cobble form is almost impossible to drill.

General Guidelines

When considering a HDD project it is usually best to consult with an experienced contractor and a qualified engineer. Here are some general considerations that should be considered:

- Select the HDD path that provides the shortest reasonable distance
- Find routes and sites where the pipeline can be constructed in one continuous length
- Compound bends are possible, but it is best to use as straight a drill path as possible
- Avoid entry and exit elevation differences in excess of 50 feet. Both points should be as close as possible to the same elevation
- Locate buried structures and utilities within 10 feet of the drill-path

- Observe and avoid aboveground structures, such as power lines, which might limit the height available for construction equipment
- Long crossings with large diameter pipe require a more powerful drill rig
- The larger the pipe diameter, the larger the volume of drilling fluids that must be pumped, requiring larger pumps and mud-cleaning and storage equipment
- Develop as-built drawings based on the final drill path. The as-built drawings are essential to knowing the exact pipeline location and to avoid future third party damage

HDD SITE CHARACTERIZATION

Site Survey

Conducting a site survey is a primary part of the HDD site characterization. The survey should include both surface and subsurface investigations. Based on the information from the site survey, a plan and profile drawing is developed. The drawing is used for contract documents and to make a working profile that will be used for navigation of the bore and developing as-built drawings. See Figure 8.1 for an example of a HDD plan and profile drawing.

Geotechnical Factors

The decision to use the HDD method for a crossing should be based on an understanding of the HDD process and the crossing site's characteristics. The natural and man-made features in the area will dictate the design for the HDD crossing. A vital element of the HDD design process is a comprehensive geotechnical survey to identify soil formations at the potential HDD site. The purpose of the geotechnical investigation is to determine if directional drilling is feasible, and to determine the most efficient way to accomplish it. Using the information gathered from the geotechnical survey, the best crossing route can be determined, drilling tools and procedures selected, and the drill path designed. The extent of the geotechnical investigation depends on several factors such as the pipe diameter, bore length, and the nature of the HDD crossing. During the survey, the geotechnical consultant will identify a number of relevant items including the following:

- Soil identification to locate rock, rock inclusions, gravelly soils, loose deposits, discontinuities and hardpan
- Soil strength and stability characteristics
- Groundwater

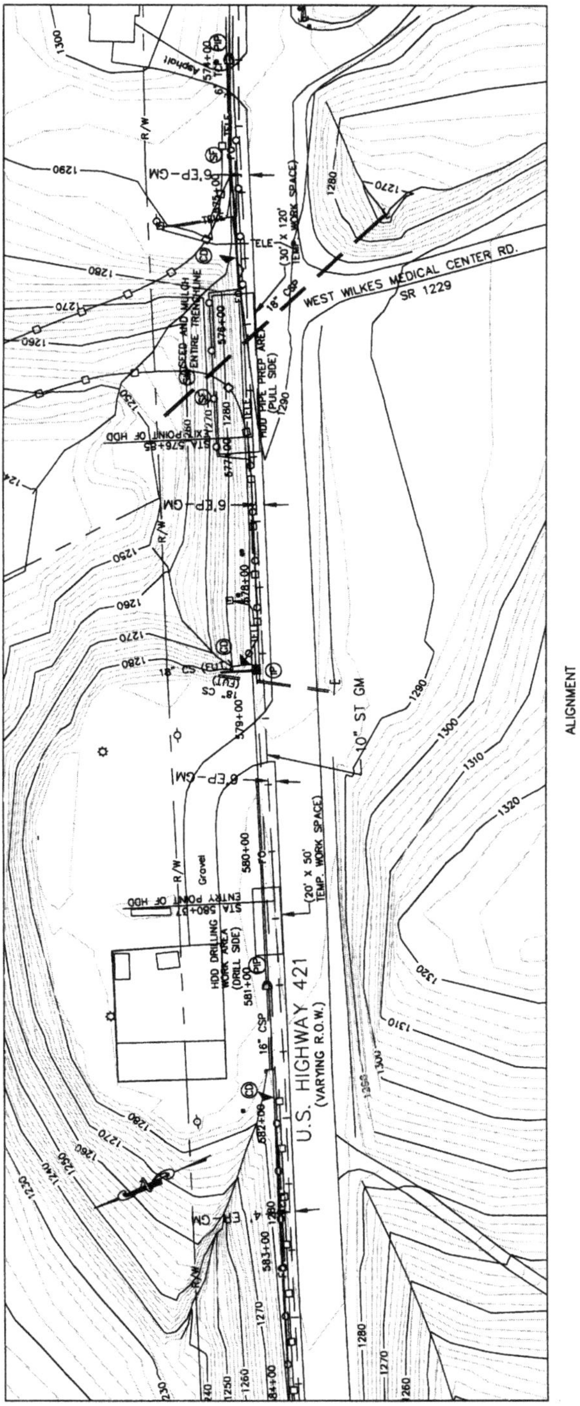

FIGURE 8.1 HDD plan and profile (alignment).

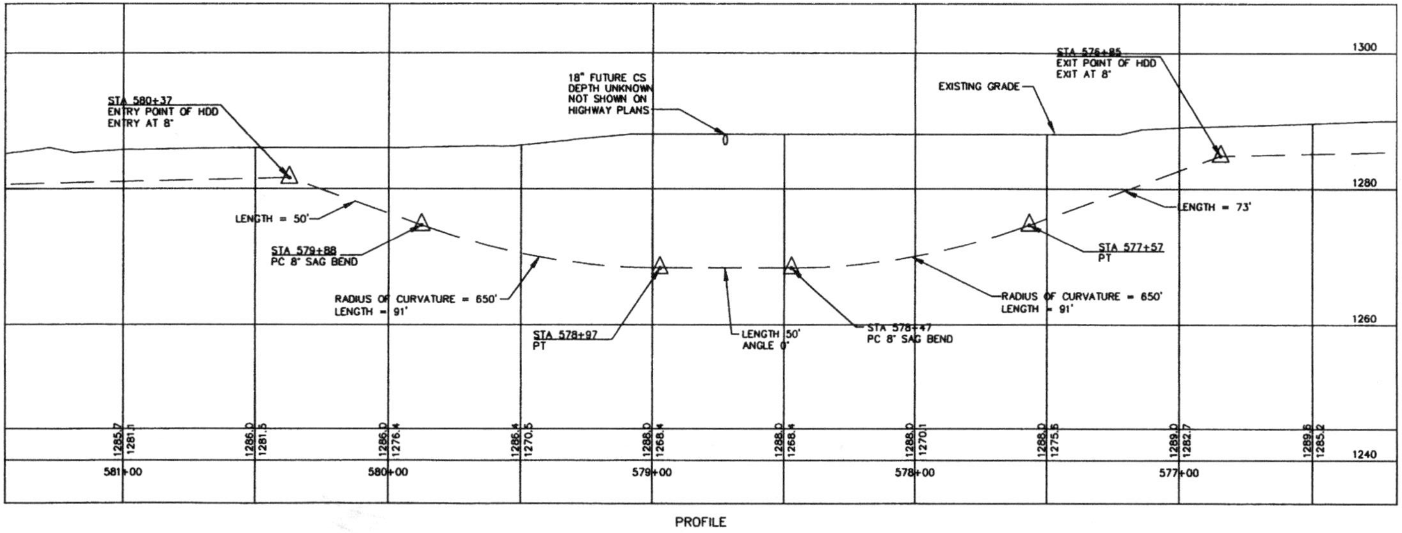

FIGURE 8.1 *(continued)* HDD plan and profile (profile).

The length of the drill and the complexity of the strata determine the number of explorations holes required for the geotechnical survey. For drill paths shorter than 1000 feet, two soil test borings (one on each end of the bore) may be adequate. If the data from these test borings indicate that the conditions are likely to be homogeneous on both sides of the bore, it may not be necessary to conduct any more test borings. If the test data indicates anomalies or discontinuity in the soils conditions, or items such as large concentrations of gravel, further tests should be conducted. For drill paths longer than 1000 feet, soil test borings are typically taken at 700 foot intervals. The soil test bores should be near the drill-path to provide accurate soil data, but far enough from the drill path bore hole to avoid pressurized mud from following natural ground fissures and rupturing to the ground surface through the soil-test bore hole. A general rule is to take soil test bores at least 30 feet of either side of drill path. River crossings require additional information such as a study to identify riverbed depth, stability, and river width. Typically, pipes are installed to a depth of at least 20 feet below the river bottom. Soil test bores for geotechnical surveys under water are usually taken to a depth that is 20 feet deeper that the pipe drill path.

The earth material and subsurface stratification are the geotechnical classifications of interest in HDD projects. Earth material is the type and conditions of the soils material at the site. Subsurface stratification defines how the earth material is distributed throughout the site. For HDD projects, soil and rock are the broad categories for earth material. Soil particles vary in size and may contain water or air in the interstitial spaces and may be excavated without drilling or blasting. Rock is a hard, consolidated material, which may require drilling or blasting. The groundwater proximity (above or blow the pipe) may have an effect on the HDD construction as well as the in-service performance of the pipe. The groundwater table and potential fluctuation should be determined during the geotechnical survey.

Soil Type Classification

A qualified technician or geologist should classify the soil material. The Unified Soil Classification System is normally used in classifying soils for HDD projects. This soils classification system is described in detail in ASTM Standard D 2487. The ASTM Standard bases the soil classifications on laboratory tests performed on soil samples passing the 3 inch sieve. Some definitions from ASTM D 2487 that are relevant to HDD operations are:

Cobbles. Particles of rock that will pass a 12 inch square opening and be retained on a 3 inch U.S. standard sieve.

Boulders. Particles of rock that will not pass a 12 inch square opening.

Gravel. Particles of rock that will pass a 3 inch sieve and be retained on a No. 4 U.S. standard sieve with the following subdivisions: Coarse will pass a 3 inch sieve and be retained on a ¾ inch sieve and Fine, will pass a ¾ inch sieve and be retained on a No. 4 sieve.

Sand. Particles of rock that will pass a No. 4 sieve and be retained on a No. 200 U.S. standard sieve with the following subdivisions: Coarse will pass a No. 4 sieve and be retained on a No. 10 sieve; Medium will pass a No. 10 sieve and be retained on a No. 40 sieve; and Fine will pass a No. 40 sieve and be retained on a No. 200 sieve.

Clay. Soil that will pass a No. 200 U.S. standard sieve that can be made to exhibit plasticity (putty-like properties) within a range of water contents and that exhibits considerable strength when air dry.

Silt. Soil that will pass a No. 200 U.S. standard sieve that is non-plastic or slightly plastic and that exhibits little or no strength when air dry.

A general discussion of the Unified Soil Classification System is in *Geotechnical Engineers Portable Handbook* by Robert W. Day. Soil types have different factors that aid in the soils classification. The unit weight and moisture content are key factors for clay soils, while the density and grain size distribution are factors for granular soils. The standard penetration test (SPT) is used to define the density of granular materials. This is a field test that drives a 2 inch split spoon sampler into the soil by dropping a hammer of a specific weight (normally 140 lbs.) a specific distance (normally 30 inches) to determine the number of blows required to drive the sampler 12 inches. In dense soils the number of blows can be lowered (i.e. 3 inches). The number is called the standard penetration resistance value (N) and is used to estimate the relative density of cohesionless soils. Sometimes these penetration test are conducted in cohesive materials and rock, and to a lesser extent, the consistency of cohesive soils and the hardness of rock can be determined.

For rock soil conditions, the unit weight and hardness are key factors. If rock is located during the test borings it is important to determine the type, the relative hardness, and the unconfined compressive strength. The geotechnical firm collects this data by core drilling with a diamond bit core barrel. The geologist will classify the rock and determine the *rock quality designator* (RQD), which rates the quality of the rock based on the length of core retrieved in relation to the total length of the core. The hardness of the rock is determined by comparing the rock to ten materials of known hardness. Measuring the core and then compressing the core to failure determines the compressive strength.

The following are a few suggested guidelines concerning soil test bores to consider during the HDD planning stage:

- Get accurate locations and elevations for each exploration borehole
- Do not spot exploration boreholes directly over the proposed HDD path—offset each 25 feet to 30 feet laterally

TABLE 8-3 Rock Quality Descriptions

RQD (%)	Rock Quality
90-100	Excellent
75-90	Good
50-75	Fair
25-50	Poor
0-25	Very Poor

- Request complete lithologic and geotechnic descriptions for all geologic strata encountered
- Require descriptions of boring techniques and all equipment used
- Test bores depths should go at least 20 feet below the lowest anticipated elevation of the horizontal directional borehole
- SPT's should be taken at 5 foot intervals for all strata (cohesive and non-cohesive units)
- Grain size analyses are helpful—either good field estimates or laboratory sieve tests
- It is important to get representative unconfined compression tests (UC) for all "harder" rock units that fail SPT tests (generally at "auger refusal"). Record RQD's and retain cores for visual inspection
- Have the driller or geologist record free water levels in the borings and note all significant observations during the actual drilling process. These notes should become part of the final log
- Have all boreholes thoroughly plugged and/or grouted upon completion of the exploration program

Surface Working Space

The surface impact associated with construction of a pipeline crossing by HDD is significantly less than the impact associated with construction by open cut excavation. However, the HDD construction is not without some surface impact. Working areas for the entry and exit points must be cleared and graded to allow for the HDD equipment and pipe pulling. The size of the area depends on the size of the drill and the drilling equipment to be used. Preferably each HDD site would allow at least 50 by 100 feet workspace for the entry side (drill rig) and adequate space on the exit side (pipe side) to pull the pipe as one continuous length. In urban areas this is often not possible. Because of restrictions such as lane closures for roads, or the need to work in alleyways, sidewalks, landscaped areas, or utility corridors, HDD equipment must often be configured in a linear arrangement. Other workspace considerations are the presence of overhead utilities and the possibility of restricted work-hours due to peak travel times on roadways.

A typical large (Maxi) rig can require as many as seven tractor-trailer loads to transport all the equipment to the HDD site. A workspace of approximately 150 by 250 feet is normally adequate for most large HDD operations. If necessary, a rig may be installed in a workspace of 60 by 150 feet. However, a workspace this small restricts the size and capability of the drilling rig. A typical HDD site plan is shown in Figure 8.2.

DRILLED PATH DESIGN

When planning and designing a HDD crossing, a key factor is properly defining the obstacle that is to be crossed. A HDD project under a river can be significantly different that crossing a city road or other utilities. The site characterization will determine the entry and exit points, the length and depth of the drill, and the horizontal and vertical movements. The designed drill path will consist of a series of straight lines and curves. The straight lines are called tangents and the curves are typically sag bends, over bends, or side bends, depending on their axial plane. Compound bends may be used but are generally avoided to simplify the drill. The entry and exit points, entry and exit angles, radius of curvature, points of curvature, and tangency define the location and configuration of a drilled profile.

The entry and exit points are the end points of the drilled profile. The drill rig is positioned at the entry point and the drill path progresses from the entry to exit point. The product pipe is positioned at the exit point and is pulled into the hole at the exit point and pulled back to the entry point. The specific locations of the entry and exits points are determined based on the site characteristics and the drill path. Steering accuracy and drilling effectiveness are greater close to the drill rig (entry point). Whenever possible, the entry point should be located close to any anticipated adverse subsurface conditions.

Another factor is the workspace for pipe fabrication for the pull section. The preferred exit point will have enough workspace to assemble and pull the entire length of pipe in one continuous pull. If space is not available, the pipe may be fabricated in sections and pulled one section at a time. This is a slow installation that will increase cost and increase the chances of the pipe getting stuck during the pullback.

Entry angles are normally designed between 8-20 degrees with horizontal. This restriction is based on equipment limitations. When designing a HDD project, ensure that the entry and exit angles can be performed with the drill rig being used. Exit angles are designed to allow easy break-over support and to not over stress the product pipe. The exit angle should not be so steep that the pull section must be severely elevated in order to guide it into the hole.

CONSTRUCTION MONITORING

This section describes some of the construction-monitoring requirements relative to HDD operations. The primary objectives of construction monitoring on an HDD

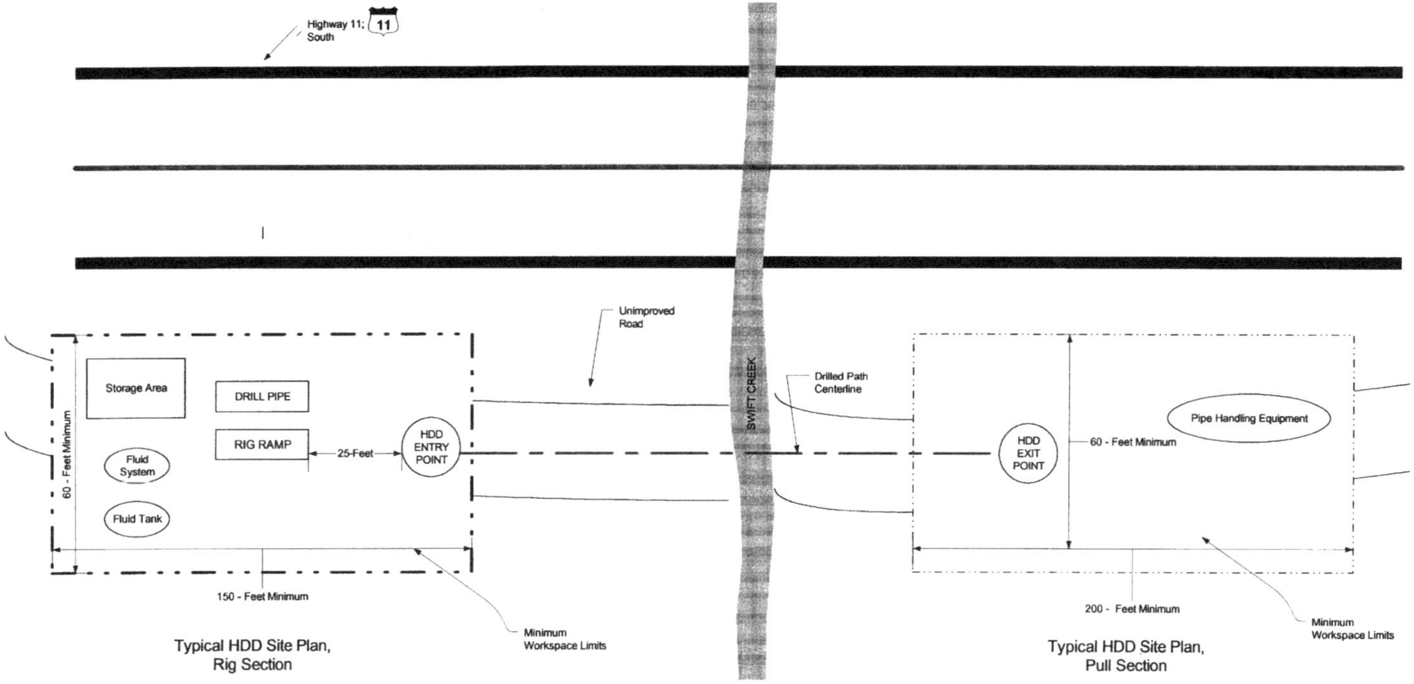

FIGURE 8.2 HDD site plan.

installation are to ensure that the contractor interprets the contract and design documents properly, and to ensure that the actual drill is documented. In doing this, it is important for the inspector to document his observations and actions. Should a question or dispute arise during or after the installation, the inspector's notes will provide the only source of confirming data. Since a drilled installation is typically buried with deep cover, its installed condition cannot be confirmed by visual examination. Figure 8.3 is an example of HDD daily report.

Drilled Path

The drilling contractor will typically rely on the owner's staking to locate the drilled segment. Two locations, the entry and exit points, should be staked. The elevations of the staked locations as well as the distance between them should be checked against the values on which the design is based. The contractor's pilot hole accuracy depends on the accuracy of the relative location, both horizontally and vertically, of these two points. The exit point coordinates will also provide a benchmark for measuring downhole survey error. If possible, the contractor should have a clear line of sight between the entry and exit points for use in orienting the downhole survey instrument. If a clear line of sight is not possible, the owner should stake points so that the drilled path centerline, or a reference line, can be established for survey instrument orientation.

Pilot Hole

Monitoring of the drilled path is accomplished during pilot hole drilling. Initially, a reading of the magnetic heading is taken to establish a reference line on which

<table>
<tr><th colspan="3">Daily Report</th></tr>
<tr><td>Date:</td><td>Client:</td><td>Time In:</td></tr>
<tr><td>Percent Complete:</td><td>Project:</td><td>Time Out:</td></tr>
<tr><td colspan="3">Description of Activities:

</td></tr>
<tr><td colspan="3">Comments:

</td></tr>
</table>

FIGURE 8.3 HDD daily report.

all drilled path data and calculations will be based. Other pertinent data which is needed to accurately locate the pilot hole drilling bit includes the bottom hole assembly length, the length from the drilling bit to the downhole probe, and the drilling rig setback distance from the entry point.

The actual path of the pilot hole is monitored during drilling by taking periodic readings of the inclination and azimuth of the downhole probe. Readings are typically taken after drilling a single joint, or approximately 30 feet. These readings are used to calculate the horizontal and vertical coordinates of the downhole probe as it progresses along the pilot hole. Data and calculations from the readings typically include the following items:

Survey. Points at which readings are taken by the downhole probe; surveys are usually tracked in a numerical sequence (1,2,3…) corresponding to the number of joints drilled.

Course Length. The distance between two downhole surveys as measured along the drilled path.

Measured Length. The total distance of a downhole survey from the entry point as measured along the drilled path; also the summation of the course lengths.

Inclination. The angle at which the downhole probe is projecting from the vertical axis at a particular downhole survey point; vertically downward corresponds to zero degrees.

Azimuth. The angle at which the downhole probe is projecting in the horizontal plane at a particular downhole survey point; magnetic north corresponds to zero degrees.

Station. The horizontal position of a downhole survey measured from an established horizontal control system.

Elevation. The vertical position of a downhole survey measured from an established vertical control system.

Right. The distance of a downhole survey from the design path reference line. Positive values indicate right of the reference line while negative values indicate left of the reference line.

Bit to Probe. The distance from the drilling bit (leading edge) to the downhole probe.

Heading. The magnetic line of azimuth to which the drilled path reference line corresponds.

Rig Setback. The distance from the drill bit when first placed on the drilling rig as measured from the staked entry point.

BHA Length. The length of the bottom hole assembly.

HDD PERFORMANCE

There are two basic areas of concern with HDD performance, the position and curvature of the pipeline. First, the pipeline must be installed so that the drilled length, depth of cover, and entry/exit angles specified by the design are achieved. Second, the installation must not curve the drilled path in such a way that the pipeline will be damaged during installation or over stressed during operation. The actual position of the drilled path cannot be readily confirmed by an independent survey. Therefore, it is necessary to have a basic understanding of the downhole survey system being used and be able to interpret the readings. It is not necessary to observe and approve the drilling of each joint. However, progress should be monitored on a routine basis and problems addressed so that remedial action can be taken as soon as possible. The inspector should insure that bends are not drilled which have a radius of curvature less than the design (minimum allowable). Figure 8.4 illustrates an example of a HDD survey report. Figure 8.5 is an an example of a radius of curvature anylsis. If a tight radius occurs, the joint or joints should be re-drilled or reviewed with the design engineers as soon as possible to insure that the codes and specifications governing design of the pipeline are not violated.

Downhole Survey Calculations

Downhole survey calculation methods are discussed in detail in API Bulletin D20. Three different methods from this bulletin are presented here for use on HDD pipeline installations. These are:

1. Average angle
2. Balanced tangential method
3. Minimum curvature

The equations for these three methods are used to calculate the horizontal and vertical distances from the entry point, as well as the distance from the reference line. Symbols used in the equations are defined below:

CL = Course length.
I_1 = Inclination angle of the previous survey point.
I_2 = Inclination angle of the current survey point.
A_1 = Deflection angle from the heading of the previous survey point.
A_2 = Deflection angle from the heading of the current survey point.

Survey Tabulation Sheet									
Date:		Bit to Probe:		Azimuth:			Rig Setback:	BHA Length:	
Raw Data				TruTracker Data		Calculations			
Survey #	Joint Length	Inclination Angle	Azimuth	Right	Elevation	Measured Distance	Station	Right	Elevation

FIGURE 8.4 HDD survey tabulation sheet.

Radius of Curvature Analysis					
Date:	Project:				
Notes:					
Radius of Curvature Per N Surveys					
Survey #	1	3	5	7	10

FIGURE 8.5 HDD radius of curvature analysis.

HD = Horizontal distance between the previous and current survey points.

RT = Differential distance from the reference line between the previous and current survey points. Also called "RIGHT" to indicate the distance right (positive value) or left (negative value) of the original reference line.

VT = Vertical distance between the previous and current survey points.

Average Angle Method

This method uses the average of the previous and current azimuth/inclination angles to project the measured distance along a path that is tangent to this average angle. The equations are:

$$HD = CL \times \cos\frac{(A_1 + A_2)}{2} \times \sin\frac{(I_1 + I_2)}{2} \tag{8.1}$$

$$RT = CL \times \sin\frac{(A_1 + A_2)}{2} \times \sin\frac{(I_1 + I_2)}{2} \tag{8.2}$$

$$VT = CL \times \cos\frac{(I_1 + I_2)}{2} \tag{8.3}$$

Balanced Tangential Method

This method assumes that half of the measured distance is tangent to the current inclination/azimuth projections and that the other half is tangent to the previous inclination/azimuth projections.

$$HD = \sqrt{\frac{CL}{2}}(\sin I_1 \cos A_1 + \sin I_2 \sin A_2) \tag{8.4}$$

$$RT = \sqrt{\frac{CL}{2}}(\sin I_1 \sin A_1 + \sin I_2 \sin A_2) \tag{8.5}$$

$$VT = \sqrt{\frac{CL}{2}}(\cos I_1 + \cos A_2) \tag{8.6}$$

Minimum Curvature Method

This method is similar to the Balanced Tangential Method; however, the tangential segments produced from the previous and current inclination/azimuth angles are smoothed into a curve using a ratio factor (RF). This ratio factor is defined by a dogleg angle (DL), which is a measure of the change in inclination/azimuth.

$$DL = \cos^{-1}\{\cos(I_2 - I_1) - \sin I_1 \times \sin I_2[1 - \cos(A_2 - A_1)]\} \tag{8.7}$$

$$RF = \sqrt{\frac{2}{DL}} \tan \sqrt{\frac{DL}{2}}; \text{ RF} = 1 \text{ for small angles (DL < 0.25 degrees)} \tag{8.8}$$

$$HD = \sqrt{\frac{CL}{2}} (\sin I_1 \cos A_1 + \sin I_2 \sin A_2) RF \tag{8.9}$$

$$RT = \sqrt{\frac{CL}{2}} (\sin I_1 \sin A_1 + \sin I_2 \sin A_2) RF \tag{8.10}$$

$$VT = \sqrt{\frac{CL}{2}} (\cos I_1 + \cos A_2) RF \tag{8.11}$$

Any one of these three methods may be used to track the downhole probe position and ensure conformance to the directional tolerances of the design. To track the probe over a specified measured distance the values from these equations must be summed over the specified length.

Radius of Curvature Calculations

The same angle readings used in the previous calculations are also used to determine the radius of curvature of the drilled path. The radius of curvature calculations are based on the relationship:

$$R = \frac{s}{Q}$$

where s = arc length in feet
Q = angular distance in radians

For a specific drilled length, the radius of curvature is calculated using the following formula:

$$R_{\text{drilled}} = \sqrt{\frac{I_{\text{drilled}}}{q_{\text{drilled}}}} \times \sqrt{\frac{180}{\pi}} \tag{8.12}$$

where R_{drilled} = the radius of curvature over a specified drill length in feet
I_{drilled} = drilled length in feet
q_{drilled} = change in angle over the drilled length in degrees

Typically, the radius of curvature is checked for conformance over any three joint course lengths using the following equation:

$$R_3 = \sqrt{\frac{L_3}{Q_3}} \times \sqrt{\frac{180}{\pi}} \quad (8.13)$$

where R_3 = the radius of curvature in feet over L_3.
L_3 = course length in feet over any 3 joint, no less than 75-feet and no greater than 100-feet.
Q_3 = total change in angle in degrees over L_3.

PIPE INSTALLATION

The inspector should review the contractor's operations to insure that the pull section is adequately supported during pull back. Roller stands should be provided as well as lifting equipment capable of moving the string into the drill hole. The section should not be dragged on the ground. All break over bends should be made with a radius long enough to insure that the pipe is not over stressed.

The inspector should always bear in mind the possibility of inadvertent drilling fluid returns. The right of way should be examined regularly for inadvertent returns. Particular attention should be paid to locations of underground utilities and pile foundations. If inadvertent returns are found, they should be cleaned up immediately and the location monitored for continuing problems (particularly during pull back).

HDD STRESSES AND FORCES

After completion of the geotechnical investigation and determination that HDD is feasible, the designer turns attention to designing a proper drill and selecting the proper pipe. The product pipe must satisfy all the requirements of the project including flow capacity, working pressure rating, and surge or vacuum capacity. These considerations have to be met regardless of the method of installation. For HDD applications, in addition to the service life operating requirements, the pipe must be able to withstand pull-back loads, which include tensile pull forces, external hydrostatic pressure, and tensile bending stresses. The pipe must also withstand external service loads which consist of post-installation soil, groundwater, and surcharge loads occurring over the life of the pipeline. Often the load the pipe sees during installation, such as the combined pulling force and external pressure will be the largest load experienced by the pipe during its life.

Load and stress analysis for a HDD project are different from similar analyses for buried pipelines. Pipe properties such as wall thickness and material grade must be selected so the pipeline can be installed and operated as planned. Directional drilling is an evolving technology and industry-wide design standards are still developing. Proper design requires considerable professional judgment. The most frequently asked questions are:

- What force is required to pull the pipe through the borehole?
- How much pulling force can be applied to the product pipe?

The pipe manufacturer can usually answer the second question. However, the force required to pull the pipe through the borehole has many factors such as:

- Borehole diameter
- Product pipe diameter
- Soil conditions
- Drilling fluid characteristics
- HDD and pullback procedures being used
- Drill path profile and radius of curvature

These factors are often site and job specific, and the pulling force required should be determined by experienced engineers, drillers, and geotechnical personnel that are familiar with the site conditions and HDD procedures.

The bore path has a significant impact in the forces acting on the pipe during pullback. The straighter the profile and alignment, the less the pulling force. Curvature in the pipe is necessary to get under or around the obstacles, but curvature should be minimized. The curvature causes bending stresses in the drill rods and the product pipe, and due to the increased friction, increases the pullback forces. When selecting the bore path, the designer has to consider the site characteristics and the allowable bending stresses in the product pipe. Based on many factors, the designer then develops a preliminary drill path that will successfully negotiate the obstacle without imposing excessive stresses or pullback forces on the product pipe.

Another key factor in the drill path is the depth of the drill profile. The profile must be deep enough to ensure any obstacles are cleared by the desired distance. Also, a minimum depth for the drill path is required to ensure that no drilling fluid breakout occurs. For mini-HDD operations this depth is typically 3 feet or more, but this depends on the site-specific soil conditions. Normally, the bore profile arcs down from the entry point, then straightens out before it finally arcs back up to the exit point.

During installation, the product pipe is subjected to:

- Tension that is required to pull the pipe into the pilot hole and around the curved sections that make up the bore path. Frictional drag due to the wetted friction between the pipe and wall of the borehole
- Fluidic drag of the pipe as it is pulled through the viscous drilling fluid trapped in the annulus
- Unbalanced gravity (weight) effects of pulling the pipe into and out of a borehole at different elevations

- Bending as the pipe is forced to negotiate the curves in the borehole
- External hoop from the pressure exerted by the drilling mud in the annulus around the pipe (unless the pipe is filled with a fluid at a similar pressure)

The stresses and potential failure of the pipe are a result of the relations of these loads. As a result, calculations of the individual effects do not accurately reflect the combined stresses. These combined stresses require specific calculations and design checks. The loads imposed on a pipe during operation are significantly different than the loads the pipe will see during installation by HDD.

HDD DESIGN CONSIDERATIONS FOR PLASTIC PIPE

After determining the pipe properties required for long-term service, the designer must determine if the pipe properties are sufficient for installation. Since installation forces are so significant, a stronger pipe may be required because of the installation stresses. Proper installation procedures may reduce some of these forces to an inconsequential level. During pullback, the pipe is subjected to axial tensile forces caused by the frictional drag between the pipe and the borehole or slurry, the frictional drag on the ground surface, the capstan effect around drill-path bends, and hydrokinetic drag. In addition, the pipe may be subjected to external hoop pressures due to net external fluid head and bending stresses. The axial pulling force reduces the pipe collapse resistance to external pressure. Furthermore, the pipe-bending radius may limit the drill path curvature. Torsional forces occur but are usually negligible when back-reamer swivels are properly designed. As discussed previously, considerable judgment is required to predict the pullback force because of the complex interaction between pipe and soil. Sources for information include experienced drillers, engineers, and publications. Typically, pullback force calculations are approximations that depend on considerable experience and judgment. Because of the large number of variables involved and the sensitivity of pullback forces to installation techniques, the formulas presented in this section are for guidelines only, and are given only to familiarize the designer with the interaction that occurs during pullback. Pullback values obtained should be considered only as qualitative values and used only for preliminary estimates. The designer is advised to consult with an experienced driller or with an engineer familiar with calculating these forces.

Pull Back Force

Large HDD rigs can exert between 100,000 to 200,000 pounds of pull force. The majority of this power is applied to the cutting face of the reamer, which precedes the pipeline segment into the borehole. It is difficult to predict what portion of the total pullback force is actually transmitted to the pipeline being inserted. The pulling force that overcomes the combined frictional drag, capstan effect, and hydrokinetic drag, is applied to the pull-head and first joint of pipe. The axial tensile stress grows in intensity over the length of the pull. The duration of the pull load is longest at the pull-nose. The tail end of the pipe segment has zero applied tensile stress for zero time. The incremental time duration of stress intensity along the length of the pipeline from nose to tail causes a varying degree of recoverable elastic strain and viscoelastic stretch per foot of length along the pipe. The pipe wall thickness must be selected so that the tensile stress due to the pullback force does not exceed the permitted tensile stress for the pipe. Increasing the pipe wall thickness will allow for a greater total pull-force, but the thicker wall also increases the weight per foot of the pipe in direct proportion. As a result, thicker wall pipe may not necessarily reduce stress. It may only increase the absolute value of the pull force or tonnage. The designer should carefully check all proposed pipe wall thicknesses and properties.

Frictional Drag Resistance

Pipe resistance to pullback in the borehole depends primarily on the frictional force created between the pipe and the borehole, the pipe and the ground surface in the entry area, the frictional drag between pipe and drilling slurry, the capstan effect at bends, and the weight of the pipe. The following equation gives the frictional resistance or required pulling force for pipe pulled in straight, level bores, or across level ground.

$$F_p = cW_BL \tag{8.14}$$

where F_p = pulling force, lbs
c = coefficient of friction between pipe and slurry (typically 0.25) or between pipe and ground (typically 0.40).
W_B = net downward (or upward) force on pipe, lb/ft
L = length, ft

When a slurry is present, W_B, is the upward buoyant force of the pipe and its contents. Filling the pipe with fluid significantly reduces the buoyancy force and thus the pulling force. If the pipe is installed empty using a closed nose-pull head, the pipe will want to float on the crown of the borehole leading to the sidewall loading, increasing frictional drag through the buoyancy-per-foot force and the wetted soil to pipe coefficient of friction. If the pipe is installed full of water the net buoyant force is drastically reduced. During pullback, the moving drill mud lubricates the contact zone. If the drilling stops, the pipe stops, or the mud flow stops, the pipe can push up and squeeze out the lubricating mud.

Capstan Force

For curves in the borehole, the force can be factored into horizontal and vertical components. When drilling with steel pipe there is an additional frictional force that occurs due to the pressure required by the borehole to keep the steel pipe curved. When drilling with plastic pipe using a radius of curvature similar to that used for steel pipe, these forces are likely insignificant. However, when using tight bends, these forces should be taken into consideration. The frictional resistance during a pull is compounded by the capstan effect. Compounding forces caused by the direction of the pulling vectors are created as the pipe is pulled around a curve or bend creating an angle. The pulling force due to the capstan effect is given in the following equation. This equation and the preceding one are applied recursively to the pipe for each section along the pullback distance. This method is credited to Larry Slavin of Bellcore (Middletown, NJ).

$$F_c = e^{\mu\phi}(\mu W_B L) \tag{8.15}$$

where e = Natural logarithm base ($e = 2.71828$)
μ = Coefficient of friction between the pipe and slurry or between the pipe and ground
ϕ = Angle of bend in pipe, radians
W_B = Weight of pipe or buoyant force on pipe, lbs/ft
L = Length of pull, ft

The majority of HDD bore paths consist of a series of straight sections and bends. To estimate the total pull forces, the above equations must be applied to each straight section and bend, and totaled. The following is a method developed by Larry Slavin for estimating the loads along a bore path with no horizontal bends. This method is an estimate only. The designer or engineer must determine their suitability for any application

$$T_D = (v_b\beta)(T_C + T_{HK} + v_b|w_b|L_4 - w_bH - \exp(v_b\alpha)(v_a w_a L_4 \exp(v_a\alpha)) \tag{8.16}$$

$$T_A = \exp(v_a\alpha)(v_a w_a(L_1 + L_2 + L_3 + L_4)) \tag{8.17}$$

$$T_B = \exp(v_b\alpha)(T_A + T_{HK} + v_b|w_b|L_2 + w_bH - v_a w_a L_2 \exp(v_a\alpha)) \tag{8.18}$$

$$T_C = T_B + T_{HK} + v_b|w_b|L_3 - \exp(v_a\alpha)(v_a w_a L_3 \exp(v_a\alpha)) \tag{8.19}$$

where T_A = pull force on pipe at point a, lbf
T_B = pull force on pipe at point b, lbf
T_C = pull force on pipe at point c, lbf
T_D = pull force on pipe at point d, lbf
T_{HK} = Hydrokinetic force, lbf (see below equation for this factor
L_1 = additional length of pipe required for handling and thermal expansion/contraction, feet
L_2 = horizontal distance to desired depth, feet
L_3 = addtional distance traversed at desired depth, feet
L_4 = horizontal distance to rise to the surface, feet
H = depth of borehole from ground surface, feet
$\exp(x)$ = e^x, where e = natural logarithm base ($e = 2.71828$)

v_a = coefficient of friction applicable at the surface before the pipe enters borehole, typically 0.5
v_b = coefficient of friction applicable within the lubricated borehole or after the wet pipe exits, typically 0.3
w_a = weight of empty pipe, lbf/ft
w_b = net upward buoyant force on the pipe in borehole, lbf/ft
a = borehole angle at pipe entry (drill exit angle), radians
b = borehole angle at pipe exit (drill entry angle), radians

The previous equations are approximations. They do not accurately account for the resistance due to pipe stiffness at curves along the bore path. The pull forces will be reduced for large radius curves and larger clearance within the borehole. The designer or engineer should ensure that the estimated pull forces do not over stress the product pipe.

GUIDELINES AND SPECIFICATIONS FOR INSTALLATION BY HDD TECHNOLOGY

This section contains practices, which are summarized and categorized chronologically beginning at the pre-qualification stage and ending with post-construction evaluation. Special sections are devoted to selected topics including segment joining, tie-ins, and the handling and disposal of drilling mud slurry.

Contract Considerations

Sharing risk with the contractor can significantly reduce the average bid price on a project. This is particularly true in underground construction. Items that should be addressed in contracts involving HDD work may include:

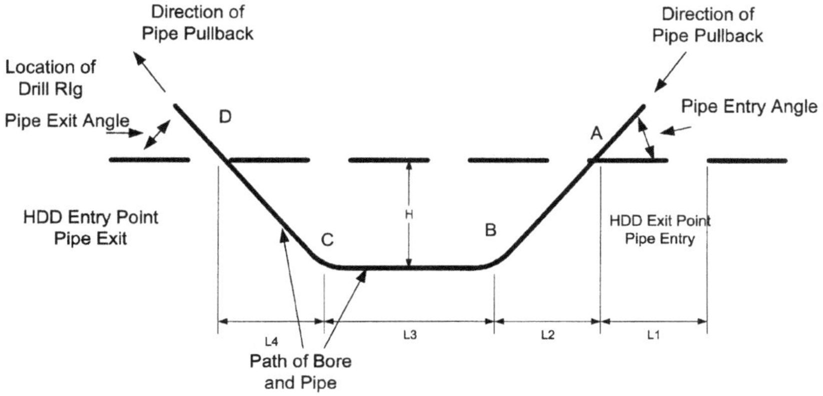

FIGURE 8.6 HDD bore path.

Different Ground Conditions and Walk-Away Provision. Adequate geo-technical information is invaluable in underground construction and can help to reduce the contractor's risk. However, even with good geo-technical data unexpected ground conditions may be encountered. These conditions can make it difficult, or even impossible, to complete the crossing using the HDD method. The contract should include provisions for if a project encounters unexpected ground conditions. Walk-away provisions in the contract entitle the contractor to stop working and walk-away from the job without penalty provided the contractor demonstrated a diligent effort to complete the project and it was decided to abandon the method.

Turbidity of Water and Inadvertent Returns. Difficult to predict these events may lead to work stoppage and loss of equipment. The contract should offer a mechanism to mutually address and mitigate these problems if and when they arise. For example, contingency plans for containment and disposal of inadvertent returns can be priced as a separate bid price and agreed prior to construction.

Contractor Proposal/Bid

As part of the bid, each HDD contractor should provide the following items:

- Construction plan
- Site layout plan
- Project schedule
- Communication plan
- Safety manual/procedures
- Emergency procedures
- Company experience record
- List of subcontractors on the project
- Drilling fluid management plan

Construction Plan

The following information should be submitted with respect to the construction plan:

- Access requirements to the site
- Type and capacity of drilling rig to be use on the project including thrust and rotary torque. The size of the drilling equipment should be adequate for the job. An industry rule of thumb is that the drilling rig's pull/push capacity should be at least equal to twice the weight of the product to be pulled or the weight of the drilling rod in the hole, which ever is greater. It should be noted that the range of a particular rig for a particular product type can vary significantly depending on soil conditions, drill path profile (i.e., radius of curvature) and crew experience

- Type and capacity of the mud mixing system. This is of particular importance if at least part of the bore path is suspected to consist of solid rock or the final ream has a diameter of 14 inches or greater
- A listing of any specialized support equipment required
- Project schedule indicating the various tasks and their expected duration
- Drawing of work site indicating the location and footprints of all equipment, location of entry and exit pits, and location of slurry containment pits
- Construction method including: diameter of pilot hole; number and size of pre-reams; use of rollers, baskets and side booms to suspend and direct pipe during pull back; number of sections in which product is to be installed
- Type, operating range and degree of accuracy of tracking equipment

Drilling Fluids Management Plan

The following information should be provided as part of the drilling fluid management plan:

1. Identify source of fresh water for mixing the drilling mud. Necessary approvals and permits are required for sources such as streams, rivers, pounds or fire hydrants
2. Method of slurry containment
3. Method of recycling drilling fluid and spoils
4. Method of transporting drilling fluids and spoils off the site
5. Approved disposal site for drilling mud and spoils

Previous Experience

The bidder should provide a list of similar projects completed by his company, including name of owner, location, project environment (e.g., urban work, river crossing), product diameter, length of installation, contact name and telephone number.

The bidder should also provide a list of key personnel assign to the project including their title, experience record and personal references.

Safety

Each bidder should submit a copy of the company safety manual including:

- Operating procedures that comply with applicable regulations, including shoring of pits and excavations when required
- Emergency procedures for inadvertently boring into natural gas line, live power cable, water mains, sewer lines, or fiber-optic cables which comply with applicable regulations
- Emergency evacuation plan in case of an injury

The drilling unit must be equipped with an electrical strike safety package. The package must include:

- Warning sound alarm
- Grounding mats
- Protective gear.

Contingency Plans

Contingency plans should consist of the following:

- Contingency plan in case of spill (e.g., drilling fluids, hydraulic fluids), including measures to contain and clean the affected area
- A contingency plan for the clean up of surface seepage of drilling fluids and spoils
- Specific action(s) required to be taken in the event that the installed pipe fails the post-installation leak test

Communication Plan

The communication plan should address the following items:

- The form and frequency of communication with owner or his representative on the site
- Identification of key person(s) which will be responsible to ensure that the communication plan is followed
- Issues to be communicated including safety, progress, and unexpected technical difficulties

Traffic Control

Traffic control considerations are as follows:

- When required, the contractor at his cost shall be responsible for supplying and placing warning signing, barricades, safety lights and flags or flagmen, as required for the protection of pedestrians and vehicle traffic
- Obstruction of the roadway should be limited to off-peak hour on major roads

List of Subcontractors

Subcontractors and their designated tasks should be identified. Possible tasks to performed by subcontractors include:

- Utility location
- Hydro-excavation

- Pipe suppliers
- Leak testing
- Fusion / welding
- Tie-ins to services and / or mains
- Mud mix disposal
- Excavation of entry / exit pits
- Surface restoration such as pavement, sidewalks and lawns

Other Considerations

Other requirements to be addressed in the bid include permits requirements:

- Permits necessary for a contractor to carry on a business
- Street opening (cut) permits
- Use of hydrants for water
- Permits for storage, piling, and disposal of material
- Permits for water/bentonite disposal
- Any other permits required to carry out the work

SITE EVALUATION

The HDD location should be inspected prior to the commencing of the project. The following should be addressed:

- Establishing whether or not there is sufficient room at the site for, entrance and exit pits, HDD equipment and its safe and unimpeded operation, support vehicles, fusion machines, and stringing out the pipe to be pulled back in a single, continuous operation
- Establishing suitability of soil conditions for HDD operations. The HDD method is ideally suited for soft sub-soils such as clays and compacted sands. Subgrade soils consisting of large grain materials like gravel and cobble and boulders make HDD difficult to use and may contribute to damaging the pipe
- Check the site for evidence of substructures such as manhole covers, valve box covers, meter boxes, electrical transformers, conduits or drop lines from utility poles, and pavement patches. HDD may be a suitable method in areas where the substructure density is relatively high

Pre-Construction

The followings are steps that should be undertaken by the contractor in order to ensure a safe and efficient construction with minimum interruption to normal every day activities at the site:

- Notify owners of subsurface utilities along and on either side of the proposed drill path of the impending work through the one-call program. Locate all utilities along and on either side of the proposed drill path
- Obtain all necessary permits or authorizations to carry construction activities near or across all such buried obstructions
- All utility crossings shall be exposed using a hydro-excavation, hand excavation or another approved method, to confirm depth
- Construction schedule should be arranged as to minimize disruption
- The proposed drill path should be determined and documented, including its horizontal and vertical alignments, location of buried utilities and substructures along the path
- Size of excavation for entrance and exit pits are to be of sufficient size as to avoid a sudden radius change of the pipe, and consequently excessive deformation at these locations. Sizing the pits is a function of the pipe depth, diameter and material. All pits must be shored as required by the relevant regulations

Drilling Operations

The following list provides general remarks and rules of thumb related to the directional boring method, as well as specific details regarding various stages along the installation process:

- Only trained operators should be permitted to operate the drilling equipment. They should always follow the manufacture's operating instructions and safety practices
- Drilling mud pressure in the borehole should not exceed that which can be supported by the overburden to prevent heaving or a hydraulic fracturing of the soil (i.e., 'Frac-out'). Allowing for a sufficient cover depth helps accomplishes this
- The drill path alignment should be as straight as possible to minimize the fractional resistance during pullback and maximize the length of the pipe that can be installed during a single pull
- It is preferable that straight tangent sections will be drilled before the introduction of a long radius curve. Under all circumstances, a minimum of one complete length of drill rod should be utilized before starting to level out the borehole path
- The radius of curvature is determined by the bending characteristics of the product line
- Entrance angle of the drill string should be between 8 and 20 degrees, with 12 degrees being considered optimal. Shallower angles may reduce the penetrating capabilities of the drilling rig, while steeper angles may result in steering difficulties, particularly in soft soils. A recommended value for the exit angle of the drill string is in the range of 5 to 10 degrees

- Whenever possible, HDD installation should be planned so that back reaming and pulling for a leg can be completed on the same day. It is permissible to drill the pilot hole and pre-ream one day, and complete both the final ream and the pull back on the next day
- If a drill hole beneath a road must be abandoned, the hole should be filled with grout or bentonite to prevent future subsidence
- Pipe installation should be performed in a manner that minimizes the overstressing and straining of the pipe. This is of particular importance in the case of a polyethylene pipe

Equipment Setup and Site Layout

- Sufficient space is required on the rig side to safely setup and operate the equipment. The workspace required depends on the type of rig used. A mini rig may require as little as 10x10 feet of working space, while a large river crossing unit requires a minimum of 100x150 feet of working area. A working space of similar dimensions to that on the rig side should be allocated on the pipe side, in case there is a need to move the rig and attempt drilling from this end of the crossing.
- If at all possible the crossing should be planned to ensure that drilling proceeds downhill, allowing the drilling mud to remain in the hole and minimizing inadvertent return.
- Sufficient space should be allocated to fabricate the product pipeline into one string, thus enabling the pull back to be conducted in a single continuous operation. Tie-ins of successive strings during pullback may considerably increase the risk of unsuccessful installation.

Drilling and Back-Reaming

- Drilling mud should be used during drilling and back reaming operations. Using water only may cause collapse the borehole in unconsolidated soils, while in clays the use of water may cause swelling, and subsequent jamming of the product
- Heaving may occur when attempting to back reaming too large of a hole. This can be avoided by using several pre-reams to gradually enlarge the hole to the desired diameter
- A swivel should be attached to the reamer, or drill rod, to prevent rotational torque been transferred to the pipe during pullback
- In order to prevent over-stressing of the product during pullback, a weak link, or breakaway pulling head, may be used between the swirl and the leading end of the pipe

- The pilot hole must be back-reamed to accommodate and permit free sliding of the product inside the borehole. A rule of thumb is to have a borehole 1.5 times the product outer diameter. This rule of thumb should be observed particularly in the larger diameter installations
- The conduit must be sealed at either end with a cap or a plug to prevent water, drilling fluids, and other foreign materials from entering the pipe as it is being pulled back
- Pipe rollers, skates, or other protective devices should be used to prevent damaging the pipe from the edges of the pit during pull-back, eliminate ground drag, and reduce pulling force and subsequently the stress on the product
- The drilling mud in the annular region should not be removed after installation, but permitted to solidify and provide support for the pipe and neighboring soil

Segment Jointing (Butt-Fusion/Welding)

- The contractor shall perform a leak test on the pipeline (particularly on fused joints) prior to pipe pull back
- It may be necessary to remove the lip on the butt fusion connections to prevent snagging on potential obstructions
- A qualified fusion technician should do all joining in accordance with the pipe manufacturers specifications
- Standard inspection techniques should be followed for assessing quality of butt and electro fusion joints

Tie-Ins and Connections

- Trenching should be used to join sections of conduits installed by the directional boring method. Trenching to join conduits shall be at the contractor's expense and should be included in the unit rate
- An additional pipe length, sufficient for joining to the next segment, should be pulled into the entrance pit. This length of the pipe should not be damaged or interfere with the subsequent drilling of the next leg. The contractor should leave a minimum of 3 feet of conduit above the ground on both sides of the borehole
- In the case of a PE pipe, tie-ins and connections should only be made after a suitable time period in order to allow the pipe to recover. Ideally, the pipe should be allowed to recover overnight. If this is not possible, the recovery period should be equal to at least twice the pull back time

Alignment and Minimum Separation

In all cases, the product shall be installed to the alignment and elevations as shown on the drawings within pre-specified tolerances. However, tolerance values are application dependent. For example, in a major river crossing a tolerance of 12 feet from the exit location along the drill path centerline may be an acceptable value. However, this tolerance is not acceptable when installing a product line between manholes. Similarly, grade requirements for a water force-main are significantly different than these on a gravity sewer project. It is recommended that a study will be initiated in order to establish tolerance limits for various applications that are acceptable from the design point of view and at the same time achievable using current tracking and steering capabilities of HDD equipment.

When a product line is installed in a crowded right-of-way, the issue of safe minimum separation distance arises. Many utilities companies have establish regulations for minimum separation distances between various utilities. These distances needed to be adjusted to account for possible minor deviation when a line product is installed using HDD technology. As a rule of thumb, if the separation distance between the proposed alignment and the existing line is 10 feet or more, normal installation procedures can be followed. If the separation is 5 feet or less, special measures, such as observation boreholes are required. The range between 5 and 10 feet is a gray area, typically subject to engineering judgment. A natural gas transmission line is likely to be treated more cautiously than a storm water drainage line.

Break-Away Pulling Head

Recent reports from several natural gas utility companies reveals concerns regarding failure experiences on HDPE pipes installed by horizontal directional drilling. These failures were attributed to deformation of the pipe due to the use of excessive pulling force during installation. A mitigation measure adopted by some gas companies involves the use of break-a-way swivels to limit the amount of force used when pulling PE products. The weak link used can be either a small diameter pipe (but same SDR) or specially manufactured breakaway links. The latter consists of a breaking pin with a defined tensile strength incorporated in a swivel. When the strength of the pin is exceeded, it will break causing the swivel to separate. The use of breakaway swivels is warranted particularly when installing small diameter PE pipes (up to 4-inches OD).

Application of such devices in the installation of larger diameter products is currently not a common practice. If the drilling equipment rated, pulling capacity is less than the safe load and the use of a weak link may not be required. Exceeding the product elastic limit can be avoided simply by following the following good drilling practices:

- Regulating pulling force
- Regulating pulling speed

- Proper ream sizing
- Appropriate amounts of drilling slurry fluid

Drilling Fluid—Collection and Disposal Practices

The collection and handling of drilling fluids and inadvertent returns is perhaps one of the most debated topics in the HDD community in North America. On one side the industry realizes the need to keep drilling fluids out of streams, streets and municipal sewer lines. On the other hand new tough regulations in some states (i.e., California) faced HDD contractors with escalating drilling fluid disposal expenses. Owners need to adopt an approach which address environment concerns while at the same time avoid unnecessary expenses and escalating drilling rates. The following clauses can be used as a guideline for the development of such an approach.

1. Drilling mud and additives to be use on a particular job should be identified in the proposal, and their Material Safety Data Sheets (MSDS) provided to the owner
2. Excess drilling mud slurry shall be contained in a lined pit or containment pond at exit and entry points, until recycled or removed from the site. Entrance and exit pits should be of sufficient size to contain the expected return of drilling mud and spoil
3. Methods to be used in the collections, transportation, and disposal of drilling fluids and spoils are to be provided as part of the pre-qualification. Excess drilling fluids should be disposed in compliance with local ordinances, regulations, and environmentally sound practices in an approved disposal site
4. In working in an area of contaminated ground, the slurry should be tested for contamination, and disposed in a manner that meets government requirements
5. Precautions should be taken to keep drilling fluids out of the streets, manholes, sanitary and storm sewers, and other drainage systems, including streams and rivers
6. Recycling drilling fluids is an acceptable alternative to disposal
7. The contractor shall make all diligent efforts to minimize the amount of drilling fluids and cuttings spilled during the drilling operation, and shall provide complete clean-up of all drilling mud overflows or spills

Site Restoration and Post Construction Evaluation

Site restoration and evaluations are the critical closing elements of HDD projects.

- It is recommended that the pipe be inspected for damage at every excavation pit as it being pulled back and after the installation is complete

- All surfaces affected by the work shall be restored to their pre-construction conditions. Performance criteria for restoration work are to be similar to these employed in traditional open excavation work
- Performance specifications should be developed as to hold the contractor responsible for settlement/heave damage that may occur along the drill path
- It is recommended that an additional length of the pipe that is one percent of the length, or 5 feet, whichever is greater, be pulled through the entrance pit, exposed, and examined for scratches, scores, cuts or other forms of damage. If excessive damage is found, a second additional pipe length, equal to the first length, should be pulled through the entrance pit
- A final leak test should be performed on the installed pipe
- The contractor shall provide a set of as-built drawings including both alignment and profile. Drawing should be constructed from actual field reading. Raw data should be available for submission at any time upon owner request. As part of the 'As-Built' document the contractor should specify the tracking equipment used, including methods or confirmatory procedure used to ensure the data was captured

GLOSSARY

Closing this chapter is a glossary of terms that are pertinent, and often specific to the application of horizontal directional drilling.

annuls: In drilling, the annulus refers to the place that surrounds the drill pipe and is enclosed by the borehole wall.

API: American Petroleum Institute.

ASTM: American Society for Testing and Materials.

azimuth: Horizontal direction expressed as an angle measured clockwise from any meridian. In drilling, azimuths are typically measured from magnetic north.

Barite: Natural barium sulfate used for increasing the density of drilling fluids.

barrel reamer: An enclosed cylindrical soft soil reaming tool with cutting teeth and fluid nozzles arrayed on the end faces. Barrel reamers may be designed with specific buoyancies to aid in borehole enlargement.

bedding plane: Any of the division planes that separate the individual strata or depositional layers in sedimentary or stratified rock.

bent sub: A short threaded piece of pipe manufactured with an axial offset or angel. In directional drilling, a bent sub is used to produce a leading edge asymmetry in a non-rotating directional drill string.

bentonite: Colloidal clay, composed primarily of montmorillonite, that swells when wet. Because of its gel-forming properties, bentonite is a major component of drilling mud.

bentonite extenders: Group of polymers that can maintain or increase the viscosity of bentonite while flocculating other clay solids in the mud. With bentonite extenders, desired viscosity can often be maintained using only half the amount of bentonite that would otherwise be required.

Bottom Hole Assembly (BHA): The combination of bit, downhole motor, subs, survey probe, and non magnetic collars assembled at the leading edge of a drill string.

boulder: Particle of rock that will not pass through a 12-inch square opening.

breakover: The over bend required to change the vertical orientation of a pipeline without inducing plastic deformation or unacceptable flexural stresses in the pipe.

bullet nose: An enclosed cylindrical soft soil reaming tool similar to a barrel reamer but with minimal cutting teeth and fluid nozzles. A bullet nose functions more as a centralized, expander, and fluid discharge point than a cutting tool and is typically used during pull back.

buoyancy control: The act of modifying the unit weight of a pipeline to achieve desired buoyancy. In HDD installation, this may be accomplished by placing water in the pipe during pull back.

Carboxymethyl Cellulose (CMC): A non-fermenting cellulose product used in drilling fluids to lower the water loss of the mud and produce viscosity.

carriage: Component of a horizontal drilling rig that travels along the frame and rotates the drill pipe. It is analogous to a top drive swivel on a vertical drilling rig.

centrifuge: Device used for the mechanical rotation to impart a centrifuge force to the fluid and achieve separation.

clay: Soil made up of particles passing a No. 200 U.S. standard sieve that can be made to exhibit plasticity (putty-like properties) within a range of water contents. Clay exhibits considerable strength when air dry.

cobble: Particle of rock that will pass through a 12 inch square opening and be retained on a 3 inch U.S. standard sieve.

density: The mass or weight of a substance per unit volume. For instance, the density of a drilling mud may be 10 pounds per gallon or 74.8 pounds per cubic foot.

desander: Centrifugal device (hydrocyclone) for removing sand from drilling fluid. Desanders are hydrocyclones larger than 5-inches in diameter.

desilter: Centrifugal device (hydrocyclone) for removing very fine particles, or silt, from drilling fluid. Desilters are hydrocyclones typically 4 or 5-inches in diameter.

diamond bit: Drilling bit that has a steel body surfaced with industrial diamonds, i.e. Polycrystalline Diamond Compact (PDC) bit.

downhole motor: Device that uses hydraulic energy contained in a drilling fluid flow stream to achieve mechanical bit rotation.

downhole survey probe: Device containing instruments that read inclination, azimuth, and tool face. A downhole survey probe is placed at the leading of a directional drill string and provides data that the drill uses to steer the string.

entry point: The point on a drilled segment where the pilot hole bit initially penetrates the surface. The horizontal drilling rig is positioned at the entry point.

exit point: The point on a drill segment where the pilot hole bit finally penetrates the surface. The pipeline pull section is positioned at the exit point.

filter cake: Layer of concentrated solids from the drilling mud or cement slurry that forms on the walls of the borehole opposite permeable formations; also called wall cake or mud cake.

filtration: The process of separating solids from liquid by forcing the liquid through a porous medium.

flocculation: The coagulation of solids in a drilling fluid, produced by special additives or by contaminants.

fluid: Substance that will flow and readily assumes the shape of the container in which it is placed. The term includes both liquids and gases.

fluid loss: Measure of the relative amount of fluid lost by filtration of the drilling fluid into a permeable formation.

flycutter: Open circular, cylindrical, or radial blade soft soil reaming tool with cutting teeth and fluid nozzles arrayed on the circumference and blades.

fracture zone: Zone of naturally occurring fissures or fractures that can pose problems with lost circulation.

frame: Component of a horizontal drilling rig on which the carriage travels. It is generally set at an angle of 6 degrees to 20 degrees from horizontal. It is analogous to the mast on a vertical drilling rig.

gel: Semi-solid, jelly state assumed by some colloidal dispersions at rest. When agitated, the gel converts to a fluid state. Gel is also used as a name for bentonite.

gel strength: Measure of the ability of a colloidal dispersion to develop and retain a gel form, based on its resistance to shear. The gel strength, or shear strength, of a drilling mud determines its ability to hold solids in suspension. Sometimes bentonite and other colloidal clays are added to drilling fluid to increase its gel strength.

gravel: Particles of rock that will pass a 3-inch sieve and be retained on a No. 4 U.S. standard sieve.

hole opener: A rock reaming tool utilizing roller cutters to cut harder material than be penetrated with a flycutter.

hole sizing: The act of moving a bit or reamer along a drilled hole one or more times to insure that the hole is open and annular drilling fluid flow can take place.

Horizontal Directional Drilling (HDD): A two-phase trench-less excavation method for installing buried pipelines and conduits. The first phase consists of drilling a directionally controlled pilot hole along a predetermined path grade at one end of a drilled segment to grade at the opposite end. The second phase consists of enlarging the pilot hole to a size, which will accommodate a pipeline or conduit, and pulling the pipeline or conduit into the enlarged hole. The method is accomplished using a horizontal drilling rig.

hydrocyclone: Conical device which directs drilling fluid flow in a spiraling manner thereby setting up centrifugal forces which aid in separating solids from the fluid. Hydrocyclones are also referred to as cyclones or cones.

hydrostatic head: Hydrostatic pressure.

hydrostatic pressure: The force exerted by a body of fluid at rest; it increases directly with the density and the depth of the fluid and is expressed in psi. The hydrostatic pressure of fresh water is 0.433 psi per foot of depth. In drilling, the term refers to the pressure exerted by the drilling fluid in the well bore.

inadvertent return: Uncontrolled flow of drilling fluid to the surface at locations other than the entry or exit points.

inclination: Angular deviation from true vertical or horizontal. In drilling, inclination is typically measured from vertical.

jetting: Advancing a drilled hole by using the hydraulic cutting action generated when drilling fluid is exhausted at high velocity through the leading edge of a drill string.

laminar flow: Flow in which fluid elements move along fixed streamlines which are parallel to the walls of the channel of flow.

LCM: Lost circulation material.

lost circulation: The quantities of whole mud lost to a formation, usually in cavernous, fissured, or coarsely permeable beds, evidenced by the complete or partial failure of the mud to return to the surface as it is being circulated in the hole; also called lost returns.

low clay solids mud: Heavily weighted mud whose high solids content (as a result of the large amounts of barite added) necessitates the reduction of clay solids.

low-solids mud: A drilling mud that contains minimum amount of soil material (sand, silt, etc.) and is used in rotary drilling when possible because it can provide fast drilling rates.

lubricity: The capacity of a fluid to reduce friction.

marsh funnel: A calibrated funnel used in field tests to indicate the viscosity of drilling mud.

metastable structure: Soil structure that is stable only because of the existence of some supplementary influence. Sand simultaneously with silt may exhibit a metastable structure due to the silt particles interfering with the intergranular contact between the sand particles. A shock or sudden loading may cause the structure to break down and liquefy.

montmorillonite: Clay mineral often used as an additive to drilling mud. It is a hydrous aluminum silicate capable of reacting with such substances as magnesium and calcium.

mud: The liquid circulated through the well bore during rotary drilling operations. Although it was originally a suspension of earth solids (especially clays) in water, the mud used in modern drilling is a more complex, three-phase mixture of liquids, reactive solids, and inert solids. The liquid phase may be fresh water, diesel oil, or crude oil and may contain one or more conditioners.

mud balance: Instrument consisting of a cup and a graduated arm with a sliding weight and rest on a fulcrum. It is used to measure the unit weight of the mud.

mud cleaner: Item of equipment combining vibratory screens and hydrocyclones to achieve effective solids control.

mud-up: To add solid materials (such as bentonite or other clay) to a drilling fluid composed mainly of clear water to obtain certain desirable properties.

Newtonian fluid: The basic and simplest fluid (from the standpoint of viscosity consideration) in which the shear stress is directly proportional to the shear rate. These fluids will immediately begin to flow when pressure or force in excess of zero is applied.

non-Newtonian fluid: Fluid in which the shear force is not directly proportional to the shear rate. Non-Newtonian fluids do not have a constant viscosity.

organic clay: Clay with sufficient organic content to influence the soil properties.

organic silt: Silt with sufficient organic content to influence the soil properties.

over bend: Vertical bends in a pipeline which progresses downward.

peat: Soil composed of vegetable tissue in various stages of decomposition usually with an organic odor, a dark brown to black color, a spongy consistency, and a texture ranging from fibrous to amorphous.

pilot hole: Small diameter hole directionally drilled along path in advance of reaming operations and pipe installation.

plastic: Capable of being shaped or formed; pliable.

plasticity index: Numerical difference between the liquid limit and the plastic limit of a soil.

plastic limit: The water content at which a soil begins to break apart and crumble when rolled into threads ⅛ inch in diameter.

plastic viscosity: Absolute flow property indicating the flow resistance of certain types of fluids. Plastic viscosity is a measure of shearing stress.

plunger effect: The sudden increase in borehole pressure brought about by the rapid movement of a larger pipe or cutting tool along a drilled or reamed hole.

polyanionic cellulose: Chemical compound used to reduce water loss in mud that are affected by salt contamination.

polymer: Substance that consists of large molecules formed from smaller molecules in repeating structural units. In petroleum refining, heat and pressure are used to polymerize light hydrocarbons into larger molecules, such as those that make up high-octane gasoline. In drilling operations, various types of organic polymers are used to thicken drilling mud, fracturing fluid, acid, and other liquids. In petrochemical production, polymer hydrocarbons are used as the basis for plastics.

polymer mud: A drilling mud to which has been added a polymer, a chemical that consists of larger molecules that were formed from small molecules in repeating structural units, to increase the viscosity of the mud.

preream: The act of enlarging a pilot hole by pulling or pushing cutting tools along the hole prior to commencing pipe installation.

pull back: The act of installing a pipeline along a horizontally drilled hole by pulling it to the drilling rig from the end of the hole opposite the drilling rig.

pull back swivel: Device placed between the rotating drill string and tools and the pipeline pull section to minimize torsion transmitted to the section during pull back installation.

pull section: String of pipeline prefabricated at an adjacent location prior to being pulled into its final position.

rock: Any indurated material that requires drilling, wedging, blasting, or other methods of brute force for excavation.

roller cone bit: Drilling bit made of two, three, or four cones, or cutters, that are mounted on extremely rugged bearings. The surface of each cone is made of rows of steel teeth or tungsten carbide inserts.

rotational viscometer: Instrument used for assessing mud properties, returning values for both plastic viscosity and yield point.

Rock Quality Designation (RQD): An indication of the fractured nature of rock determined by summing the total length of core recovered counting only those pieces which are 4 inches or more in length and which are hard and sound. RQD is expressed as a percentage of the total core run.

sag bend: Vertical bends in a pipeline which progresses upward.

sand: Particles of rock that will pass a No. 4 U.S. standard sieve and be retained on a No. 200 U.S. standard sieve.

shale shaker: Device that utilizes vibrating screens to remove larger solid particles from circulating drilling fluid. The fluid passes through the screen openings while solids are retained and moved off of the shaker by the vibrating motion.

side bend: Horizontal bend in a pipeline.

silt: Soil passing a No. 200 U.S. standard sieve that is non-plastic or very slightly plastic and that exhibits little or no strength when air dry.

slickenside: A smooth surface produced in rock or clay by movement along a fault or joint.

soil: Any unconsolidated material composed of discrete solid particles with gases or liquids between.

spoil: Excavated soil or rock.

Standard Penetration Test (SPT): An indication of the density or consistency of soils given by counting the number of blows required to drive a 2 inch OD split spoon sampler 12 inches using a 140 pound hammer falling 30 inches. The sampler is driven in three 6 inch increments. The sum of the blows required for the last two increments is referred to as the "N" value, blow count, or Standard Penetration Resistance.

spud in: To begin drilling; to start the hole.

sub: Short threaded piece of pipe used in a drill string to perform a special function.

surfactant: Substance that affects the properties of the surface of a liquid or solid by concentrating on the surface layer. Surfactants are useful in that their use can ensure that the surface of one substance or object is in thorough contact with the surface of another substance.

suspension: Mixture of small non setting particles of solid material within a gaseous or liquid medium.

swabbing effect: Phenomenon characterized by formation fluids being pulled or swabbed into the well bore when the drill stem and bit are pulled up the well bore fast enough to reduce the hydrostatic pressure of the mud below the bit.

tool face: Direction of the asymmetry of a drilling string. A directional drilling string will progress in the direction of the tool face. Tool face is normally expressed as a angle measured clockwise from the top of the drill pipe in a plane perpendicular to the axis of the drill pipe.

transition velocity: Velocity at which the flow in a particular fluid flowing in a particular channel shifts between laminar and turbulent.

trip: Act of withdrawing (tripping out) or inserting (tripping in) the drill string.

turbulent flow: Fluid flow in which the velocity at a given point changes constantly in magnitude and direction.

twist off: To part or split drill pipe or drill collars, primarily because of metal fatigue in the pipe or because of mishandling.

velocity: Rate of linear motion per unit of time.

vices: Devices mounted on the frame of a horizontal drilling rig which grip the drill pipe and allow it to be made up (screwed together) or broken (unscrewed).

viscometer: Apparatus to determine the viscosity of a fluid.

viscosity: Measure of the resistance of a liquid flow. The internal friction resulting from the combined effects of cohesion and adhesion bring about resistance.

wall cake: Solid material deposited along the wall of a drilled hole resulting from filtration of the fluid part of the mud into the formation.

wash pipe: Drill pipe that is run, or rotated, concentrically over a smaller drill pipe so that the smaller (internal) pipe can be freely moved or rotated.

water-back: To reduce the weight or density of a drilling mud by adding water or to reduce the solids content of a mud by adding water.

weight-up: To increase the weight or density of drilling by adding weighting material.

yield point: Maximum stress that a solid can withstand without undergoing permanent deformation either by plastic flow or by rupture.

CHAPTER 9

CHEMICAL RESISTANCE OF PLASTICS AND ELASTOMERS

Chemical Resistance of Plastics and Elastomers Used in Pipeline Construction

1 . Introduction

It is now inconceivable to construct pipelines without the use of plastics. Pipes made from plastics are used not only for drinking water, water for general use and waste water, but also for the conveyance of aggressive liquids and gases. Expensive pipe materials such as lined metal, ceramic or glass, have been largely superseded by plastic pipes. It is, however, important that the most suitable plastic material is selected for each application. This "Chemical Resistance List" serves as a useful guide in this respect. The list is periodically revised to include the latest findings. It contains all plastics and elastomers in the George Fischer product range which can come into direct contact with the media.

The information is based on experiments, immersion and, when available, on data from tests which include temperature and pressure as stress factors. The results achieved in immersion experiments cannot be applied without reservation to pipes under stress, i.e. internal pressure, as the factor "stress corrosion cracking" is not taken into consideration. In certain cases it can be of advantage to test the suitability under the planned working conditions. The tests referred to have been carried out partly by George Fischer and partly by the International Standardization Organization (ISO) or national standards organizations.

Pure chemicals were used for the tests. If a mixture of chemicals is to be conveyed in practice this may affect the chemical resistance of the plastic. It is possible in special cases to carry out appropriate tests with the specific mixture. Suitable test equipment is available at George Fischer for this purpose, which we regard as part of our service to the customer. It goes without saying that we are willing to give individual advice at any time. In this connection it is worth mentioning that George Fischer already possesses information concerning the behavior towards plastics of a number of chemicals or mixtures of chemicals which are not yet included in this list. The "Chemical Resistance List" gives valuable assistance in the planning of plastic pipelines. Please refer to the following instructions, which are important for the application and evaluation of this list.

2. Instructions for the Use of the Chemical Resistance List

2.1 General

As stated in the introduction, the "Chemical Resistance List" is only intended as a guide. Changes in the composition of the medium or special working conditions could lead to deviations. If there is any doubt, it is advisable to test the behavior of the material under the specific working conditions, by means of a pilot installation. No guarantees can be given in respect of the information contained in this booklet. The data shown is based upon information available at the time of printing, but it may, however, be revised from time to time in the light of subsequent research and experience.

2.2 Classification

The customary classifications: **resistant, conditionally resistant** and **not recommended** are depicted by the signs: +, O, and –, which allow simple presentation and application. These classifications are defined as:

Resistant: +
Within the acceptable limits of pressure and temperature the material is unaffected or only insignificantly affected.

Conditionally Resistant: O
The medium can attack the material or cause swelling. Restrictions must be made in regard to pressure and/or temperature, taking the expected service life into account. The service life of the installation can be noticeably shortened. Further consultation with George Fischer is recommended.

Not recommended: –
The material cannot be used with the medium at all, or only under special conditions.

(Courtesy George Fischer Engineering Handbook)

2.3 Pipe Joints

2.3.1 Solvent Cement Joints (PVC)

Solvent cement joints made with standard PVC cement and primer systems are generally as resistant as the PVC material itself. The following chemicals are, however, an exception:

- Sulphuric acid H_2SO_4 in concentrations above 70 percent
- Hydrochloric acid HCl in concentrations above 25 percent
- Nitric acid HNO_3 in concentrations above 20 percent

Hydrofluoric acid in any concentration In conjunction with the above media the solvent cement joining is classified as "conditionally resistant". Previously recommended solvent cement (Dytex, by Henkel, Germany) used for pipe and fittings to carry concentrated acids, can no longer be brought into the United States because of its methylene chloride solvent system being classified as a carcinogen. There is no known domestically available substitute. Special consideration should be given to the possible attack of the cemented joints by these concentrated acids.

2.3.2 Fusion Joints

In the case of PE, PP, and PVDF (SYGEF®) heat fusion joints have practically the same chemical resistance as the respective material. In conjunction with media which could cause stress cracking, the fused joints can be subjected to an increased risk due to residual stress from the joining process.

2.4 Sealing Materials

Depending upon the working conditions and the stress involved, the life span of the sealing materials can differ from that of the pipeline material. Seals in PTFE, which are not included in this list, are resistant to all the chemicals indicated. The greater permeability of PTFE should, however, be considered. Under certain working conditions, for example when conveying highly aggressive media such as hydrochloric acid, this material characteristic must be taken into account.

2.5 General Summary and Limits of Application

The following table includes all the materials contained in the George Fischer product range, and their abbreviations. The summary gives preliminary information regarding the general behavior of the materials and the temperature limits.

2.6 Standards

This list has been compiled with reference to the following ISO standards:

ISO/TR 7473
Unplasticized polyvinyl chloride pipes and fittings – Chemical resistance with respect to fluids.

ISO/TR 7474
High density polyethylene pipes and fittings – Chemical resistance with respect to fluids to be conveyed.

ISO/TR 7471
Polypropylene (PP) pipes and fittings – Chemical resistance with respect to fluids.

ISO TR 10358
Plastic pipes and fittings – Combined chemical resistance classification table.

DVS 2205 Part I
Calculations for thermoplastic containers and appliances.

DIN 8080 Supplement 1 «Pipes of chlorinated polyvinyl chloride (PVC-C), PVC-C 250 – Chemical Resistance».

(Courtesy George Fischer Engineering Handbook)

Material	Abbre-viation	Remarks	Maximum Permissible Temperature (Water) °C	
			Constant	Short Term
Polyvinyl Chloride	PVC	Resistant to most solutions of acids, alkalis and salts and to organic compounds miscible with water. Not resistant to aromatic and chlorinated hydrocarbons	60°	60°
Chlorinated Polyvinyl Chloride	CPVC	Can be used similarly to PVC but at higher temperatures. Consult factory for specific applications.	90°	110°
High-density Polyethylene	PE 50	Resistant to hydrous solutions of acids, alkalis and salts as well as to a large number of organic solvents. Unsuitable for concentrated oxidizing acids.	60°	80°
Polypropylene, heat stabilized	PP	Chemical resistance similar to that of PE but suitable for higher temperatures	90°	110°
Polyvinylidene Fluoride	PVDF (SYGEF®)	Resistant to acids, solutions of salts, aliphatic, aromatic and chlorinated hydrocarbons, alcohols and halogens. Conditionally suitable for ketones, esters, organic bases and alkaline solutions	140°	150°
Polybutylene-1	PB	Similar to PE 50, but can be used up to 90°C	90°	100°
Polyoxymethylene	POM	Resistant to most solvents and hydrous alkalis. Unsuitable for acids	60°	80°
Polytetrafluoroethylene (e.g. Teflon®)	PTFE	Resistant to all chemicals in this list	250°	300°
Nitrile Rubber	NBR	Good resistance to oil and petrol. Unsuitable for oxidizing media	90°	120°
Butyl Rubber Ethylene Propylene Rubber	IIR EFDM	Good resistance to ozone and weather. Especially suitable for aggressive chemicals. Unsuitable for oils and fats	90°	120°
Chloroprene Rubber (e.g. Neoprene®)	CR	Chemical resistance very similar to that of PVC-U and between that of Nitrile and Butyl Rubber	80°	110°
Fluorine Rubber (e.g. Viton®)	FPM	Has best chemical resistance to solvents of all elastomers	150°	200°
Chlorine Sulphonyl Polyethylene (e.g. Hypalon®)	CSM	Chemical resistance similar to that of EPDM	100°	140°

®Registered trade name

The abbreviations listed below are found throughout the listings and have the following definition:

Q/E (Quellung/Erweichung) = swelling/softening
D/P (Diffusion/Permeation) = diffusion/permeation
SpRB (Spannungsrissbildung) = environmental stress cracking

(Courtesy George Fischer Engineering Handbook)

Aggressive Media					Chemical Resistance											
Medium	Formula	Boiling point °C	Concentration	Temperature °C	PVC	CPVC	ABS	PE	PP-H	PVDF (SYGEF)		EPDM	FPM	NBR	CR	CSM
Acetaldehyde	CH_3-CHO (C_2H_4O)	21	technically pure	20	-	-	-	+	O	-		+	O	-	-	O
				40				O	-			O	-			-
				60								-				
				80												
				100												
				120												
				140												
Acetaldehyde			40%, aqueous solution	20	O	-	-	+	+	-		+	+	-	+	+
				40	-			+	+			+	+		+	+
				60				O	+			+	O		O	+
				80					O			+	-		-	+
				100					-							
				120												
				140												
Acetic acid (SpRB)	CH_3COOH	118	technically pure, glacial	20	O	-	-	+	+	+		O	-	-	O	O
				40	-			+	+	O						
				60				O	O	-						
				80					-							
				100												
				120												
				140												
Acetic acid (SpRB)	CH_3COOH		10%, aqueous	20	+	+	+	+	+	+		+	O	+	+	O
				40	+	+	+	+	+	+		+	-	O	+	-
				60	O	+	O	+	+	+		O			O	
				80		+			+	+						
				100					+	+						
				120						+						
				140												
Acetic acid (SpRB)			50%, aqueous	20	+	+	-	+	+	+		+	O	-	O	O
				40	+			+	+	+						
				60	O			+	+	+						
				80						O						
				100						O						
				120												
				140												
Acetic acid (SpRB)	CH_3COOH		60%	20	+	-	-	+	+	+		+				
				40												
				60												
				80												
				100												
				120												
				140												
Acetic acid (SpRB)		118	98%	20	-	-	-	+	+	+		O	-			
				40												
				60												
				80												
				100												
				120												
				140												
Acetic acid anhydride (SpRB)	(CH_3-CO)$_2$O	139	technically pure	20	-	-	-	+	+	-		O	-	-	-	+
				40				O	O							
				60												
				80												
				100												
				120												
				140												

(Courtesy George Fischer Engineering Handbook)

Aggressive Media					Chemical Resistance											
Medium	Formula	Boiling point °C	Concentration	Temperature °C	PVC	CPVC	ABS	PE	PP-H	PVDF (SYGEF)		EPDM	FPM	NBR	CR	CSM
Acetic acid isobutyl ester	$(CH_2)_2$-CH-$(CH_2)_2$-CO_2H		technically pure	20						-						
				40												
				60												
				80												
				100												
				120												
				140												
Acetone	CH_3-CO-CH_3	56	technically pure	20	-	-	-	+	+	-		+	-	-	-	O
				40				+	+			+				O
				60				+	+			+				O
				80												
				100												
				120												
				140												
Acetone			up to 10%, aqueous	20	-	-	O	+	+	O		+	O	-	+	O
				40				+	+	O		+	O		O	O
				60				+	+	O		+	-		-	O
				80												
				100												
				120												
				140												
Acetonitrile	CH_3CN	81.6	100%	20	-	-	-			-						
				40												
				60												
				80												
				100												
				120												
				140												
Acetophenone	CH_3-CO-C_6H_5		100 %	20	-	-	-			-		+	-	-	-	
				40												
				60												
				80												
				100												
				120												
				140												
Acrylic acid methyl ester	CH_2=$CHCOOCH_3$	80.3	technically pure	20	-	-	-			+		O				
				40												
				60												
				80												
				100												
				120												
				140												
Acrylic ester	CH_2=CH-COO CH_2CH_3	100	technically pure	20	-	-	-		-	-		O	-	-	O	+
				40												
				60												
				80												
				100												
				120												
				140												
Acrylonitrile	CH_2=CH-CN	77	technically pure	20	-	-	-	+	+	-		+	O	-	+	O
				40				+	O			+	O		+	O
				60				+				O	-		+	-
				80												
				100												
				120												
				140												

(Courtesy George Fischer Engineering Handbook)

Aggressive Media					Chemical Resistance											
Medium	Formula	Boiling point °C	Concentration	Temperature °C	PVC	CPVC	ABS	PE	PP-H	PVDF (SYGEF)		EPDM	FPM	NBR	CR	CSM
Adipic acid	$HOOC\text{-}(CH_2)_4\text{-}COOH$	Fp 153	saturated, aqueous	20	+	+	-	+	+	+		+	+	+	+	+
				40	+	+		+	+			+	+	+	+	+
				60	-	+		+	+			+	+	+	+	+
				80		+			+							
				100												
				120												
				140												
Aluni	see Potassium/ aluminium sulphate															
Alcoholic spirits (Gin, Whisky,etc.)			approx. 40% ethyl alcohol	20	+	O	-	+	+	+		+	+	+	+	+
				40												
				60												
				80												
				100												
				120												
				140												
Allyl alcohol	$H_2C{=}CH\text{-}CH_2\text{-}OH$	97	96%	20	O	O	-	+	+			O	O	+	O	+
				40	-			+	+			O	-	+	-	+
				60				+	+			O		+		+
				80								-		+		-
				100												
				120												
				140												
Aluminium chloride	$AlCl_3$		10%, aqueous	20	+	+	+	+	+	+		+	+	+	+	+
				40	+	+	+	+	+	+		+	+	+	+	+
				60	+	+	+	+	+	+		+	+	O	+	+
				80		+				+		+	+			+
				100						+		+	+			
				120						+						
				140												
Aluminium chloride	$AlCl_3$	115	saturated	20	+	+	+	+	+	+		+	+	+	+	+
				40	+	+	+	+	+	+		+	+	+	+	+
				60	+	+	+	+	+	+		+	+	+	+	+
				80		+			+	+			+	O	+	+
				100					O	+			+	-		+
				120						+						
				140												
Aluminium fluoride	AlF_3		saturated	20		+				+						
				40		+				+						
				60		+				+						
				80						+						
				100												
				120												
				140												
Aluminium hydroxide	$Al(OH)_3$		Suspension	20		+						+				
				40		+						+				
				60		+						+				
				80		+										
				100												
				120												
				140												
Aluminium nitrate	$Al(NO_3)_3$		saturated	20		+				+		+				
				40		+				+		+				
				60		+				+		+				
				80		+										
				100												
				120												
				140												

(Courtesy George Fischer Engineering Handbook)

Aggressive Media					Chemical Resistance											
Medium	Formula	Boiling point °C	Concentration	Temperature °C	PVC	CPVC	ABS	PE	PP-H	PVDF (SYGEF)		EPDM	FPM	NBR	CR	CSM
Aluminium sulphate	$Al_2(SO_4)_3$		10%, aqueous	20	+	+	+	+	+	+		+	+	+	+	+
				40	+	+	+	+	+	+		+	+	+	+	+
				60	O	+	+	+	+	+		+	+	+	+	+
				80		+			+	+			+			O
				100					+	+						
				120						+						
				140												
Aluminium sulphate			cold saturated, aqueous	20	+	+	+	+	+	+		+	+	+	+	+
				40	+	+	+	+	+	+		+	+	+	+	+
				60	+	+	+	+	+	+		+	+	+	+	O
				80		+			+	+			+			O
				100						+						
				120						+						
				140												
Ammonia (SpRB)	NH_3	-33	gaseous, technically pure	20	+	-	-	+	+	+		+	+	+	+	+
				40	+			+	+	+			O			
				60	+			+	+	O						
				80						O						
				100						O						
				120												
				140												
Ammonium acetate	CH_3COONH_4		aqueous, all	20	+	+	O	+	+	+		+	+	+	+	+
				40	+	+		+	+	+		+	+	O	+	+
				60	O	+		+	+	+		+	+		O	
				80		+			+	+		O				
				100					+	+						
				120												
				140												
Ammonium aluminium sulfate				20						+						
				40						+						
				60						+						
				80						+						
				100						+						
				120												
				140												
Ammonium bromide				20						+						
				40						+						
				60						+						
				80												
				100												
				120												
				140												
Ammonium carbonate	$(NH_4)_2CO_3$		50%, aqueous	20	+	+	+	+	+	+		+	+	+	+	+
				40	+	+	+	+	+	+		+	+	+	+	+
				60	O	+	+	+	+	+		+	+	+	+	+
				80		+			+	+		+	+			+
				100					+	+						
				120						+						
				140												
Ammonium chloride	NH_4Cl	115	aqueous, cold saturated	20	+	+	+	+	+	+		+	+	+	+	+
				40	+	+	+	+	+	+		+	+	+	+	+
				60	O	+	+	+	+	+		+	+	+	+	+
				80		+			+	+		+	+		+	+
				100					+	+		+	+			+
				120						+			+			
				140												

(Courtesy George Fischer Engineering Handbook)

Aggressive Media					Chemical Resistance											
Medium	Formula	Boiling point °C	Concentration	Temperature °C	PVC	CPVC	ABS	PE	PP-H	PVDF (SYGEF)		EPDM	FPM	NBR	CR	CSM
Ammonium citrate				20	+					+						
				40	+					+						
				60	+					+						
				80						+						
				100						+						
				120												
				140												
Ammonium dicromate	$(NH_4)_2 Cr_2 O_7$		saturated	20		+										
				40		+										
				60		+										
				80		+										
				100												
				120												
				140												
Ammonium dihydrogenphosphate				20	+			+	+							
				40	+			+	+							
				60	+			+	+							
				80					+							
				100												
				120												
				140												
Ammonium fluoride	NH_4F			20	+	+		+	+	+						
				40	+			+	+	+						
				60	+			+	+	+						
				80					+	+						
				100						+						
				120												
				140												
Ammonium formiate				20						+						
				40						+						
				60						+						
				80						+						
				100						+						
				120												
				140												
Ammonium hexafluorosuilicate				20						+						
				40						+						
				60						+						
				80						+						
				100						+						
				120												
				140												
Ammonium hydrogen fluoride	NH_4HF_2		50%, aqueous	20	+	+	-	+	+	+		+	+			
				40	+	+		+	+	+						
				60	O	+		+	+	+						
				80												
				100												
				120												
				140												
Ammonium hydrogencarbonate				20	+			+	+							
				40	+			+	+							
				60	+			+	+							
				80					+							
				100												
				120												
				140												

(Courtesy George Fischer Engineering Handbook)

Aggressive Media					Chemical Resistance											
Medium	Formula	Boiling point °C	Concentration	Temperature °C	PVC	CPVC	ABS	PE	PP-H	PVDF (SYGEF)		EPDM	FPM	NBR	CR	CSM
Ammonium hydrogenphosphate				20	+			+	+							
				40	+			+	+							
				60	+			+	+							
				80					+							
				100												
				120												
				140												
Ammonium hydrogensulfite				20						+						
				40						+						
				60						+						
				80						+						
				100						+						
				120												
				140												
Ammonium hydroxide	NH_4OH		aqueous, cold saturated	20	+	-	+		+	-		+	-	+	+	+
				40	+		+		+			+		O	+	+
				60	O		O		+			O		O	+	O
				80											O	-
				100												
				120												
				140												
Ammonium nitrate	NH_4NO_3	112	aqueous, saturated	20	+	+	+	+	+	+		+	+	+	+	+
				40	+	+	+	+	+	+		+	+	+	+	+
				60	+	+	+	O	+	+		+	+	O		O
				80		+			O	+			+			
				100						+						
				120						+						
				140												
Ammonium oxalate	$H_4NOOC\text{-}COONH_4$			20						+		+				
				40						+						
				60						+						
				80						+						
				100						+						
				120												
				140												
Ammonium persulphate	$(NH_4)_2S_2O_8$			20		+				+						
				40		+				+						
				60		+				+						
				80		+				+						
				100						+						
				120												
				140												
Ammonium phosphate	$(NH_4)_3PO_4$		saturated	20	+	+	+	+	+	+		+	+	+	+	+
				40	+	+	+	+	+	+		+	+	+	+	+
				60	+	+	+	+	+	+			+	O	+	O
				80					+	+			+			
				100						+			+			
				120												
				140												
Ammonium sulphate	$(NH_4)_2SO_4$		aqueous, saturated	20	+	+	+	+	+	+		+	+	+	+	+
				40	+	+	+	+	+	+		+	+	+	+	+
				60	+	+	+	+	+	+		+	+	O	+	O
				80		+			+	+		+	+			
				100					+	+			+			
				120						+						
				140												

(Courtesy George Fischer Engineering Handbook)

Aggressive Media					Chemical Resistance											
Medium	Formula	Boiling point °C	Concentration	Temperature °C	PVC	CPVC	ABS	PE	PP-H	PVDF (SYGEF)		EPDM	FPM	NBR	CR	CSM
Ammonium sulphide	$(NH_4)_2S$		aqueous, all	20	+	O	+	+	+	+		+	+	+	+	+
				40	+	O	+	+	+	+		+	O	+	+	+
				60	O	O	+	+	+	+		+	-	+	+	+
				80		-			+							
				100												
				120												
				140												
Ammonium tetrafluoroborate				20						+						
				40						+						
				60						+						
				80						+						
				100						+						
				120												
				140												
Ammonium thiocyanate	NH_4SCN		saturated	20		+				+						
				40		+				+						
				60		+				+						
				80		+										
				100												
				120												
				140												
Amyl acetate	$CH_3(CH_2)_4\text{-}COOCH_3$	141	technically pure	20	-	-	-	+	O	+		O	-	-	-	-
				40				+	O	O						
				60				+	-	O						
				80												
				100												
				120												
				140												
Amyl alcohol (SpRB)	$CH_3(CH_2)_3\text{-}CH_2\text{-}OH$	137	technically pure	20	+	+	-	+	+	+		+	O	+	+	O
				40	+	+		+	+	+		+		+	+	
				60	O	+		+	+	+		+		+	+	
				80					+	+						
				100						+						
				120						O						
				140												
Aniline	$C_6H_5NH_2$	182	technically pure	20	-	-	-	O	O	+		-	O	-	-	-
				40						O			O			
				60						-			O			
				80												
				100												
				120												
				140												
Aniline hydrochloride	$C_6H_7N{+}HCl$	245	aqueous, saturated	20	+	+	-	+	+	+		+	O	O	-	+
				40	O	+		+	+			+	-	-		+
				60				O	O			+				+
				80												O
				100												
				120												
				140												
Antimony thiocyanate				20						+						
				40						+						
				60						+						
				80						+						
				100						+						
				120												
				140												

(Courtesy George Fischer Engineering Handbook)

Aggressive Media					Chemical Resistance											
Medium	Formula	Boiling point °C	Concentration	Temperature °C	PVC	CPVC	ABS	PE	PP-H	PVDF (SYGEF)		EPDM	FPM	NBR	CR	CSM
Antimony trichloride (SpRB)	$SbCl_3$		90%, aqueous	20	+	+	-	+	+	+		+	+	-	+	+
				40	+			+	+	+						
				60				+	+	+						
				80												
				100												
				120												
				140												
Aqua regia (SpRB)	HNO_3+HCl			20	+	+	-	-	-	O		-	O	-	-	O
				40	O											
				60												
				80												
				100												
				120												
				140												
Arsenic acid	H_3AsO_4		80%, aqueous	20	+	+	+	+	+	+		+	+	+	+	+
				40	+	+	+	+	+	+		+	+	+	+	+
				60	O	+	+	+	+	+		+	+	+	+	+
				80		+			+	+		+	+	O	+	+
				100						+			+			
				120						+						
				140												
Barium carbonate	$BaCO_3$			20	+	+	+	+	+	+		+	+	+	+	+
				40	+	+		+	+	+		+	+	+		
				60	+	+		+	+	+		+				
				80		+			+							
				100												
				120												
				140												
Barium chloride	$BaCl_2$		saturated	20	+	+	+	+	+	+		+	+	+	+	+
				40	+	+		+	+	+		+	+	+		
				60	+	+		+	+	+		+	+			
				80		+			+	+			+			
				100						+						
				120												
				140												
Barium hydroxide	$Ba(OH)_2$	102	aqueous, saturated	20	+	+	+	+	+	-		+	+	+	+	+
				40	+	+	+	+	+			+		+	+	O
				60	O	+	+	+	+			+		+	+	
				80		+			+			+				
				100												
				120												
				140												
Barium salts			aqueous, all	20	+	+	+	+	+	+		+	+	+	+	+
				40	+	+	+	+	+	+		+	+	+	+	+
				60	+	+	+	+	+	+		+	+	+	+	+
				80		+			+	+		+	+			
				100						+						
				120												
				140												
Barium sulfate	$BaSO_4$			20	+			+	+	+		+				
				40	+			+	+	+		+				
				60	+			+	+	+		+				
				80					+	+						
				100						+						
				120												
				140												

(Courtesy George Fischer Engineering Handbook)

Aggressive Media					Chemical Resistance											
Medium	Formula	Boiling point °C	Concentration	Temperature °C	PVC	CPVC	ABS	PE	PP-H	PVDF (SYGEF)		EPDM	FPM	NBR	CR	CSM
Barium sulfide	BaS		suspension	20	+	+		+	+	+		+				
				40	+	+		+	+	+		+				
				60	+	+		+	+	+		+				
				80		+			+	+						
				100						+						
				120												
				140												
Battery acid	see Sulphuric acid 40%															
Beef tallow emulsion, sulphonated (SpRB)			usual commercial	20	+	o	+	+	+	+		-	+	+	+	+
				40						+						
				60						+						
				80												
				100												
				120												
				140												
Beer			usual commercial	20	+	+	+	+	+	+			+	+	+	+
				40	+	+	+	+	+	+						
				60	+	+	+	+	+	+						
				80		+				+						
				100												
				120												
				140												
Benzaldehyde	C_6H_5-CHO	180	saturated, aqueous	20	-	-	-	+	+	+		+	+	O	-	-
				40				+		O		+	+			
				60				+		-		+	+			
				80												
				100												
				120												
				140												
Benzene	C_6H_6	80	technically pure	20	-	-	-	O	O	+		-	+	O	-	-
				40				O	-	O						
				60						-						
				80												
				100												
				120												
				140												
Benzenesulfonic acid	$C_6H_5SO_3H$		technically pure	20						+			+			
				40						+			+			
				60						+						
				80						+						
				100						+						
				120												
				140												
Benzoic acid	C_6H_5-COOH	Fp.*, 122	aqueous, all	20	+	+	+	+	+	+		+	+	-	-	-
				40	+	+	+	+	+	+		+	+			
				60	O	+		+	+	+			+			
				80		O			+	+			+			
				100					+	+			O			
				120						+						
				140												
Benzoyl chloride	$C_6H_5CHCl_2$		technically pure	20						+						
				40						+						
				60						+						
				80						O						
				100						-						
				120												
				140												

(Courtesy George Fischer Engineering Handbook)

Aggressive Media					Chemical Resistance											
Medium	Formula	Boiling point °C	Concentration	Temperature °C	PVC	CPVC	ABS	PE	PP-H	PVDF (SYGEF)		EPDM	FPM	NBR	CR	CSM
Benzyl alcohol (SpRB)	C_6H_5-CH_2-OH	206	technically pure	20	O	-	-	+	+	+		-	+	-	+	O
				40				+	+	+					+	
				60				O	O	O					+	
				80						-						
				100												
				120												
				140												
Beryllium chloride				20						+						
				40						+						
				60						+						
				80												
				100												
				120												
				140												
Beryllium sulfate				20						+			+			
				40						+			+			
				60						+			+			
				80						+						
				100						+						
				120												
				140												
Borax	$Na_2B_4O_7$		aqueous, all	20	+	+	+	+	+	+		+	+	+	+	+
				40	+	+	+	+	+	+		+	+	+	+	+
				60	O	+		+	+	+		+	+	+	+	O
				80		+			+	+		+	+			
				100					+	+						
				120												
				140												
Boric acid	H_3BO_3		all, aqueous	20	+	+	+	+	+	+		+	+	+	+	+
				40	+	+	+	+	+	+		+	+	+	+	+
				60	O	+	+	+	+	+		+	+	+	+	+
				80		+			+	+		+	+			
				100					+	+			+			
				120						+						
				140												
Brine, containing chlorine				20	+	+	-	+	O	+		O	+	O	O	O
				40	+	+				+						
				60	+	+				O						
				80												
				100												
				120												
				140												
Brombenzene	C_6H_5Br			20	-	-				+			+			
				40												
				60												
				80												
				100												
				120												
				140												
Bromine, liquid	Br_2	59	technically pure	20	-	-	-	-	-	+		-	+	-	-	-
				40						+						
				60						+						
				80						+						
				100						O						
				120												
				140												

(Courtesy George Fischer Engineering Handbook)

Aggregive Media					Chemical Resistance											
Medium	Formula	Boiling point °C	Concentration	Temperature °C	PVC	CPVC	ABS	PE	PP-H	PVDF (SYGEF)		EPDM	FPM	NBR	CR	CSM
Bromine, vapours	Br_2		high	20	-	-	-	-	-	+		-	+	-	-	-
				40						+						
				60						+						
				80						+						
				100						O						
				120												
				140												
Bromine water	$Br.H_2O$		saturated, aqueous	20	+	o	-	-	-	+		-	+	-	-	-
				40						+						
				60						+						
				80						+						
				100												
				120												
				140												
Butadiene (Q/E)	$H_2C{=}CH{-}CH{=}CH_2$	-4	technically pure	20	+	+	-	+	+	+		-	O	-	+	+
				40					+	+					+	O
				60					+	+					O	-
				80						+						
				100						+						
				120												
				140												
Butane	C_4H_{10}	0	technically pure	20	+	+	+	+	+	+		-	+	+	+	+
				40												
				60												
				80												
				100												
				120												
				140												
Butanediol (SpRB)	$HO{-}(CH_2)_4{-}OH$	230	aqueous, 10%	20	+	+	-	+	+			+	+	+	O	+
				40	O	+		+	+			+	+	+	-	+
				60				+	+			+	+	+		+
				80												
				100												
				120												
				140												
Butanol (SpRB)	C_4H_9OH	117	technically pure	20	+	+	-	+	+	+		+	+	+	+	+
				40	+	+		+	+	+		+	O	+	+	+
				60	O	+		+	O	+		+	-	+	O	+
				80		O			-	+						
				100						O						
				120												
				140												
Butyl acetate	$CH_3COOCH_2CH_2CH_2CH_3$	126	technically pure	20	-	-	-	+	O	+		+	O	-	O	O
				40						O		-	-		-	-
				60						-						
				80												
				100												
				120												
				140												
Butyl phenol, p-tertiary	$(CH_3)_3C{-}C_6H_4{-}OH$	237	technically pure	20	O	O	-	O	+	+		-	O	-	-	-
				40	-	-				+						
				60						+						
				80						+						
				100												
				120												
				140												

(Courtesy George Fischer Engineering Handbook)

Aggressive Media					Chemical Resistance											
Medium	Formula	Boiling point °C	Concentration	Temperature °C	PVC	CPVC	ABS	PE	PP-H	PVDF (SYGEF)		EPDM	FPM	NBR	CR	CSM
Butylene glycol (SpRB)	$HO\text{-}CH_2\text{-}CH{=}CH\text{-}CH_2\text{-}OH$	235	technically pure	20	+	+	+	+	+	+		+	+	-	+	O
				40	+	+	+	+	+	+		+	+		+	-
				60	O	+	+	+	+	+		+	O		+	
				80						+						
				100												
				120												
				140												
Butylene liquid	C_4H_8	51	technically pure	20	+			-	-	+		O	+	+	+	O
				40												
				60												
				80												
				100												
				120												
				140												
Butyric acid (SpRB)	$CH_3\text{-}CH_2\text{-}CH_2\text{-}COOH$	163	technically pure	20	+	+	-	+	+	+		O	O	-	O	O
				40				+		+						
				60				O		+						
				80						+						
				100						O						
				120												
				140												
Cadmium bromide	$CdBr_2$			20	+	+		+	+			+	+			
				40	+	+		+	+			+	+			
				60	+	+		+	+			+	+			
				80		+							+			
				100												
				120												
				140												
Cadmium chloride	$CdCl_2$			20	+	+		+	+			+	+			
				40	+	+		+	+			+	+			
				60	+	+		+	+			+	+			
				80					+							
				100												
				120												
				140												
Cadmium cyanide	Cd(CN)2			20	+			+	+							
				40	+			+	+							
				60	+			+	+							
				80					+							
				100												
				120												
				140												
Cadmium sulfate	$CdSO_4$			20	+	+		+	+			+	+			
				40	+	+		+	+			+	+			
				60	+	+		+	+			+	+			
				80		+			+				+			
				100												
				120												
				140												
Calcium acetate	$(CH_5COO)_2Ca$		saturated	20	+	+	+	+	+	+		+	+			
				40	+	+		+	+	+		+				
				60		+		+	+	+		+				
				80		+										
				100												
				120												
				140												

(Courtesy George Fischer Engineering Handbook)

Aggressive Media					Chemical Resistance											
Medium	Formula	Boiling point °C	Concentration	Temperature °C	PVC	CPVC	ABS	PE	PP-H	PVDF (SYGEF)		EPDM	FPM	NBR	CR	CSM
Calcium bisulphite	$Ca(HSO_3)_2$		cold saturated, aqueous	20	+	+	+			+		+	+	-	O	+
				40		+	+			+			+			
				60		+				+			+			
				80		+				+			+			
				100						+			+			
				120						+						
				140												
Calcium carbonate	$CaCO_3$			20	+	+		+	+	+		+	+			
				40	+	+		+	+	+		+	+			
				60	+	+		+	+	+		+	+			
				80		+			+	+		+	+			
				100						+						
				120												
				140												
Calcium chlorate	$Ca(ClO_3)_2$			20	+	+		+	+	+						
				40	+	+		+	+	+						
				60	+	+		+	+	+						
				80		+										
				100												
				120												
				140												
Calcium chloride	$CaCl_2$	125	saturated, aqueous, all	20	+	+	+	+	+	+		+	+	+	+	+
				40	+	+	+	+	+	+		+	+	+	+	+
				60	O	+		+	+	+		+	+	+	+	+
				80		+			+	+		+	+	O	O	+
				100					+	+		O	+			+
				120						+						
				140												
Calcium fluoride	CaF_2			20	+			+	+				+			
				40	+			+	+				+			
				60	+			+	+				+			
				80					+							
				100												
				120												
				140												
Calcium hydrogencarbonate				20						+		+	+			
				40						+		+	+			
				60						+		+	+			
				80						+		+	+			
				100						+			+			
				120												
				140												
Calcium hydrogensulfide	$Ca(SH)_2$			20		+				+		+				
				40		+				+		+				
				60		+				+		+				
				80		+				+						
				100						+						
				120												
				140												
Calcium hydrosulfite	$Ca(HSO_3)_2$		saturated	20						+						
				40						+						
				60						+						
				80						+						
				100						+						
				120												
				140												

(Courtesy George Fischer Engineering Handbook)

Aggressive Media					Chemical Resistance											
Medium	Formula	Boiling point °C	Concentration	Temperature °C	PVC	CPVC	ABS	PE	PP-H	PVDF (SYGEF)		EPDM	FPM	NBR	CR	CSM
Calcium hydroxide	$Ca(OH)_2$	100	saturated, aqueous	20	+	+	+	+	+	O		+	+	+	+	+
				40	+	+	+	+	+	-		+	+	+	+	+
				60	+	+	+	+	+			+	+	O	+	+
				80		+			+			+	+			+
				100									+			+
				120												
				140												
Calcium nitrate	$Ca(NO_3)_2$	115	50%, aqueous	20	+	+	+	+	+	+		+	+	+	+	+
				40	+	+	+	+	+	+		+	+	+	+	+
				60		+		+	+	+		+	+			+
				80		+				+			+			
				100						+						
				120												
				140												
Calcium phosphate	$Ca(H_2PO_4)_2$			20						+						
	$CaHPO_4$			40						+						
	$Ca_3(PO_4)_2$			60						+						
				80						+						
				100						+						
				120												
				140												
Calcium sulfide	Ca_5			20	+				+			+				
				40	+				+							
				60	+				+							
				80					+							
				100												
				120												
				140												
Calcium sulphate	$CaSO_4$		suspensions	20	+	+				+		+				
				40	+	+				+		+				
				60	+	+				+		+				
				80		+				+						
				100						+						
				120												
				140												
Calcium sulphite	$Ca(HSO_3)_2$		aqueous, cold saturated	20	+			+	+			+				
				40	+			+	+							
				60	+			+	+							
				80					+							
				100												
				120												
				140												
Calcium tungstate				20						+						
				40						+						
				60						+						
				80												
				100												
				120												
				140												
Calciumbromide	$CaBr_2$			20	+	+		+	+			+	+			
				40	+	+		+	+			+	+			
				60	+	+		+	+			+	+			
				80		+			+							
				100												
				120												
				140												

(Courtesy George Fischer Engineering Handbook)

Aggressive Media					Chemical Resistance											
Medium	Formula	Boiling point °C	Concentration	Temperature °C	PVC	CPVC	ABS	PE	PP-H	PVDF (SYGEF)		EPDM	FPM	NBR	CR	CSM
Calciumlactate	$(CH_3COO)_2Ca$		saturated	20				+	+	+		+				
				40				+	+	+		+				
				60				+	+	+		+				
				80					+	+						
				100						+						
				120												
				140												
Caprolactam	$C_6H_{11}NO$			20		-										
				40												
				60												
				80												
				100												
				120												
				140												
Caprolactone	$C_6H_{10}O_2$			20		-										
				40												
				60												
				80												
				100												
				120												
				140												
Carbon dioxide -carbonic acid	CO_2		technically pure, anhydrous	20	+	+	+	+	+	+		+	+	+	+	+
				40	+	+	+	+	+	+		+	+	+	+	+
				60	+	+	+	+	+	+			+	+	+	+
				80		+			+	+			+			+
				100						+						
				120												
				140												
Carbon disulphide	CS_2	46	technically pure	20	-	-	-	O	O	+			+	-	-	-
				40												
				60												
				80												
				100												
				120												
				140												
Carbon tetrachloride	CCl_4	77	technically pure	20	-	-	-	-	-	+		-	+	-	-	-
				40												
				60												
				80												
				100												
				120												
				140												
Carbonic acid				20	+	+		+	+	+		+	+			
				40	+	+		+	+	+		+	+			
				60	+	+		+	+	+		+	+			
				80		+			+	+		+	+			
				100						+			+			
				120												
				140												
Caro's acid	see Peroxomonosulfuric acid															
Casein				20						+						
				40						+						
				60						+						
				80												
				100												
				120												
				140												

(Courtesy George Fischer Engineering Handbook)

Aggressive Media					Chemical Resistance											
Medium	Formula	Boiling point °C	Concentration	Temperature °C	PVC	CPVC	ABS	PE	PP-H	PVDF (SYGEF)		EPDM	FPM	NBR	CR	CSM
Cäsium chloride	ClCs			20						+						
				40						+						
				60						+						
				80												
				100												
				120												
				140												
Cäsiumhydroxide	CsOH			20						+						
				40						+						
				60						+						
				80												
				100												
				120												
				140												
Caustic potash solution (potassium hydroxide)	KOH	131	50%, aqueous	20	+	+	+	+	+	-		+	-	O	-	+
				40	+	+	+	+	+			+		-		O
				60	O	+	+	+	+			+				O
				80		+			+			O				-
				100					+							
				120												
				140												
Caustic soda solution	NaOH		50%, aqueous	20	+	+	+	+	+	O		+	-	O	-	+
				40	+	+		+	+			+		-		O
				60		+		+	+			+				-
				80					+							
				100												
				120												
				140												
Cerium (III) -chloride	$CeCl_3$			20						+						
				40						+						
				60						+						
				80												
				100												
				120												
				140												
Chloral hydrate	$CCl_3\text{-}CH(OH)_2$	98	technically pure	20	-		-	+	O	-		O	O	-	O	+
				40				+								
				60				+	-							
				80												
				100												
				120												
				140												
Chloric acid (SpRB)	$HClO_3$		10%, aqueous	20	+	+	-	+	-	+		+	-	-	-	+
				40	+	+		+		+		+				+
				60	O	+						+				+
				80												
				100												
				120												
				140												
Chloric acid (SpRB)	$HClO_3$		20%, aqueous	20	+	+	-	O	-	+		+	-	-	-	+
				40	+	+						+				+
				60	O	+										
				80												
				100												
				120												
				140												

(Courtesy George Fischer Engineering Handbook)

Aggressive Media					Chemical Resistance											
Medium	Formula	Boiling point °C	Concentration	Temperature °C	PVC	CPVC	ABS	PE	PP-H	PVDF (SYGEF)		EPDM	FPM	NBR	CR	CSM
Chlorosulphonic acid	$ClSO_3H$	158	technically pure	20	O	-	-	-	-	O		-	-	-	-	-
				40						-						
				60												
				80												
				100												
				120												
				140												
Chrome alum (chromium potassium sulphate)	$KCr(SO_4)_2$		cold saturated, aqueous	20	+	+	+	+	+	+		+	+	+	+	+
				40	+	+	+	+	+	+		+	+	+	+	+
				60	+	+		+	+			+	+	+	+	+
				80								+	+			
				100									+			
				120												
				140												
Chromic acid (SpRB)	CrO_3+H_2O		up to 50%, aqueous	20	O	+	-	O	O	+		O	+	-	-	O
				40	O	O		-	-	+		O	+			O
				60	-	O				+		O	+			O
				80		-				+						
				100						O						
				120						O						
				140												
Chromic acid (SpRB)			all, aqueous	20	O	O	-	O	O	+			+	-	-	O
				40						+			+			O
				60						+			O			O
				80						O						
				100						O						
				120												
				140												
Chromic acid + sulphuric acid + water (SpRB)	CrO_3 H_2SO_4 H_2O		50 g 15 g 35 g	20	+	+	-	-	-	+		O	+	-	-	O
				40	+	+				+		O	+			O
				60	O	+				+			+			
				80		O				O						
				100												
				120												
				140												
Chromium (II) -chloride				20	+					+						
				40	+					+						
				60	+					+						
				80						+						
				100						+						
				120												
				140												
Chromium (II) -fluoride	CrF_3			20						+						
				40						+						
				60						+						
				80												
				100												
				120												
				140												
Chromium (III) -chloride	$CrCl_3$			20	+					+						
				40	+					+						
				60	+					+						
				80						+						
				100						+						
				120												
				140												

(Courtesy George Fischer Engineering Handbook)

Aggressive Media					Chemical Resistance											
Medium	Formula	Boiling point °C	Concentration	Temperature °C	PVC	CPVC	ABS	PE	PP-H	PVDF (SYGEF)		EPDM	FPM	NBR	CR	CSM
Chlorine	Cl_2		moist, 97%, gaseous	20	-	-	-	-	-	-		-	+	-	-	O
				40												
				60												
				80												
				100												
				120												
				140												
Chlorine	Cl_2		anhydrous, technically pure	20	-	-	-	O	-	+		O	+	-	-	O
				40				O		+						
				60				-		+						
				80						+						
				100						O						
				120												
				140												
Chlorine	Cl_2		liquid, technically pure	20	-	-	-	-	-	+		-	O	-	-	-
				40												
				60												
				80												
				100												
				120												
				140												
Chlorine water (SpRB)	Cl_2 H_2O		saturated	20	+	+	O	O	O	O		O	O	-	O	-
				40	+	+		O								
				60	O	O										
				80		-										
				100												
				120												
				140												
Chloroacetic acid, mono (SpRB)	$ClCH_2COOH$		50%, aqueous	20	+	-	-	+	+	+		O	-	-	-	O
				40	+			+	+	O						
				60				+	+	-						
				80												
				100												
				120												
				140												
Chloroacetic acid, mono (SpRB)	$ClCH_2COOH$	188	technically pure	20	+	-	-	+	+	-		O	-	-	-	O
				40	+			+	+							
				60	O			+	+							
				80												
				100												
				120												
				140												
Chlorobenzene	C_6H_5Cl	132	technically pure	20	-	-	-	O	+	+		-	-	-	-	O
				40						+						
				60						O						
				80						-						
				100												
				120												
				140												
Chloroethanol	$ClCH_2$-CH_2OH	129	technically pure	20	-		-	+	+	+		O	-	+	-	O
				40				+	+	O						
				60				+	+	O						
				80						-						
				100												
				120												
				140												

(Courtesy George Fischer Engineering Handbook)

Aggressive Media					Chemical Resistance											
Medium	Formula	Boiling point °C	Concentration	Temperature °C	PVC	CPVC	ABS	PE	PP-H	PVDF (SYGEF)		EPDM	FPM	NBR	CR	CSM
Chromium (III) -nitrate	$Cr(NO_3)_3$			20	+					+						
				40	+					+						
				60	+					+						
				80						+						
				100						+						
				120												
				140												
Chromium (III) -sulfate	$Cr_2(SO_4)_3$			20	+					+						
				40	+					+						
				60	+					+						
				80						+						
				100						+						
				120												
				140												
Cider				20	+	+	+	+	+	+		+	+	+	+	+
				40			+			+						
				60						+						
				80												
				100												
				120												
				140												
Citric acid		Fp. *153	10%, aqueous	20	+	+	+	+	+	+		+	+	+	+	+
				40	+	+	+	+	+	+		+	+	+	+	+
				60	O	+	+	+	+	+		+	+	O	+	+
				80					+	+						
				100					+	+						
				120												
				140												
Citric acid				20	+			+	+							
				40	+			+	+							
				60				+	+							
				80					+							
				100												
				120												
				140												
Citric acid up to 10 %				20						+						
				40						+						
				60						+						
				80												
				100												
				120												
				140												
Coal gas, benzene free				20	+	+	+	+	+	+			+	+	O	+
				40												
				60												
				80												
				100												
				120												
				140												
Coconut fat alcohol (SpRB)			technically pure	20	+	-	-	+	+	+			+	+	+	+
				40	+			O	+	+			+	+	O	O
				60	O				O	+			+	+		
				80												
				100												
				120												
				140												

(Courtesy George Fischer Engineering Handbook)

Aggressive Media					Chemical Resistance											
Medium	Formula	Boiling point °C	Concentration	Temperature °C	PVC	CPVC	ABS	PE	PP-H	PVDF (SYGEF)		EPDM	FPM	NBR	CR	CSM
Compressedair, containing oil (SpRB)				20	-	-	-	+	O	+		-	+	+	+	+
				40				+		+						
				60						+						
				80												
				100												
				120												
				140												
Copper salts	$CuCl$, $CuCl_2$, CuF_2, $Cu(NO_3)_2$, $CuSO_4$, $Cu(CN)_2$		all, aqueous	20	+	+	+	+	+	+		+	+	+	+	+
				40	+	+	+	+	+	+		+	+	+	+	+
				60	O	+	+	+	+	+			+	O	+	O
				80		+			+	+			+			
				100						+						
				120												
				140												
Corn oil (SpRB)			technically pure	20	O	O	O	+	+	+			+	+	O	+
				40		O		+	+	+			+	+	-	+
				60				O	O	+			+	+		O
				80						+						
				100												
				120												
				140												
Cresol	$HO-C_6H_4-CH_3$		cold saturated, aqueous	20	O	-	-	+	+	+			+	O	-	O
				40				+	+	+			+	O		
				60						+						
				80						O						
				100												
				120												
				140												
Crotonic aldehyde	$CH_3-CH=CH-CHO$	102	technically pure	20	-	-	-	+	+	+		+	+	+	+	+
				40						O						
				60						-						
				80												
				100												
				120												
				140												
Cyclohexane (Q/E)	C_6H_{12}	81	technically pure	20	-	-	-	+	+	+		-	+	+	-	-
				40				+		+						
				60				+		+						
				80						+						
				100												
				120												
				140												
Cyclohexanol (SpRB)	$C_6H_{12}O$	161	technically pure	20	+	+	-	+	+	+		-	+	O	+	+
				40	+	+		+	+	+						
				60	+	+		+	O	O						
				80		O				O						
				100						-						
				120												
				140												
Cyclohexanone	$C_6H_{10}O$	155	technically pure	20	-	-	-	+	+	+		O	-	-	-	-
				40				O	O	O						
				60				O	O	-						
				80												
				100												
				120												
				140												

(Courtesy George Fischer Engineering Handbook)

Aggregive Media					Chemical Resistance											
Medium	Formula	Boiling point °C	Concentration	Temperature °C	PVC	CPVC	ABS	PE	PP-H	PVDF (SYGEF)		EPDM	FPM	NBR	CR	CSM
Densodrine W				20	+	+	O			+			+	+	+	
				40	+	+										
				60	+	+										
				80												
				100												
				120												
				140												
Detergents (SpRB)	see washing powder		for usual washing lathers													
Dextrine	$(C_6H_{10}O_5)_n$		usual commercial	20	+	+	+	+	+	+		+	+	+	+	+
				40	+	+	+	+		+		+	+	+	+	+
				60	+	+	+	+		+		+	+	+	+	+
				80		+				+						
				100						+						
				120						+						
				140												
Dextrose	siehe Glucose			20	+	+	+	+	+	+		+	+			
				40	+	+		+	+	+		+	+			
				60	+	+		+	+	+		+	+			
				80		+			+	+		+	+			
				100						+			+			
				120												
				140												
Dibuttyl ether	$C_4H_9OC_4H_9$	142	technically pure	20	-	-	-	O	O			-	+	+	-	O
				40									+	O		O
				60				-	-				O	-		O
				80												
				100												
				120												
				140												
Dibutyl phthalate	$C_6H_4(COOC_4H_9)_2$	340	technically pure	20	-	-	-	+	+	+		O	O	-	-	-
				40				O	O	+						
				60				O	O	O						
				80												
				100												
				120												
				140												
Dibutyl sebacate	$C_8H_{16}(COOC_4H_9)_2$	344	technically pure	20	-	-	-	+	+	+		+	+	-	-	-
				40												
				60												
				80												
				100												
				120												
				140												
Dichlorbenzol	$C_6H_4Cl_2$	180	technically pure	20	-	-	-									
				40												
				60												
				80												
				100												
				120												
				140												
Dichloroacetic acid	$Cl_2CHCOOH$	194	technically pure	20	+	-	-	+	+	+		+	O	-	O	+
				40	+			+	+	+		+	-		-	O
				60	O			O	O	O		+				-
				80						-						
				100												
				120												
				140												

(Courtesy George Fischer Engineering Handbook)

Aggressive Media					Chemical Resistance											
Medium	Formula	Boiling point °C	Concentration	Temperature °C	PVC	CPVC	ABS	PE	PP-H	PVDF (SYGEF)		EPDM	FPM	NBR	CR	CSM
Dichloroacetic acid (SpRB)	$Cl_2CHCOOH$		50%, aqueous	20	+	-	-	+	+	+		+	O	-	+	+
				40	+			+	+	+		+	O		O	+
				60	O			+	+	+		+	-		-	O
				80						O						
				100						-						
				120												
				140												
Dichloroacetic acid methyl ester	$Cl_2CHCOOCH_3$	143	technically pure	20	-	-	-	+	+	O		+	-	-	-	+
				40				+	+			+				+
				60				+	+			O				O
				80												
				100												
				120												
				140												
Dichloroethan	Ethylene chloride															
Dichloroethylene	$ClCH{=}CHCl$	60	technically pure	20	-	-	-	-	O	+		-	O	-	-	-
				40						+						
				60												
				80												
				100												
				120												
				140												
Dichloromethane				20	-	-	-									
				40												
				60												
				80												
				100												
				120												
				140												
Diesel oil (SpRB, Q/E)				20	+	+	O	+	O	+		-	+	+	O	O
				40	+	+				+			+	+		-
				60				O		+						
				80						+						
				100						+						
				120						+						
				140												
Diethyl ether				20	-	-	-									
				40												
				60												
				80												
				100												
				120												
				140												
Diethylamine	$(C_2H_5)_2NH$	56	technically pure	20	O	-	-	+	+	+		O	-	-	-	-
				40						O						
				60						-						
				80												
				100												
				120												
				140												
Diethylene glycol butyl ether				20	-	-	-									
				40												
				60												
				80												
				100												
				120												
				140												

(Courtesy George Fischer Engineering Handbook)

Aggressive Media					Chemical Resistance											
Medium	Formula	Boiling point °C	Concentration	Temperature °C	PVC	CPVC	ABS	PE	PP-H	PVDF (SYGEF)		EPDM	FPM	NBR	CR	CSM
Diglycolic acid (SpRB)	HOOC-CH_2-O-CH_2-COOH	Fp*., 148	30%, aqueous	20	+	+	+	+	+	+		+	O	+	+	O
				40	+	+		+	+							
				60	O	+		+	+							
				80												
				100												
				120												
				140												
Di-isobutyl ketone	$[(CH_3)2CHCH_2]_2CO$	124	technically pure	20	-	-	-	+	+	+		O	-	-	-	-
				40						+						
				60				-	-	O						
				80												
				100												
				120												
				140												
Dimethyl formamide	$(CH_3)_2CHNO$	153	technically pure	20	-	-	-	+	+	-		O	-	O	+	+
				40				+	+							
				60				O	+							
				80												
				100												
				120												
				140												
Dimethylamine	$(CH_3)_2NH$	7	technically pure	20	O	-	-	+	+	O		O	-	-	-	-
				40						-						
				60				O								
				80												
				100												
				120												
				140												
Dimethylphthalate (DMP)	$C_6H_4(CH_3)_2$			20	-	-	-									
				40												
				60												
				80												
				100												
				120												
				140												
Dinonylphthalate (DNP)			technically pure	20	-	-	-	O	+			O	+	-	-	-
				40												
				60												
				80												
				100												
				120												
				140												
Dioctylephthalate (SpRB) (DOP)			technically pure	20	-	-	-	O	+			O	+	-	-	-
				40												
				60					-							
				80												
				100												
				120												
				140												
Dioxane	$C_4H_8O_2$	101	technically pure	20	-	-	-	+	O	-			-	O	-	-
				40				+	O							
				60				+	O							
				80					-							
				100												
				120												
				140												
Drinking water	see water															

(Courtesy George Fischer Engineering Handbook)

Aggregssive Media					Chemical Resistance											
Medium	Formula	Boiling point °C	Concentration	Temperature °C	PVC	CPVC	ABS	PE	PP-H	PVDF (SYGEF)		EPDM	FPM	NBR	CR	CSM
Ethanolamine	see Annino ethanol															
Ethyl acetate	CH_3COOCH_2-CH_3	77	technically pure	20	-	-	-	+	+	O		+	-	-	-	-
				40				O	O							
				60				O	O							
				80												
				100												
				120												
				140												
Ethyl alcohol + acetic acid (fermentation mixture)			technically pure	20	+	O	-	+	+	+		O	O	O	+	+
				40	+			+		+		O	O	O	+	+
				60	O			+		+			O		+	+
				80						O						
				100												
				120												
				140												
Ethyl alcohol (Ethnoc) (SpRB)	CH_3-CH_2-OH	78	technically pure, 96%	20	+	O	-	+	+	+		+	O	O	+	+
				40	+			+	+	O		+	O			
				60	O			+	+	-		+	O			
				80					+							
				100												
				120												
				140												
Ethyl benzene	C_6H_5-CH_2CH_3	136	technically pure	20	-	-	-	O	O	O		-	+	-	-	-
				40												
				60					-							
				80												
				100												
				120												
				140												
Ethyl chloride	CH_3-CH_2Cl	12	technically pure	20	-	-	-	O	O	O		-	O	-	-	-
				40												
				60												
				80												
				100												
				120												
				140												
Ethyl ether	CH_3CH_2-O-CH_2CH_3	35	technically pure	20	-	-	-	+	O	+		-	-	-	-	-
				40												
				60												
				80												
				100												
				120												
				140												
Ethylenchloride (1,2-Dichloroethane)				20	-	-	-									
				40												
				60												
				80												
				100												
				120												
				140												
Ethylene chloride	$ClCH_2$-CH_2Cl	83	technically pure	20	-	-	-	O	O	+		-	+	O	O	-
				40						+			+	-	-	
				60						+			O			
				80						O						
				100						-						
				120												
				140												

(Courtesy George Fischer Engineering Handbook)

Aggregive Media					Chemical Resistance											
Medium	Formula	Boiling point °C	Concentration	Temperature °C	PVC	CPVC	ABS	PE	PP-H	PVDF (SYGEF)		EPDM	FPM	NBR	CR	CSM
Ethylene diamine	$H_2N\text{-}CH_2\text{-}CH_2\text{-}NH_2$	117	technically pure	20	○	-	-	+	+	○		+	○	+	+	○
				40				+	+	○			○	○	○	○
				60				+	+	-			-	-	-	-
				80												
				100												
				120												
				140												
Ethylene glycol (SpRB)	$HO\text{-}CH_2\text{-}CH_2\text{-}OH$	198	technically pure	20	+	○	-	+	+	+		+	+	+	+	+
				40	+			+	+	+		+	+	+	+	+
				60	+			+	+	+		+	+	○	○	+
				80					+	+		+	○			○
				100					+	+						
				120						+						
				140												
Ethylene glycol	CH_2OHCH_2OH	198	technically pure	20	+		-	+	+	+		+	+	+	+	+
				40	+			+	+	+		+	+	+	+	+
				60	+			+	+	+		+	+	○	○	+
				80					+	+			○			○
				100						+						
				120												
				140												
Ethylene oxide	$CH_2\text{-}CH_2$	10	technically pure, moist	20	-	-	-	-	○	+		○	-	-	-	-
				40												
				60												
				80												
				100												
				120												
				140												
Ethylenediaminetetra-acetic acid (EDTA)				20				+	+	+		+				
				40												
				60												
				80												
				100												
				120												
				140												
Fatty acids $>C_6$ (SpRB)	R-COOH		technically pure	20	+	+	-	+	+	+		+	+	○	○	-
				40	+	+		+	+	+						
				60	+	+		○	+	+						
				80						+						
				100												
				120												
				140												
Fatty alcohol sulphonates (SpRB)			aqueous	20	+	+		+	+	+		+	+	+	+	+
				40	+	+		+	+	+		+	+	+	+	+
				60	○	○		+	○	+		+	+	+	+	+
				80						+						
				100						+						
				120												
				140												
Fertilizers			aqueous	20	+	+	○	+	+	+		+	+	+	+	+
				40	+	+		+	+	+		+	+	+	+	+
				60	○	+		+	+	+		+	+	+	+	+
				80		+				+			+			+
				100									+			
				120												
				140												

(Courtesy George Fischer Engineering Handbook)

Aggressive Media					Chemical Resistance											
Medium	Formula	Boiling point °C	Concentration	Temperature °C	PVC	CPVC	ABS	PE	PP-H	PVDF (SYGEF)		EPDM	FPM	NBR	CR	CSM
Fluorine	F_2		technically pure	20	-	-	-	-	-	-		-	-	-	-	-
				40												
				60												
				80												
				100												
				120												
				140												
Fluorosilicic acid (Q/E)	H_2SiF_6		32%, aqueous	20	+	+	+	+	+	+		+	O	O	O	+
				40	+	+	+	+		+				-	-	O
				60	+	+	+	+		+						-
				80						+						
				100						+						
				120												
				140												
Formaldehyde (SpRB)	HCHO		40%, aqueous	20	+	+	+	+	+	+		+	+	+	+	+
				40	+	+	+	+	+	+		+	+	+	+	+
				60			+	+		+		+	+	O	O	O
				80						+						
				100												
				120												
				140												
Formamide	$HCONH_2$	210	technically pure	20	-	-	-	+	+			+	O	+	+	
				40				+	+							
				60				+	+							
				80												
				100												
				120												
				140												
Formic acid (SpRB)	HCOOH		up to 50%, aqueous	20	+	-	O	+	+	+		+	+	-	+	+
				40	+			+	+	+		+	+		+	+
				60	O			+	O	+		O	O		O	+
				80						+			-			O
				100						+						
				120												
				140												
Formic acid (SpRB)	HCOOH	101	technically pure	20	+	-	-	+	+	+		+	+	-	+	+
				40	O			+	O	+		+			O	+
				60	-			+	-	+		+			-	+
				80						+		O				O
				100						+						
				120												
				140												
Formic acid (SpRB)			25%	20	+	+		+	+	+		+				
				40	+	+		+	+	+		+				
				60	+	+		+	+	+		+				
				80					+	+						
				100												
				120												
				140												
Freon 113	see trifluoro, trichlorethane	48														
Frigen 12 (D/P)	see Freon 12	-30	technically pure													

(Courtesy George Fischer Engineering Handbook)

Aggressive Media					Chemical Resistance											
Medium	Formula	Boiling point °C	Concentration	Temperature °C	PVC	CPVC	ABS	PE	PP-H	PVDF (SYGEF)		EPDM	FPM	NBR	CR	CSM
Fruit juices (SpRB)				20	+	+	+	+	+	+		+	+	+	+	+
				40	+	+	+	+	+	+		+	+	+	+	+
				60	+	+	+	+	+	+		+	+	+	+	+
				80		+			+	+		+	+	+	+	+
				100						+		+	+	+		+
				120						+			+			
				140												
Fruit pulp				20	+	+	+	+	+			+	+	+	+	+
				40			+	+	+							
				60				+	+							
				80												
				100												
				120												
				140												
Fuel oil				20	+	+		O	O	+		-	+	+	+	-
				40	O	+		-	-	+			+	+	+	
				60						+			+	+	O	
				80						+						
				100						+						
				120												
				140												
Furfuryl alcohol (SpRB)	$C_5H_6O_2$	171	technically pure	20	-	-	-	+	+	+		O	-	-	O	O
				40				+		+						
				60				+	O	O						
				80						-						
				100												
				120												
				140												
Gasoline (SpRB)	C_5H_{12} to $C_{12}H_{26}$	80-130	free of lead and aromatic compounds	20	+	+	-	+	O	+		-	+	+	-	O
				40	+	+		+		+			+	+		-
				60	+	+		O	-	+			+	+		
				80						+						
				100						+						
				120						+						
				140												
Gelatin			all, aqueous	20	+	+	+	+	+	+		+	+	+	+	+
				40	+	+	+	+	+	+		+	+	+	+	+
				60		+	+	+	+	+						
				80						+						
				100												
				120												
				140												
Glucose	$C_6H_{12}O_6$	Fp*., 148	all, aqueous	20	+	+	+	+	+	+		+	+	+	+	+
				40	+	+	+	+	+	+		+	+	+	+	+
				60	O	+	+	+	+	+		+	+	+	+	+
				80		+			+	+		+	+	+	+	+
				100						+			+			
				120												
				140												
Glycerol	HO-CH_2-CH(OH)-CH_2OH	290	technically pure	20	+	+	+	+	+	+		+	+	+	+	+
				40	+	+	+	+	+	+		O	+	+	+	+
				60	+	+		+	+	+		O	O	+	+	+
				80		+			+	+			-	O	+	+
				100					+	+					O	O
				120						+						
				140												

(Courtesy George Fischer Engineering Handbook)

Aggressive Media					Chemical Resistance											
Medium	Formula	Boiling point °C	Concentration	Temperature °C	PVC	CPVC	ABS	PE	PP-H	PVDF (SYGEF)		EPDM	FPM	NBR	CR	CSM
Glycocoll (SpRB)	NH_2-CH_2-COOH	Fp.* 233	10%, aqueous	20	+	+	+	+	+	+			+	+	+	+
				40	+	+	+	+	+	+			+	O	+	O
				60		+				+						
				80						+						
				100												
				120												
				140												
Glycol	see Ethylene glycol															
Glycolic acid	HO-CH_2-COOH	Fp.*, 80	37%, aqueous	20	+	-	+	+	+	+			+	+	+	+
				40			+	+		+						
				60				+		+						
				80						+						
				100						+						
				120												
				140												
Heptane (SpRB)	C_7H_{16}	98	technically pure	20	+	O	-	+	+	+		-	+	+	+	+
				40						+			+	+	+	O
				60				O	O	+			+	+	+	-
				80						+						
				100						+						
				120												
				140												
Hexane (SpRB)	C_6H_{14}	69	technically pure	20	+	O	-	+	+	+		-	+	+	+	+
				40						+			+	+	+	O
				60				O	O	+			+	+	+	-
				80						+						
				100						+						
				120												
				140												
Hydrazine hydrate (SpRB)	H_2N-NH_2 . H_2O	113	aqueous	20	+	-	-	+	+	-		+	O	-	-	+
				40				+	+							
				60				+	+							
				80												
				100												
				120												
				140												
Hydrobromic acid (SpRB)	HBr	124	aqueous, 50%	20	+	+	+	+	+	+		+	+	O	+	+
				40	+	+	+	+	+	+		+	+	-	+	+
				60	+	+		+	+	+		O	+		O	+
				80		O				+		-	O		-	O
				100						+			-			-
				120												
				140												
Hydrochloric acid (Q/E, D/P)	HCl		up to 38%	20	+	+	-	+	O	+		+	+	-	O	+
				40	+	+		+		+			+			
				60	+	+				+						
				80						+						
				100						+						
				120												
				140												
Hydrochloric acid (Q/E, D/P)	HCl		5%, aqueous	20	+		+	+	+	+		+	+	O	O	+
				40	+		+	+	+	+		+	+	-	-	O
				60	O			+	+	+		+	+			-
				80					O	+			+			
				100						+						
				120						+						
				140												

(Courtesy George Fischer Engineering Handbook)

Aggressive Media					Chemical Resistance											
Medium	Formula	Boiling point °C	Concentration	Temperature °C	PVC	CPVC	ABS	PE	PP-H	PVDF (SYGEF)		EPDM	FPM	NBR	CR	CSM
Hydrochloric acid (Q/E, D/P)	HCl		10%, aqueous	20	+	+	+	+	+	+		+	+	O	O	+
				40	+	+	+	+	+	+		+	+	-	-	O
				60	O	+		+	O	+		+	+			-
				80		+			O	+			+			
				100						+						
				120						+						
				140												
Hydrochloric acid (Q/E, D/P)	HCl		up to 30%, aqueous	20	+	+	O	+	+	+		+	+	-	-	+
				40	+	+	-	+	O	+		+	+			O
				60	O	+		+	O	+		O	O			-
				80		+			-	+						
				100						+						
				120												
				140												
Hydrochloric acid (Q/E, D/P)	HCl		36%, aqueous	20	+	+	-	+	+	+		O	+	-	-	O
				40	+	+		+	O	+		O	O			-
				60	O	O		+	-	+		-	-			
				80		O				+						
				100						+						
				120												
				140												
Hydrocyanic acid	HCN	26	technically pure	20	+	+	-	+	+	+		+	+	O	O	+
				40	+	+		+	+	+		O	O	-	-	O
				60	O	+		+	+	+						
				80						+						
				100												
				120												
				140												
Hydrofluoric acid	HF			20	+	-	-	+	+	+		-	+	-	-	+
				40	O			+	+	+			+			+
				60	O			O	+	+			O			O
				80						+						
				100						+						
				120												
				140												
Hydrogen	H_2	-253	technically pure	20	+	+	+	+	+	+		+	+	+	+	+
				40	+	+	+	+	+	+		+	+	+	+	+
				60	+	+	+	+	+	+		+	+	+	+	+
				80		+				+		+	+	+	+	+
				100					-	+		+	+	+		
				120												
				140												
Hydrogen chloride (Q/E)	HCl	-85	technically pure, gaseous	20	+	+	-	+	+	+		+	+	O	O	O
				40	+	+		+	+	+		+	+	-	-	O
				60	O	+		+	+	+		+	+			-
				80		O				+						
				100						+						
				120												
				140												
Hydrogen perocide			70%	20	+	+	-	+	+	O		O	+			
				40						O						
				60												
				80												
				100												
				120												
				140												

(Courtesy George Fischer Engineering Handbook)

Aggressive Media					Chemical Resistance											
Medium	Formula	Boiling point °C	Concentration	Temperature °C	PVC	CPVC	ABS	PE	PP-H	PVDF (SYGEF)		EPDM	FPM	NBR	CR	CSM
Hydrogen peroxide (SpRB)	H_2O_2		50%, aqueous	20	+	+	-	+	+	O		O	+			
				40												
				60												
				80												
				100												
				120												
				140												
Hydrogen peroxide (SpRB)	H_2O_2		10%, aqueous	20	+	+	-	+	+	O		O	+	O	-	+
				40	+			+	+	O		O	O	-		+
				60	O			+	+	-		-	-			O
				80												-
				100												
				120												
				140												
Hydrogen peroxide (SpRB)	H_2O_2	139	90%, aqueous	20	+		-	+	-	O			O	-	-	O
				40												
				60				-								
				80												
				100												
				120												
				140												
Hydrogen peroxide (SpRB)	H_2O_2	105	30%, aqueous	20	+	+	-	+	+	O		O	+	-	-	+
				40												O
				60												-
				80												
				100												
				120												
				140												
Hydrogen sulphide	H_2S		technically pure	20	+	+	+	+	+	+		+	+	+	O	+
				40	+	+	+	+	+	+		-	+	O	-	O
				60	+	+		O	+	+			O	-		O
				80						+			-			-
				100						+						
				120						+						
				140												
Hydrogen sulphide	H_2S		saturated, aqueous	20	+	+	+	+	+	+		+	+	-	+	+
				40	+	+	+	+	+	+		-	+	-	-	+
				60	O	+		+	+	+			+			O
				80						+			O			-
				100						+						
				120						+						
				140												
Hydroquinone	$C_6H_4(OH)_2$		saturated	20	+	+		+	+			+				
				40	+	+		+	+							
				60				+	+							
				80					+							
				100												
				120												
				140												
Hydrosulphite	see Sodium dithione															
Hydroxylamine sulfate				20	+			+				+				
				40	+			+								
				60	+			+								
				80												
				100												
				120												
				140												

(Courtesy George Fischer Engineering Handbook)

Aggressive Media					Chemical Resistance											
Medium	Formula	Boiling point °C	Concentration	Temperature °C	PVC	CPVC	ABS	PE	PP-H	PVDF (SYGEF)		EPDM	FPM	NBR	CR	CSM
Hydroxylamine sulphate	$(NH_3OH)_2SO_4$		all, aqueous	20	+	+	-	+	+			+	+	+	O	+
				40	+	+		+	+				+	O		+
				60				+	+							
				80												
				100												
				120												
				140												
Iodine-potassium iodide solution (lugol's solution)				20	+	-	-			+			+			
				40												
				60												
				80												
				100												
				120												
				140												
Iodium	I_2	185	100%	20	-	-	-			+			+			
				40												
				60												
				80												
				100												
				120												
				140												
Iron (II) -chloride			saturated	20	+	+	+	+	+	+		+	+			
				40	+	+		+	+	+		+	+			
				60	+	+		+	+	+		+	+			
				80		+			+	+			+			
				100						+			+			
				120												
				140												
Iron (II) -chloride	$FeCl_2$		saturated	20	+	+	+	+	+	+		+	+			
				40	+	+		+	+	+		+	+			
				60	+	+		+	+	+		+	+			
				80		+			+	+		+	+			
				100						+			+			
				120												
				140												
Iron (II) -nitrate	$Fe(NO_3)_2$		saturated	20	+	+	+	+	+	+		+	+			
				40	+	+		+	+	+		+	+			
				60	+	+		+	+	+		+	+			
				80		+			+	+			+			
				100						+			+			
				120												
				140												
Iron (III) -chloride	$FeCl_3$		saturated	20	+	+	+	+	+	+		+	+			
				40	+	+		+	+	+		+	+			
				60	+	+		+	+	+		+	+			
				80		+			+	+		+	+			
				100						+			+			
				120												
				140												
Iron (III) -chloride			saturated	20	+	+	+	+	+	+		+	+			
				40	+	+		+	+	+		+	+			
				60	+	+		+	+	+		+	+			
				80		+			+	+		+	+			
				100						+			+			
				120												
				140												

(Courtesy George Fischer Engineering Handbook)

Aggressive Media					Chemical Resistance											
Medium	Formula	Boiling point °C	Concentration	Temperature °C	PVC	CPVC	ABS	PE	PP-H	PVDF (SYGEF)		EPDM	FPM	NBR	CR	CSM
Iron (III) -chloridsulfate			saturated	20	+	+	+	+	+	+		+	+			
				40	+	+		+	+	+		+	+			
				60	+	+		+	+	+		+	+			
				80		+			+	+		+	+			
				100						+			+			
				120												
				140												
Iron (III) -nitrate			saturated	20	+	+	+	+	+	+		+	+			
				40	+	+		+	+	+		+	+			
				60	+	+		+	+	+		+	+			
				80		+			+	+		+	+			
				100						+			+			
				120												
				140												
Iron (III) -nitrate	$Fe(NO_3)_3$		saturated	20	+	+	+	+	+	+		+	+			
				40	+	+		+	+	+		+	+			
				60	+	+		+	+	+		+	+			
				80		+			+	+		+	+			
				100						+			+			
				120												
				140												
Iron (III) -sulfate	$Fe_2(SO_4)_3$		saturated	20	+	+	+	+	+	+		+	+			
				40	+	+		+	+	+		+	+			
				60	+	+		+	+	+		+	+			
				80		+			+	+		+	+			
				100						+			+			
				120												
				140												
Iron (III) -sulfate			saturated	20	+	+	+	+	+	+		+	+			
				40	+	+		+	+	+		+	+			
				60	+	+		+	+	+		+	+			
				80		+			+	+		+	+			
				100						+			+			
				120												
				140												
Iron (III) -nitrate	$FE(NO_3)3$		saturated	20	+					+		+	+			
				40	+					+		+	+			
				60	+					+		+	+			
				80						+		+	+			
				100						+			+			
				120												
				140												
Iron (II) -sulfate	$FeSO_4$		saturated	20	+	+	+	+	+	+		+	+			
				40	+	+		+	+	+		+	+			
				60	+	+		+	+	+		+	+			
				80		+			+	+		+	+			
				100						+			+			
				120												
				140												
Iron (III) -sulfate			saturated	20						+						
				40						+						
				60						+						
				80						+						
				100						+						
				120												
				140												

(Courtesy George Fischer Engineering Handbook)

Aggressive Media					Chemical Resistance											
Medium	Formula	Boiling point °C	Concentration	Temperature °C	PVC	CPVC	ABS	PE	PP-H	PVDF (SYGEF)		EPDM	FPM	NBR	CR	CSM
Iron salts			all, aqueous	20	+	+	+	+	+	+		+	+	+	+	+
				40	+	+	+	+	+	+		+	+	+	+	+
				60	O	+		+	+	+		+	+	+	+	+
				80		+			+	+		+	+			+
				100						+			+			
				120						+						
				140												
Isooctane (SpRB)	$(CH_3)_3$-C-CH_2-CH-$(CH_3)_2$	99	technically pure	20	+		-	+	+	+			+	+	+	O
				40						+						
				60				O	O	+						
				80						+						
				100						+						
				120												
				140												
Isophorone (SpRB)	$C_9H_{14}O$		technically pure	20						-						
				40												
				60												
				80												
				100												
				120												
				140												
Isopropyl alcohol (SpRB)	$(CH_3)_2$-CH-OH	82	technically pure	20				+	+	+		+				
				40						+						
				60						+						
				80						O						
				100												
				120												
				140												
Isopropyl ether	$(CH_3)_2$-CH-O-CH-$(CH_3)_2$	68	technically pure	20	-	-	-	O	O	+		O	-	-	-	-
				40						+						
				60				-	-	+						
				80												
				100												
				120												
				140												
Isopropylbenzene				20	-	-	-									
				40												
				60												
				80												
				100												
				120												
				140												
Jam, Marmalade				20	+	+	+	+	+	+		+	+	+	+	+
				40	O	+	+	+	+	+		+	+	+	+	+
				60	O	+		+	+	+		+	+	+	+	+
				80					+	+		+	+			
				100					+	+						
				120						+						
				140												
Lactic acid (SpRB)	$CH_3CHOHCOOH$		10%, aqueous	20	+	+	+	+	+	+		O	+	-	-	O
				40	O	+	O	+	+	+		O	O			O
				60	-	+	-	+	+	O		O	O			O
				80		+			+	O		-	O			
				100					+	-						
				120												
				140												

(Courtesy George Fischer Engineering Handbook)

Aggressive Media					Chemical Resistance											
Medium	Formula	Boiling point °C	Concentration	Temperature °C	PVC	CPVC	ABS	PE	PP-H	PVDF (SYGEF)		EPDM	FPM	NBR	CR	CSM
Lanolin (SpRB)			technically pure	20	+	o	+	+	+	+			+	+	+	O
				40	O		+	+	+	+			+	+	O	-
				60			+	+	+	+			+	+	-	
				80						+						
				100						+						
				120						+						
				140												
Lead acetate	$Pb(CH_3COO)_2$		aqueous, saturated	20	+	+	+	+	+	+		+	+	+	+	+
				40	+	+	+	+	+	+		+	+	+	+	+
				60	+	+	+	+	+	+		+	+	+	+	+
				80		+				+						
				100						+						
				120												
				140												
Lead salts	$PbCl_2$, $Pb(NO_3)_2$, $PbSO_4$		saturated	20		+										
				40		+										
				60		+										
				80		+										
				100												
				120												
				140												
Leadcarbonate				20	+	+		+	+	+		+				
				40						+						
				60						+						
				80												
				100												
				120												
				140												
Leadnitrate	$Pb(NO_3)2$			20		+										
				40		+										
				60		+										
				80		+										
				100												
				120												
				140												
Leadnitrate				20	+					+						
				40	+					+						
				60	+					+						
				80						+						
				100						+						
				120												
				140												
Leadtetrafluoroborate				20						+						
				40						+						
				60						+						
				80						+						
				100						+						
				120												
				140												
Linoleic acid				20						+						
				40						+						
				60						+						
				80						+						
				100						+						
				120												
				140												

(Courtesy George Fischer Engineering Handbook)

Aggressive Media					Chemical Resistance											
Medium	Formula	Boiling point °C	Concentration	Temperature °C	PVC	CPVC	ABS	PE	PP-H	PVDF (SYGEF)		EPDM	FPM	NBR	CR	CSM
Linseed oil (SpRB)			technically pure	20	+	+	+	+	+	+			+	+	O	+
				40	+	+	-	+	+	+			+	+	-	O
				60	O			+	+	+			+	+		-
				80					+	+						
				100					+	+						
				120						+						
				140												
Liqueurs				20	+			+	+	+		+	+	+	+	+
				40	+			+		+						
				60						+						
				80						+						
				100												
				120												
				140												
Liquid fertilizers				20				+	+			+				
				40				+	+			+				
				60				+	+			+				
				80					+							
				100												
				120												
				140												
Lithiumbromide	LiBr			20	+	+		+	+	+		+	+			
				40						+						
				60						+						
				80						+						
				100						+						
				120												
				140												
Lithiumsulfate				20	+	+		+	+	+		+	+			
				40						+						
				60						+						
				80												
				100												
				120												
				140												
Lubricating oils				20	+	O	-	+	O	+		-	+	+	+	+
				40	+			+		+			+	+	O	O
				60	+			O		+			+	O	-	-
				80						+			O			
				100						+			-			
				120						+						
				140												
Magnesium salts	$MgCl_2$, $MgCO_3$, $Mg(No3)_2$, $Mg(OH)_2$, $MgSO_4$		all, aqueous,saturated	20	+	+	+	+	+	+			+	+	+	+
				40	+	+	+	+	+	+			+	+	+	+
				60	O	+		+	+	+		+	+	+	+	+
				80		+			+	+		+	+			+
				100					+	+			+			
				120						+						
				140												
Magnesiumhydrogen-carbonate				20	+			+	+			+				
				40	+			+	+			+				
				60	+			+	+			+				
				80												
				100												
				120												
				140												

(Courtesy George Fischer Engineering Handbook)

Aggressive Media					Chemical Resistance											
Medium	Formula	Boiling point °C	Concentration	Temperature °C	PVC	CPVC	ABS	PE	PP-H	PVDF (SYGEF)		EPDM	FPM	NBR	CR	CSM
Maleic acid (SpRB)	$(CH\text{-}COOH)_2$	Fp.	cold saturated,	20	+	+	+	+	+	+			+	-	-	-
		*131	aqueous	40	+	+		+	+	+			+			
				60	O			+	+	+			+			
				80						+			-			
				100						+						
				120						+						
				140												
Media water or similar				20	+	+	+	+	+	+						
media				40	+	+	+	+	+	+						
				60	+	+	+	+	+	+						
				80		+			+	+						
				100					+	+						
				120						+						
				140						+						
Mercury	Hg	357	pure	20	+	+	+	+	+	+		+	+	+	+	+
				40	+			+	+	+		+	+	+	+	+
				60	+			+	+	+		+	+	+	+	+
				80						+			+			
				100						+						
				120						+						
				140												
Mercury (II) -chloride	$HgCl_2$			20	+	+	+	+	+	+		+	+	+	+	+
				40	+	+		+	+	+		+	+	+	+	+
				60	+	+		+	+	+		+	+	+	+	+
				80		+			+	+						
				100						+						
				120												
				140												
Mercury (II) -cyanide	Hg(CN)2			20	+	+	+	+	+	+		+	+	+	+	+
				40	+	+		+	+	+		+	+	+	+	+
				60	+	+		+	+	+		+	+	+	+	+
				80		+				+						
				100						+						
				120												
				140												
Mercury (II) -cyanide	$Hg(NO_3)_2$			20	+	+		+	+	+		+	+	+	+	+
				40	+	+		+	+	+		+	+	+	+	+
				60	+	+		+	+	+		+	+	+	+	+
				80		+			+	+			+			
				100						+						
				120												
				140												
Mercury (II) -sulfate				20	+	+	+	+	+	+		+	+	+	+	+
				40	+	+		+	+	+		+	+	+	+	+
				60	+	+		+	+	+		+	+	+	+	+
				80		+			+	+			+			
				100						+						
				120												
				140												
Mercury salts	$HgNO_3$, Hg Cl_2, $Hg(CN)_2$		cold saturated,	20	+	+	+	+	+	+		+	+	O	O	O
			aqueous	40	+	+		+	+	+		+	+	O	O	O
				60	O	+		+	+	+			+	-	-	-
				80						+						
				100						+						
				120						+						
				140												
Methane	see natural gas	-161	technically pure													

(Courtesy George Fischer Engineering Handbook)

Aggressive Media					Chemical Resistance											
Medium	Formula	Boiling point °C	Concentration	Temperature °C	PVC	CPVC	ABS	PE	PP-H	PVDF (SYGEF)		EPDM	FPM	NBR	CR	CSM
Methanol (SpRB)	CH_3OH	65	all	20	+	-	-	+	+	+		+	O	+	+	+
				40	+			+	+	O		+	O	+	+	+
				60	O			+	+	-		+	O	+	O	+
				80												
				100												
				120												
				140												
Methyl acetate	CH_3COOCH_3	56	technically pure	20	-	-	-	+	+	+			-	-	-	-
				40					+	O						
				60					O							
				80												
				100												
				120												
				140												
Methyl amine	CH_3NH_2	-6	32%, aqueous	20	O	-	-	+	+	O			+	-	+	+
				40												
				60												
				80												
				100												
				120												
				140												
Methyl bromide	CH_3Br	4	technically pure	20	-	-	-	O	-	+			O	-	-	O
				40						+						
				60						+						
				80												
				100												
				120												
				140												
Methyl chloride	CH_3Cl	-24	technically pure	20	-	-	-	O	-	+			-	-	-	-
				40						+						
				60						+						
				80												
				100												
				120												
				140												
Methyl ethyl ketone	$CH_3COC_2H_5$	80	technically pure	20	-	-	-	+	+	-			-	-	-	-
				40				O	O							
				60				-	O							
				80												
				100												
				120												
				140												
Methylene chloride	CH_2Cl_2	40	technically pure	20	-	-	-	O	O	+			O	-	-	-
				40						O						
				60						O						
				80												
				100												
				120												
				140												
Methylisobutylketone	$C_6H_{12}O$			20	-	-	-									
				40												
				60												
				80												
				100												
				120												
				140												

(Courtesy George Fischer Engineering Handbook)

Aggressive Media					Chemical Resistance											
Medium	Formula	Boiling point °C	Concentration	Temperature °C	PVC	CPVC	ABS	PE	PP-H	PVDF (SYGEF)		EPDM	FPM	NBR	CR	CSM
Methylmethacrylate	$C_5H_8O_2$			20	-	-	-									
				40												
				60												
				80												
				100												
				120												
				140												
Methylphenylketone	C_8H_8O			20	-	-	-									
(Acetophenon)				40												
				60												
				80												
				100												
				120												
				140												
Milk (SpRB)				20	+	+	+	+	+	+			+	+	+	+
				40	+	+	+	+	+	+						
				60	+	+	+	+	+	+						
				80		+			+	+						
				100					+	+						
				120						+						
				140												
Mineral oils, free of				20	+	+	-	+	+	+			+	+	O	O
aromatics				40	+			+	+	+			+	+	-	-
				60	+			O	O	+			+	+		
				80						+						
				100						+						
				120						+						
				140												
Mineral water				20	+	+	+	+	+	+		+	+	+	+	+
				40	+	+	+	+	+	+		+	+	+	+	+
				60	+	+	+	+	+	+		+	+	+	+	+
				80		+			+	+		+	+	+		+
				100					+	+		+	+			+
				120						+			+			
				140												
Mixed acids	15% HNO_3		3 parts	20	O	O	-	O	-	+			+	-	-	+
- nitric	15% HF		1 part	40						+			O			O
- hydrofluoric	18% H_2SO_4		2 parts	60						+						
- sulphuric				80												
				100												
				120												
				140												
Mixed acids	H_2SO_4		48%	20	+	+	-	-	-	+			-	-	-	-
- sulphuric	HNO_3		49%	40	O											
- nitric	H_2O		43%	60	-											
- water				80												
				100												
				120												
				140												
Mixed acids	H_2SO_4		50%	20	O	O	-	-	-	+			-	-	-	-
- sulphuric	HNO_3		50%	40	-											
- nitric	H_2O		40%	60												
- water				80												
				100												
				120												
				140												

(Courtesy George Fischer Engineering Handbook)

Aggressive Media					Chemical Resistance											
Medium	Formula	Boiling point °C	Concentration	Temperature °C	PVC	CPVC	ABS	PE	PP-H	PVDF (SYGEF)		EPDM	FPM	NBR	CR	CSM
Mixed acids	H_2SO_4		10%	20	O	O	-	-	-	O			-	-	-	-
- sulphuric	HNO_3		87%	40												
- nitric	H_2O		43%	60												
- water				80												
				100												
				120												
				140												
Mixed acids	H_2SO_4		50%	20	+	+	-	-	-	+			+	-	-	O
- sulphuric	HNO_3		33%	40	O											
- nitric	H_2O		17%	60												
- water				80												
				100												
				120												
				140												
Mixed acids	H_2SO_4		10%	20	+	+	-	O	-	+			+	-	O	+
- sulphuric	HNO_3		20%	40	+					+			+			O
- nitric	H_2O		70%	60						+			+			
- water				80						+						
				100												
				120												
				140												
Mixed acids	H_2SO_4		50%	20	+		-	-	-	+			+	-	O	O
- sulphuric	HNO_3		31%	40												
- nitric	H_2O		19%	60												
- water				80												
				100												
				120												
				140												
Mixed acids	H_2SO_4		30%	20	+	+	-	+	+	+			+	-	+	+
- sulphuric	H_3PO_4		60%	40	+	+		O	O	+			+		O	O
- phosphoric	H_2O		10%	60		+				+			+			
- phosphoric				80						+						
				100												
				120												
				140												
Molasses				20	+	+	+	+	+	+		+	+	+	+	+
				40	+	+	+	+	+	+		+	+	+	+	+
				60	O	+	+	+	+	+		+	+	+	+	+
				80		+				+		+	+	+	O	+
				100												
				120												
				140												
Molasses wort				20	+	+	+	+	+	+		+	+	+	+	+
				40	+	+	+	+	+	+		+	+	+	+	+
				60	+	+	+	+	+	+		+	+	+	+	+
				80		+				+						
				100												
				120												
				140												
Monochloroacetic acid ethyl ester	$ClCH_2COOC_2H_5$	144	technically pure	20	-	-	-	+	+	O			O	-	-	-
				40				+	+	-						
				60				+	+							
				80												
				100												
				120												
				140												

(Courtesy George Fischer Engineering Handbook)

Aggressive Media					Chemical Resistance											
Medium	Formula	Boiling point °C	Concentration	Temperature °C	PVC	CPVC	ABS	PE	PP-H	PVDF (SYGEF)		EPDM	FPM	NBR	CR	CSM
Morpholin	C_4H_9NO	129	technically pure	20	-	-	-	+	+	+			+	-	O	O
				40				+	+	+						
				60				+	+	O						
				80												
				100												
				120												
				140												
Mowilith D			usual commercial	20	+	+		+	+	+			+	+	+	+
				40												
				60												
				80												
				100												
				120												
				140												
Naphthalene		218	technically pure	20	-	-	-	+	+	+		-	+	+	-	O
				40						+			+	+		-
				60				O		O			+	+		
				80												
				100												
				120												
				140												
Natriumhydrogensulfite	$NaHSO_3$			20	+	+		+	+	+		+	+			
				40	+	+		+	+	+		+				
				60	+	+		+	+	+		+				
				80		+			+	+						
				100						+						
				120												
				140												
Natriumsulfate				20	+	+	+	+	+	+		+	+			
				40	+	+		+	+	+		+	+			
				60	+	+		+	+	+		+	+			
				80		+			+	+		+	+			
				100						+			+			
				120												
				140												
Natriumtetraborate (Borax)				20	+	+	+	+	+	+		+	+			
				40	+	+		+	+	+		+				
				60	+	+		+	+	+		+				
				80		+			+							
				100												
				120												
				140												
Nickel salts	$(CH_3COO)_2Ni$, $NiCl_2$, $Ni(NO_3)_2$, $NiSO_4$		cold saturated, aqueous	20	+	+	+	+	+	+		+	+	+	+	+
				40	+	+	+	+	+	+		+	+	+	+	+
				60	O	+	+	+	+	+			+	+	+	+
				80						+			+			
				100						+			+			
				120						+						
				140												
Nitrating acid	H_2SO_4 HNO_3 H_2O		65% 20% 15%	20						+						
				40												
				60												
				80												
				100												
				120												
				140												

(Courtesy George Fischer Engineering Handbook)

Aggressive Media					Chemical Resistance											
Medium	Formula	Boiling point °C	Concentration	Temperature °C	PVC	CPVC	ABS	PE	PP-H	PVDF (SYGEF)		EPDM	FPM	NBR	CR	CSM
Nitric acid (SpRB)	HNO_3			20	+	+	-	+	O	+		+	+			
				40	+	+				+			+			
				60	+	+				+			+			
				80						+						
				100						+						
				120												
				140												
Nitric acid (SpRB)	HNO_3			20	+	+	-	+	+	+		+	+			
				40	+	+		+	O	+			+			
				60	+	+		+		+						
				80		+				+						
				100						+						
				120												
				140												
Nitric acid up to 55% (SpRB)				20	+	+	-	+	-	+			+			
				40	+	+				+						
				60						+						
				80												
				100												
				120												
				140												
Nitric acid (see note 2.3.1 on jointing) (SpRB)	see Salpetre		6,3%, aqueous													
Nitric acid (see note 2.3.1 on jointing) (SpRB)	see Salpetre		up to 40%, aqueous													
Nitric acid (see note 2.3.1 on jointing) (SpRB)	see Salpetre		65%, aqueous													
Nitric acid (see note 2.3.1 on jointing) (SpRB)	see Salpetre		100%													
Nitric acid (see note 2.3.1 on jointing) (SpRB)	see Salpetre		85%													
Nitric oxide	see Nitrous gases															
Nitrilotriacetic acid	$N(CH_2\text{-}COOH)_3$			20				+	+			+				
				40												
				60												
				80												
				100												
				120												
				140												
Nitrobenzene	$C_6H_5\text{-}NO_2$	209	technically pure	20	-	-	-	+	+	+		-	O	-	-	-
				40				O	+	O						
				60				O		-						
				80												
				100												
				120												
				140												

(Courtesy George Fischer Engineering Handbook)

Aggressive Media					Chemical Resistance											
Medium	Formula	Boiling point °C	Concentration	Temperature °C	PVC	CPVC	ABS	PE	PP-H	PVDF (SYGEF)		EPDM	FPM	NBR	CR	CSM
Nitrotoluene (o-, m-, p-)		222-238	technically pure	20	-	-	-	+	+	+		-	O	O	-	-
				40				O	O	+			-	-		
				60						+						
				80						+						
				100						O						
				120												
				140												
Nitrous acid	HNO_2			20	+	+	-	+	-	+		+	+			
				40	+	+				+						
				60						+						
				80												
				100												
				120												
				140												
Nitrous gases	see Nitric oxide		diluted, moist, anhydrous													
N-Methylpyrrolidon				20	-	-	-									
				40												
				60												
				80												
				100												
				120												
				140												
N,N-Dimethylaniline	$C_6H_5N(CH_3)_2$		technically pure	20	-	-	-	+	+			+				
				40												
				60												
				80												
				100												
				120												
				140												
n-Pentylacetate				20	-	-	-									
				40												
				60												
				80												
				100												
				120												
				140												
Oleic acid (SpRB)	$C_{17}H_{33}COOH$		technically pure	20	+	O	-	+	+	+		-	+	O	-	-
				40	+			+	+	+			O	-		
				60	+			O	O	+			-			
				80						+						
				100						+						
				120						+						
				140												
Oleum (SpRB)	$H_2SO_4+SO_3$		10% SO_3	20	-	-	-	-	-	-		-	-	-	-	-
				40												
				60												
				80												
				100												
				120												
				140												
Oleum vapours (SpRB)			traces	20	+	-	-	-	-	-		-	+	-	-	O
				40												
				60												
				80												
				100												
				120												
				140												

(Courtesy George Fischer Engineering Handbook)

Aggressive Media					Chemical Resistance											
Medium	Formula	Boiling point °C	Concentration	Temperature °C	PVC	CPVC	ABS	PE	PP-H	PVDF (SYGEF)		EPDM	FPM	NBR	CR	CSM
Olive oil (SpRB)				20	+	-	-	+	+	+		-	+	+	+	+
				40	+			+	+	+			+	+	+	+
				60	+			O	+	+			+	+	+	O
				80					+	+			+			-
				100												
				120												
				140												
Oxalic acid (SpRB)	$(COOH)_2$		cold saturated, aqueous	20	+	+	+	+	+	+		+	+	O	O	O
				40	+	+	+	+	+	+			+	-	-	O
				60	+	+		+	+	+			O			-
				80		O				+			-			
				100						+						
				120												
				140												
Oxygen	O_2		technically pure	20	+	+	+	+	+	+		+	+	+	+	+
				40	+		+	+	+	+		+	+		+	+
				60	+		+	O	O	+		+	+		+	+
				80						+			+		+	+
				100						O			+			+
				120						O			+			
				140												
Ozone (SpRB)	O_3		up to 2%, in air	20	+	+	-	O	O	O		O	+	-	O	+
				40				-	-							
				60												
				80												
				100												
				120												
				140												
Ozone (SpRB)	O_3		cold saturated, aqueous	20	+	+	-	O	O	O		-	+	-	O	+
				40	+			-	-				O		-	+
				60									-			O
				80												
				100												
				120												
				140												
Palm oil, palm nut oil (SpRB)				20	+	O	+	+	+	+		-	+	+	+	O
				40	-			+	+	+			+	+	O	-
				60				O	O	+			+	O	-	
				80						+						
				100						+						
				120												
				140												
Palmitic acid (SpRB)	$C_{15}H_{31}COOH$	390	technically pure	20	+	-	+	O	O	+		O	+	O	+	O
				40						+			O	-	-	-
				60					-	+			-			
				80						+						
				100						+						
				120						+						
				140												
Paraffin emulsions			usual commercial, aqueous	20	+	+	O	+	+	+		-	+	+	+	+
				40	+	+		+	+	+			+	+	O	-
				60				O	O	+			+	O	-	
				80						+			+			
				100						+						
				120												
				140												

(Courtesy George Fischer Engineering Handbook)

Aggressive Media					Chemical Resistance										
Medium	Formula	Boiling point °C	Concentration	Temperature °C	PVC	CPVC	ABS	PE	PP-H	PVDF (SYGEF)	EPDM	FPM	NBR	CR	CSM
Paraffin oil				20	+	+	O	+	+	+	-	+	+	+	O
				40	+			+	+	+		+	O	O	-
				60	O			+	O	+		+	O	-	
				80						+		O			
				100						+					
				120						+					
				140											
p-Dibromo benzene	$C_6H_5Br_2$		technically pure	20	-	-	-	O	O	+	-	+	-	-	-
				40						+					
				60						+					
				80						+					
				100						+					
				120											
				140											
Perchlorethylene (tetrachlorethylene)	$Cl_2C{=}CCl_2$	121	technically pure	20	-	-		O	O	+		+	O	-	-
				40						+		+	-		
				60						+		+			
				80						O					
				100						-					
				120											
				140											
Perchlorid acid (SpRB)	$HClO_4$		10%, aqueous	20	+	+	O	+	+	+	+	+	-	-	+
				40	+	+		+	+	+	O	+			+
				60	O	+		+	+	+		+			O
				80						+		O			-
				100						+					
				120											
				140											
Perchlorid acid (SpRB)			70%, aqueous	20	O	O	-	+	O	+	-	+	-	-	+
				40				O	-	+		+			+
				60				-		+		+			O
				80						+		O			
				100						+					
				120											
				140											
Petroleum			technically pure	20	+	-	-	+	+	+	-	+	+	O	-
				40				+	O	+		+	+	-	
				60				O	O	+		O	+		
				80						+					
				100						+					
				120						+					
				140											
Petroleum ether (SpRB)		40-70	technically pure	20	+	-	-	+	+	+	-	+	+	-	-
				40	+			O	O	+		+	O		
				60	+			O	O	+		O	-		
				80						+					
				100						+					
				120											
				140											
Phenol (SpRB)	C_6H_5-OH	182	up to 10%, aqueous	20	+	+	-	+	+	+		+	-	-	-
				40	O	+		+	+	+		+			
				60				O	+	+	+	+			
				80						+	O	O			
				100						+					
				120											
				140											

(Courtesy George Fischer Engineering Handbook)

Aggressive Media					Chemical Resistance											
Medium	Formula	Boiling point °C	Concentration	Temperature °C	PVC	CPVC	ABS	PE	PP-H	PVDF (SYGEF)		EPDM	FPM	NBR	CR	CSM
Phenol (SpRB)			up to 5%	20	+	+	-	+	+	+		+	+	-	-	
				40		+		+	+	+		O	+			
				60				+	+	+			+			
				80						+			O			
				100						O						
				120												
				140												
Phenol (SpRB)	C_6H_5-OH		up to 90%, aqueous	20	O	-	-	+	+	+		-	+	-	-	-
				40				+	+	+			O		O	
				60				O	+	O			-			
				80												
				100												
				120												
				140												
Phenylhydrazine	C_6H_5-NH-NH_2	243	technically pure	20	-	-	-	O	O	O		-	+	-	-	-
				40						-			+			
				60									O			
				80												
				100												
				120												
				140												
Phenylhydrazine hydrochloride	C_6H_5-NH-NH_2.HCl		aqueous	20	O	O	-		+	+			+	O	O	+
				40					O	+			+	-	-	+
				60					O	+		O	O			O
				80									-			-
				100												
				120												
				140												
Phosgene (SpRB)	$COCl_2$	8	liquid, technically pure	20	-	-	-	-	-	-		-	+	O	+	+
				40												
				60												
				80												
				100												
				120												
				140												
Phosgene (SpRB)			gaseous, technically pure	20	+	-	-	O	O	+		+	+	+	+	+
				40	O					+			+	+	O	O
				60	O								O	+	-	
				80												
				100												
				120												
				140												
Phosphate disodique	see d'isodiumphosphate		saturated													
Phosphoric acid	H_3PO_4		up to 30%, aqueous	20	+	+	+	+	+	+		+	+	O	+	+
				40	+	+	+	+	+	+		+	+	O	+	+
				60	O	+	O	+	+	+		O	+	-	+	+
				80		+			+	+			+		O	O
				100						+			+			
				120						+						
				140												
Phosphoric acid			50%, aqueous	20	+	+	+	+	+	+		+	+	O	+	+
				40	+	+	+	+	+	+		O	+	-	+	+
				60	+	+	O	+	+	+			+		O	+
				80		+			+	+			+			+
				100						+			O			O
				120						+						
				140												

(Courtesy George Fischer Engineering Handbook)

Aggressive Media					Chemical Resistance											
Medium	Formula	Boiling point °C	Concentration	Temperature °C	PVC	CPVC	ABS	PE	PP-H	PVDF (SYGEF)		EPDM	FPM	NBR	CR	CSM
Phosphoric acid			85%, aqueous	20	+	+	+	+	+	+		+	+	-	+	+
				40	+	+	+	+	+	+		+	+		+	O
				60	+	+	O	O	+	+		O	+		O	-
				80		+			+	+			+			
				100						+			O			
				120						+						
				140												
Phosphoric acid	H_3PO_4			20	+	+	-	+	+	+		O	+	-	-	-
				40	+	+		+	+	+			+			
				60		+				+			O			
				80						+						
				100												
				120												
				140												
Phosphoric acid	H_3PO_4			20	+	+	-	+	+	+			+			
				40	+	+		+	+	+			+			
				60	+	+			+	+			+			
				80		+			+	+			+			
				100						+			O			
				120												
				140												
Phosphoric acid tributyl ester	$(HaC_4O)3P=O$			20	-	-	-	+	+	-		+	-			
				40												
				60												
				80												
				100												
				120												
				140												
Phosphorous chlorides:		175	technically pure	20	-	-	-	+	O	-			+	-	-	+
- Phosphorous trichloride	PCl_3	162		40												
- Phosphorous	PCl5	105		60				O	O							
pentachloride	$POCl_3$			80												
- Phosphorous oxichloride				100												
(SpRB)				120												
				140												
Photographic developer (SpRB)			usual commercial	20	+	+	+	+	+	+		+	+	O	+	+
				40	+	+	+	+	+	+		+	+	O	+	+
				60	O	+	O	O		+						
				80		O										
				100												
				120												
				140												
Photographic emulsions (SpRB)				20	+	+	+	+	+	+		+	+	O	+	+
				40	+	+	+	+	+	+		+	+		+	+
				60		O				+						
				80												
				100												
				120												
				140												
Photographic fixer (SpRB)			usual commercial	20	+	+	+	+	+	+		+	+	+	+	+
				40	+	+	+	+	+	+		+	+	+	+	+
				60	O	+	O			+						
				80												
				100												
				120												
				140												

(Courtesy George Fischer Engineering Handbook)

Aggressive Media					Chemical Resistance											
Medium	Formula	Boiling point °C	Concentration	Temperature °C	PVC	CPVC	ABS	PE	PP-H	PVDF (SYGEF)		EPDM	FPM	NBR	CR	CSM
Phthalic acid (SpRB)	$C_6H_5(COOH)_2$	Fp.*, 208	saturated, aqueous	20	+	-	-	+	+	+		+	-	-	+	+
				40	O			+	+	+		O			+	+
				60	-			+	+	+					O	
				80						+						
				100						+						
				120												
				140												
Phthalic acid dioctyl ester	$C_{24}H_{38}O_4$			20	-	-	-	+	+	-		+	-	-		
				40												
				60												
				80												
				100												
				120												
				140												
Picric acid (SpRB)	$C_6H_3N_3O_7$	FP. 122	1%, aqueous	20	+	-	-	+	+	+		+	+	O	O	+
				40						+		+	+	-	-	O
				60						+		+	+			-
				80						+		+	+			
				100						+		+	+			
				120												
				140												
Potash	see potassium carbonate		cold saturated, aqueous													
Potash lye	KOH		50%	20	+	+		+	+	-		+				
				40	+	+		+	+			+				
				60	+	+		+	+			+				
				80												
				100												
				120												
				140												
Potassium (SpRB)	$KMnO_4$		cold saturated, aqueous	20	+	+		+	+	+		+	+	O	O	+
				40	+	+		+	+	+			+	-	-	+
				60	O			O		+			+			+
				80						+			+			
				100						+						
				120												
				140												
Potassium acetate (SpRB)	CH_3COOK		saturated	20	+	+	+	+	+	+		+	-			
				40	+	+		+	+	+		+				
				60	+	+		+	+	+		+				
				80		+			+	+		+				
				100						+						
				120												
				140												
Potassium bichromate (SpRB)	$K_2Cr_2O_7$	107	saturated, aqueous	20	+	+	+	+	+	+		+	+	+	O	+
				40	+	+	+	+	+	+			+	O	-	+
				60	O	+	+	+	+	+			+			+
				80					+	+						
				100					+	+						
				120						+						
				140												
Potassium borate	K_3BO_3		10%, aqueous	20	+	+	+	+	+	+		+	+	+	+	+
				40	+	+	+	+	+	+			+	+	+	+
				60	O	+	+	+	+	+			+	+	+	+
				80		+										
				100												
				120												
				140												

(Courtesy George Fischer Engineering Handbook)

Aggressive Media					Chemical Resistance											
Medium	Formula	Boiling point °C	Concentration	Temperature °C	PVC	CPVC	ABS	PE	PP-H	PVDF (SYGEF)		EPDM	FPM	NBR	CR	CSM
Potassium bromate	$KBrO_3$		cold saturated, aqueous	20	+	+	+	+	+	+		+	+	+	+	+
				40	+	+	+	+	+	+		+	+	+	+	+
				60	O	+	+	O	+	+		+	+	+	+	+
				80		+			+	+			+	+	O	+
				100					+	+				+	O	+
				120						+						
				140												
Potassium bromide	KBr		all, aqueous	20	+	+	+	+	+	+		+	+	+	+	+
				40	+	+	+	+	+	+		+	+	+	+	+
				60	O	+	+	+	+	+		+	+	+	+	+
				80		+				+			+	O	+	+
				100						+			+	O	O	+
				120						+						
				140												
Potassium carbonate (potash)				20	+	+	+	+	+	O		+				
				40	+	+		+	+			+				
				60	+	+		+	+			+				
				80		+			+							
				100												
				120												
				140												
Potassium chlorate (SpRB)	K ClO_3		cold saturated, aqueous	20	+	+	+	+	+	O		+	+	+	+	+
				40	+	+	+	+	+	-		+	+	O	O	+
				60	+	+	+	+	+				+			+
				80		+							+			O
				100									+			
				120												
				140												
Potassium chloride	KCl		all, aqueous	20	+	+	+	+	+	+		+	+	+	+	+
				40	+	+	+	+	+	+		+	+	+	+	+
				60	+	+	+	+	+	+		+	+	+	+	+
				80		+			+	+			+			+
				100					+	+			+			
				120						+						
				140												
Potassium chromate (SpRB)	K_2CrO_4		cold saturated, aqueous	20	+	+	+	+	+	+		+	+	+	+	+
				40	+	+	+		+	+			+	O	+	+
				60	+	+	+		+	+			+	-	O	O
				80						+						
				100						+						
				120												
				140												
Potassium cyanide	KCN		cold saturated, aqueous	20	+	+	+	+	+	+			+	+	+	+
				40	+	+	+	+	+	+			O	+	+	+
				60	+	+	+	+	+	O		+	-	+	O	+
				80		+			+			+		+	-	+
				100												
				120												
				140												
Potassium dichromate	$K_2Cr_2O_7$		saturated	20		+		+	+	+		+	+			
				40		+		+	+	+						
				60		+		+	+	+						
				80		+				+						
				100						+						
				120												
				140												

(Courtesy George Fischer Engineering Handbook)

Aggressive Media					Chemical Resistance											
Medium	Formula	Boiling point °C	Concentration	Temperature °C	PVC	CPVC	ABS	PE	PP-H	PVDF (SYGEF)		EPDM	FPM	NBR	CR	CSM
Potassium fluoride	KF		saturated	20	+	+		+	+	+			+			
				40	+	+		+	+	+						
				60	+	+		+	+	+						
				80		+			+	+						
				100						+						
				120												
				140												
Potassium Hexacyanoferrate -(II)	$K_4[Fe(CN)_6].3H_2O$			20	+	+		+	+	+		+	+			
				40	+	+		+	+	+						
				60	+	+		+	+	+						
				80		+			+	+						
				100						+						
				120												
				140												
Potassium hydrogen carbonate	$KHCO_3$		saturated	20	+	+		+	+	+		+	+			
				40	+	+		+	+	+		+				
				60	+	+		+	+	+		+				
				80		+			+							
				100												
				120												
				140												
Potassium hydrogen sulphate	$KHSO_4$		saturated	20	+	+		+	+			+				
				40	+	+		+	+			+				
				60	+	+		+	+			+				
				80		+			+			+				
				100												
				120												
				140												
Potassium iodide	KJ		cold saturated, aqueous	20	+	+	+	+	+	+		+	+	+	+	+
				40	+	+	+	+	+	+			+	O	O	+
				60	+	+	+	+	+	+			+	-	-	+
				80		+				+			+			
				100						+						
				120												
				140												
Potassium nitrate	KNO_3		50%, aqueous	20	+	+	+	+	+	+		+	+	+	+	+
				40	+	+	+	+	+	+			+	+	+	+
				60	+	+	+	+	+	+			+	+	+	+
				80		+				+						
				100						+						
				120												
				140												
Potassium perchlorate (SpRB)	$KClO_4$		cold saturated, aqueous	20	+	+		+		+		+	+	+	+	+
				40	+	+				+			+	O	O	+
				60	O	+				+			+			+
				80		+				+			+			O
				100												
				120												
				140												
Potassium persulphate (SpRB)	$K_2S_2O_8$		all, aqueous	20	+	+	+	+	+	+		+	+	-	+	+
				40	+	+	+	+	+	+		+	+		+	+
				60	O	+	+	+	+	+			+		+	+
				80						+			+		O	O
				100									+			
				120												
				140												

(Courtesy George Fischer Engineering Handbook)

Aggressive Media					Chemical Resistance											
Medium	Formula	Boiling point °C	Concentration	Temperature °C	PVC	CPVC	ABS	PE	PP-H	PVDF (SYGEF)		EPDM	FPM	NBR	CR	CSM
Potassium sulphate	K_2SO_4		all, aqueous	20	+	+	+	+	+	+		+	+	+	+	+
				40	+	+	+	+	+	+		+	+	+	+	+
				60	O	+	+	+	+	+			+	+	+	+
				80		+				+			+			
				100						+			+			
				120												
				140												
Potassium sulphide	K_2S		saturated	20	+	+		+	+	O		+				
				40	+	+		+	+	O		+				
				60	+	+		+	+	O		+				
				80		+			+			+				
				100												
				120												
				140												
Potassium sulphite	K_2SO_3		saturated	20	+	+		+	+			+				
				40	+	+		+	+							
				60	+	+		+	+							
				80		+										
				100												
				120												
				140												
Potassium-aluminiumsulfate (alum)			50%	20	+	+		+	+	+		+				
				40	+	+		+	+	+		+				
				60	+	+		+	+	+		+				
				80		+			+	+		+				
				100						+						
				120												
				140												
Pottasium hexacyanoferrate -(III)	$K_3[Fe(CN)_6]$.			20	+	+		+	+	+		+	+			
				40	+	+		+	+	+						
				60	+	+		+	+	+						
				80		+			+	+						
				100						+						
				120												
				140												
Pottasium tartrat				20	+			+	+	+		+				
				40	+			+	+	+		+				
				60				+	+	+		+				
				80					+	+						
				100						+						
				120												
				140												
Pottasiumhydrogensulfite				20	+					+		+				
				40	+					+		+				
				60						+						
				80						+						
				100						+						
				120												
				140												
Pottasiumhypochlorite	KOCl			20	+	O		+	+	O		+	O			
				40												
				60												
				80												
				100												
				120												
				140												

(Courtesy George Fischer Engineering Handbook)

Aggressive Media					Chemical Resistance											
Medium	Formula	Boiling point °C	Concentration	Temperature °C	PVC	CPVC	ABS	PE	PP-H	PVDF (SYGEF)		EPDM	FPM	NBR	CR	CSM
Pottasiumperoxodisulfate	$K_2S_2O_8$		saturated	20	+	+										
				40	+	+										
				60	+	+										
				80		+										
				100												
				120												
				140												
Pottasiumphosphate	KH_2PO_4 und $K_2H\ PO_4$		all, aqueous	20	+	+	O	+	+	+		+	+	+	+	+
				40	+	+		+	+	+		+	+	O	O	+
				60	O	+		+	+	+			+	-	-	+
				80		+			+	+			+			O
				100									+			
				120												
				140												
Pottasiumphosphate				20	+				+							
				40	+				+							
				60	+				+							
				80					+							
				100												
				120												
				140												
Propane	C_3H_8	-42	technically pure, liquid	20	+	-	-	+	+	+		-	+	+	-	-
				40						+						
				60						+						
				80												
				100												
				120												
				140												
Propane			technically pure, gaseous	20	+	+	-	+	+	+		-	+	+	+	O
				40						+						
				60						+						
				80												
				100												
				120												
				140												
Propanol, n- and iso- (SpRB)	C_3H_7OH	97 bzw. 82	technically pure	20	+	-	-	+	+	+		+	+	+	+	+
				40	O			+	+	+		+	+	O	+	O
				60	O			+	+	+			+	-	+	O
				80						O						
				100												
				120												
				140												
Propargyl alcohol (SpRB)	CHčC-CH_2-OH	114	7%, aqueous	20	+	-	-	+	+	+		+	+	+	+	+
				40	+			+	+	O		+	+	+	+	+
				60	+			+	+	O			+	+	O	O
				80												
				100												
				120												
				140												
Propionic acid (SpRB)	CH_3CH_2COOH	141	50%, aqueous	20	+	O	-	+	+	+		+	+	-	O	O
				40	+			+	+	+		+	+		-	-
				60	O			+	+	+			O			
				80												
				100												
				120												
				140												

(Courtesy George Fischer Engineering Handbook)

Aggressive Media					Chemical Resistance											
Medium	Formula	Boiling point °C	Concentration	Temperature °C	PVC	CPVC	ABS	PE	PP-H	PVDF (SYGEF)		EPDM	FPM	NBR	CR	CSM
Propionic acid (SpRB)		141	technically pure	20	+	O	-	+	+	+		+	+	-	-	-
				40	O			O	O	+		O	+			
				60				O	O	+			+			
				80									O			
				100												
				120												
				140												
Propylene glycol (SpRB)	$C_3H_8O_2$	188	technically pure	20	+	-	O	+	+	+		+	+	+	+	+
				40	+			+	+	+		+	+	O	+	+
				60	+			+	+	+			O	-		+
				80												
				100												
				120												
				140												
Propylene oxide	C_3H_6O	35	technically pure	20	O		-	+	+	+		O	-	-	-	-
				40						O						
				60												
				80												
				100												
				120												
				140												
Pyridine	C_5H_5N	115	technically pure	20	-	-	-	+	O	+		O	-	-	-	-
				40				O	O	-						
				60				O	O							
				80												
				100												
				120												
				140												
Pyrogallol	$C_6H_3(OH)_3$		100%	20						+			+			
				40						+						
				60												
				80												
				100												
				120												
				140												
Ramsit fabric waterproofing agents			usual commercial	20	+			+	+	+		+	+	+	+	+
				40	+					+						
				60	+					+						
				80												
				100												
				120												
				140												
Salicylic acid	$C_6H_4(OH)COOH$		saturated	20	+	+	O	+	+	+		+	+	+	+	+
				40	+			+	+	+		+				
				60				+	+	+		+				
				80						+						
				100												
				120												
				140												
Sea water	see Brine															
Silicic acid	$Si(OH)_4$			20	+	+	+	+	+			+				
				40	+	+		+	+			+				
				60	+	+		+	+			+				
				80					+							
				100												
				120												
				140												

(Courtesy George Fischer Engineering Handbook)

Aggressive Media					Chemical Resistance											
Medium	Formula	Boiling point °C	Concentration	Temperature °C	PVC	CPVC	ABS	PE	PP-H	PVDF (SYGEF)		EPDM	FPM	NBR	CR	CSM
Silicone oil				20	+	+	+	+	+	+		+	+	+	+	+
				40	O	+		+	+	+		+	+	+	+	+
				60	-			+	+	+		+	+	+	O	
				80					+	+						
				100						+						
				120												
				140												
Silver	AgCn		saturated	20	+	+	+	+	+	+		+	+	+	+	+
				40	+	+		+	+	+		+	+			
				60	+	+		+	+	+		+	+			
				80												
				100												
				120												
				140												
Silver salts	$AgNO_3$, AgCN, AgCl		cold saturated, aqueous	20	+	+	+	+	+	+		+	+	+	+	+
				40	+	+	+	+	+	+		+	+	+	+	+
				60	O	+	+	+	+	+		+	+	+	+	+
				80		+			+	+						
				100						+						
				120												
				140												
Silvercyanide				20	+	+	+	+	+	+		+	+	+	+	+
				40	+	+		+	+	+		+	+			
				60	+	+		+	+	+		+	+			
				80												
				100												
				120												
				140												
Soap solution (SpRB)			all, aqueous	20	+	+	+	+	+	+		+	+	+	+	+
				40	+	+	+	+	+	+		+	+	+	+	+
				60	O	+		+	+	+		+	+	+	+	+
				80						+						
				100						+						
				120												
				140												
Soda	see Sodium carbonate															
Sodium acetate	CH_3COONa		all, aqueous	20	+	+	+	+	+	+		+	-	+	+	O
				40	+	+		+	+	+		+		+	+	
				60	+	+		+	+	+		+				
				80		+			+	+						
				100					+	O						
				120												
				140												
Sodium aluminium sulfate				20	+			+	+							
				40	+			+	+							
				60	+			+	+							
				80					+							
				100												
				120												
				140												
Sodium arsenite	Na_3AsO_3		saturated	20	+	+		+	+			+				
				40	+	+		+	+			+				
				60	+	+		+	+			+				
				80		+			+							
				100												
				120												
				140												

(Courtesy George Fischer Engineering Handbook)

Aggressive Media					Chemical Resistance											
Medium	Formula	Boiling point °C	Concentration	Temperature °C	PVC	CPVC	ABS	PE	PP-H	PVDF (SYGEF)		EPDM	FPM	NBR	CR	CSM
Sodium benzoate	C_6H_5-COONa		cold saturated, aqueous	20	+	+	-	+	+	+		+	+	+	+	+
				40	+	+		+	+	+		O	+	+	+	+
				60	O	+		+	+	+			+			O
				80		+				+			O			
				100						O						
				120												
				140												
Sodium bicarbonate	$NaHCO_3$		cold saturated, aqueous	20	+	+	+	+	+	+		+	+	+	+	+
				40	+	+	+	+	+	+		+	+	+	+	+
				60	+	+	+	+	+	+		+	+	+	+	+
				80					+	+		+	+			
				100						+						
				120												
				140												
Sodium bisulphate	$NaHSO_4$		10%, aqueous	20	+	+	+	+	+	+		+	+	+	+	+
				40	+	+	+	+	+	+		O	+	O	+	+
				60	O	+	+	+	+	+			+	-	O	
				80						+			+			
				100						+			+			
				120						+						
				140												
Sodium bisulphite	$NaHSO_3$		all, aqueous	20	+	+		+	+	+		+	O	O	+	+
				40	O	+		+	+	+		O	-	-	+	+
				60	-	+		+	+	+		-			O	+
				80						+					-	O
				100						+						
				120												
				140												
Sodium borate	Na_3BO_3		saturated	20	+	+		+	+			+	+			
				40	+	+		+	+			+	+			
				60	+	+		+	+			+	+			
				80		+			+							
				100												
				120												
				140												
Sodium bromate	$NaBrO_3$		all, aqueous	20	+	+		+	+	+		+	+	+	+	+
				40	O	+		O	O	+			+	O	+	+
				60						+			+	-	O	+
				80						+						
				100						+						
				120												
				140												
Sodium bromide	NaBr		all, aqueous	20	+	+	+	+	+	+		+	+	+	+	+
				40	+	+	+	+	+	+			+	O	+	+
				60	O	+	+	+	+	+			+		O	O
				80		+			+	+			+			
				100						+						
				120						+						
				140												
Sodium carbonate	see soda		cold saturated, aqueous													
Sodium chlorate (SpRB)	$NaClO_3$		all, aqueous	20	+	+	+	+	+	O		+	+	+	+	+
				40	+	+	+	+	+			+	+	O	+	+
				60	O	+	+	+	+			+	+	-	O	+
				80		+						O	+		-	O
				100								-				
				120												
				140												

(Courtesy George Fischer Engineering Handbook)

Aggressive Media					Chemical Resistance											
Medium	Formula	Boiling point °C	Concentration	Temperature °C	PVC	CPVC	ABS	PE	PP-H	PVDF (SYGEF)		EPDM	FPM	NBR	CR	CSM
Sodium chlorite (SpRB)	$NaClO_2$		diluted, aqueous	20	O	+		+	+	+		+	+	-	O	+
				40		+		O	O	O		+	+		-	+
				60		+				O		+	+			+
				80		+										
				100												
				120												
				140												
Sodium chromate (SpRB)	Na_2CrO_4		diluted, aqueous	20	+	+	+	+	+	+		+	+	+	+	+
				40	+	+	+		+	+		+	+	O	+	+
				60	O		+			+		+	+	-	O	O
				80						+						
				100						+						
				120												
				140												
Sodium disulphite	$Na_2S_2O_5$		all, aqueous	20	+	+		+	+	+		+	+	O	+	+
				40	+	+				+		+	+	-	+	+
				60	O	+				+		+	+		+	O
				80						+						
				100						+						
				120												
				140												
Sodium dithionite	see hyposulphite		up to 10%, aqueous													
Sodium fluoride	NaF		cold saturated, aqueous	20	+	+	+	+	+	+		+	+	+	+	+
				40	+	+	+			+		+	+	+	+	+
				60	+	+	+			+		+	+	O	+	+
Sodium hydroxide (see Caustic soda)				80						+						
				100						+						
				120												
				140												
Sodium hypochlorite (SpRB)	NaOCl		12,5% active chlorine, aqueous	20	+			O	O	O		+	+	-	-	+
				40	+			-	-							
				60	O											
Sodium iodide	NaI		all, aqueous	20	+	+	+	+	+	+		+	+	+	+	+
				40	+	+	+			+			+	+	+	+
				60	O	+	+			+			+	O	+	O
				80		+				+						
				100						+						
				120												
				140												
Sodium nitrate	$NaNO_3$		cold saturated, aqueous	20	+	+	+	+	+	+		+	+	+	+	+
				40	+	+	+	+	+	+		+	+	+	+	+
				60	O	+		+	+	+		+	+	+	+	+
				80					+	+						
				100						+						
				120						+						
				140												
Sodium nitrite	$NaNO_2$		cold saturated, aqueous	20	+	+	+	+	+	+		+	+	+	+	+
				40	+	+	+	+	+	+			+	O	+	+
				60	+	+		+	+	+			+	-	+	+
				80		+			+	+						
				100						+						
				120						+						
				140												

(Courtesy George Fischer Engineering Handbook)

Aggressive Media					Chemical Resistance											
Medium	Formula	Boiling point °C	Concentration	Temperature °C	PVC	CPVC	ABS	PE	PP-H	PVDF (SYGEF)		EPDM	FPM	NBR	CR	CSM
Sodium oxalate	$Na_2C_2O_4$		cold saturated, aqueous	20	+	+	+	+	+	+			+	+	+	+
				40	+	+				+						
				60	O	+				O						
				80												
				100												
				120												
				140												
Sodium perborate	$NaBO_3\ 4H_2O$		saturated	20	+	+		+	+	+		+	+			
				40	+	+				+						
				60	+	+				+						
				80		+										
				100												
				120												
				140												
Sodium perchlorate	$NaClO_4$		saturated	20	+	+		+	+			+	+			
				40	+	+										
				60	+	+										
				80		+										
				100												
				120												
				140												
Sodium persulphate (SpRB)	$Na_2S_2O_8$		cold saturated, aqueous	20	+			+	+	+		+	+	-	+	+
				40	+			+	+	+		+	+		+	+
				60	O			+	+	+			+		+	+
				80									+		O	O
				100									+			
				120												
				140												
Sodium phosphate	Na_3PO_4		cold saturated, aqueous	20	+	+	+	+	+	+		+	+	+	+	+
				40	+	+		+	+	+		+	+	+	+	+
				60	O	+		+	+	+		+	+	+	+	+
				80		+			+	O						
				100					+	-						
				120												
				140												
Sodium silicate	Na_2SiO_3		all, aqueous	20	+	+	+	+	+	+		+	+	+	+	+
				40	+	+	+	+	+	+		+	+	+	+	+
				60	O	+		+	+	O		+	+	+	+	+
				80					+	-						
				100												
				120												
				140												
Sodium Sulfide	Natriumsulfid															
Sodium sulphate	Na_2SO_4, $NaHSO_4$		cold saturated, aqueous	20	+	+	+	+	+	+		O	+	+	+	+
				40	+	+	+	+	+	+		O	+	+	+	+
				60	O	+	+	+	+	+		O	+	+	+	+
				80					+	+			+			
				100						+						
				120						+						
				140												
Sodium sulphide	Na_2S		cold saturated, aqueous	20	+	+	+	+	+	O		+	+	+	-	+
				40	+	+	+	+	+	O		+		+		+
				60	O	+	+	+	+	O		+		+		+
				80		+			+							
				100												
				120												
				140												

(Courtesy George Fischer Engineering Handbook)

Aggressive Media					Chemical Resistance											
Medium	Formula	Boiling point °C	Concentration	Temperature °C	PVC	CPVC	ABS	PE	PP-H	PVDF (SYGEF)		EPDM	FPM	NBR	CR	CSM
Sodium sulphite	Na_2SO_3		cold saturated, aqueous	20	+	+	+	+	+	+		+	+	+	+	+
				40	+	+		+	+	+		+	+	O	+	+
				60	O	+		+	+	+		+	+	-	O	+
				80		+			+	+						
				100						+						
				120												
				140												
Sodium thiosulphate	$Na_2S_2O_3$		cold saturated, aqueous	20	+	+	+	+	+	+		+	+	+	+	+
				40										O	+	+
				60										-	O	O
				80												
				100												
				120												
				140												
Sodiumchloride	NaCl		each, aqueous	20	+	+	+	+	+	+		+	+			
				40	+	+	+	+	+	+		+	+			
				60	+	+		+	+	+		+	+			
				80		+			+	+		+	+			
				100						+			+			
				120												
				140												
Sodiumcyanide	NaCN			20	+	+		+	+	+		+	+			
				40	+	+		+	+	+		+	+			
				60	+	+		+	+	+		+	+			
				80		+			+							
				100												
				120												
				140												
Sodiumdichromate	Na_2Cs2O_7			20	O	+		+	+			+	+			
				40		+		+				+	+			
				60		+							+			
				80		+										
				100												
				120												
				140												
Sodiumhydrogen-carbonate	$NaHCO_3$			20	+	+	+	+	+	+		+	+			
				40	+	+		+	+	+		+				
				60	+	+		+	+	+		+				
				80		+			+							
				100												
				120												
				140												
Sodiumhydrogensulfate	$NaHSO_4$			20	+	+	+	+	+	+		+	+			
				40	+	+		+	+	+		+	+			
				60	+	+		+	+	+		+	+			
				80		+			+	+						
				100						+						
				120												
				140												
Spindle oil				20	O	O	-	O	+	+		-	+	+	O	-
				40					O	+			O	+	-	O
				60				O	-	+			-	O		-
				80						+				-		
				100												
				120												
				140												

(Courtesy George Fischer Engineering Handbook)

Aggressive Media					Chemical Resistance											
Medium	Formula	Boiling point °C	Concentration	Temperature °C	PVC	CPVC	ABS	PE	PP-H	PVDF (SYGEF)		EPDM	FPM	NBR	CR	CSM
Spinning bath acids containing carbon disulphide (SpRB)			100 mg CS_2/l	20	+			+	+	+			+	-	-	O
				40	+					+						
				60												
				80												
				100												
				120												
				140												
Spinning bath acids containing carbon disulphide (SpRB)			200 mg CS_2/l	20	O			+	+	+		-	+	-	-	-
				40						+						
				60												
				80												
				100												
				120												
				140												
Spinning bath acids containing carbon disulphide (SpRB)			700 mg CS_2/l	20	-			+	+	+		-	+	-	-	-
				40						+						
				60												
				80												
				100												
				120												
				140												
Stannous chloride	see Tin II chloride		cold saturated, aqueous													
Stannous chloride - Tin IV chloride	$SnCl_4$		cold saturated, aqueous	20				+	+							
				40				+	+							
				60				+	+							
				80					+							
				100												
				120												
				140												
Starch solution	$(C_6H_{10}O_5)n$		all, aqueous	20	+	+	+	+	+	+		+	+	+	+	+
				40	+	+	+	+	+	+		+	+	+	+	+
				60	+	+		+	+	+		+	+	+	+	+
				80		+				+						
				100						+						
				120												
				140												
Starch syrup			usual commercial	20	+	+	+	+	+	+		+	+	+	+	+
				40	+	+	+	+	+	+		+	+	+	+	+
				60	+	+		+	+	+		+	+	+	+	+
				80		+			+	+		+	+	+	+	+
				100						+		+	+	+	+	+
				120												
				140												
Stearic acid (SpRB)	$C_{17}H_{35}COOH$	Fp. 69	technically pure	20	+	O	+	+	+	+		+	+	+	+	O
				40	+		+			+		+	+	+	+	O
				60	+			O	O	+		O	O	O	O	-
				80						+						
				100						+						
				120						+						
				140												
Styrol				20	-	-	-			+			+			
				40												
				60												
				80												
				100												
				120												
				140												

(Courtesy George Fischer Engineering Handbook)

Aggressive Media					Chemical Resistance											
Medium	Formula	Boiling point °C	Concentration	Temperature °C	PVC	CPVC	ABS	PE	PP-H	PVDF (SYGEF)		EPDM	FPM	NBR	CR	CSM
Succinic acid	HOOC-CH_2-CH_2-COOH	Fp*., 185	aqueous, all	20	+	+	+	+	+	+		+	+	+	+	+
				40	+	+		+	+	+		+	+	+	+	+
				60	+	+		+	+	+		+	+	+	+	+
				80									+			
				100												
				120												
				140												
Sugar syrup			usual commercial	20	+	+	+	+	+	+		+	+	+	+	+
				40	+	+	O	+	+	+		+	+	+	+	+
				60	O	+		+	+	+		+	+	+	+	+
				80					+	+		+	+			+
				100					+	+						
				120						+						
				140												
Sulfur	S	Fp.*, 119	technically pure	20	O	O	-	+	+	+		+	+	-	+	+
				40	-			+	+	+			+			+
				60				+	+	+			+			+
				80					+	+			+			+
				100						+						
				120						+						
				140												
Sulfur dioxide	SO_2	-10	technically pure, anhydrous	20	+	+	-	+	+	O		+	+	-	-	O
				40	+	+		+	+	O		O	O			-
				60	+	+		+	+	-		-				
				80												
				100												
				120												
				140												
Sulfur dioxide	SO_2		technically pure, moist	20	-	-	-	-	-	-		-	O	-	-	O
				40												
				60												
				80												
				100												
				120												
				140												
Sulfur dioxide	SO_2		all, moist	20	+	+	-	+	+	+		+	+	-	-	O
				40	+	+		+	+	O		O	O			-
				60	O			+	+	-		-	-			
				80												
				100												
				120												
				140												
Sulfur trioxide	SO_3			20	-	-	-	-	-	-		-	-	-	-	-
				40												
				60												
				80												
				100												
				120												
				140												
Sulfuric acid saturated by Chlorine	H_2SO_4+Cl_2		60%	20						+						
				40						+						
				60						+						
				80						+						
				100						+						
				120												
				140												

(Courtesy George Fischer Engineering Handbook)

Aggressive Media					Chemical Resistance											
Medium	Formula	Boiling point °C	Concentration	Temperature °C	PVC	CPVC	ABS	PE	PP-H	PVDF (SYGEF)		EPDM	FPM	NBR	CR	CSM
Sulfuric acid	H_2SO_4	120	up to 40%,	20	+	+	+	+	+	+		+	+	O	+	+
(see note 2.3.1 on jointing)			aqueous	40	+	+	O	+	+	+		+	+	-	O	+
				60	O	+		+	+	+		+	+			O
				80		+				+		O	O			O
				100						+		-	-			-
				120						+						
				140												
Sulfuric acid	H_2SO_4	140	up to 60%,	20	+	+	-	+	+	+		+	+	-	-	+
(see note 2.3.1 on jointing)			aqueous	40	+	+		+	O	+		+	+			O
(SpRB)				60	+	+		+	-	+		O	+			O
				80		+				+		-	O			-
				100						+						
				120						+						
				140												
Sulfuric acid	H_2SO_4	195	up to 80%,	20	+	+	-	+	+	+		O	+	-	-	+
(see note 2.3.1 on jointing)			aqueous	40	+	+		+	+	+		O	+			O
(SpRB)				60	+	+		O	O	+		-	O			-
				80		+				+			-			
				100						+						
				120						O						
				140												
Sulfuric acid	H_2SO_4	250	90%, aqueous	20	+	+	-	O	O	+		-	+	-	-	-
(see note 2.3.1 on jointing)				40	+	+				+			+			
(SpRB)				60		+				+						
				80						+						
				100						O						
				120						O						
				140												
Sulfuric acid	H_2SO_4		96%, aqueous	20	+	+	-	-	-	+		-	+	-	-	-
(see note 2.3.1 on jointing)				40	+	+				+			+			
(SpRB)				60	O	+				-						
				80												
				100												
				120												
				140												
Sulfuric acid	H_2SO_4		97%	20	+	+	-	-	-	O		-	+	-	-	-
(see note 2.3.1 on jointing)				40		+										
(SpRB)				60												
				80												
				100												
				120												
				140												
Sulfuric acid	H_2SO_4	340	98%	20	+	+	-	-	-	-		-	O	-	-	-
(see note 2.3.1 on jointing)				40	O	+										
(SpRB)				60		O										
				80												
				100												
				120												
				140												
Sulfurous acid	H_2SO_3		saturated,	20	+	+	O	+	+	+		+	+	-	-	O
			aqueous	40	+	+	+	+	+	+		-	+			O
				60	O			+	+	+			O			-
				80						+			-			
				100						+						
				120												
				140												

(Courtesy George Fischer Engineering Handbook)

Aggressive Media					Chemical Resistance											
Medium	Formula	Boiling point °C	Concentration	Temperature °C	PVC	CPVC	ABS	PE	PP-H	PVDF (SYGEF)		EPDM	FPM	NBR	CR	CSM
Sulfuryl chloride	SO_2Cl_2	69	technically pure	20	-	-	-	-	-	O			+	-	O	+
				40												
				60												
				80												
				100												
				120												
				140												
Surfactants (SpRB)			up to 5%, aqueous	20	O	-	-	+	+	+		+	+	+	+	+
				40	O				O	O						
				60	O				O	O						
				80						O						
				100												
				120												
				140												
Surfactants (ESC)				20	O	O	O	O	O	O		O	O	O	O	
				40												
				60												
				80												
				100												
				120												
				140												
Tallow (SpRB)			technically pure	20	+	-	-	+	+	+		+	+	+	+	+
				40	+			+	+	+		+	+	+	+	+
				60	+			+	+	+		+	+	+	+	+
				80						+						
				100						+						
				120												
				140												
Tannic acid (SpRB)			all, aqueous	20	+	+	+	+	+				+	+	+	+
				40		+	+	+	+							
				60		+		+	+							
				80												
				100												
				120												
				140												
Tanning extracts form plants (SpRB)			usual commercial	20	+	+	+	+	+	+		+	+	+	+	+
				40		+	+									
				60												
				80												
				100												
				120												
				140												
Tartaric acid				20						+						
				40						+						
				60												
				80												
				100												
				120												
				140												
Tartaric acid	$HO_2C\text{-}CH(OH)\text{-}CH(OH)\text{-}CO_2H$		all, aqueous	20	+	+	+	+	+	+		+	+	+	+	+
				40	+		+	+	+	+		+	+	+	+	+
				60	O			+	+	+		+	+	O	+	+
				80						+						
				100						+						
				120						+						
				140												

(Courtesy George Fischer Engineering Handbook)

Aggressive Media					Chemical Resistance											
Medium	Formula	Boiling point °C	Concentration	Temperature °C	PVC	CPVC	ABS	PE	PP-H	PVDF (SYGEF)		EPDM	FPM	NBR	CR	CSM
Tartaric acid up to 10%				20						+						
				40						+						
				60						+						
				80												
				100												
				120												
				140												
Tetrachlorethylene				20	-	-	-	-	-	+		-	+			
				40												
				60												
				80												
				100												
				120												
				140												
Tetrachloroethane	$Cl_2CH\text{-}CHCl_2$	146	technically pure	20	-	-	-	O	O	+		-	O	-	-	-
				40						+						
				60						O						
				80												
				100												
				120												
				140												
Tetrachloroethylene	see Perchloroethylene	121														
Tetraetylene lead (SpRB)	$(C_2H_5)_4Pb$		technically pure	20	+	+	-	+	+	+		O	+	+	O	+
				40						+						
				60						+						
				80						+						
				100						+						
				120						+						
				140												
Tetrahydrofurane	C_4H_8O	66	technically pure	20	-	-	-	O	O	-		O	-	-	-	-
				40												
				60												
				80												
				100												
				120												
				140												
Tetrahydronaphthalene	Teralin	207	technically pure													
Thionyl chloride	$SOCl_2$	79	technically pure	20	-	-	-	-	-	-		O	+	-	-	-
				40												
				60												
				80												
				100												
				120												
				140												
Tin (IV) -chloride				20	+	+				+		+	+			
				40	+	+				+		+	+			
				60	+	+				+		+	+			
				80		+				+			+			
				100						+						
				120												
				140												

(Courtesy George Fischer Engineering Handbook)

Aggressive Media					Chemical Resistance											
Medium	Formula	Boiling point °C	Concentration	Temperature °C	PVC	CPVC	ABS	PE	PP-H	PVDF (SYGEF)		EPDM	FPM	NBR	CR	CSM
Tin-(II)-chloride	$SnCl_2$			20				+	+							
				40				+	+							
				60				+	+							
				80					+							
				100												
				120												
				140												
Toluene	C_6H_5-CH_3	111	technically pure	20	-	-	-	O	O	+		-	+	-	-	-
				40												
				60												
				80												
				100												
				120												
				140												
Triacetin (Glycerintriacetat)	$C_9H_{14}O_6$			20	-	-	-	+	+	+		+				
				40						+						
				60												
				80												
				100												
				120												
				140												
Tributylphosphate	$(C_4H_9)_3PO_4$	289	technically pure	20	-	-	-	+	+	+		+	-	-	-	-
				40				+	+							
				60				+	+							
				80												
				100												
				120												
				140												
Trichloroacetic acid	Cl_3C-COOH	196	technically pure	20	O	-	-	+	+	O		O	-	-	-	-
				40				O	+							
				60				-	O							
				80												
				100												
				120												
				140												
Trichloroacetic acid	Cl_3-C-COOH		50%, aqueous	20	+	-	-	+	+	+		O	-	-	-	-
				40	O			+	+	+						
				60				+	O	O						
				80						-						
				100												
				120												
				140												
Trichloroethane	Methylchloroform	74	technically pure													
Trichloroethylene	Cl_2C=CHCl	87	technically pure	20	-	-	-	-	O	+		-	+	-	-	-
				40						+						
				60						+						
				80						O						
				100												
				120												
				140												
Trichloromethane	Chloroform	61														

(Courtesy George Fischer Engineering Handbook)

Aggressive Media					Chemical Resistance											
Medium	Formula	Boiling point °C	Concentration	Temperature °C	PVC	CPVC	ABS	PE	PP-H	PVDF (SYGEF)		EPDM	FPM	NBR	CR	CSM
Tricresyl phosphate (SpRB)	H_3C-C_6H_5-O $)_3PO_4$		technically pure	20	-	-	-	+	+			+	-	O	-	-
				40				+						-		
				60				+	O							
				80												
				100												
				120												
				140												
Triethanolamine (SpRB)	$N(CH_2$-CH_2-$OH)_3$	Fp. *21	technically pure	20	O	-	-	+	+	+		O	-	O	-	-
				40						+						
				60												
				80												
				100												
				120												
				140												
Triethylamine (SpRB)	$N(CH_2$-$CH_3)3$	89	technically pure	20	-	-	-	+	+	O		-	-	-	-	-
				40						-						
				60												
				80												
				100												
				120												
				140												
Trifluoro acetic acid (SpRB)	F_3C-COOH		up to 50%	20	-	-	-	+	+	+		O	-	-	-	-
				40						O						
				60												
				80												
				100												
				120												
				140												
Trioctyl phosphate (SpRB)	$(C_8H_{17})_3$ PO_4		technically pure	20	-	-	-	+	+	O		+	-	O	-	-
				40				+	O							
				60												
				80												
				100												
				120												
				140												
Turpentine oil (SpRB)			technically pure	20	+	-	-	O	-	+		-	+	O	-	-
				40	O			O					+			
				60									+			
				80												
				100												
				120												
				140												
Urea (SpRB)	H_2N-CO-NH_2	Fp.*, 133	up to 30%, aqueous	20	+	+	+	+	+	+		+	+	+	+	+
				40	+	+	+	+	+	+		+	+	+	+	+
				60	O	+		+	+	+		+	+	+	+	+
				80		O				+						
				100						O						
				120												
				140												
Urine				20	+	+	+	+	+	+		+	+	+	+	+
				40	+	+	+	+	+	+		+	+	+	+	+
				60	O	+		+	+	+		+	+	+	+	+
				80						+						
				100						+						
				120												
				140												

(Courtesy George Fischer Engineering Handbook)

Aggressive Media					Chemical Resistance											
Medium	Formula	Boiling point °C	Concentration	Temperature °C	PVC	CPVC	ABS	PE	PP-H	PVDF (SYGEF)		EPDM	FPM	NBR	CR	CSM
Vaseline			technically pure	20	O	O	-	+	O	+		-	+	+	-	-
				40	-					+			+	+		
				60				-	-	+			+	+		
				80						+			+	+		
				100						+			+	+		
				120						+			+			
				140												
Vegetable oils				20	O	-	-	+	+	+		-	+	+	O	O
				40						+						
				60						+						
				80												
				100												
				120												
				140												
Vegetable oils and fats (SpRB)				20	+	O	-	+	+	+		-	+	+	O	O
				40	O			O	+	+			+	+	O	O
				60					O	+			+	+	-	-
				80						+						
				100						+						
				120												
				140												
Vinegar	see wine vinegar															
Vinyl acetate	$CH_2{=}CHOOCCH_3$	73	technically pure	20	-	-	-	+	+	+		+	-	-	-	-
				40				+		-						
				60					O							
				80												
				100												
				120												
				140												
Vinyl chloride	$CH_2{=}CHCl$	-14	technically pure	20	-	-	-	-	-	+		-	+	-	-	
				40						+						
				60												
				80												
				100												
				120												
				140												
Viscose spinning solution				20	+	-	-	+	+	+			+	-	O	+
				40	+			+	+	+		+	+		O	+
				60	+			+	+	+		+	+		-	+
				80												
				100												
				120												
				140												
Waste gases containing - Alkaline				20	+	+		+	+	+		+	+	+	+	+
				40	+	+		+	+	O		+	+	+	+	+
				60	+	+		+	+	-		+	+	+	+	O
				80		+			+			+	O			-
				100									-			
				120												
				140												
Waste gases containing - Carbon oxides			all	20	+	+		+	+	+		+	+	+	+	+
				40	+	+		+	+	+		+	+	+	+	+
				60	+	+		+	+	+		+	+	+	+	+
				80		+			+	+		+	+			+
				100						+			+			
				120						+						
				140												

(Courtesy George Fischer Engineering Handbook)

Aggressive Media					Chemical Resistance											
Medium	Formula	Boiling point °C	Concentration	Temperature °C	PVC	CPVC	ABS	PE	PP-H	PVDF (SYGEF)		EPDM	FPM	NBR	CR	CSM
Waste gases containing - Hydrochloric acid			all	20	+	+		+	+	+		+	+	O	+	+
				40	+	+		+	+	+		+	+	-	+	+
				60	+	+		+	O	+		+	+		+	+
				80		+				+		O	+			+
				100						+			+			
				120						+						
				140												
Waste gases containing - Hydrogen fluoride (SpRB)			traces	20	+	+		+	+	+		+	+	+	+	+
				40	+	+		+	+	+		+	+	O	+	+
				60	+	+		+	+	+		O	+	-	O	+
				80		+				+			+			
				100						+						
				120												
				140												
Waste gases containing - Nitrous gases			traces	20	+	+		+	+	+		+	+	O	+	+
				40	+	+		+	+	+		+	+	-	+	+
				60	+	+		+	O	+		+	+		O	+
				80		+				+		O	+			O
				100						+			O			
				120												
				140												
Waste gases containing - Sulphur dioxide			traces	20	+	+		+	+	+		+	+	O	+	+
				40	+	+		+	+	+		+	+	-	+	+
				60	+	+		+	+	+		+	+		+	+
				80		+			+	+		+	+			+
				100						+			+			
				120						+						
				140												
Waste gases containing - Sulphur trioxide (SpRB)			traces	20	+	+		+	+	+		+	+	O	+	+
				40	+	+		+	+	+		+	+	-	+	+
				60	+	+		+	O	+		+	+		+	+
				80		+				+		O	+			
				100						+						
				120												
				140												
Waste gases containing - Sulphuric acid			all	20	+	+		+	+	+		+	+	O	+	+
				40	+	+		+	+	+		+	+	-	+	+
				60	+	+		+	+	+		+	+		+	+
				80		+			O	+		O	+			+
				100						+			+			
				120						+						
				140												
Water - distilled - deionised	H_2O	100		20	+	+	+	+	+	+			+	+	+	+
				40	+	+	+	+	+	+		+	+	+	+	+
				60	+	+	+	+	+	+		O	+	+	+	+
				80		+			+	+		-	+	+		+
				100					+	+			+	+		
				120						+			+			
				140												
Water, condensed				20	+	+	+	+	+	+			+	+	+	+
				40	+	+	+	+	+	+		+	+	+	+	+
				60	O	+	+	+	+	+		O	+	+	+	+
				80		+			+	+			+	O		+
				100						+						
				120						+						
				140												

(Courtesy George Fischer Engineering Handbook)

Aggressive Media					Chemical Resistance											
Medium	Formula	Boiling point °C	Concentration	Temperature °C	PVC	CPVC	ABS	PE	PP-H	PVDF (SYGEF)		EPDM	FPM	NBR	CR	CSM
Water, drinking, chlorinated				20	+	+	+	+	+	+			+	+	+	+
				40	+	+	+	+	+	+		+	+	+	+	+
				60	+	+	+	+	+	+		O	+	+	+	+
				80					+	+			+	O		+
				100					+	+			+			
				120						+						
				140												
Water, waste water without organic solvent and surfactants				20	+	+	+	+	+	+			+	+	+	+
				40	+	+	+	+	+	+		+	+	+	+	+
				60		+	+	+	+	+		O	+	+	+	+
				80		+			+	+			+	+		+
				100						+				O		
				120						+						
				140												
Wax alcohol (SpRB)	$C_{31}H_{63}OH$		technically pure	20	+	O	-	O	O	+		+	+	+	+	-
				40	+			-	-	+		+	+	+	+	
				60	+					+		+	+	+	+	
				80												
				100												
				120												
				140												
Wine vinegar (SpRB)			usual commercial	20	+	o	O	+	+	+		+	O	-	O	+
				40	+			+	+	+			-		-	O
				60	+			+	+	+						-
				80					+	+						
				100						+						
				120												
				140												
Wines, red and white			usual commercial	20	+	O	+	+	+	+		+	+	+	+	+
				40			+	+	+	+						
				60				+	+	+						
				80						+						
				100						+						
				120												
				140												
Xylene	$C_6H_4(CH_3)_2$	138?	technically pure	20	-	-	-	-	-	+		-	+	-	-	-
		144		40						+			O			
				60						O			-			
				80						-						
				100												
				120												
				140												
yeasts			all, aqueous	20	+	+	+	+	+	+		+	+	+	+	+
				40	+	+		+	+	+		+	+	+	+	+
				60		+		+	+	+						
				80		+				+						
				100												
				120												
				140												
Zinc salts	$ZnCl_2$, $ZnCO_3$, $Zn(NO_3)_2$, $ZnSO_4$		all, aqueous	20	+	+	+	+	+	+			+	+	+	+
				40	+	+	+	+	+	+		+	+	O	+	+
				60	O	+		+	+	+		+	+	-	+	+
				80						+						
				100						+						
				120						+						
				140												

(Courtesy George Fischer Engineering Handbook)

Aggressive Media					Chemical Resistance											
Medium	Formula	Boiling point °C	Concentration	Temperature °C	PVC	CPVC	ABS	PE	PP-H	PVDF (SYGEF)		EPDM	FPM	NBR	CR	CSM
Zinccarbonate				20	+	+	+	+	+	+		+	+			
				40	+	+		+	+	+		+	+			
				60	+	+		+	+	+		+	+			
				80		+			+	+		+	+			
				100						+						
				120												
				140												
Zincchloride			saturated	20	+	+	+	+	+	+		+	+			
				40	+	+		+	+	+		+	+			
				60	+	+		+	+	+		+	+			
				80		+			+	+			+			
				100						+						
				120												
				140												
Zincnitrate	$Zn(NO_3)_2$		saturated	20	+	+	+	+	+	+		+	+			
				40	+	+		+	+	+		+	+			
				60	+	+		+	+	+		+	+			
				80		+			+	+			+			
				100						+						
				120												
				140												
Zincoxide			Suspension	20						+						
				40						+						
				60						+						
				80						+						
				100						+						
				120												
				140												
Zincphosphate			saturated	20	+	+	O	+	+	+		+	+			
				40	+	+		+	+	+		+	+			
				60	+	+		+	+	+		+	+			
				80		+			+	+			+			
				100						+						
				120												
				140												
Zincstearate			Suspension	20	-	-	-	+	+	+		+	O			
				40				+	+	+		+				
				60				+	+	+		+				
				80					+	+						
				100						+						
				120												
				140												
Zincsulfate	$ZnSO_4$			20	+	+		+	+	+		+	+			
				40	+	+		+	+	+		+	+			
				60	+	+		+	+	+		+	+			
				80		+			+	+		+	+			
				100						+			+			
				120												
				140												
1-Chloropentan	$C_5H_{11}Cl$			20	-	-	-									
				40												
				60												
				80												
				100												
				120												
				140												

(Courtesy George Fischer Engineering Handbook)

Aggressive Media					Chemical Resistance												
Medium	Formula	Boiling point °C	Concentration	Temperature °C	PVC	CPVC	ABS	PE	PP-H	PVDF (SYGEF)		EPDM	FPM	NBR	CR	CSM	
1,1,2-Trifluoro, 1,2,2-Trichloroethane (Freon 113) (SpRB)	FCl_2C-$CClF_2$	47	technically pure	20	+		-			+			+	+	+	+	
				40	+												
				60													
				80													
				100													
				120													
				140													

(Courtesy George Fischer Engineering Handbook)

CHAPTER 10
SLIPLINING SEWERS

The sliplining of sewers with PE pipe is a fine way to rehabilitate old sewers. By inserting PE pipe in old sewers you can create a new, smooth, watertight interior surface that is far less prone to problems. Standard procedure is to have a continuous length of watertight lining extend from one manhole to the next.

Polyethylene material used to manufacture pipe and fittings for sliplining is either a high-density polyethylene or a minimum polyethylene. These materials must meet or exceed the minimum requirements for cell class 334433 or cell class 234333, respectively, in accordance with ASTM D 3350. Check all ASTM standards prior to making an installation.

There are several considerations that must be reviewed prior to the sliplining procedure. You will need to establish the minimum anticipated clearance that will exist between the lining material and the existing sewer pipe. Flow capacity is another factor to establish. Will there be external loads on the pipe? What will the earth load be? Is hydrostatic pressure expected? Internal pressure and construction loads are additional considerations. How will structural support of the pipe be creating with grouting?

When large sewers are to be lined, they can be examined with video cameras. When you can't gather visual data, it is generally acceptable to plan on a lining diameter that is approximately 10 percent smaller than the diameter of the existing sewer pipe. A difference of five percent is sometimes acceptable when the sewer pipe has a minimum diameter of 24 inches. A liner may float if its diameter is too small. This is a especially true when a force main is lined.

Solid wall PE made in standard outside diameters and to standard dimension ratios (SDR). ASTM has a standard series of rations from which SDRs are developed. The formula for finding a SDR is fairly simple. You divide the specified outside diameter by the specified minimum wall thickness for pipe of solid wall construction. Evaluation of the SDR will show that the lower the SDR, the stiffer the pipe wall. When working with force mains, the SDR must be based on pressure requirements for the system.

When profile wall pipe is to be used, it's important to remember that it is based on standard inside diameters and ring stiffness constants (RSC). This, of course, is in accordance with ASTM requirements. Profile wall pipe can have reinforced walls to strengthen the pipe against diametrical deformation. To obtain a RSC value, you divide the parallel plate load in pounds per foot of pipe length by the deflection. Standard inside diameters for this type of pipe range from 18 to 120 inches.

Sufficient flow capacity for the anticipated hydraulics of a rehabilitated sewer system can be ensured by choosing the proper inside diameter for the pipe liner. Wall stiffness of a liner pipe must be adequate to withstand external pressure from such things as ground water.

PREPARATION

In preparation of lining a sewer there are steps to be taken on the job site. As you might imagine, the existing sewer pipe has to be cleaned. When feasible, a video inspection should be made of the interior of the sewer. In many cases the flow of sewage must be diverted. This can be done by plugging an upstream manhole and pumping sewage to a manhole that is downstream of the work area. A pipe liner that is joined with gaskets can be installed in a flow stream. The decision to divert sewage flow is one that has to be made on a case-by-case basis. Jobs that require more than one day to complete must be equipped to make a temporary tie-in during the time that work is stopped.

BLOCKAGES

All blockages must be removed. Any obstruction is a risk and must be removed. Depending upon the circumstances, excavation may be required to remove blockages. Care should be taken when making decisions on excavation plans. For example, plan an excavation that will minimize traffic congestion. It's not uncommon for sliplining to be inserted in two directions from a single excavation site. This can greatly reduce the disturbance of traffic and other forms of complications related to excavations.

When digging a pit to work in, the excavation should be sloped at a 2.5 to 1 slope. All safety procedures should be followed to prevent personal injury. If the liner pipe will be of a gasket type, the excavation will need to be long enough to allow the length of pipe to be worked with to be inserted in the sewer. In addition, the pit must be long enough to allow workers to join and insert the liner. When fused pipe is used, the length of the excavation should be at least 12 times the diameter of the liner pipe, plus the sloping ends of the work pit.

How wide should a work pit be? At a minimum, it should be as wide as the diameter of the existing sewer, plus 12 inches. Sewers with diameters of between 18 to 48 inches call for a work pit that is as wide as the pipe diameter, plus 18 inches. A sewer with a diameter of 48 inches, or more, will require a width of the outside diameter, plus 24 inches. All bracing and sheeting must be established as required by local conditions and pertinent regulations.

INSTALLATION

Before installation, the top of the existing sewer has to be exposed. It should be exposed to the spring line and the crown of the pipe must be removed for the full length of the work pit. Be careful when removing the top of the existing sewer. Try not to disturb the bottom of the sewer. If the bottom surface of the existing sewer can be maintained it will serve as a solid support for the liner.

In most cases, fused-joint liners are joined outside of the existing sewer. Thermal-butt-fusion or the thermal extrusion welding method is used to join the liner. If SDR solid wall PE pipe is being used as a liner, it may be joined with the use of a stainless steel full encirclement clamp and the proper gasket material. This method is used when the work pit is not at a manhole. Sections of PE pipe with gasket joints can be assembled in the work pit with flow passing through the previously inserted sections.

A power winch and steel cable that is connected to the end of the liner by use of a pulling head is used to insert fused or welded liners. The length of the liner pipe is determined by the capacity of the winch drum and how much power the winch has. There is some risk that the liner pipe will be damaged during insertion. Take this into consideration and make all reasonable efforts to assure a smooth insertion. Once the winch is turned on and the slipliner is being inserted, the process should not be stopped until the liner is in place. Sometimes a pushing-and-pulling process is used to insert a liner. The choice is largely that of the contractor.

There will be a period of relaxation and thermal equilibrium before the liner is ready to be grouted. Check manufacturer recommendations on how long you should wait before sealing the annular space between the liner and the existing sewer.

When lining a sewer with gasket-joined pipe, the pipe is lowered into the work pit one piece at a time. The pipe sections are joined in the pit and then inserted into the sewer. Don't allow the liner to float within the existing sewer. Equal force should be used as the insertion is made.

GROUTING

Grouting is used to stabilize and support some liner installations. The liner must be protected during grouting. There are two options. Enter the grouting material at a very low pressure, or fill the liner with water to pressurize it during the grouting procedure. The liner manufacturer can provide complete specifications on how much grouting pressure the lining material can be subjected to.

In addition to grouting between the liner and the sewer, you will also have to seal the annular space between the liner and where it enters and exists a manhole. A rule of thumb is to install a sealant for a distance equal to 1½ times the pipe diameter of the liner. Wait until relaxation and thermal equilibrium has settled in. Normally, this will not take more than 24 hours, but check manufacturers' recommendations to confirm a satisfactory waiting period.

If a foam sealant is used to seal the manhole locations, the foam should not protrude into the manhole. It should be finished off flush and covered with a quick-setting, non-shrinking cement.

CONNECTIONS

Connections to the new liner will be needed once the liner installation is complete. This requires a portion of the old sewer be removed at the service connections. Once the liner is exposed, connection can be made. The connection to the liner is made with a heat-fusion saddle or a strap-on saddle. If a strap-on saddle is used, it must be held in place with stainless steel clamps. A neoprene gasket is inserted between the liner and the strap-on saddle. Connections made to the saddle fittings are made with boots, full-encirclement clamps, or some other approved method.

The next step is sealing the open space between the old sewer and the grout or liner. This requirement prevents invasion of ground water, backfill material, and other debris. Once this is done, you are ready to backfill the work area.

BACKFILLING

Before backfilling, all exposed PE pipe and components must be protected. This is done by encasing them with a cement-stabilized sand or some other suitable high-density material. All soil and debris should be removed from the area to be encased. Once the encasement material is in place and approved, the backfilling process may begin. All laterals must be supported prior to being backfilled. Normal backfilling procedures are then used to repair the work area.

COST EFFECTIVENESS

Sliplining is a cost-effective means of refurbishing old sewers. The process is much less costly than full sewer replacement. In addition to the financial savings, the sliplining process is far less intrusive on local traffic and other activities than a full replacement would be. PE pipe is a very versatile material that can be used for numerous cost-saving applications.

CHAPTER 11
TESTING

The inspection and testing of a new installation is essential to a safe installation. Suitable testing is required by various codes. Even if official testing and inspection is not required, common sense would call for careful testing and inspection. Testing methods differ between pressurized pipe and nonpressurized pipe.

Air pressure is often used on nonpressurised installations during testing. But, air and gasses should never be used when testing pressurized systems. There is far too much risk of property and personal damage when a pressurized system is tested with air. Potable water is the preferred medium for testing pressurized systems.

Test methods can also vary between underground installations and aboveground installations. Additionally, some leakage is sometimes allowed for certain types of underground pipelines. Leakage is not allowed in aboveground installations. Another consideration is the type of pipe and joints being tested. For example, systems constructed with solvent-welded joints require plenty of curing time before being tested. Piping installed with mechanical connectors can be tested right away.

When testing a plastic piping system, the test section should be restrained from sudden uncontrolled movement in the event of rupture. If expansion joints are in the system, they should be temporarily restrained, isolated, or removed during the pressure test. When testing a piping system, the test may be conducted on the entire system, or in sections. The test section size is determined y test equipment capability. If the pressurizing equipment is too small, it may not be possible to complete the test within allowable testing time limits. If so, higher capacity test equipment, or a smaller test section may be necessary.

A key point in determining the length of the test section is that the longer the section, the harder it may be to find a leak. If possible, test medium and test section temperatures should be less than 100°F. At temperatures above 100°F, reduced test pressure is required. Before applying test pressure, allow the required time for the test medium and the test section to temperature equalize. If in doubt about the testing procedures for a particular application, contact the pipe manufacturer for technical assistance. Pressure testing procedures may or may not be applicable depending upon piping products and/or piping applications.

Valves or other devices may limit the test pressure. Such components may not be able to withstand the required test pressure, and should be either removed from, or isolated from the section being tested to avoid possible damage to these components. For continuous pressure systems where test pressure limiting components or devices have been isolated, or removed, or are not present in the test section, the maximum allowable test pressure is 1½ times the system design pressure at the lowest elevation in the section under test. If the test pressure limiting device or component cannot be removed or isolated, then the limiting section or system test pressure is the maximum allowable test pressure of that device or component. For non-pressure, low pressure, or gravity flow systems, consult the piping manufacturer for the maximum allowable test pressure.

The time of a pressure test is often dictated by the applicable code. If there is no code requirement, a good rule of thumb is to conduct a four-hour test. Pressure testing time should not exceed eight hours. If the pressure test is not completed due to leakage, equipment failure, etc., the test section should be depressurized and allowed to "relax" for at least eight hours before bringing the test section up to test pressure again.

Test equipment and the pipeline should be examined before pressure is applied to ensure that connections are tight, necessary restraints are in place and secure, and components that should be isolated or disconnected are in fact, isolated or disconnected. All low pressure filling lines and other items not subject to the test pressure should be disconnected or isolated.

The two types of pressure testing are hydrostatic and pneumatic. Hydrostatic pressure testing with clean water is the preferred method. The test section should be completely filled with the test medium, taking care to bleed off any trapped air. Venting at high points may be required to purge air pockets while the test section is filling.

Pneumatic testing of the plastic piping system may be conducted with compressed air or any pressurized gas. However, pneumatic pressure testing my present severe hazards to personnel in the vicinity of lines being tested. Extra personnel protection precautions should be observed when a gas under pressure is used as the test medium. Pneumatic testing should not be used unless the Owner and responsible Project Engineer specify pneumatic testing or approve its use as an alternative to hydrostatic testing.

The testing medium should be non-flammable and non-toxic. Leaks may be detected using mild soap solutions or other not-deleterious leak detecting fluids applied to the joint. Bubbles indicate leakage. After leak testing, all soap solutions or leak-detecting fluids should be rinsed of the system with clean water.

After pressure testing, cleaning the piping sysem is often required. If cleaning is required, sedimentation dpepostis can often be flushed from the system using high-pressure water. Water-jet cleaning uses high-pressure water from a nozzle that is pulled through the pipe system with a cable. Pigs may also be used to clean pressure piping systems. Pigging involves forcing a plastic plug (soft pig) through the pipeline. The pig is forced down the pipeline by hydrostatic or pneumatic pressure that is applied behind the pig. When using pigs to clean a pipeline, a pig launcher and a pig catcher is required. A pig launcher is a wye or a removable spool. In the wye, the pig is fitted into the branch, then the branch behind the pig is pressurized to move the pig into the pipeline and downstream. In the removable pipe spool, the pig is loaded into the spool, the spool is installed into the pipeline, and then the pig is forced downstream. A pig may discharge from the pipeline with considerable velocity and force. The pig catcher is a basket or other device at the end of the line designed to receive the pig when is discharges from the pipeline. The pig catcher provides a means of safe pig discharge from the pipeline. Soft pigs must be used with plastic pipe. Brush type pigs may severely damage a plastic pipe and must not be used.

LOW-PRESSURE TEST

A low-pressure test is normally conducted prior to a high-pressure test. The low-pressure test is done prior to the piping being covered or concealed. Generally, the test pressure should never exceed 50 pounds per square inch (PSI). Initial testing should be done to detect any leaks that will be evident at low pressure. The duration of an initial test varies and should always be maintained long enough to allow the detection of any leaks.

A standard pressure gauge can be used to monitor any pressure drops. However, it can take hours of observation to see if the pressure gauge indicates any leaks. In the meantime, a visual inspection of all joints can be helpful in locating leaks. It's rare for pipe to leak along its length, but this can happen. For example, PVC pipe that was dropped on a hard surface can crack and leak when tested. PE pipe that has been punctured can also leak along a section of pipe.

When conducting a test, maintain the test conditions long enough to find all leaks. Any leaks missed on an initial test can be much more costly to repair when an installation is complete and concealed. It is wise to wait at least 24 hours before testing pipelines that include solvent-welded joints. In some cases with larger pipe and fittings, the drying time will be longer. This is especially true in cold weather. Waiting up to 48 hours to conduct a test may be wise when working with large pipe in cold temperatures.

HIGH-PRESSURE TEST

A high-pressure test is needed before a piping system is put into service. It is often recommended that a high-pressure test be maintained for at least 12 hours. Pipe rupture is a risk during high-pressure testing. Therefore, nonessential personnel should vacate the test area during testing.

Water hammer can be a problem when performing a hydrostatic, high-pressure test. If water is introduced into a piping system too quickly a water hammer can result. This can damage the piping and should be avoided. Water hammer can be avoided by introducing water into a system at lower velocities.

UNDERGROUND PIPING

Underground piping must be secured prior to testing. This is done with bracing and partial backfilling. If proper precautions are not taken in advance, the testing procedure could cause pipe movement and potential damage. Thrust blocks at fittings should be in place prior to testing.

Air will need to be purged from a pipeline that is being tested with water. Provisions for air relief must be installed on the pipeline prior to testing. Ends of the pipe must be capped to prevent leakage during testing. Bracing may be needed to keep the test caps in place when they are under pressure. For example, the thrust at the end of a 100-foot run of 2-inch pipe is 455 pounds. The thrust on a 100-foot length of 8-inch pipe is 5,930 pounds.

Nonpressurized, underground piping is often tested by filling vertical risers with water. A ten-foot head of water over a pipeline can be all that is needed. If the pipeline will have manholes or cleanouts at grade level, the risers are already in place. It's common for several risers to be installed along a pipeline. This is a simple way of testing, but fluctuations in air temperature can cause the static head of water in the risers to move. Don't let this movement fool you when looking for leaks.

It can take a lot of water to fill a pipeline. This generally is not a problem, but there are times when water must be hauled to a site and pumped or poured into a pipeline for testing. Reaching 160 psi in a 2-inch pipe that is 100 feet long will take 20 gallons of water. The same conditions with an 8-inch pipe would require

259 gallons of water. When the water is released it must have a place to go. Keep this in mind when you are planning your test procedures. The test pressure should never be more than 1.5 times the designed maximum operating pressure of the pipe, fittings, valves, or other elements included in the test. Usually, a test pressure that is 50 psi over the designed maximum operating pressure will be sufficient. The best advice on test pressure is to consult code requirements and manufacturers' recommendations.

TEST CAPS

The removal of test caps can be dangerous. Over the years I've seen several workers come close to serious accidents when removing test gear. Create a safety procedure for removing test equipment and enforce the safety rules. Inspectors will sometimes require test equipment to be removed in their presence so that they can see firsthand that a full test has been applied. There are many tricks of the trade that have been used from time to time to fool inspectors. Seasoned inspectors know most of the tricks and may require proof of testing. Removing an inflatable test ball or a cap can result in a lot of pressure at the point of removal. Make sure that no one is in front of that force.

ABOVE GROUND TESTS

Above ground tests are very similar to those used with underground systems. However, while limited leakage may be allowed in some situations with an underground system, no leakage is allowed in an above ground system. Before conducting any tests, you should consult the designer of the system to be tested. If there are no engineered specifications for testing, contact the manufacturers of materials installed in the piping system to obtain ideal test pressure.

Don't cut corners when testing and inspecting a system. Any time saved in rushing a test will likely be lost in making additional tests and repairs later. In addition to potential financial losses, the risk of personal injury is always a serious consideration. Play by the rules and make your tests and inspections effective.

CHAPTER 12
PROTECTING PUBLIC SAFETY THROUGH EXCAVATION DAMAGE PREVENTION

Source: National Transportation Safety Board Safety Study NTSB/SS-97/01, Washington, D.C.

INTRODUCTION

Pipeline accidents result in fewer fatalities annually than accidents in the other modes of transportation; however, a single pipeline accident has the potential to cause a catastrophic disaster that can injure hundreds of persons, affect thousands more, and cost millions of dollars in terms of property damage, loss of work opportunity, community disruption, ecological damage, and insurance liability. In March 1994, a pipeline accident in Edison, New Jersey, injured 112 persons, destroyed eight buildings, and resulted in the evacuation of 1,500 apartment residents [1]. Accident damage exceeded $25 million. The National Transportation Safety Board's investigation determined that the probable cause.of the accident was excavation damage to the exterior of a 36-inch gas pipe. Less than 3 months later, a gas explosion in Allentown, Pennsylvania, resulted in 1 fatality, 66 injuries, and more than $5 million in property damage [2]. The Safety Board concluded that the accident was caused by a service line that had been exposed during excavation and had subsequently separated at a compression coupling.

A propane gas explosion on November 21, 1996, in the Rio Piedras shopping district of San Juan, Puerto Rico, resulted in 33 fatalities and 69 injuries. This accident, one of the deadliest in pipeline history, made 1996 a record year for pipeline fatalities. The San Juan accident accounted for more fatalities than occurred the entire previous year, and it vividly illustrates the tragic potential of a single excavation damaged pipe.

The Safety Board determined that the probable cause of the propane gas.explosion, fueled by an excavation-caused gas leak in the basement of the Humberto Vidal, Inc., office building, was the failure of San Juan Gas Company, Inc., to oversee its employees' actions to ensure timely identification and correction of unsafe conditions and strict adherence to operating practices; and to provide adequate training to employees [3].

Also contributing to the explosion was the failure of the Research and Special Programs Administration/Office of Pipeline Safety to effectively oversee the pipeline safety program in Puerto Rico; the failure of the Puerto Rico Public Service Commission to require San Juan Gas Company, Inc., to correct identified safety deficiencies; and the failure of Enron Corporation to adequately oversee the operation of San Juan Gas Company, Inc. Contributing to the loss of life was the failure of San Juan Gas Company, Inc., to adequately inform citizens and businesses of the dangers of propane gas and the safety steps to take when a gas leak is suspected or detected.

In 1994, a tragic pipeline accident occurred in Caracas, Venezuela. A 22-ton trenching device, working on a road construction project, struck a 10-inch gas transmission line. An occupied bus and cars stopped by the road construction were engulfed in flames. Fifty-one persons were killed and 34 injured. The next year, in April 1995, construction work on a subway system in Taegu, Korea, ruptured a gas line, killing 103 persons. These accidents occurred in systems that do not operate under U.S. regulations, but they illustrate the catastrophic consequences that can result from excavation damage to underground facilities [4].

Excavation and construction activities are the largest single cause of accidents to pipelines. Data maintained by the U.S. Department of Transportation (DOT), Research and Special Programs Administration (RSPA), Office of Pipeline Safety (OPS), indicate that damage from outside force is the leading cause of leaks and ruptures to pipeline systems, accounting for more than 40 percent of the reported failures [5]. According to the data, two-thirds of these failures are the result of third-party damage; that is, damage caused by someone other than the pipeline operator. Reports from the 20 th World Gas Congress confirm that excavator damage is also the leading cause of pipeline accidents in other countries [6].

According to the Network Reliability Steering Committee (NRSC), an industry group appointed by the Federal Communications Commission, excavation damage is also the single largest cause of interruptions to fiber cable service. Network reliability data, compiled since 1993 by NRSC, show that more than half of all facility outages are the result of excavation damage (53 percent), and in more than half of those cases (51 percent), the excavator failed to notify the facility owner or provided inadequate notification [7]. The Safety Board's review of NRSC first quarter data for 1997 indicates that this relationship has not changed. In addition to being expensive and inconvenient, disruption of the telecommunications network can have significant safety implications, such as impact on traffic control systems, health services, and emergency response activities. The Federal Aviation Administration's (FAA) study of cable cuts in 1993 documented 1,444 equipment outages or communications service disruptions resulting from 590 cable cuts nationwide over a 2-year period. The majority of cable cuts were relatedto construction and excavation activities [8]. For 1995, the FAA's National Maintenance Control Center documented cable cuts that affected 32 air traffic control facilities, including five en route control centers. Cable cuts for the first 8 months of 1997 affected air traffic control operations for a total of 158 hours [9].

The Safety Board has long been concerned about the number of excavation-caused pipeline accidents. Because of several excavation-caused pipeline accidents that occurred between 1968 and 1972, the Safety Board sponsored a symposium on pipeline damage prevention [10]. Many of the proposals developed at that April 1972 symposium led to a Safety Board special study on damage prevention and recommendations that resulted in many of the concepts and systems that have now been implemented to minimize excavation-caused damage to pipelines; for example, the local utility location and coordinating councils (ULCCs) established by the American Public Works Association (APWA) [11]. Since that symposium, the Safety Board has continued to support the initiatives of the APWA, the States, and the national organizations to reduce excavation damage to pipelines. The Safety Board has been an advocate of strong damage prevention programs through its recommendation process and through testimony before Congress and State legislatures, and before groups and trade associations interested in pipeline safety, such as the Interstate Natural Gas Association of America (INGAA), the American Public Gas Association (APGA), the Association of Oil Pipe Lines (AOPL), the American. Gas Association (AGA), and the American Petroleum Institute (API).

The combined efforts of industry, the States, the Safety Board, and other Federal agencies led to a decrease in the number of accidents during the 1980s. Nevertheless, excavation-caused damage remains the largest single cause of pipeline accidents. The Board is currently investigating three other accidents that involved excavation: Gramercy, Louisiana; Tiger Pass, Louisiana; and Indianapolis, Indiana [12].

In response to six serious pipeline accidents during 1993 and 1994 that were caused by excavation damage and to foster improvements in State excavation damage prevention programs, the Safety Board and RSPA jointly sponsored a workshop in September 1994 [13, 14]. This workshop brought together about 400 representatives from pipeline operators, excavators, trade associations, and local, State, and Federal government agencies to identify and recommend ways to improve prevention programs.

On May 20, 1997, the Safety Board updated its "Most Wanted" list of safety improvements to include excavation damage prevention [15]. The Board's recommendations on this issue address requirements for excavation damage prevention programs; comprehensive education and training for operators of buried facilities and the public; and effective government monitoring and enforcement.

This safety study, "Protecting Public Safety Through Excavation Damage Prevention," was initiated to analyze the findings of the 1994 workshop, to discuss industry and government actions undertaken since the workshop, and to formalize recommendations aimed at further advancing improvements in excavation damage prevention programs. Chapter 2 of the study provides some background information on the subsurface infrastructure and an overview of pertinent regulatory and legislative initiatives. Chapter 3 discusses the various components of a damage prevention program, Chapter 4 discusses the accuracy of information regarding buried facilities, Chapter 5 addresses system performance measures, and the last sections contain the Safety Board's conclusions and recommendations.

OVERVIEW OF SUBSURFACE INFRASTRUCTURE AND REGULATORY AND LEGISLATIVE INITIATIVES

Subsurface Infrastructure

The term "underground facilities" generally refers to the buried pipelines and cables that transport petroleum, natural gas, electricity, communications, cable television, steam, water, and sewer. These subsurface networks are constructed of cast iron, steel, fiberglass, copper wire, concrete, clay, plastic, or optical fiber depending on the age of the system and its product content. In addition to being categorized by product type and structural components, underground networks are further grouped according to function (gathering, transmission, distribution, service lines); owner (public utility or private industry);.or jurisdiction (municipalities, State, and Federal agencies). The U.S. underground infrastructure comprises about 20 million miles (32.2 million kilometers) of pipe, cable, and wire [16].

Pipeline regulation and oversight by DOT distinguishes between the transport of carbon dioxide, hazardous liquids, and gas. Hazardous liquid lines carry petroleum, petroleum products, or anhydrous ammonia. Their functions include gathering lines that transport petroleum from a field production facility to the primary pumping stations. Trunk lines differentiate a line-haul function for the transport of crude oil to refineries and product from the refineries. Gas lines are categorized as gathering, transmission, or distribution. Gathering lines transport gas from a current production facility to a transmission line or processing facility; transmission lines transport gas to distribution centers, storage facilities, or large volume customers; and distribution service lines transport gas to end users [17].

The diverse and segmented nature of underground facilities is evident from the variety of organizational interests that work with the subsurface infrastructure:

- Facility owners design, install, and maintain the underground network..Owners are a diverse group with varied interests; they include private.corporations, municipal public works systems, private and public utility.companies, telecommunication providers, and State transportation traffic.control systems.
- Construction crews engage in excavation activities for a variety of reasons, and they use an assortment of government permits. Excavation activities are carried out by building trades, farmers, homeowners, State and local transportation departments, and others.

- States regulate actions to protect safety.
- Insurance companies insure the underground facilities, property, and construction business activities.
- Locators work at excavation sites to identify and mark underground facilities. This work may be conducted by the operators of the underground facility or by locating contractors who specialize in providing underground locating services.
- One-call communication centers coordinate notifications about digging activities. These centers may be an operating unit of a facility owner or they may be independent entities that provide notification service to several facility owners.

The number of times people dig into the underground infrastructure illustrates the sheer frequency of excavation: there were an estimated 13 million excavation notices issued to utility operators across the United States in 1996, though the actual number is higher because some excavators do not use one-call notification services [18].

Urbanization of lands through which utility lines are routed, combined with an increase in the number of users of the underground, has created competition for the underground space. A recent study by the American Farmland Trust states that the rate of farmland lost to development is 2 acres per minute, or 1 million acres per year (0.81 hectare per minute, or 405 000 hectares per year) Increased construction activity, which results in increased excavation, is directly related to population growth, demographic shifts, and a growing national economy [19]. New building construction requires that additional services—more utility lines and communication services—be placed in the underground. There is also a trend in current suburban development to remove aboveground utilities to reduce clutter and storm damage. Additions to the underground infrastructure are installed within the underground space occupied by the existing systems. Thus, increased construction can be considered a corollary of increased excavation. This relationship affects the approach to excavation damage prevention because the desire to avoid damage is a genuine interest of everyone, but the success of damage prevention depends on systematic safeguards.

REGULATORY AND LEGISLATIVE INITIATIVES

Underground facilities and pipelines are addressed by various Federal regulations. The Federal regulations issued by RSPA are contained in Title 49 Code of Federal Regulations (49 CFR) Parts 190-199. Parts 191 and 192 address natural gas regulations; Part 195 covers hazardous liquids and carbon dioxide; and Part 198 prescribes regulations for grants to aid State safety programs. Federal regulations establishing minimum standards for excavation damage prevention programs for gas pipeline operators (49 CFR 192.614) were extended to the operators of hazardous liquid pipelines (49 CFR 195.442) effective April 1995 [20, 21].

Federal regulations mandate companies to develop and participate in damage prevention.programs when those companies transport gas and hazardous liquids subject to DOT jurisdiction [22]. Participation in a one-call notification system satisfies parts of this requirement; consequently, one-call centers have become a key element in damage prevention programs. As the Safety Board's accident investigations and 1994 workshop have pointed out, however, one-call notification programs do not ensure damage prevention. Protection from excavation damage will occur only when facility owners, excavators, locators, and one-call operators—people working with the underground facilities—share responsibilities to protect underground facilities from excavation damage. (These responsibilities are discussed in the next two chapters.)

Both government and industry have, in the past, prepared model statutes that would serve as a framework for individual State legislation of damage prevention programs. The Office of Pipeline Safety Operations (OPSO) prepared model statutes in 1974 and 1977 and encouraged State and local governments to enact model legislation [23]. The APWA prepared guidelines for damage prevention laws that were not substantially different from the OPSO model. The AGA also developed elements for damage prevention legislation; these elements were documented in a 1988 report issued by the Transportation Research Board [24]. Several features of the OPSO model statute overlapped with features in regulations issued by the Occupational Safety and Health Administration.(OSHA) of the Department of Labor.

OSHA regulations require excavators to notify utility owners of planned excavation and to request that the estimated location of underground facilities be marked prior to the start of excavation (29 CFR 1926.651(b)). The regulations also require excavators to determine the exact location "by safe and acceptable means" when they approach the estimated location during excavation [25].

Model legislation was introduced in the 103 d Congress following the 1994 accident in Edison, New Jersey [26]. That bill, which strongly recommended rather than mandated State participation in onecall systems, passed the House of Representatives but not the U.S. Senate. A different version of the bill (HR431/S164) was introduced in, but not adopted by the 104 th Congress. Industry representatives worked with the 105 th Congress.to again develop legislation. The Comprehensive One-Call Notification Act of 1997 (S1115) was introduced into the Senate in July 1997, and the Surface Transportation Safety Act of 1997 (HR1720) was introduced into the House in May 1997. The Senate Committee on Commerce, Science, and Transportation held a public hearing on S1115 in September 1997; the Senate passed the measure on November 9, 1997. The issues related.to the currently proposed legislation are not substantially different from earlier versions; a comparison of the features indicates the following:

- The current bills are advisory in nature rather than prescriptive.
- The House bill recommends participation by all facility owners and excavators; the Senate bill recommends appropriate participation by underground facility operators and excavators. Both contain incentives for.compliance based on providing grant monies for State use.

- Both bills recommend general components to be included in the State programs; the House bill calls these elements, the Senate bill calls them minimum standards. There is no specific guidance for States concerning the organizational structure and funding mechanisms of one-call centers, or the administration of enforcement provisions.
- Both bills include a mechanism for recommending effective damage prevention practices; specifically, the Secretary of Transportation shall study existing one-call systems to determine practices that are most effective in preventing excavation damage.

The recently introduced legislation makes no specific requirements on the States because the Federal government has not exercised jurisdiction over one-call operations, and because States cannot be required to pass legislation. This has led one industry trade publication to characterize both bills as "toothless in terms of being able to require states, excavators, facility operators, or one-call centers" to modify existing practices to achieve the objectives set forth in the legislation [27]. Table 12.1 shows a comparison of the two bills that appeared in a recent trade association newsletter. The Safety Board's position regarding certain provisions of the proposed legislation is discussed in the next chapter.

House Bill HR1720	Senate Bill S1115[a]
1. Shall consider the establishment of a nationwide toll-free telephone number system.	1. (Telephone number language deleted because OCSI already has a system in operation.)
2. Elements of a State program shall include: (a) All excavators and facility operators; (b) 24-hour coverage for emergency notice; (c) Public education about the program; (d) Proper excavation procedures training; (e) Excavators must contact the One-Call Center; (f) Facility operators must mark their facilites; (g) Effective enforcement; (h) Fair free schedule for operating the State program.	2. Minimum standards of a State program to include: (a) appropriate participation by excavators and facility operators considering risks, cost: benefit ratios, and allowing voluntary participation where risk is low; (b) Administrative system of variable penalties.
3. Funding of $1M in fiscal year 1998, "as necessary" in FY 1999, and 2000 to be used for improving damage prevention. Money to come from fees collected from pipeline operators.	3. Funding of: (a) grants of $1M in FY 1999 and $5 M in FY 2000 to be available until expended; (b) funds "as necessary" in FY 1998–2000 for administration. Money to come from general revenues. States would get money when the Secretary of Transportation has determined, from a grant application, that the State is in compliance with the minimum program standards.
4. Secretary of Transportation to develop a model program in consultation with all affected parties, conduct workshops and public education.	4. Secretary to review the one-call system "best practices" and issue a report.

[a] The U.S. Senate passed S1115 on November 9, 1997.

Source: The Conduit 3(4): 1, 4. August 1997. Spooner, WI: National Utility Locating Contractors Association. (The full text of the bills is available on the Internet at www.underspace.com.)

FIGURE 12.1 Comparison by a trade association newsletter of the currently proposed federal legislation for one-call systems.

DAMAGE PREVENTION PRACTICES

In its report of the accident in Allentown, Pennsylvania, on June 9, 1994, the Safety Board highlighted the common elements of effective State excavation damage prevention programs that have been recognized in the industry and that were discussed in detail at the Safety Board's 1994 workshop [28]. The elements include:

1. Mandatory participation by all affected parties, whether private or public
2. A true one-call notification system in which excavators can alert all operators of buried systems
3. Swift, effective sanctions against violators of State damage prevention laws
4. An effective education program for the public, contractors, excavation machine operators, and operators of underground systems that stresses the importance of notifying before excavating, accurately marking buried facilities, and protecting marked facilities when excavating.

Other elements that have been deemed critical to an effective damage prevention program and that have been the subject of past Safety Board recommendations include accurate mapping, employee training, and emergency response planning. This chapter discusses the various aspects of these elements and summarizes the reports and conclusions of the 1994 workshop participants as they relate to these elements.

MANDATORY PARTICIPATION

Every State except Hawaii and the District of Columbia has a damage prevention law to govern the activities of operators and excavators of most buried facilities. Texas, the most recent State to enact legislation, passed the Underground Facility Damage Prevention and Safety Act in June 1997 to establish a non-profit corporation to oversee the State's three one-call systems. The Governor of Puerto Rico is preparing damage prevention legislation for introduction in the Legislative Assembly. In the interim, he has issued an Executive Order that establishes an excavations notice center, requires government facility operators to use the center, and encourages its use by private entities.

Individual States have developed a variety of program approaches to handling the problem of excavation damage of underground facilities. A key finding in a 1995 OPS study was that there were "significant variations among state statutes, among excavators, and among facility operators in the ways that excavation around underground facilities is done " [29]. Table 12.2 provides an overview of the variations among State programs.

More than half the States (30) have mandatory one-call participation programs and most (39) are intended to protect all utilities. However, all but seven States (Connecticut, Iowa, Massachusetts, Maryland, Maine, New Hampshire, and Vermont) have granted exemptions to a variety of organizations. State laws

Overview of the characteristics of State programs to prevent excavation damage.

Program characteristic	Alabama	Alaska	Arizona	Arkansas	California	Colorado	Con-necticut	Delaware	District of Columbia	Florida	Georgia	Hawaii	Idaho
Protects all utilities[a]	✓		✓	✓	✓	✓	✓	✓	✓	✓	✓		✓
Mandatory participation for facility owners			✓	✓	✓	✓	✓	✓	✓	✓	✓		✓
No exemptions for excavation notification							✓						
24-hour access[b]						✓	✓						
Emergency response procedures	✓	✓	✓	✓	✓	✓	✓	✓	✓	✓	✓		✓
Permits issued		✓			✓				✓				
Penalty clause	✓	✓	✓	✓	✓	✓	✓	✓	✓	✓	✓		✓
Notification time 2 days or less		✓				✓				✓			
Positive response			✓	✓	✓		✓	✓	✓	✓			✓
Standard marking color code	✓	✓	✓	✓	✓	✓	✓	✓	✓	✓	✓		✓
Tolerance zone within 24 inches	✓	✓	✓	✓	✓	✓	✓	✓	✓	✓	✓		✓
Year of State legislation	1994	1987	1995	1995	1984	1993	1987	1979	1979	1997	1986		1991
Legislative activity in 1997						✓				✓	✓		
Number of one-call service centers	1	1	1	1	2	1	1	1	1	1	1		6

FIGURE 12.2 Overview of the characteristics of state programs to prevent excavation damage.

Overview of the characteristics of State programs to prevent excavation damage.

Program characteristic	Illinois	Indiana	Iowa	Kansas	Kentucky	Louisiana	Maine	Maryland	Massa-chusetts	Michigan	Minne-sota	Missis-sippi	Missouri
Protects all utilities[a]	✓		✓		✓	✓		✓		✓	✓	✓	✓
Mandatory participation for facility owners	✓		✓	✓		✓	✓	✓	✓	✓	✓		
No exemptions for excavation notification			✓				✓	✓	✓			✓	
24-hour access[b]	✓	✓	✓	✓			✓		✓	✓	✓		✓
Emergency response procedures	✓	✓	✓	✓	✓	✓	✓	✓	✓	✓	✓	✓	✓
Permits issued							✓		✓				
Penalty clause	✓	✓	✓	✓	✓	✓	✓	✓	✓	✓	✓	✓	✓
Notification time 2 days or less	✓	✓	✓		✓		✓	✓			✓	✓	✓
Positive response	✓	✓	✓	✓		✓		✓		✓		✓	✓
Standard marking color code	✓	✓	✓	✓	✓	✓	✓	✓	✓		✓	✓	✓
Tolerance zone within 24 inches	✓	✓	✓	✓	✓	✓		✓		✓	✓		
Year of State legislation	1991	1990	1993	1993	1995	1995	1991	1974	1980	1989	1987	1985	1992
Legislative activity in 1997	✓		✓								✓	✓	
Number of one-call service centers	2	1	1	1	1	1	1	2	1	1	1	1	1

FIGURE 12.2 *continued* Overview of the characteristics of state programs to prevent excavation damage.

Overview of the characteristics of State programs to prevent excavation damage.

Program characteristic	Montana	Nebraska	Nevada	New Hamp-shire	New Jersey	New Mexico	New York	North Carolina	North Dakota	Ohio	Oklahoma	Oregon	Pennsyl-vania
Protects all utilities[a]	✓	✓	✓		✓		✓	✓	✓	✓	✓	✓	✓
Mandatory participation for facility owners	✓	✓	✓	✓	✓		✓		✓	✓		✓	✓
No exemptions for excavation notification	✓			✓		✓		✓					
24-hour access[b]	✓	✓	✓	✓					✓	✓	✓	✓	✓
Emergency response procedures	✓	✓	✓	✓	✓	✓	✓	✓	✓	✓	✓	✓	✓
Permits issued				✓					✓				✓
Penalty clause	✓	✓	✓	✓	✓	✓	✓		✓			✓	✓
Notification time 2 days or less	✓					✓			✓				
Positive response	✓	✓	✓		✓		✓		✓		✓	✓	✓
Standard marking color code	✓	✓	✓	✓	✓	✓	✓	✓	✓	✓	✓	✓	✓
Tolerance zone within 24 inches	✓	✓			✓	✓	✓		✓	✓	✓	✓	✓
Year of State legislation	1991	1995	1987	1983	1994	1991	1997	1986	1995	1990	1982	1997	1996
Legislative activity in 1997				✓		✓		✓		✓			
Number of one-call service centers	2	1	1	1	1	2	2	1	1	2	1	1	1

FIGURE 12.2 *continued* Overview of the characteristics of state programs to prevent excavation damage.

Overview of the characteristics of State programs to prevent excavation damage.

Program characteristic	Rhode Island	South Carolina	South Dakota	Tennes-see	Texas	Utah	Vermont	Virginia	Wash-ington	West Virginia	Wis-consin	Wyoming
Protects all utilities[a]		✓	✓	✓		✓		✓	✓	✓	✓	✓
Mandatory participation for facility owners	✓		✓			✓	✓	✓	✓		✓	✓
No exemptions for excavation notification				✓			✓					
24-hour access[b]	✓		✓	✓	✓		✓			✓		
Emergency response procedures	✓	✓	✓	✓	✓	✓	✓	✓	✓	✓	✓	✓
Permits issued	✓						✓					✓
Penalty clause	✓	✓		✓		✓	✓	✓	✓		✓	✓
Notification time 2 days or less	✓		✓			✓	✓			✓		✓
Positive response		✓	✓	✓		✓		✓			✓	✓
Standard marking color code	✓	✓		✓	✓	✓	✓	✓	✓	✓	✓	✓
Tolerance zone within 24 inches	✓			✓	✓	✓	✓	✓	✓	✓		✓
Year of State legislation	1984	1978	1993	1978	1997	1977	1988	1994	1990	1996	1995	1996
Legislative activity in 1997		✓	✓	✓	✓							
Number of one-call service centers	1	1	1	1	3	1	1	3	8	1	1	10

[a] Subject to each State's definition of activity.

[b] Although notification calls may be recorded on a 24-hour basis, most States process notifications only during normal business hours.

Source: One-Call Systems International Directory, 1997–1998. The criteria identified for program characteristics were taken from the proceedings of the National Transportation Safety Board's workshop held in 1994 (NTSB/RP-95/01).

FIGURE 12.2 *continued* Overview of the characteristics of state programs to prevent excavation damage.

specifically qualify their exemptions, but, in general, exempt organizations are not required to participate in the State's excavation damage prevention program. Exemptions have been granted to State transportation departments, railroads, mining operations, city/State/Federal governments, cemeteries, water utilities, military bases, and Native American Lands. Although underground facilities frequently follow road rights-of-way, nine States have current damage prevention legislation that specifically exempts State transportation departments; another dozen States exempt substantial State highway maintenance activities. State highway administrators have argued that they do not have the resources for participating in notification and marking. Several States (Arizona, Arkansas, Delaware, Oregon, Mississippi, and Washington) exempt agricultural activities, home owners, and tilling operations less than 12 inches (30.5 centimeters) deep. By receiving exemptions, these entities are not required to inform utilities or underground facility owners of their digging activities, nor are the underground facilities operated by these exempt entities marked or protected in advance of scheduled excavations.

In the 1994 Green River, Wyoming, accident investigated by the Safety Board, a highway contractor operating excavation equipment struck a 10-inch-diameter natural gas gathering line [30]. The accident resulted in three fatalities. The pipeline operator did not participate in the local excavation notification one-call program, though the operator was required by the State of Wyoming to belong to the one-call system. The highway contractor notified the one-call center prior to excavation but did not notify one-call concerning project modifications that expanded the geographic area of work. Neither the Wyoming Department of Transportation nor its contractor made telephone notification directly to the pipeline operator. Had these parties participated in the one-call notification program, the gas line would have been marked and the accident likely would not have happened.

In April 1996, excavation damage of a water main in Buffalo, New York, flooded the downtown area. The municipal water department was not a member of the local one-call system. In fact, at that time four separate city utilities in Buffalo had to be notified to coordinate excavation work; none of those utilities participated in the local one-call system. This situation existed even though State law required participation and made it free for municipalities.

Panelists at the 1994 damage prevention workshop agreed that all owners/operators of buried facilities should participate in damage prevention programs; there should be no exceptions. Some States have realized the value of full participation and have taken legislative action to ensure participation. For example, according to Pennsylvania law, underground facility owners who are not one-call members cannot collect damage costs from excavators who hit their lines. A similar requirement became effective in Florida in October 1997. Oregon has mandatory one-call membership provisions for all facility owners with lines that cross public rights-of-way [31].

The Safety Board agrees that the failure of all parties to participate in damage prevention programs can substantially undermine the effectiveness of the programs. When parties such as State transportation departments and railroads are given exemptions to participation in excavation damage prevention programs, these parties, in essence, are no longer obligated to use one-call notification centers to protect their facilities or to protect the facilities of others that use their rights-of-way. Nor are they obligated to inform other parties of their intent to dig or excavate. In addition to public safety interests, the Board is concerned that taxpayers ultimately bear the burden for these exemptions by paying for the cost of fixing excavation damage, particularly damage caused by State agencies that are not protecting their facilities. The Safety Board concludes that full participation in excavation damage prevention programs by all excavators and underground facility owners is essential to achieve optimum effectiveness of these programs.

Because of the number of State transportation department activities that are exempt from participating in excavation damage prevention programs, the Safety Board believes that the Federal Highway Administration should require State transportation departments to participate in excavation damage prevention programs and consider withholding funds to States if they do not fully participate in these programs.

Although railroad rights-of-way are not as prevalent as those of highways, they frequently serve as ideal routes for underground facilities, particularly for gas and oil transmission lines. Contractual provisions for underground facilities to use railroad rights-of-way are a revenue source for the railroads. However, railroads are also exempt from participating in some State excavation damage prevention programs. For the larger, Class 1 railroads, there are usually internal operating procedures for notification of excavation work on railroad property. However, recent trends in contracting out construction and maintenance services suggest that not all work is controlled through internal operations. Additionally, the number of small, short line railroad companies is increasing. The Association of American Railroads estimates that there are 450–475 short line railroads; 424 are members of the American Short Line Railroad Association (ASLRA). ASLRA membership has doubled in the past 25 years. These smaller companies often do not have the resources to operate internal excavation notification systems. Consequently, the Safety Board believes that the Association of American Railroads and the American Short Line Railroad Association should urge their members to fully participate in statewide excavation damage prevention programs, including one-call notification centers.

ONE-CALL NOTIFICATION SYSTEM

Function and Structure of the Centers

A cornerstone of current damage prevention programs involves the use of one-call notification centers. One-call notification centers function as communication systems established by two or more utilities, government agencies, or other operators of underground facilities to provide one telephone number for notification of excavating, tunneling, demolition, or any other similar work [32]. The system is designed so that excavation contractors, other facility owners, or the general public can notify the one-call center of the location of intended digging or construction activity. The intended area of excavation may be premarked, generally with white spray paint, to specifically indicate the digging or construction area [33]. Based on that one call, the center, in turn, notifies its members that digging or construction will occur in a given location. The facility owners, or their contract locator service, go to the excavation site and mark the location of any of their underground facilities in that area. By avoiding the use of power-driven tools in the vicinity of marked facilities, there is a decreased risk of damage to underground facilities.

Notification services use a variety of names and logos to create meaningful associations in the public's mind: Miss Dig System in Michigan, Underground Service Alert in California, Utility Protection Center in Georgia, and Digger for the Chicago Utility Alert Network. There are 84 one-call centers in the U.S. covering almost all areas of the country [34]. Of these, 55 are members of One-Call Systems International (OSCI). OSCI members recorded over 13 million excavation notifications in 1996 [35].

In 1996, a nationwide referral number, 1-888-258-0808, was established by the APWA and administered by Sprint. In the fall of 1997, this number was automated by the APWA and is handled through the Georgia One-Call Center. A placard containing this number is placed on all newly manufactured construction equipment; placarding resulted from coordination with the Equipment Manufacturers Institute. Ideally, a call to that nationwide referral number would result in an automatic transfer to the appropriate one-call center. This automatic transfer exists on a small scale for the State of California, which uses two one-call systems but uses only one statewide phone number. However, because automated switching of calls to the referral number would result in substantial expenses for long distance telephone charges and billing, the existing referral service informs the caller of the correct one-call number, based on the caller's identification of the geographic location of the excavation, and the caller must then place a second phone call to that center.

The organizational structure of one-call centers varies: some are functioning units of the local ULCC; others are joint efforts of a few facility owners. Statutory language in some States stipulates the composition of the Board of Directors (for example, Minnesota, North Dakota, South Dakota, Nebraska, and Oregon), but government involvement varies by State and by one-call center [36]. Many one-call centers have been organized as not-for-profit corporations that operate with a limited degree of State oversight. Their administrative framework, funding arrangements, and operating procedures also vary. For example, California allows local government agencies to recover all costs of one-call membership through the permit fees that it charges contractors. Several States make participation by municipalities free. According to the OCSI, its member organizations were "developed to best suit the needs of the underground facility owners in that state" and "state laws do not govern the operation of a one-call system. The laws generally set out who is required to belong to a one-call system and who must call a one-call system as well as enforcement provisions. No two State laws are alike" [37].

Methods of Operation

The differences in State involvement translate into very practical distinctions between one-call centers. An assortment of communication methods are used to receive excavators' calls and to issue notification tickets to the centers' participants; centers may use telephone staff operators, voice recorded messages, e-mail, fax machines, Internet bulletin boards, or a combination of methods. Service hours may be seasonally limited to a few hours a day or extend to 24 hours a day. Some locations operate only seasonally because of construction demand; most operate year-round. Most centers have statewide coverage but may not strictly follow State boundaries. A center may cover portions of several States (Miss Utility in Virginia, Maryland, and the District of Columbia) or there may be several centers within a State (Idaho has six different one-call systems; Washington and Wyoming each have nine). Centers may provide training to the construction community, conduct publicity campaigns to educate the public to excavation notification requirements, and work with facility operators to protect their underground facilities. Other centers do little work in these areas.

Some centers use positive response procedures—members who do not mark facilities in the construction area confirm that they have no facilities in the area rather than just not mark a location; other centers do not have this requirement. A part of the Miss Utility program in the Richmond, Virginia, area uses positive response procedures to notify the excavator when the marking is complete. Facility owners directly inform a voice messaging system of the status of a notification ticket. (Notification tickets are identified and discussed in the following section.) As a timesaving alternative, the contractor can call the information system anytime to receive an up-to-date status of their marking request. Information indicating that marking has been completed, or that no facilities are located in the area of excavation, allows construction work to proceed as soon as marking is completed rather than waiting the full time period for which marking activity is allowed.

The important elements of an effective one-call notification center have been generally identified by industry organizations. For example, the position of the Associated General Contractors of America on one-call systems is summarized in six elements: mandatory participation; statewide coverage; 48-hour marking response; standard marking requirements; continuing education; and a fair system of liability [38]. Participants at the Safety Board's 1994 workshop, on the other hand, developed detailed lists of elements they believed are essential for an effective one-call notification center, other elements a center should have, and elements it could have. All agreed, however, that first and foremost was the need for mandatory participation and use of notification centers by all parties. The Safety Board concludes that many essential elements and activities of a one-call notification center have been identified but have not been uniformly implemented.

Excavation Notification Tickets

A record of a locate request is generally called an excavation notification ticket, but there is no standard format for one-call excavation notification tickets. One-call centers track excavation activity based on the number of notification tickets they handle for their members, but they do not necessarily track how many of those digging activities result in excavation hits. For the centers that do maintain a record of hits, one-call members must report their hits to the center; the center then compiles the information.

The OCSI Committee on Communication Standards is developing a universal ticket format to address the problem of underground facility owners who work in different States and who receive tickets from more than one notification center [39]. For large companies working in different one-call areas, information that is organized into different formats can be confusing and can lead to unsafe activities at the excavation site. According to discussions at the Safety Board's 1994 workshop, the format needs to be consistent between centers, both in terms of ticket information and the work unit represented by a ticket. For example, a ticket from one center might encompass work for all utilities at a given two-block construction site, whereas another might separate tickets for each utility, or by smaller geographic areas. Damage reports must also be consistent, and OCSI is considering the feasibility of including damage information in the universal ticket format. The committee expects to finalize a universal ticket format in January 1998. The Safety Board encourages the OCSI members and all other notification centers to adopt a universal ticket format and to maintain ticket data. Standard ticket information would be an essential first step in developing performance measures for damage prevention programs.

EFFECTIVE SANCTIONS

Penalties for failure to act in accordance with State damage prevention programs vary depending on location; provisions for oversight and timely enforcement can be quite different from State to State. Administrative enforcement of State excavation damage prevention laws does not require State court actions and has been shown to be effective in several States. For example, in Pennsylvania's new legislation (Act 187 enacted in 1996), the process of enforcement includes $100 and $200 citations for minor infractions to the State's excavation damage prevention law and $2,500 and $25,000 civil penalties for more serious infractions. The Department of Labor and Industry is responsible for administrative enforcement; the State's Attorney General handles civil penalties.

The use of administrative enforcement is a characteristic of several State programs for excavation damage prevention. The following examples from Massachusetts, Minnesota, and Connecticut illustrate three programs that operate differently but use administrative enforcement to effect safer excavation practices.

The structure and cost of penalties for nonparticipation in the excavation damage prevention programs varies, but the States' common goal is to foster safe practices.

Massachusetts originally passed damage prevention legislation in 1959; it required all excavators to notify utilities before they began to dig. Legislation in 1980 empowered the Department of Public Utilities (DPU) with enforcement authority under the State's Administrative Procedures Act. Beginning in 1986, the Dig Safe Law enforcement was delegated to the chief engineer in the pipeline engineering and safety division of the DPU. A staff of one person handles the administrative enforcement of damage prevention for the State of Massachusetts. That person has authority to issue notices of probable violation with fines that range from $200 to $1,000. In administering the program, the DPU keeps fines at a reasonable level compared to many other States. Since 1986, the State has issued over 3,000 notices; third-party damages dropped from 1,138 in 1986 to 412 in 1993. The Department requires utility companies to report any third party damages within 30 days, and excavators are encouraged to send the Department violation notices for State adjudication if they find fault with the utility companies. State utility owners and excavators are provided books of violation tickets to document infractions of damage prevention rules to the DPU. Using this mechanism, involved parties can notify the State of problems, such as when facilities are mismarked or not marked within the required time, when excavators do not use the notification system, or when line hits are not reported. The State has found that its readiness to dispense small penalties has resulted in awareness of damage prevention throughout the industry. The State's administrative enforcement process does not rely on the Attorney General's office for execution; thus it keeps State pipeline safety actions from being in direct competition with all other State actions.

Other States have also found benefits from administrative rather than court enforcement of their regulations. Minnesota's Department of Public Safety, Office of Pipeline Safety, takes complaints, investigates, and issues penalties of up to $500. The State Office of Pipeline Safety, MnOPS, is the enforcement entity. Minnesota focuses strongly on education as a key to the success of damage prevention. Violators of the regulation often are allowed to institute training actions instead of paying fines.

Connecticut also uses an administrative process to enforce its damage prevention program. Its Call-Before-You-Dig law, first passed in 1978, had only one penalty provision: if an excavator failed to call before digging and subsequently damaged facilities, the excavator could be fined up to $10,000. A representative from Connecticut Call Before You Dig has stated, however, because of the severity, the penalty was not used [40]. An accident on December 6, 1985, in Derby, Connecticut, occurred when an excavator struck a gas line; the excavator had used the notification system [41]. Natural gas from the broken main migrated into the basement of a restaurant, exploded, and killed seven people. The severity of this accident focused attention on the shortcomings of the existing law and resulted in a change in the penalties, fines, and overall structure for enforcement. Connecticut created a position of compliance supervisor, an employee of the one-call center, who serves as a field investigator and expert witness at that State's DPU hearings. The compliance supervisor receives incident reports and maintains case files on noncompliance. Connecticut law allows for fines on the first or second offense if the severity of the offense, injuries, or past performance warrants. Otherwise, the compliance supervisor sends a letter to the party explaining the damage prevention program and stating what compliance actions are needed. The DPU may send letters of inquiry or interrogatories, proceed with a docket for penalties, or schedule a "show cause" hearing. Offending parties have 30 days to appeal by requesting a hearing. In 1994, excavation damage to underground lines in Connecticut declined 28.5 percent compared to earlier years before administrative enforcement. Of the 436 incidents of damage in that year, 223 were gas lines, 91 were water lines, 78 were electric lines, 40 were communication lines, and 4 were sewer lines [42].

Other States have implemented stringent programs for excavation damage prevention, with severe penalties for noncompliance. Because of its small area and concentrated population, New Jersey has a dense network of pipelines: 30,000 miles (48 270 kilometers) of intrastate lines and 1,000 miles (1609 kilometers) of interstate lines [43]. As a result of the 1994 accident in Edison, New Jersey, the State implemented heavy fines and strong enforcement, effective in 1995. Digging near a gas line without calling for the facility owner to mark the location can result in a $25,000 fine, and the company involved in underground facility damage is required to provide a written plan for remediation and training. The number of one-call notifications has increased 30 percent between 1995 and 1996. The New Jersey Board of Public Utilities recorded 2.2 million notifications for 1996. Even though there were 17 percent fewer hits in 1996, they still totaled 3,961. As previously mentioned, however, the New Jersey State Department of Transportation is exempt from participating in the one-call notification process.

Participants at the Safety Board's 1994 workshop generally agreed that penalties need to be enforced in order to recover the costs of the damage prevention programs; however, the participants also believed that self-policing partnerships between operators and excavators were essential and that the administration of the program should be as simple and streamlined as possible with a minimum of government oversight. The participants believed that by doing so, costs to stakeholders would be minimized and there would be a greater potential for success. The participants also indicated that State programs should have enough flexibility to be able to implement alternative procedures that still meet the intent of the program.

The administrative approach to enforcement of damage prevention programs is designed to promote compliance rather than punishment, and to create awareness of good damage prevention practices rather than to collect fines or to put small companies out of business. Administrative enforcement has been accomplished without creating an additional bureaucracy, and the cost of the enforcement program has been covered even with the small fines and penalties that are imposed. The Safety Board concludes that administrative enforcement has proven effective in some State excavation damage prevention programs.

Excavation Marking

Excavation occurs frequently. The excavation notification system in Illinois recorded over 100,000 calls during the month of April 1997 [44]. In New Jersey, its one-call system records 2.2 million excavation markings per year, an average of more than 6,000 per day [45]. With this rate of occurrence, the frequency of hits would be dramatically higher if some information about line locations were not available.

An entire industry of underground utility locating businesses have developed in the last two decades. Primarily, these businesses serve utility companies by performing the marking services associated with one-call notification. Referred to as locators, these technicians visit construction sites and mark the location of underground facilities using both mapping technology and electronic tools. Practices for marking the underground facilities can have an impact on the risk of excavation damage. Good practices include pre-marking the intended excavation site by the excavator to clearly identify to the facility locator the area of digging; positive response by the utility owner to confirm that underground facilities have been marked or to verify that no marking was necessary; the use of industry-accepted marking standards to unambiguously communicate the type of facility and its location; marking facility locations within the specified notification time; and responding to requests for emergency markings, when necessary.

The timeframe for excavation marking is usually specified by State damage prevention laws. Twenty States require underground facility marking to be accomplished within 48 hours of excavation notification. Construction work planning is not evenly distributed throughout the week; consequently, one-call centers may schedule three or four times the number of locates for some days compared to other days. This, in turn, creates variable workloads for utility locators.

Pre-Marking. Participants at the 1994 workshop agreed that pre-marking the proposed excavation area has been demonstrated to enhance the safety of excavation activities. Pre-marking allows the excavators to specifically tell facility owners where they intend to dig. Some States require the use of white marking to indicate the boundaries of planned excavations. Maine was one of the first States to have mandatory pre-marking for non-emergency excavations. Connecticut has also adopted a pre-marking requirement; the law provides for face-to-face meetings between operators and excavators for projects that are too large for or not conducive to pre-marking.

According to workshop participants, pre-marking an excavation site helps to ensure that owners of underground facilities are aware of the specific area that is to be excavated. Facility owners avoid unnecessary work locating underground facilities that are not associated with the planned excavation. Excavators can be certain that underground facilities within their intended area of excavation are well marked. Because pre-marking defines the physical boundary of the excavation site, it removes ambiguity about what underground facilities need to be located. Marking the intended excavation area creates a greater likelihood that affected underground facilities will be identified to the excavator. The Safety Board concludes that pre-marking an intended excavation site to specifically indicate the area where underground facilities need to be identified is a practice that helps prevent excavation damage.

Marking Standards. Most State laws on damage prevention call for facility owners to follow the standards for temporary marking developed by the ULCC. Figure 12.1 identifies the color codes. Local one-call centers often distribute pocket-size flash cards with these color codes to excavators. The use of standard marking colors informs the excavator about the type of underground facility whose location has been marked.

Markings of the appropriate color for each facility are placed directly over the centerline of the pipe, conduit, cable, or other feature. There are procedures for offset markings when direct marking cannot be accomplished. For most surfaces, including paved surfaces, spray paint is used for markings; however, stakes or flags may be used if necessary. In addition to uniform color codes used to transmit standard information about the type of facility marked, the National Utility Locating Contractors Association (NULCA) has developed a proposal for standard marking symbols. The proposal is currently available only for internal use but is being designed for distribution to members in the future. NULCA's proposed standard addresses conventions for marking the width of the facility, change of direction, termination points, and multiple lines within the same trench. The standard symbology indicates how to mark construction sites to ensure that excavators know important facts about the underground facilities; for example, hand-dig areas, multiple lines in the same trench, and line termination points. The Safety Board recognizes the benefit of industry efforts to standardize marking practices. Such conventions help to avoid misinterpretation between locators who designate the position of underground facilities and excavators who work around those facilities. Participants at the workshop recommended that uniform marking include the facility owners' identification.

NULCA's work to define standard marking symbols incorporates the use of facility owner's identification marks along with conventions for identifying underground system configurations.

EMPLOYEE QUALIFICATIONS AND TRAINING

Training to prevent excavation damage to the underground infrastructure is not limited to the pipeline industry and operating personnel: locators need training in locating techniques, equipment technology, and marking procedures; excavators need training to fully participate in the notification process and to understand locator marking symbols; one-call operators need training to efficiently and effectively transmit information between excavators and underground system operators; and the general public needs to be aware of the one-call notification process when they dig for private projects. In addition, anyone working to oper-

Color	Feature identified
red	electric power lines, cables, conduit and lighting cables
yellow	gas, oil, steam, petroleum, gaseous materials
orange	communications, alarm or signal lines, cable or conduit
blue	water, irrigation, and slurry lines
green	sewers, drain lines
pink	temporary survey markings
purple	cable television
white	proposed excavation

FIGURE 12.3 Uniform color code of the American Public Works Association, Utility Location and Coordinating Council.

ate underground systems or whose work requires underground digging needs to be trained in emergency response procedures. This diversity of training needs presents a challenge to both system regulators and the industry.

Training and Educating Excavation Personnel. Excavators need to be trained and educated about safe work conditions, good excavation practices, relevant State laws, and one-call procedures. In this context, the excavator is not only the backhoe operator at the construction site, but also the project manager, the scheduler, company officials—anyone connected to excavation work. In an effort to ensure that excavators are aware of their responsibilities to protect underground facilities, some States have licensing requirements that assess professional knowledge. For example, Florida law (in Section 556.104 of the Florida Statutes) requires contractors who work near buried facilities to be licensed, a process that involves passing a written examination. Excavators should fully understand the one-call notification process: the meaning of facility markings, requirements for hand digging near underground facilities, notification responsibilities when the scope of work changes, and emergency response procedures. Many one-call centers offer outreach training programs designed for excavators. Some one-call center personnel have met with local union organizations and industry associations to explain the notification process and State damage prevention laws.

Because marking the position of an underground line is a safety-critical job, training is necessary to ensure that locators are well prepared to perform this function. NULCA has defined a set of minimum standards for its members to adopt as part of their training programs [46]. The program includes 118 hours of structured training in the topics of system design, construction standards, equipment techniques, recognition of line type, locating theory, and safety procedures. In addition to recommending the use of written tests, the program recommends field training and annual retesting.

The NULCA has also developed guidelines for excavation practices and procedures for damage prevention. These guidelines, which were revised in September 1997, incorporate OSHA requirements and identify best practices applicable to excavation work. Use of the guidelines is voluntary, but NULCA's brochure explains that legislation in most States requires contractors who plan to excavate to notify the appropriate one-call center and non-member facility owners before the job begins. The guidelines address preplanning and job site activities for both large and small projects. Instructions for handling damage, along with a construction facility damage report form, are also included. The Safety Board commends NULCA's efforts in promulgating good practices among its members and the excavation community.

Title 29 CFR 1926, Subpart P, contains several worker safety requirements on excavation activities. In 1990, OSHA developed and issued a booklet, Excavation, to assist excavation firms and contractors in protecting workers from excavation hazards. The booklet is based on the requirements of Part 1926 and gives specific advice on preventing cave-ins and providing protective support systems. OSHA employs several methods of providing information to persons subject to its regulations; its latest information system uses the Internet via the World Wide Web to provide assistance to excavators and contractors on complying with OSHA requirements. Responses to frequently asked questions, statistical data, news releases, OSHA pamphlets and publications, and a listing of available training materials can be obtained via the computer.

Federal training requirements for the transport of hazardous liquids are stated in 49 CFR 195.403. These are general requirements that do not specifically discuss excavation activities, and there are no comparable general training requirements for gas operator employees. RSPA has a joint industry and government working group that periodically meets to develop proposed requirements for employee qualification and training. That committee, the Negotiated Rulemaking Advisory Committee on Pipeline Personnel Qualifications, completed its fourth meeting in August 1997. It has prepared three drafts of a proposed operator qualification regulation for committee consideration. The committee has not reached consensus and is still considering draft regulatory language.

Participants at the Safety Board's workshop recommended that excavator associations work in conjunction with operators of buried facilities and one-call notification centers to provide buried-facility damage-prevention training as part of their safety training programs. The participants acknowledged that the Associated General Contractors of America and many contractor organizations are very safety conscious and have produced several videotapes about safety issues. Few of these education efforts, however, include testing. The current negotiated regulation process at RSPA has addressed the issue of training verification and testing, but the scope of that work is limited to only oil and gas operators subject to Federal regulations.

The Safety Board has long been concerned that all personnel involved in excavation activity be properly trained and qualified and has issued several recommendations in this area as a result of its accident investigations. Following the investigation of an accident in Derby, Connecticut, in December 1985, the Safety Board recommended that Northeast Utility Service Company

> *Emphasize in its training of operating personnel the importance of following the company procedures for patrolling and protecting its gas mains in proximity to excavation projects. (P-86-19) [47].*

The Safety Board's investigation of an accident that occurred 3 months later in Chicago Heights, Illinois, also generated a recommendation concerning training. The Board recommended that Northern Illinois Gas Company

> *Emphasize in company training the importance of following company procedures for making areas near gas pipeline leaks safe for the public by evacuation or other means. (P-87-38) [48].*

As a result of an explosion and gas-fueled fire that occurred on July 22, 1993, when a backhoe of the city of St. Paul Department of Public Works hooked and pulled apart a high-pressure gas service line, the Safety Board asked the American Public Works Association to

> *Advise its members of the circumstances of the July 22, 1993, explosion in St. Paul, Minnesota, and urge them to develop and implement written procedures and training to prevent excavation-caused pipeline damages. (P-95-24) [49].*

In 1987, RSPA first issued a notice of proposed rulemaking (NPRM) to improve the competency of operator personnel and to set minimum training and testing standards for employees of pipeline operators. A notice issued in October 1991 stated that a second proposal, based on comments received earlier, would be forthcoming. By 1993, RSPA still had not acted to implement any employee qualification and testing standards, and the Safety Board urged that this issue become a priority in the regulatory agenda. Ten years after its original NPRM in 1987, RSPA has entered into negotiated rulemaking. Action on this issue is long overdue. The Safety Board concludes that employee qualification and training is an integral component of an effective excavation damage prevention program, and industry has recognized the need for employee training but has not implemented training uniformly. Inadequate employee training was highlighted in the Safety Board's report of the San Juan accident [50]. In that report, the Board recommended (P-97- 7) that RSPA complete a final rule on operator employee qualification, training, and testing standards within 1 year. The Board further stated in that recommendation that the final rule should require operators to test employees on the safety procedures they are expected to follow and to demonstrate that they can correctly perform the work.

Because RSPA's rulemaking would cover only those employees of oil and gas operators subject to Federal regulations, additional efforts are needed by industry to provide training materials to those employees not covered by the regulations. The OCSI's Training Committee—which develops educational materials for use by notification center employees, facility owners, and excavators—would appear to be the appropriate organization to accomplish this goal. Therefore, the Safety Board believes that the APWA should review existing training programs and materials related to excavation damage prevention and develop guidelines and materials for distribution to one-call notification centers.

Emergency Response Planning. Pipeline operators are required by law to establish written emergency procedures for classifying events that require immediate response, communicating with emergency response officials, and responding to each type of emergency [51]. Although the extent of emergency response planning may vary depending on the type of excavation activity, emergency response planning should involve a definition of responsibilities, a flow chart of actions, execution criteria, systems inventory and resource information, coordination procedures (internal and external), and simulation exercises of response actions.

Federal regulations require no emergency response plan for excavators; however, these are the very people that often have responsibility for first response at an excavation disaster. The Safety Board has addressed the need for emergency response plans and procedures in many of its reports of accidents involving excavation damage. One such accident was an explosion in Cliffwood Beach, New Jersey, on June 9, 1993, that occurred as a result of a utility contractor's trenching operation. The Safety Board's investigation determined that a failure in training was causal to the accident [52]. The utility operator did not brief or determine whether the contractor knew what procedures to follow should the crew damage a main or service line. In addition, the Safety Board found no record or evidence of the contractor being properly trained in emergency procedures, and the facility operator's procedures did not include emergency response training for contractors. As a result of its investigation, the Safety Board recommended that the gas company take the following actions:

> *Train all gas operations construction contractors for emergencies involving struck pipelines; training should stress immediately reporting natural gas pipeline strikes to New Jersey Natural Gas's emergency phone number. (P-94-01) [53].*

As a result of the previously mentioned accident in St. Paul, Minnesota, on July 22, 1993, the Safety Board recommended that the American Public Works Association

> *Urge your members to call 911 immediately, in addition to calling the gas company, if a natural gas line has been severed. (P-95-25) [54].*

The Safety Board concludes that, at a minimum, excavators should formulate an emergency response plan appropriate for the specific construction site and ensure that employees working at that site know the correct action to take if a buried facility is damaged. The local one-call center can also play an important role in planning with local officials to define the best emergency response appropriate for its communities. The local one-call centers also are in a good position to disseminate this information on a regular basis. Therefore, the Safety Board believes that the APWA should develop guidelines and materials that address initial emergency actions by excavators when buried facilities are damaged and then distribute this information to all one-call notification centers.

Discussion

The Safety Board acknowledges that considerable progress has been made by RSPA and the industry in the area of improving excavation damage prevention programs since the Board's 1994 workshop and most likely because of it. The workshop provided a valuable forum to discuss how government and industry can work together to improve excavation damage prevention programs. The Safety Board believes that by continuing to focus attention on this important safety issue, the number of excavation-caused accidents to the Nation's underground facilities will ultimately decrease. Therefore, the Safety Board believes that RSPA should conduct at regular intervals, joint government and industry workshops on excavation damage prevention that highlight specific safety issues, such as full participation, enforcement, good marking practices, the importance of mapping, and emergency response planning.

Specific progress has been made to standardize marking symbols, to develop a uniform notification ticket, to develop guidelines for excavation practices and procedures, and to develop minimum standards for training programs. The importance of mandatory participation has been advocated by industry as well as government, yet many entities are granted exemptions to participation in State excavation damage prevention programs. Although many elements of an effective State excavation damage prevention program have been identified, the Safety Board is concerned that these elements have not been uniformly implemented. Some States have realized the benefit of swift, effective sanctions through the administrative process, yet many States are lacking effective enforcement programs. The practices and activities of one-call notification centers have also been identified, but these practices have not been uniformly implemented. The Safety

Board concludes that although considerable progress has been made to improve State excavation damage prevention programs, additional efforts are needed to uniformly develop and implement programs that are most effective.

In 1996, RSPA established a joint government/industry Damage Prevention Quality Action Team. Participants include the American Petroleum Institute (API), the American Gas Association (AGA), the American Public Gas Association (APGA), the Interstate Natural Gas Association of America (INGAA), One-Call Systems International (OCSI) of the APWA, the National Telecommunications Damage Prevention Council, the National Association of Regulatory Utility Commissioners (NARUC), the Associated Electrical and Gas Insurance Services, the National Association of Pipeline Safety Representatives, and industry participants. As stated in its charter, "the purpose of that team is to assess the status of current excavation damage prevention efforts and their effectiveness, and to identify additional efforts that would lead to reduction of excavation damage." However, rather than assessing the status of damage prevention efforts, the group set as its goal to "conduct a national pipeline awareness campaign." As of June 1997, the team had developed and distributed surveys to assess the awareness of one-call systems. Because the critical elements of an effective excavation damage prevention program have not been uniformly implemented at the State level, the Safety Board believes there is a need to review and evaluate existing damage prevention programs and to highlight deficiencies in existing programs so that corrective action can be taken. The Safety Board supports current legislative interest in provisions for a review of existing excavation damage prevention programs but does not believe there is a need to await Congressional action before such an evaluation is undertaken. The Damage Prevention Quality Team appears to be an appropriate mechanism for accomplishing a detailed evaluation of existing programs. Therefore, the Safety Board believes that RSPA, in conjunction with the APWA, should initiate and periodically conduct detailed and comprehensive reviews and evaluations of existing State excavation damage prevention programs and recommend changes and improvements, where warranted, such as full participation, administrative enforcement of the program, pre-marking requirements, and training requirements for all personnel involved in excavation activity.

ACCURACY OF INFORMATION REGARDING BURIED FACILITIES

Underground Detection Technologies

Both facility owners and excavators have genuine interest in identifying the location of underground facilities. But with current locating equipment technologies and mapping records, there remains a variety of errors that can potentially affect the ability to positively identify the position of underground facilities. There is no one procedure or tool that can provide accurate location information for all types of facilities in all types of situations. Location work is a combination of operator experience and the correct use of technology.

A variety of remote sensing technologies can be used for detecting underground facilities. Different types of locating equipment and techniques are needed depending on structural composition of the buried materials, soil composition, and surface access [55]. A brief description of the types and attributes of locator tools is shown in Table 12.3. In addition to equipment choice, there are situational variables that affect detection accuracy. The more conductive the soil, the more shallow the conductor will appear. Sandy, loose soil with a high mineral content will give sensitive readings; pipe locations under these conditions may be deeper than the locator equipment readings indicate. Moisture content or water table levels can also affect depth readings.

For equipment types that determine location by sensing an electronic signal that has been introduced into the underground system, strength of the locating signal depends on where the signal was introduced into the system, the proximity of structural uprights connected to the underground system, and nearby surface obstructions that dissipate the signal. Selection of radio signal frequency can also affect signal clarity. Equipment readings cannot be taken as absolute values; they depend on situational effects associated with locator equipment calibration, field conditions, and the operators familiarity with the particular operating characteristics. Many water and sewer lines are made of plastic or concrete pipe, gas systems commonly use plastic pipe, and fiber optic cable is often used in telecommunication lines. These systems are difficult to detect with common locator tools because they do not contain metal. A metal tracer wire can be buried with the pipe to facilitate future locating work. Typically, pipe is laid in the trench and covered by a shallow layer of fill dirt. The tracer wire is then placed over the pipe and trench filling is completed. Detectable warning tape—aluminum foil covered with color-coded polyester—can be buried with nonmetallic underground facilities to permanently mark the lines. Varieties of tracer wire and detectable warning tape are designed to be sturdy enough to be plowed into the trench during backfill operations.

The Safety Board recognizes industry efforts to inform locators about issues relevant to locator technology. Underground Focus magazine sponsors an annual utility locating technology seminar. This training event, currently in its 6 th year, provides information on utility locating techniques, equipment, and new technology. Participants include locators, equipment manufacturers, engineers, trade association representatives, and academic interests. Topics related to locator equipment are also regularly addressed at conferences such as the annual OCSI symposium and the Underground Safety Association forum [56].

VERTICAL/DEPTH LOCATION

The only certain method of determining facility depth is to expose the pipe, conduit, or cable through hand digging or through vacuum excavation. Southwestern

Types of locator equipment.

Equipment type	Functional description	Attributes
Radio frequency (RF) detection techniques	Conventional underground line detection method. Requires a transmitter and a receiver. Conductive tracing attaches the transmitter directly to the line or tracer wire. Inductive tracing does not require direct line connection.	Oldest, most widely used technology. Inductive signal detection is quicker, but conductive signal reading is more accurate.
Electromagnetic techniques	Records signal differentials of magnetic fields. Similar to radio frequency technology.	Useful for detecting metal objects or structures that exhibit strong magnetic fields at the ground surface. This type of detector is affected by obstructions between the transmit signal and the locating equipment.
Magnetic methods	Useful for detecting iron and steel facilities.	Magnetic flux methods are easy to use and inexpensive, but they are subject to interference from metal surface structures.
Vacuum extraction	Small test holes are dug from the surface by vacuuming out the soil. The activity, usually referred to as "potholing," follows more preliminary locating work to identify the general facility location. The pothole then confirms the location and verifies a depth for that specific site.	Requires preliminary records search to approximate location for potholing and special vacuum equipment. Process can be expensive and labor intensive.
Ground penetrating radar	Radar wave reflections from underground surfaces of different dielectric constants are used to identify subsurface structures.	This method is relatively expensive compared to other locator methods and does not work well in clay or saltwater.
Terrain conductivity	Detects current measures that differ from average ground surface conductivity.	This method can be useful in areas of high conductivity, such as marine clay soils, particularly for locating underground storage tanks.
Global positioning system (GPS)	Uses triangulated satellite telemetry to identify latitude/longitude location of ground unit.	While not a detection technology, GPS coordinates are frequently used to define geographic location.

Adapted from: Anspach, J.H. 1994. Locating and Evaluating the Conditions of Underground Utilities. In: RETROFIT '94. Washington, DC: National Science Foundation. Sponsored by: Stanford University and the National Science Foundation.

FIGURE 12.4 Types of locater equipment.

Bell's use of vacuum excavation to expose and document exact facility locations is credited with decreasing cable damages by 50 percent in Texas during 1996 [57]. This method positively identifies both the horizontal and vertical location of the pipe at a specific site. But certainty about the line's position is inversely related to its distance from the test hole. Depth depends on how the line was installed and on the changes in surface grade caused by erosion or construction since installation.

For selected models of locating equipment, manufacturers claim that the units can accurately determine depth [58]. Accurate depth measurements are a highly desirable attribute of locating equipment. Based on equipment manufacturers' claims, States have begun to consider adding requirements for depth location information to their damage prevention legislation. Wyoming's Underground Facilities Notification Act of 1996 requires construction project owners to furnish information on the nature, location, and elevation of underground facilities.59 Minnesota is considering a similar requirement.

Remote locating devices that measure depth are susceptible to calibration problems, antenna misalignment, and electronic fields that are combined from more than one surface conductor [60]. The capability for accurate depth measurement may exist under ideal situations, but given field conditions, depth measures may lack a high rate of reliability. Participants in the 1994 damage prevention workshop concluded that remote sensing methods should not be used for determining facility depth location. More recently, at the 1997 One-Call Systems and damage prevention symposium, a session on depth perception concluded that remote locator equipment was available that could provide elevation readings but not with a degree of accuracy that warrants placing the liability with the locating service [61].

The capability of locator equipment needs to be incorporated into damage prevention practices. The Safety Board concludes that more research and testing is needed to determine the accuracy of depth detection by remote locating equipment. Therefore, the Safety Board believes that RSPA should sponsor independent testing of locator equipment performance under a variety of field conditions. Further, the Board believes that as a result of the testing, RSPA should develop uniform certification criteria of locator equipment. Finally, once locator equipment performance has been evaluated and defined by certification criteria, RSPA should review State requirements for location accuracy and hand-dig tolerance zones to determine that they can be accomplished with commercially available technology.

DIRECTIONAL BORING/TRENCHLESS TECHNOLOGY

Excavation work is frequently for the purpose of installing additional facilities. General practices require digging an open trench from the surface down to the installation depth. However, trenchless technology offers a different method for installing underground facilities. Directional boring "snakes" a new line that follows a drill bit horizontally through the subsurface. This method is particularly advantageous for traversing below waterways, ecologically sensitive wet lands, or major traffic arteries. But there are practical limits to the depth that lines are installed. Eventually, additional depth becomes infeasible because of the cost of the extended line runs, geologic changes at lower stratum, or practical concerns for future maintenance. New lines must then go through the areas that have had line laid by directional boring.

Differences in soil density, rock formations, and variable torque on the drilling head often result in a directional line that does not run along a straight route. Drilling heads can be deflected by hard rock or unknown underground objects. The operational accuracy of directional boring depends on the accuracy of sensors located on the drill bit and the drilling unit's resolution and correlation to a common base map. Though they do not involve sensors, similar problems can be found with the use of pneumatic drills and mechanical augers.

Directional boring is not always sensitive to line hits; it is possible for a boring equipment operator to hit a facility without being aware of the hit. The drill bits, designed to go through rock, experience little change in resistance when going through plastic pipe or cable. This sets up a situation for hitting a gas line without knowing it; migrating gas can then collect, creating conditions for an explosion. The Safety Board recently investigated an accident involving directional boring in Indianapolis, Indiana [62]. The explosion resulted in one fatality, one injury, and extensive damage to a residential subdivision.

Over the past year, the trade literature has documented several accidents, not investigated by the Safety Board, that resulted from horizontal directional boring. For example:

- In Seattle, directional boring caused a gas explosion that destroyed a home
- A major traffic artery in northern New York State was closed for several days to determine if a water main break resulting from directional boring had seriously weakened the roadbed
- Two people were hospitalized in Overland Park, Kansas, when a gas explosion, caused by directional boring, destroyed four homes [63]

Equipment manufacturers have tried to address the problem of recording the position of lines installed by directional boring. Sensors, generally magnetic guidance-type sensors attached to the drill bit, record location information for mapping the line. The relative position of the drill bit is plotted on a real-time display at the drilling operator's control position [64]. Stored as an electronic data file, this information can be archived in facility data records. Conceptually, this accounts for "recording the course of a new line." Associated issues, however, can affect the accuracy of information gathered in this manner. First, accuracy depends on sensor calibration. Operators must know how to check for and correct calibration error. Second, the drill's sensor may know where it is in relation to some global positioning system (GPS) coordinates, but it may not know its location in relation to ground surface. Depth of line, an important fact, is dependent on accurately orienting the drilling activity on a topographic survey map. The accuracy of the topographic map is, in turn, affected by erosion and grade changes over time.

The Safety Board concludes that facility maps should have a standard depiction for underground facilities that were installed using directional boring techniques. The Safety Board believes that the APWA should work in conjunction with the American Society of Civil Engineers (ASCE) to develop standards for map depiction of underground facilities that were installed using directional boring techniques.

MAPPING

Maps are important to many aspects of excavation damage prevention. Rather than using a standard, common mapping system, current damage prevention programs use many different maps. An excavator usually uses a city road map to identify to the one-call center the intended area of construction activity. The one-call center refers to its coverage map (grid system coded with database information) to identify which facility owners should be notified to mark their underground facilities. Locators use a combination of utility maps to direct their field work.

Engineers and project designers are forced to use a variety of data sources from both public and private organizations to determine the structure and location of the under-ground facility network. Land use and zoning maps, tax assessor maps, easement descriptions, highway and transportation network maps, quadrangle and topographic maps of the U.S. Geologic Survey, construction permit drawings, construction plans, and aerial photographs are also used to help define the location. As the following example illustrates, map quality can vary. Excavation to install telephone cable on the University of New Haven campus in Connecticut in August 1996 hit a gas main, but the gas did not ignite. The gas crew searched for 33 minutes to find the correct shutoff valve. The director of facilities for the university said the gas line was not marked on maps of the campus [65].

Facility records maintained by the utility owners or pipeline operators are the most widely used sources of information about the underground infrastructure. As a result of the Pipeline Safety Act of 1992, OPS requires operators to identify facilities in environmentally sensitive areas and in densely populated areas, but there is no requirement for system operators to maintain a comprehensive system map of their underground facilities, though most do maintain this information to facilitate their business operations. Different utility services use different types of maps; they vary in scale, accuracy, resolution, standard notation, and data format. System records developed prior to the widespread use of computer technology most likely exist as architectural and engineering diagrams. For some systems, these diagrams have been electronically imaged so that they are easier to reference, update, and store. Digitized versions of early maps do not always reflect the uncertainty of information that may have been inherent on the hand-drafted version. Structural references and landmarks that define the relative locations of underground facilities also change over time and may not be reflected on maps.

Many system maps lack documentation of abandoned facilities. Abandoned facilities result when the use of segments of the underground system are discontinued, or when replaced lines run in new locations, or when entire systems are upgraded. Without accurate records of abandoned facilities, excavators run the risk of mistaking the abandoned line for an active one, thereby increasing the likelihood of hitting the active line. Several States have recognized the need to require facility operators to map abandoned lines; for example, Arizona requires that any line abandoned after December 1988 be mapped.

In addition to documenting the location of a facility, utility map records may also contain information on the age of the facility, type and dimensions of the material, history of leakage and maintenance, status of cathodic protection, soil content, and activity related to pending construction. However, the quality of this information varies widely. Participants at the 1994 damage prevention workshop recommended that when excavation revealed errors in mapping, operators should be required to update system maps.

Recent utility records often exist as geographical information systems in a variety of computerized software packages and electronic data storage formats. The Mapping Requirements and Standards task group of the AGA's Distribution Engineering Committee surveyed member companies in 1995 about mapping requirements and practices. Of the 27 companies that responded, 12 used computerbased mapping systems, 12 others were planning to automate their mapping systems, and 3 reported that they had no plan to automate mapping records [66].

Many automated mapping programs are not compatible, and it is difficult to merge system records developed over the years by different departments and companies. Additionally, computerized diagrams may be associated with large databases that contain entry errors that are difficult to identify. Inconsistencies between data dictionaries—similar information labeled differently in different databases—require considerable effort to correct once identified. More importantly, these differences may lead to an unknown error if they are not resolved. A good quality printed image of an electronic map can disguise the poor level of information used to generate the image.

One-call systems are beginning to use GPS receivers and mapping programs [67]. Arizona Blue Stake One-Call and Ohio Utility Protection Service are working to develop positionally accurate, map-driven software to support their notification systems. California's USA North One Call ticket locations can be displayed as GPS coordinates [68]. Excavators, locators, and utility operators can use GPS information to identify field locations (longitude and latitude coordinates), and they can use this information to navigate to the sites. With the added capability of differential GPS, objects can be located to an accuracy of better than 1 meter (1.1 yards). This degree of accuracy makes differential GPS appropriate for many aspects of mapping underground facilities. The Tennessee One-Call System is considering the feasibility of installing differential GPS antennas across the State to provide location accuracy.

Most commercial maps are based on topographically integrated geographic encoding and referencing (TIGER) files from the U.S. Census Bureau. These files often contain positional inaccuracies that can be problematic when integrated with GPS latitude and longitude coordinates. For example, many, if not most, existing underground systems are not documented by GPS coordinates. Consequently, a facility owner working on a line may want to update the positional record of that line to include the coordinates. Using a GPS receiver, the facility owner acquires the line's position and then references a TIGER-based map for that area to verify aboveground landmarks. The map can indicate that those coordinates are on the south side of the highway, yet the locator might actually be standing above the underground facility on the north side of the highway.

In 1994, the Federal Geographic Data Committee recommended a plan for the Nation's spatial data infrastructure, and Congress mandated governmental response to the plan [69]. The OPS subsequently formed a joint government/industry team to start a national pipeline mapping system. The team's 1996 report, "Strategies for Creating a National Pipeline Mapping System," made several recommendations: (1) develop, promote, and aggressively communicate pipeline data standards that are consistent with the standards of the Federal Geographic Data Committee; (2) formalize a partnership with industry, and Federal and State agencies; (3) develop a partnership with One-Call Systems International to reach a better understanding of one-call system data needs and gather support for using geographically referenced data; and (4) create a distributed mapping system with centralized quality control and decentralized access capabilities.

There are many different facility mapping systems in use by one-call systems and facility owners. Those with GPS positional accuracy lack information on landmarks and developed structures, and maps that accurately reflect current structural improvements often lack positional accuracy. The Safety Board concludes that underground facility mapping must consider the amount of detail and the accuracy of information necessary for effective use. The Safety Board recognizes RSPA's effort in creating strategies for a national pipeline mapping system and for its current Mapping Implementation Quality Action Team. The Board believes RSPA should develop mapping standards for a common mapping system, with a goal to actively promote its widespread use.

SUBSURFACE UTILITY ENGINEERING

Subsurface utility engineering (SUE) is a process for identifying, verifying, and documenting underground facilities. Depending on the information available and the technologies employed to verify facility locations, a level of the quality of information can be associated with underground facilities. These levels, shown in Table 12.4, indicate the degree of uncertainty associated with the information; level A is the most reliable and level D the least reliable. This categorization is a direct result of the source of information and the technologies used to verify the information.

A comprehensive map and automated computer diagram of a construction site is developed as a SUE product; it depicts coregistered information for all utilities in that area. The SUE process identifies all utilities during a single coordinated effort. In this way, information known about one facility can beneficially affect the mapping of other utilities, and unknown facilities are more likely to be documented. By signing the SUE product, a professional engineer warrants the maps against errors and omissions and assumes liability for the accuracy of the information.

The Federal Highway Administration (FHWA) considers SUE an integral part of preliminary engineering work on highway projects receiving Federal aid. It has the potential to reduce facility conflicts, relocation costs, construction delays, and redesign work. In 1984, the State of Virginia began a SUE program, called the Utility Designation and Locating Program, and determined that there were substantial cost savings. A highway project in the city of Richmond used SUE work costing $93,553 to avoid an estimated $731,425 worth of expenses to move utilities had the highway projects not been designed to avoid conflict with underground facilities. Virginia's estimate of cost savings, just in terms of avoiding utility relocations, was $4 saved for each dollar spent.

Additionally, Virginia credits the process with reducing design time by 20 percent [70]. The utility coordinator for Maryland's State Highway Administration estimates a savings of $18 for each dollar spent. Florida DOT found that it saved $3 in contract construction delay claims for each dollar spent on SUE. Variations in these estimates reflect different cost assumptions, geographic conditions, and system configurations. Twenty-six highway agencies have used SUE at some level on some projects; FHWA estimates a nationwide savings of $100 million a year as a result of SUE [71, 72].

Compiling comprehensive information on underground facilities can be expensive and labor intensive. Small contractors generally do not have the resources or expertise available to accomplish SUE on a regular basis; consequently, SUE is generally used on large construction projects such as those typical of highway development.

Architects, engineers, and contractors should have ready access to information on the location of underground facilities to plan construction activities. The advantage of this information was recognized at the 1994 damage prevention workshop. The Safety Board concludes that providing construction planners with information on the location of underground facilities, referred to as "planning locates," can reduce conflicts between construction activities and existing underground facilities. The Safety Board believes that the APWA should encourage one-call notification centers to work with their members to provide facility location information for the purpose of construction planning.

Subsurface utility engineering (SUE) levels of information.

Quality level of the information	Description
Level D	Information is collected from **existing utility records** without field activities to verify the information. The accuracy or comprehensiveness of the information cannot be guaranteed; consequently, this least certain set of data is the lowest quality level.
Level C	**Adds aboveground survey data** (such as manholes, valve boxes, posts, and meters) to existing utility records. The Federal Highway Administration Office of Engineering estimates that 15–30 percent of level C facility information pertinent to highway construction is omitted or plotted with an error rate of more than 2 feet.[a]
Level B	**Confirmed existence and horizontal position of facilities are mapped** using surface geophysical techniques. The two-dimensional, plan-view map is useful in the construction planning phase when slight changes to avoid conflicts can produce substantial cost savings by eliminating the relocation of utilities.
Level A	Vacuum excavation is used to **positively verify both the horizontal and vertical depth location** of facilities.

[a] Scott, Paul. 1995. "Subsurface Utility Engineering: An Alternative to Excavation Damage." In: Proceedings, Excavation Damage Prevention Workshop; 1994 September 8-9; Washington, DC. Report of Proceedings NTSB/RP-95/01. Washington, DC: National Transportation Safety Board: 186-189.

Source: Stutzman, H.G.; Anspach, J.H. 1993. "Site Investigation and Detection." In: Research Needs in Automated Excavation and Material Handling: Proceedings, National Science Foundation Symposium; 1993 April. Washington, DC: National Science Foundation.

FIGURE 12.5 Subsurface utility engineering (SUE) levels of information.

The Standards Committee of the ASCE is developing standards for depicting underground facilities on construction drawings. The Board thus believes that the APWA and the ASCE should address the accuracy of information that depicts subsurface facility location on construction drawings. Further, the Safety Board believes that the Associated General Contractors of America should promote the use of subsurface utility engineering practices among its members to minimize conflicts between construction activities and underground systems.

SYSTEM PERFORMANCE MEASURES

Few performance-based measures are available and useful for assessing excavation damage prevention programs. Those measures that are maintained are specific to selected States or are maintained by individual companies for a specific underground system. Data concerning underground damage for all types of systems are needed:

1. to determine if changes to State damage prevention programs are effective in decreasing underground facility damage
2. to assess the benefit of different practices followed by one-call notification centers

3. to identify the risks of different field practices used by facility operators, locators, and excavators
4. to allow facility operators to evaluate their company's excavation damage prevention programs
5. to assess the needs and benefits of training; and (6) to perform risk assessment for the purposes of business, insurance, and public policy decisions

RISK EXPOSURE

A critical component of excavation damage data is the total number of excavations that present a chance for damage. These data, however, are not available. The number of excavations presented in this report are industry estimates; they did not result from a national data collection system. To quantify the number of accidents in relation to how many could have occurred, it is necessary to determine some frequency of exposure. In the context of excavation damage, exposure can be measured by the number of excavations. This measure can be approximated by the number of locate tickets issued by one-call centers, although that number will capture only those excavations that were reported to one-call centers.

One-call centers offer the best opportunity for the industry as a whole to determine the rate of excavation damage. The OCSI Delegate Committee is developing a process to standardize and collect one-call center information from its members. To be useful, the information will need to be qualified by reporting criteria. Categories will need to be clearly defined: what is an excavation activity, what constitutes a facility hit, how is the level of damage categorized, what caused the damage?

Many facility operators, particularly companies that transport gas and hazardous liquids, investigate and record "line hits" in terms of damages per thousand locate requests. But because of proprietary interests, these numbers are rarely compiled across companies. The Gas Research Institute's (GRI) 1995 study made an effort to determine risk exposure for the gas industry [73]. The study surveyed 65 local distribution companies and 35 transmission companies regarding line hits. Less than half (41percent) of the companies responded, and several major gas-producing States were poorly represented (only one respondent from Texas and one from Oklahoma). The GRI estimate was determined by extrapolation and may be subject to a large degree of error because the data sample was not representative. Based on survey responses, however, GRI calculated an approximate magnitude of risk. For those companies that responded, a total of 25,123 hits to gas lines were recorded in 1993; from that, the GRI estimated total U.S. pipeline hits in 1993 to be 104,128. For a rate of exposure, this number can be compared to pipeline miles: for 1993, Gas Facts reported 1,778,600 miles (2 861 767.4 kilometers) of gas transmission, main, and service line, which calculates to a risk exposure rate of 58 hits per 1,000 line miles (1609 kilometers) [74].

Because the risk of excavation damage is associated with digging activity rather than system size, "hits per digs" is a useful measure of risk exposure. For the same year that GRI conducted its survey, one-call systems collectively received more than an estimated 20 million calls from excavators. (These calls generated 300 million worksite notifications for participating members to mark many different types of underground systems.) Using GRI's estimate of hits, the risk exposure rate for 1993 was 5 hits per 1,000 notifications to dig [75]. A comprehensive measure of hits per digs tracked over time can be a useful indicator of how well excavation damage prevention programs are performing. Because the measure is expressed as a rate rather than simply a number of hits, it can be used to compare years in which there were different levels of construction activity. The measure can also be used to compare geographic locations or utility systems of different size. Industry is beginning to use such measures of performance; for example, measures of hits per locates have been incorporated into contractual agreements between utilities and their locator services [76].

The Safety Board is encouraged that attempts are being made to calculate risk exposure data. Without this information, evaluations on the effectiveness of State damage prevention programs cannot be adequately performed. The Safety Board is concerned, however, that these isolated attempts to calculate exposure data are neither standardized nor centrally reported. A "utility" in one State may be defined differently for another State, resulting in inconsistent measures of damage.

If all digging activity were recorded through one-call systems, notification ticket volume would be a useful measure of risk exposure. The Safety Board recognizes that not all excavators currently use one-call notifications systems and that there are 84 separate one-call systems operating in the United States collecting different information in different formats. The Safety Board concludes that the one-call notification centers may be the most appropriate organizations to collect risk exposure data on frequency of digging and data on accidents. To standardize how and what information should be collected, the Safety Board believes that the APWA, in conjunction with RSPA, should develop a plan for collecting excavation damage exposure data and then work with the one-call systems to implement the plan to ensure that excavation damage exposure data are being consistently collected. The universal damage report form developed by Alberta One-Call could be considered by the OCSI. Finally, the Safety Board believes that the APWA and RSPA should use excavation damage exposure data in the periodic assessments of the effectiveness of State excavation damage prevention programs described in other safety recommendations in this report.

ACCIDENT REPORTING REQUIREMENTS OF RSPA

The requirements and criteria for reporting natural gas and hazardous liquid pipeline accidents are found in 49 CFR Part 191.3 and Part 195.50, respectively.

A natural gas incident report is required for:

1. an event that involves release of gas causing a death, or personal injury necessitating in-patient hospitalization, or property damage or loss of $50,000 [77]
2. an event that results in an emergency shutdown
3. an event that is significant in the judgment of the operator.

For hazardous liquids, an accident report is required for any of the following conditions:

1. Explosion or fire not intentionally set by the operator
2. Loss of 50 or more barrels of liquid product
3. Escape to the atmosphere of more than 5 barrels a day of volatile liquids
4. Death of any person
5. Bodily harm
6. Estimated property damage exceeding $50,000.

RSPA receives accident reports on only a small portion of the underground infrastructure, not as a result of failure to report on the part of industry, but because RSPA's oversight responsibilities are limited to only a portion of the gas and hazardous liquids systems, and of that subset, accident reports are required only when reporting thresholds are exceeded. Nonetheless, RSPA's database is important because there are few sources for national accident measures and because RSPA's experience in collecting pipeline accident data can be useful for designing future databases on excavation damage. According to the GRI study of damage prevention, gas transmission and distribution systems accident reports by RSPA account for less than 1 percent of the occurrences of underground pipeline damage [78]. Although numerous accidents and incidents do not meet the above reporting criteria and, consequently, are not recorded by RSPA, the Safety Board is concerned that many accidents that should be reported are not being reported because the cost of damage is underestimated. For example, a recent university study determined that a gas line rupture, originally reported to cost $15,000, cost substantially more [79]. Survey responses from businesses, homeowners, and emergency response units determined that the cost of the accident, not including the cost of lost gas or legal fees associated with ongoing litigation, was over $300,000. Because of the $50,000 reporting threshold, this accident, based on the original damage estimate, was not required to be reported to RSPA.

Although a determination by the operator that an incident costs less than $50,000 alleviates the operator of the requirement to report the incident to RSPA and may be a factor in the under-reporting of accidents, estimating property damage can be difficult and very subjective. The incident reports filed by operators ask for estimated property damage; however, little guidance is provided to operators on all costs that should be included to ensure accurate reporting. Dollar amounts are generally assumed to represent product loss, facility damage incurred by the operator and others, and the environmental cleanup cost; however, the exclusion of any one of these costs could reduce the estimated damage to below the reporting threshold. As a result, the accident would not be reported to RSPA. The Safety Board concludes that facility operators are provided little guidance for estimating property damage resulting from an accident, and subjective estimates of damage below the reporting threshold may account for some accidents not being reported to RSPA when they should have been. Therefore, the Safety Board believes that RSPA should develop and distribute to pipeline operators written guidance to improve the accuracy of information for reportable accidents, including parameters for estimating property damage resulting from an accident.

ACCIDENT CAUSES

The accident report form for hazardous liquid pipelines offers seven categories of cause [80]. For accidents reported between 1986 and 1995, three categories (corrosion, outside force damage, and other) accounted for 78 percent of the reported accidents. For 1996, RSPA data indicated that "outside force damage" was the leading cause of accidents (damage by outside force is primarily, though not exclusively, the category in which excavation damage is placed). The second leading cause for that year was "other." The Safety Board has previously expressed concern that the definition of accident cause is imprecise and that distinctions between categories of cause are vague. For example, in the data for hazardous liquid pipeline accidents, pipeline accidents resulting from similar events (as described by text explanations) are categorized differently. Accidents described as "lightning strike," "vandalism," "drilled into pipe," and "bullet hole" appear in both the "outside force damage" and "other" categories. Because excavation damage is not separately categorized, Safety Board staff conducted a systematic review of the accidents reported to RSPA for the years 1991 through 1996 to determine the number of excavation-related accidents (Table 12.5). The review indicated no trend toward a long-term decrease in excavation-related accidents (Figure 12.2).

Numerous accident records in the databases for distribution, transmission, and hazardous liquids systems show \$0.00 for accident costs [81]. This is particularly disturbing because in one case, a damage cost of \$0.00 was reported for an accident that injured 12 persons (a distribution system accident, July 1996 in Brooklyn, New York). A review of text comments associated with the accident records indicated that most excavation damage accidents were classified in the database as "outside force damage." However, there were many additional accidents classified as "outside force damage" that were not excavation-caused and several incidents of excavation damage were mis-categorized as "other," "corrosion," "accidentally caused by operator," or "construction/operating error."

Based on this review and previous analysis of RSPA data, the Safety Board concludes that deficiencies in RSPA accident data, particularly with respect to the cause of accidents and a record of whether those involved in pipeline accidents

Proportion of accidents that were excavation-related, 1991 through 1996.

System type and item	1991 All accidents	1991 Excavation accidents[a]	1991 Percentage	1992 All accidents	1992 Excavation accidents[a]	1992 Percentage	1993 All accidents	1993 Excavation accidents[a]	1993 Percentage
Distribution:									
Accidents	165	59	35.8	103	35	34.0	119	33	27.7
Injuries	78	29	37.2	65	20	30.8	82	27	32.9
Fatalities	14	6	42.9	7	2	28.6	16	7	50.0
Property damage (million)	$7.9	$3.1	39.0	$6.8	$1.5	21.5	$14.7	$3.7	25.1
Transmission:									
Accidents	72	28	38.9	75	11	14.7	97	16	16.5
Injuries	12	2	16.7	15	3	20.0	18	2	11.1
Fatalities	0	0	0	3	2	66.7	1	0	0
Property damage (million)	$12.0	$1.9	15.8	$24.7	$1.0	4.0	$23.0	$0.7	3.2
Hazardous liquids:									
Accidents	220	43	19.5	223	42	18.8	229	51	22.3
Injuries	9	1	11.1	38	8	21.1	10	5	50.0
Fatalities	0	0	0	5	0	0	0	0	0
Property damage (million)	$25.5	$6.5	25.4	$64.4	$29.1	45.3	$29.2	$13.4	45.8
Total:									
Accidents	457	130	28.4	401	88	21.9	445	100	22.5
Injuries	99	32	32.3	118	31	26.3	110	34	30.9
Fatalities	14	6	42.9	15	4	26.7	17	7	41.2
Property damage (million)	$45.3	$11.4	25.2	$95.9	$31.6	32.9	$67.0	$17.8	26.6

FIGURE 12.6 Proportion of accidents that were excavation-related, 1991 through 1996.

Proportion of accidents that were excavation-related, 1991 through 1996.

	1994			1995			1996		
	All accidents	Excavation accidents[a]	Percentage	All accidents	Excavation accidents[a]	Percentage	All accidents	Excavation accidents[a]	Percentage
(continued)									
Distribution:									
Accidents	125	37	29.6	103	38	36.9	120	36	30.0
Injuries	86	25	29.1	47	17	36.2	69	31	44.9
Fatalities	18	4	22.2	12	1	8.3	22	4	18.2
Property damage (million)	$50.2	$4.4	8.7	$12.6	$1.7	13.5	$12.2	$4.6	$37.8
Transmission:									
Accidents	69	7	10.1	68	10	14.7	73	19	26.0
Injuries	12	8	66.7	12	0	0	4	0	0
Fatalities	0	0	0	2	0	0	1	0	0
Property damage (million)	$43.3	$25.3	58.4	$11.1	$1.2	11.2	$12.5	$2.2	17.8
Hazardous liquids:									
Accidents	206	24	11.7	214	34	15.9	211	42	19.9
Injuries	930	0	0	13	0	0	13	7	53.8
Fatalities	1	1	100.0	3	0	0	5	3	60.0
Property damage (million)	$49.8	$2.7	5.5	$38.3	$8.7	22.8	$41.6	$6.2	14.9
Total:									
Accidents	400	68	17.0	385	82	21.3	404	97	24.0
Injuries	1,028	33	3.2	72	17	23.6	86	38	44.2
Fatalities	19	5	26.3	17	1	5.9	28	7	25.0
Property damage (million)	$143.3	$32.4	22.6	$62.2	$11.7	18.8	$66.3	$13.0	19.6

[a] Using data maintained by the Office of Pipeline Safety, Research and Special Programs Administration, the Safety Board charactertized the accidents/incidents that were determined to be excavation-related based on incident text descriptions, then calculated the proportion of injuries, fatalities, and property damage.

Source: Office of Pipeline Safety, Research and Special Programs Administration.

FIGURE 12.6 *(continued)* Proportion of accidents that were excavation-related, 1991 through 1996.

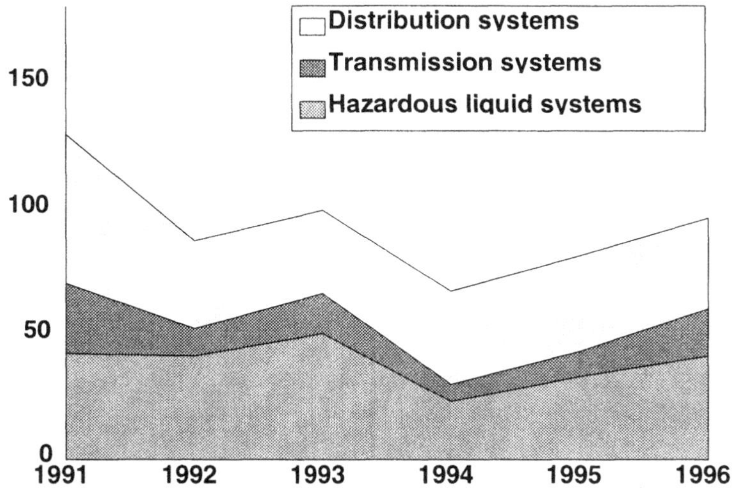

FIGURE 12.7 Number of excavation-related accidents for distribution, transmission, and hazardous liquid systems, 1991-1996.

participated in excavation damage prevention programs, precludes effective analyses of accident trends and evaluations of operator performance. Although RSPA and the industry consider excavation damage to be one of the leading causes of pipeline accidents, excavation damage is not specifically indicated on RSPA's accident form as a separate data element. A more useful analysis of accident data could also be performed if information were available on the primary, secondary, and contributing causes. The Safety Board has found through years of accident investigations that accidents are rarely the result of one event, but rather the consequence of a sequence or combination of events. Categories based on purpose of the excavation (building construction, road grading, utility maintenance); type of equipment involved (backhoe, grader, road vehicle); excavator (facility owner employee, contract employee, landowner, general public); and locator (facility owner or contract support) could provide meaningful information with which analyses of accident trends and evaluations of operator performance could be conducted.

The Safety Board has addressed deficiencies in RSPA's accident data on several previous occasions [82]. Most recently, in a 1996 special investigation report, the Safety Board evaluated RSPA's collection and analysis of accident data for petroleum product pipelines [83]. In that report, the Board concluded that RSPA's failure to fully implement the Safety Board's original 1978 safety recommendations to evaluate and analyze its accident data reporting needs has hampered RSPA's oversight of pipeline safety. Consequently, the Safety Board recommended that RSPA

> *Develop within 1 year and implement within 2 years a comprehensive plan for the collection and use of gas and hazardous liquid pipeline accident data that details the type and extent of data to be collected, to provide the Research and Special Programs Administration with the capability to perform methodologically sound accident trend analyses and evaluations of pipeline operator performance using normalized accident data. (P-96-1)*

RSPA indicated that it agreed with the Board's recommendation and was working with the pipeline industry to determine the value of industry's data to RSPA [84]. Industry and RSPA have conducted workshops to review data issues and, as recommended by the Safety Board, RSPA has obtained database information from the Federal Energy Regulatory Commission for analysis. The Safety Board believes that given the large percentage of accidents that are caused by excavation damage and the emphasis in recent years by industry to address and respond to these types of accidents, RSPA should, as part of its comprehensive plan for the collection and use of gas and hazardous liquid data, revise the cause categories on its accident report forms to eliminate overlapping and confusing categories and to clearly list excavation damage as one of the data elements, and consider developing categories that address the purpose of the excavation.

CONCLUSIONS

1. Full participation in excavation damage prevention programs by all excavators and underground facility owners is essential to achieve optimum effectiveness of these programs.
2. Many essential elements and activities of a one-call notification center have been identified but have not been uniformly implemented.
3. Administrative enforcement has proven effective in some State excavation damage prevention programs.
4. Pre-marking an intended excavation site to specifically indicate the area where underground facilities need to be identified is a practice that helps prevent excavation damage.

5. Employee qualification and training is an integral component of an effective excavation damage prevention program, and industry has recognized the need for employee training but has not implemented training uniformly.
6. At a minimum, excavators should formulate an emergency response plan appropriate for the specific construction site and ensure that employees working at that site know the correct action to take if a buried facility is damaged.
7. Although considerable progress has been made to improve State excavation damage prevention programs, additional efforts are needed to uniformly develop and implement programs that are most effective.
8. More research and testing is needed to determine the accuracy of depth detection by remote locating equipment.
9. Facility maps should have a standard depiction for underground facilities that were installed using directional boring techniques.
10. Underground facility mapping must consider the amount of detail and the accuracy of information necessary for effective use.
11. Providing construction planners with information on the location of underground facilities, referred to as "planning locates," can reduce conflicts between construction activities and existing underground facilities.
12. One-call notification centers may be the most appropriate organizations to collect risk exposure data on frequency of digging and data on accidents.
13. Facility operators are provided little guidance for estimating property damage resulting from an accident, and subjective estimates of damage below the reporting threshold may account for some accidents not being reported to the Research and Special Programs Administration when they should have been.
14. Deficiencies in the Research and Special Programs Administration's accident data, particularly with respect to the cause of accidents and a record of whether those involved in pipeline accidents participated in excavation damage prevention programs, precludes effective analyses of accident trends and evaluations of operator performance.

RECOMMENDATIONS

As a result of this safety study, the National Transportation Safety Board made the following recommendations:

To the Research and Special Programs Administration

Conduct at regular intervals joint government and industry workshops on excavation damage prevention that highlight specific safety issues, such as full participation, enforcement, good marking practices, the importance of mapping, and emergency response planning. (P-97-14)

Initiate and periodically conduct, in conjunction with the American Public Works Association, detailed and comprehensive reviews and evaluations of existing State excavation damage prevention programs and recommend changes and improvements, where warranted, such as full participation, administrative enforcement of the program, pre-marking requirements, and training requirements for all personnel involved in excavation activity. (P-97-15)

Sponsor independent testing of locator equipment performance under a variety of field conditions. (P-97-16)

As a result of the testing outlined in Safety Recommendation P-97-16, develop uniform certification criteria of locator equipment. (P-97-17)

Once locator equipment performance has been evaluated and defined by certification criteria as outlined in Safety Recommendation P-97-17, review State requirements for location accuracy and hand-dig tolerance zones to determine that they can be accomplished with commercially available technology. (P-97-18)

Develop mapping standards for a common mapping system, with a goal to actively promote its widespread use. (P-97-19)

Develop and distribute to pipeline operators written guidance to improve the accuracy of information for reportable accidents, including parameters for estimating property damage resulting from an accident. (P-97-20)

As part of the comprehensive plan for the collection and use of gas and hazardous liquid data, revise the cause categories on the accident report forms to eliminate overlapping and confusing categories and to clearly list excavation damage as one of the data elements, and consider developing categories that address the purpose of the excavation. (P-97-21)

In conjunction with the American Public Works Association, develop a plan for collecting excavation damage exposure data. (P-97-22)

Work with the one-call systems to implement the plan outlined in Safety Recommendation P-97-22 to ensure that excavation damage exposure data are being consistently collected. (P-97-23)

Use excavation damage exposure data outlined in Safety Recommendation P-97-22 in the periodic assessments of the effectiveness of State excavation damage prevention programs described in Safety Recommendation P-97-15. (P-97-24)

To the American Public Works Association

Initiate and periodically conduct, in conjunction with the Research and Special Programs Administration, detailed and comprehensive reviews and evaluations of existing State excavation damage prevention programs and recommend changes and improvements, where warranted, such as full participation, administrative enforcement of the program, pre-marking requirements, and training requirements for all personnel involved in excavation activity. (P-97-25)

In conjunction with the Research and Special Programs Administration, develop a plan for collecting excavation damage exposure data. (P-97-26)

Work with the one-call systems to implement the plan outlined in Safety Recommendation P-97-26 to ensure that excavation damage exposure data are being consistently collected. (P-97-27)

Use excavation damage exposure data outlined in Safety Recommendation P-97-26 in the periodic assessments of the effectiveness of State excavation damage prevention programs described in Safety Recommendation P-97-25. (P-97-28)

Review existing training programs and materials related to excavation damage prevention and develop guidelines and materials for distribution to one-call notification centers. (P-97-29)

Develop guidelines and materials that address initial emergency actions by excavators when buried facilities are damaged and then distribute this information to all one-call notification centers. (P-97-30)

Encourage one-call notification centers to work with their members to provide facility location information for the purpose of construction planning. (P-97-31)

Develop standards, in conjunction with the American Society of Civil Engineers, for map depiction of underground facilities that were installed using directional boring techniques. (P-97-32)

Address, in conjunction with the American Society of Civil Engineers, the accuracy of information that depicts subsurface facility locations on construction drawings. (P-97-33)

To the Federal Highway Administration

Require State transportation departments to participate in excavation damage prevention programs and consider withholding funds to States if they do not fully participate in these programs. (P-97-34)

To the Association of American Railroads

Urge your members to fully participate in statewide excavation damage prevention programs, including one-call notification centers. (P-97-35)

To the American Short Line Railroad Association

Urge your members to fully participate in statewide excavation damage prevention programs, including one-call notification centers. (P-97-36)

To the American Society of Civil Engineers

Develop standards, in conjunction with the American Public Works Association, for map depiction of underground facilities that were installed using directional boring techniques. (P-97-37)

Address, in conjunction with the American Public Works Association, the accuracy of information that depicts subsurface facility locations on construction drawings. (P-97-38)

To the Associated General Contractors of America

Promote the use of subsurface utility engineering practices among your members to minimize conflicts between construction activities and underground systems. (P-97-39)

By the National Transportation Safety Board
James E. Hall John A. Hammerschmidt
Chairman Member
Robert T. Francis II John Goglia
Vice Chairman Member
George W. Black, Jr.
Member
Adopted: December 16, 1997

REFERENCES

1. National Transportation Safety Board. 1995. Texas Eastern Transmission Corporation Natural Gas Pipeline Explosion and Fire; Edison, New Jersey; March 23, 1994. Pipeline Accident Report NTSB/PAR-95/01. Washington, DC. 104 p.
2. National Transportation Safety Board. 1996. UGI Utilities, Inc., Natural Gas Distribution Pipeline Explosion and Fire; Allentown, Pennsylvania; June 9, 1994. Pipeline Accident Report NTSB/PAR-96/01. Washington, DC. 94 p.
3. National Transportation Safety Board. 1997. San Juan Gas Company, Inc./Enron Corp. Propane Gas Explosion in San Juan, Puerto Rico, on November 21, 1996. Pipeline Accident Report NTSB/PAR-97/01. Washington, DC.
4. The National Transportation Safety Board does not have the authority to investigate pipeline accidents in other countries.
5. Transportation Research Board, National Research Council. 1988. Pipelines and Public Safety. Special Report 219. Washington, DC.
6. Dipl.Ing, Klees Alfred; Wasserfaches, e.V. 1997. The Safety Concept of Public Gas Supply in Germany. In: Proceedings, 20 th IGU World Gas Conference; Copenhagen.
7. Alliance for Telecommunications Industry Solutions/Network Reliability Steering Committee. 1996. Results and Recommendations Pertaining to Facilities Reliability. Facilities Solutions Report. Washington, DC. February.
8. Federal Aviation Administration, Safety and Quality Assurance Division, Associate Administrator for Aviation Safety. 1993. Cable Cuts: Causes, Impacts, and Preventive Measures. Special Review. Washington, DC. 30 p.
9. Federal Aviation Administration, National Maintenance Control Center AOP-100. 1997. Adhoc Report of facility/service outages associated with cable cuts, 7/1/95-8/22/97.
10. A gas explosion in Annandale, Virginia, on March 24, 1972, occurred just 1 month before the symposium.
11. National Transportation Safety Board. 1973. Prevention of Damage to Pipelines. Special Study NTSB/PSS-73/01. Washington, DC.
12. NTSB accident Nos. DCA96FP004 (Gramercy, Louisiana; May 24, 1996); DCA97FP001 (Tiger Pass, Louisiana; October 23, 1996); and DCA97FP005 (Indianapolis, Indiana; July 21, 1997).
13. The accidents occurred at Allentown, Pennsylvania; Edison, New Jersey; Green River, Wyoming; St. Paul, Minnesota; Cliffwood Beach, New Jersey; and Reston, Virginia.
14. National Transportation Safety Board. 1995. Proceedings of the Excavation Damage Prevention Workshop; September 8-9, 1994; Washington, DC. Report of Proceedings NTSB/RP-95/01. Washington, DC.
15. In October 1990, the Safety Board adopted a program to identify the "Most Wanted" safety improvements. The purpose of the Board's "Most Wanted" list, which is drawn up from safety recommendations previously issued, is to bring special emphasis to the safety issues the Board deems most critical.
16. Estimates of the total infrastructure size are difficult to verify. Bell Communications Research used 20 million miles during the Safety Board's 1994 excavation damage prevention workshop.
17. Wright, P.H.; Ashford, N.J. 1989. Transportation Engineering: Planning and Design. 3d ed. New York: Wiley & Sons (p. 25). 776 p.

18. The estimated number of excavations is based on the number of notifications received in 1996 by member organizations of One-Call Systems International.
19. American Farmland Trust. 1997. Farming on the Edge. Washington, DC.
20. In September 1988, RSPA's Technical Pipeline Safety Standards Committee unanimously supported extending Section 192.614 to cover onshore gas pipelines in Class 1 and 2 locations.
21. RSPA published the final rule, "Excavation Damage Prevention Programs for Gas and Hazardous Liquid and Carbon Dioxide Pipelines," in the Federal Register on March 20, 1995.
22. 49 CFR Parts 192.614 and 195.442.
23. Courtney, W.J.; Kalkbrenner, D.; Yie, G. 1977. Effectiveness of Programs for Prevention of Damage to Pipelines by Outside Forces. Final Report DOT/NTB/OPSO-77/12. Washington, DC: U.S. Department of Transportation, Materials Transportation Bureau, Office of Pipeline Safety Operations; contract DOT-OS-60521. 290 p. (The Office of Pipeline Safety Operations later became the Office of Pipeline Safety within the DOT's Research and Special Programs Administration.)
24. Transportation Research Board, National Research Council. 1988. Pipelines and Public Safety. Special Report 219. Washington, DC.
25. The Safety Board's 1973 review of OSHA regulations is contained in its special report entitled "Prevention of Damage to Pipelines" (NTSB/PSS-73/01).
26. National Transportation Safety Board. 1995. Texas Eastern Transmission Corporation Natural Gas Pipeline Explosion and Fire; Edison, New Jersey; March 23, 1994. Pipeline Accident Report NTSB/PAR-95/ 01. Washington, DC. 104 p.
27. The Conduit 3(4): 1, 4. August 1997. Spooner, WI: National Utility Locating Contractors Association.
28. National Transportation Safety Board. 1996. UGI Utilities, Inc., Natural Gas Distribution Pipeline Explosion and Fire; Allentown, Pennsylvania; June 9, 1994. Pipeline Accident Report NTSB/PAR-96/01, "Gas Piping Technical Committee Excavation Damage Prevention Guidelines"). 94 p.
29. U.S. Department of Transportation, Office of Pipeline Safety. 1995. Exemplary Practices and Success Stories In One-Call Systems. Washington, DC. May.
30. National Transportation Safety Board Accident Brief DCA94MP002; Green River, Wyoming; May 3, 1994.
31. Underspace Bulletin 3(9): 2. June 1997. Spooner, WI: Center for Subsurface Strategic Action (CSSA).
32. As defined by One-Call Systems International (OCSI), a subgroup of the American Public Works Association (APWA).
33. Pre-marking is discussed later in this chapter.
34. "Nationwide One-Call Directory." Pipeline & Utilities Construction. April 1996.
35. Don Evans of USA South cited 13,362,684 in the "Options in Load Management" session at the 22 d annual One Call Systems and Damage Prevention Symposium, April 20-23, 1997, New York City.
36. Kelly, Walter. 1996. Making One-Call Work for You. Constructor, November 1996: 19.
37. Correspondence dated February 10, 1997, from L.D. Shamp, representing the OCSI, to the Association of Oil Pipe Lines. The letter offered comments and suggestions regarding the provisions of the proposed one-call legislation.
38. Transportation Research Board, National Research Council. 1988. Pipelines and Public Safety. Special Report 219 (p. 133). Washington, DC.

39. Underspace Bulletin 2(11): 2. August 1996. Spooner, WI: Center for Subsurface Strategic Action (CSSA).
40. National Transportation Safety Board. 1995. Proceedings of the Excavation Damage Prevention Workshop; September 8–9, 1994. Report of Proceedings NTSB/RP-95/01 (p. 16). Washington, DC.
41. National Transportation Safety Board Accident Brief DCA86FP004.
42. "Connecticut Data Reveals Damage Causes." Underground Focus 10(5): 17. July/August 1996.
43. "New Jersey Takes a Hard Line on Underground Damage Prevention." Underground Focus 11(3): 26-27. April 1997.
44. "News Briefs." Underground Focus 11(5): 14. July/August 1997.
45. "New Jersey Takes a Hard Line on Underground Damage Prevention." Underground Focus 11(3): 26-27. April 1997.
46. National Utility Locating Contractors Association. 1996. Locator Training Standards & Practices. Spooner, WI.
47. National Transportation Safety Board. 1986. Northeast Utilities Service Co. Explosion and Fire; Derby, Connecticut; December 6, 1995. Pipeline Accident Report NTSB/PAR-86/02. Washington, DC. As a result of the Northeast Utility Service Company's positive response to Safety Recommendation P-86-19, the recommendation was classified "Closed—Acceptable Action" on May 14, 1987.
48. National Transportation Safety Board. 1987. Chicago Heights, Illinois; March 13, 1986. Pipeline Accident Summary Report NTSB/PAR-87/01-SUM. Washington, DC. Safety Recommendation P-87-38 was classified "Closed—Acceptable Action" on September 29, 1988.
49. NTSB accident DCA93FP004. Safety Recommendation P-95-24 is currently classified "Open—Acceptable Response" pending receipt of further information from the APWA.
50. National Transportation Safety Board. 1997. San Juan Gas Company, Inc./Enron Corp. Propane Gas Explosion in San Juan, Puerto Rico, on November 21, 1996. Pipeline Accident Report NTSB/PAR-97/01. Washington, DC.
51. 49 CFR Part 192.615, "Emergency plans" [for gas pipelines]; and Part 195.402, "Procedural manual for operations, maintenance, and emergencies" [for hazardous liquids].
52. NTSB accident DCA93FP008.
53. On August 1, 1995, the Safety Board classified this recommendation "Closed—Acceptable Action."
54. This recommendation is currently in an "Open—Acceptable Response" status pending further action by the APWA.
55. Anspach, J.H. 1994. Locating and Evaluating the Conditions of Underground Utilities. In: RETROFIT '94. Washington, DC: National Science Foundation. Sponsored by: Stanford University and the National Science Foundation.
56. OCSI will hold its 23 d annual symposium in March 1998, and the Underground Safety Association will hold its forum in February 1998.
57. Underspace Bulletin 3(7): 2. April 1997. Spooner, WI: Center for Subsurface Strategic Action (CSSA).
58. According to advertisements for the Sure-Lock locator by Heath Consultants, that equipment provides a continuous depth reading. Other equipment manufacturers, Fisher TW-770 and Metrotech 9800, advertise a pushbutton feature for digital display of depth.

59. "Wyoming's Unique One-Call Legislation." Constructor, November 1996:17.
60. Anspach, J.H. 1994. Locating and Evaluating the Conditions of Underground Utilities. In: RETROFIT '94. Washington, DC: National Science Foundation. Sponsored by: Stanford University and the National Science Foundation.
61. "Depth Perception" session at the 22 d annual One-Call Systems and damage prevention symposium, April 20-23, 1997, New York City. Panel participants at the moderated session represented equipment manufacturers and underground locator services.
62. NTSB accident DCA97FP005; the accident occurred on July 21, 1997.
63. (a) Underground Focus 10(6): 16-19; 22-23. September/October 1996. (b) Underground Focus 10(7): 18-19. November/December 1996.
64. Configuration of the Mole Map System developed by McLaughlin Boring Systems.
65. Underground Focus 10(6): 17. September/October 1996.
66. Place, J.C. 1996. "Gas Utility Mapping: What's Needed, What's Used." Gas Industries, January: 21-22.
67. Vista One Call Mapping Program by Kuhagen, Inc., is compatible with California's USA North One Call System and has been accepted for use by the State fire marshal as a method for digitizing pipeline mapping.
68. "One-Calls Eye New Mapping." Underground Focus 10(2): 6. Symposium Edition 1996.
69. The Federal Office of Management and Budget (OMB), under the directive of OMB Circular A16, created the Federal Geographic Data Committee, which is chaired by the Secretary of the Interior. The 1994 Plan for the National Spatial Data Infrastructure was issued in March 1994.
70. U.S. Department of Transportation, Federal Highway Administration. 1995. Subsurface Utility Engineering Handbook. FHWA-PD-96-004 (p. I-14). Washington, DC. November.
71. According to the FHWA, Maryland, Pennsylvania, Delaware, North Carolina, and Arizona use SUE on an extensive basis.
72. U.S. Department of Transportation, Federal Highway Administration. 1995. Subsurface Utility Engineering Handbook. FHWA-PD-96-004 (p. I-29). Washington, DC. November.
73. Doctor, R.H.; Dunker, N.A.; Santee, N.M. 1995. Third-Party Damage Prevention Systems. GRI-95/ 0316. Final report, contract 5094–810–2870. Chicago, IL: Gas Research Institute. 67 p., plus appendixes.
74. Calculated as total hits (104,128)¸ miles of gas line (1,778,600) = 0.0585 hits per mile or 58.5 hits per 1,000 miles. (104,128 hits, 2 861 767.4 kilometers = 0.0364 hits per kilometer or 36.4 hits per 1000 kilometers.) Note: Different categories of gas lines were added together. Transmission lines have a substantially lower rate than other gas systems: survey respondents reported 201 hits per 36,042 line miles (57 992 kilometers), for a rate of 5.5 hits per 1,000 miles (1609 kilometers). However, GRI survey numbers account for only 10 percent of the U.S. gas transmission system. If the number of transmission system hits per 1,000 miles is separated from the U.S. total, the rate for local distribution companies increases to 71 hits per 1,000 miles.
75. Calculated as total hits (104,128)¸ excavation notifications (20,000,000) = 0.0052 per notification or 5.2 per 1,000 notifications.
76. Northern Illinois Gas incorporated a performance incentive based on hits per locates into its most recent locator service contract with Kelly Cable Corporation.

77. Before 1984, $5,000 was the OPS property loss threshold for reporting natural gas and liquid pipeline failures. In July 1984, this threshold was increased, resulting in a sharp decline in reportable line failures after 1983.

78. Doctor, R.H.; Dunker, N.A.; Santee, N.M. 1995. Third-Party Damage Prevention Systems. GRI-95/ 0316. Final report, contract 5094-810-2870. Chicago, IL: Gas Research Institute. 67 p., plus appendixes.

79. North Carolina State University, Construction Automation & Robotics Laboratory. 1996. Assessment of the Cost of Underground Utility Damages. Raleigh, NC. 17 p., plus appendixes. The study was also the subject of the following article: Carver, C. 1996. "Real Costs of Utility Damages Researched by NCSU." Underground Focus 10(6): 28. September/October.

80. DOT Form 7000-1, Part D: (1) corrosion, (2) failed weld, (3) incorrect operation by operator personnel, (4) failed pipe, (5) outside force damage, (6) malfunction of control or relief equipment, (7) other— specify. Category 7 includes cases involving excavation damage, such as backhoe dug into line, and category 5 (outside force damage) includes vandalism and lightning strikes. Excavation damage is not separately categorized.

81. Accidents with $0.00 damage are included in the database because they meet one of the other criteria for reporting. For 1996, the three databases show 76 accidents with $0.00 damage costs.

82. See appendix for a list of Safety Board recommendations related to RSPA's accident data.

83. National Transportation Safety Board. 1996. Evaluation of Accident Data and Federal Oversight of Petroleum Product Pipelines. Pipeline Special Investigation Report NTSB/SIR-96/02. Washington, DC. 67 p. The special investigation was prompted by the ruptures of two petroleum product pipelines operated by the same company. Both ruptures occurred within a 15-month period.

84. Correspondence dated August 7, 1996, from the RSPA Administrator. On January 2, 1997, the Safety Board classified Safety Recommendation P-96-1 "Open—Acceptable Response" based on RSPA's response and pending a further progress report.

CHAPTER 13

BRITTLE-LIKE CRACKING IN PLASTIC PIPE FOR GAS SERVICE

Source: National Transportation Safety Board
Special Investigation Report NTCB/SIR-98/01, Washington, D.C.

INTRODUCTION

The use of plastic piping to transport natural gas has grown steadily over the years because of the material's economy, outstanding corrosion resistance, light weight, and ease of installing and joining. According to the American Gas Association (A.G.A.), the total miles of plastic piping in use in natural gas distribution systems in the United States grew from about 9,200 miles in 1965 to more than 45,800 miles in 1970 [1]. By 1982, this figure had grown to about 215,000 miles, of which more than 85 percent was polyethylene Data maintained by Office of Pipeline Safety (OPS), an office of the Research and Special Programs Administration (RSPA) within the U.S. Department of Transportation (DOT), indicate that, by the end of 1996, more than 500,000 miles of plastic piping had been installed [2]. Plastic piping as a percentage of all gas distribution piping installed each year has also grown steadily, as illustrated in Figure 13.1.

Despite the general acceptance of plastic piping as a safe and economical alternative to piping made of steel or other materials, the Safety Board notes that a number of pipeline accidents it has investigated have involved plastic piping that cracked in a brittle-like manner [3]. (see Table 13.1 for information on three recent accidents.) For example, on October 17, 1994, an explosion and fire in Waterloo, Iowa, destroyed a building and damaged other property. Six persons died and seven were injured in the accident. The Safety Board investigation determined that natural gas had been released from a plastic service pipe that had failed in a brittle-like manner at a connection to a steel main.

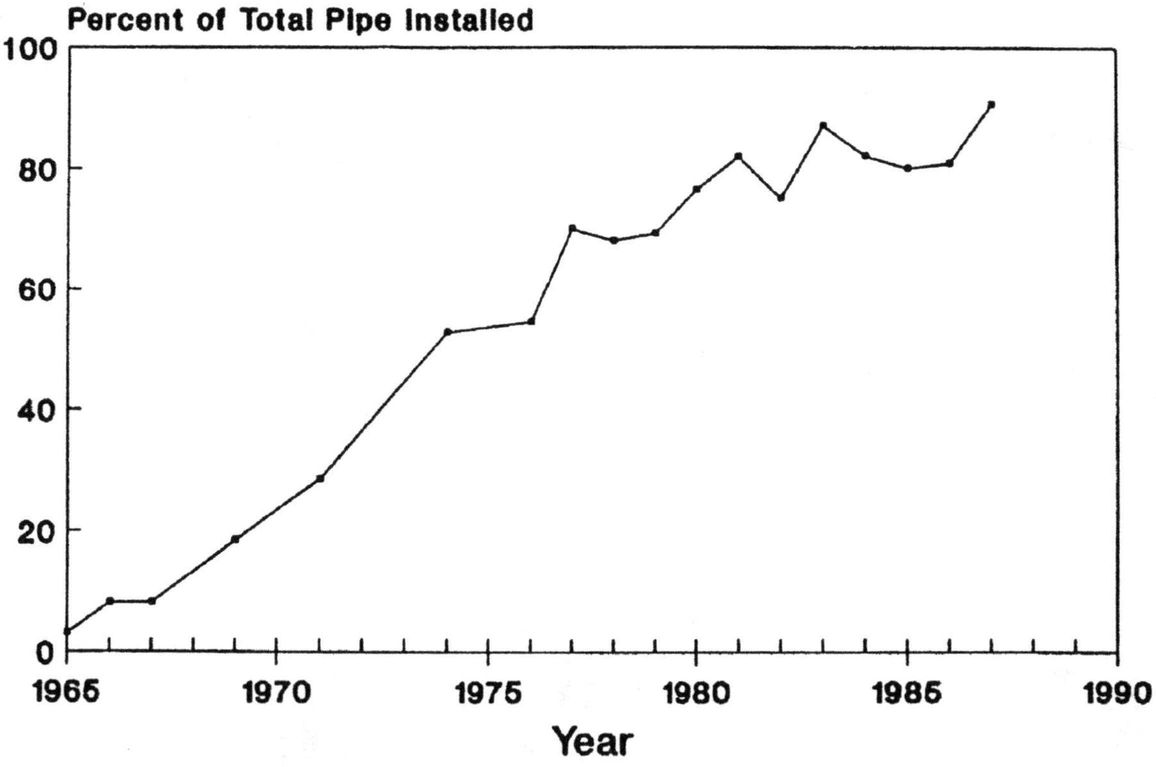

FIGURE 13.1 Plastic pipe as a percentage of all piping used in gas distribution. *(Source: Duvall, D.E., "Polyethylene Pipe for Natural Gas Distribution," presented at the Transportation Safety Institute's Pipeline Failure Investigation course, 1997. Data from Pipeline & Gas Journal surveys)*

The Safety Board also investigated a gas explosion that resulted in 33 deaths and 69 injuries in San Juan, Puerto Rico, in November 1996 [4]. The Safety Board's investigation determined that the explosion resulted from ignition of propane gas that had migrated under pressure from a failed plastic pipe. Stress intensification at a connection to a plastic fitting led to the formation of brittle-like cracks.

The Railroad Commission of Texas investigated a natural gas explosion and fire that resulted in one fatality in Lake Dallas, Texas, in August 1997 [5]. A metal pipe pressing against a plastic pipe generated stress intensification that led to a brittle-like crack in the plastic pipe.

A Safety Board survey of the accident history of plastic piping suggested that the material may be susceptible to brittle-like cracking under conditions of stress intensification. No statistics exist that detail how much and from what years any plastic piping may already have been replaced; however, as noted above, hundreds of thousands of miles of plastic piping have been installed, with a significant amount of it having been installed prior to the mid-1980s. Any vulnerability of this material to premature failure could represent a serious potential hazard to public safety.

In an attempt to gauge the extent of brittle-like failures in plastic piping and to assess trends and causes, the Safety Board examined pipeline accident data compiled by RSPA. The examination revealed that the RSPA data are insufficient to serve as a basis for assessing the long-term performance of plastic pipe.

Lacking adequate data from RSPA, the Safety Board reviewed published technical literature and contacted more than 20 experts in gas distribution plastic piping to determine the estimated frequency of brittle-like cracks in plastic piping. The majority of the published literature and experts indicated that failure statistics would be expected to vary from one gas system operator to another based on factors such as brands and dates of manufacture of plastic piping in service, installation practices, and ground temperatures, but they indicated that brittle-like failures, as a nationwide average, may represent the second most frequent failure mode for older plastic piping, exceeded only by excavation damage.

The Safety Board asked several gas system operators about their direct experience with brittle-like cracks. Four major gas system operators reported that they had compiled failure statistics sufficient to estimate the extent of brittle-like failures. Three of those four said that brittle-like failures are the second most frequent failure mode in their plastic pipeline systems. One of these operators supplied data showing that it experienced at least 77 brittle-like failures in plastic piping in 1996 alone.

As an outgrowth of the Safety Board's investigations into the Waterloo, Iowa, San Juan, Puerto Rico, and other accidents, and in view of indications that some plastic piping, particularly older piping, may be subject to premature failure attributable to brittle-like cracking, the Safety Board undertook a special investigation of polyethylene gas service pipe. The investigation addressed the following safety issues:

- The vulnerability of plastic piping to premature failures due to brittle-like cracking;
- The adequacy of available guidance relating to the installation and protection of plastic piping connections to steel mains; and
- Performance monitoring of plastic pipeline systems as a way of detecting unacceptable performance in piping systems.

As a result of its investigation, the Safety Board makes three safety recommendations to the Research and Special Programs Administration, one safety recommendation to the Gas Research Institute, three safety recommendations to the Plastics Pipe Institute, one safety recommendation to the Gas Piping Technology Committee, two safety recommendations to the American Society for Testing and Materials, one safety recommendation to the American Gas Association, two safety recommendations to MidAmerican Energy Corporation, two safety recommendations to Continental Industries, Inc., and one safety recommendation each to Dresser Industries, Inc., Inner-Tite Corporation, and Mueller Company.

INVESTIGATION

Accident History

On October 17, 1994, a natural gas explosion and fire in Waterloo, Iowa, destroyed a building and damaged other property. Six persons died and seven were injured in the accident. The Safety Board investigation determined that the source of the gas was a ½ inch diameter plastic service pipe that had failed in a brittle-like manner at a connection to a steel main [6].

Excavations following the accident uncovered, at a depth of about 3 feet, a 4-inch steel main. Welded to the top of the main was a steel tapping tee manufactured by Continental Industries, Inc. (Continental). Connected to the steel tee was a 1/2-inch plastic service pipe. (see Figure 13.2.) Markings on the plastic pipe indicated that it was a medium-density polyethylene material manufactured on June 11, 1970, in accordance with American Society for Testing and Materials (ASTM) standard D2513. The pipe had been marketed by Century Utility Products, Inc. (Century). The plastic pipe was found cracked at the end of the tee's internal stiffener and beyond the coupling nut.

The investigation determined that much of the top portion of the circumference of the pipe immediately outside the tee's internal stiffener displayed several brittle-like slow crack initiation and growth fracture sites. These slow crack fractures propagated on almost parallel planes slightly offset from each other through the wall of the pipe. As the slow cracks from different planes continued to grow and began to overlap one another, ductile tearing occurred between the planes. Substantial deformation was observed in part of the fracture; however, the initiating cracks were still classified as brittle-like.

Samples recovered from the plastic service line underwent several laboratory tests under the supervision of the Safety Board. Two of these tests were meant to roughly gauge the pipe's susceptibility to brittle-like cracking. These tests were a compressed ring environmental stress crack resistance (ESCR) test in accordance with ASTM F1248 and a notch tensile test known as a PENT test that is now ASTM F1473. Lower failure times in these tests indicate greater susceptibility to brittle-like cracking under test conditions. The ESCR testing of 10 samples from the pipe yielded a mean failure time of 1.5 hours, and the PENT testing of 2 samples yielded failure times of 0.6 and 0.7 hours. Test values this low have been associated with materials having poor performance histories characterized by high leakage rates at points of stress intensification due to crack initiation and slow crack growth typical of brittle-like cracking [7, 8].

In late 1996, the Safety Board began an investigation of a November 1996 gas explosion that resulted in 33 deaths and 69 injuries in San Juan, Puerto Rico. The investigation determined that the explosion resulted from ignition of propane gas that, after migrating under pressure from a failed plastic pipe at a connection to a plastic fitting, had accumulated in the basement of a commercial building. The Safety Board concluded that apparent inadequate support under the piping and the resulting differential settlement generated long-term stress intensification that led to the formation of brittle-like circumferential cracks on the pipe.

The Railroad Commission of Texas investigation of a fatal natural gas explosion and fire in Lake Dallas, Texas, in August 1997 determined that a metal pipe pressing against a plastic pipe generated stress intensification that led to a brittle-like crack in the plastic pipe.

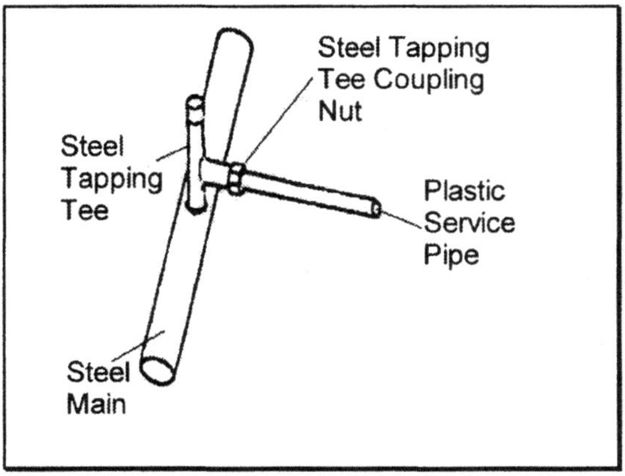

FIGURE 13.2 Typical plastic service pipe connection to steel gas main. Many connections are protected against shear and bending forces by a plastic sleeve that encloses the service pipe-to-tee connection on either side of the coupling nut.

The Waterloo, San Juan, and Lake Dallas accidents were only three of the most recent in a series of accidents in which brittle-like cracks in plastic piping have been implicated. In Texas in 1971, natural gas migrated into a house from a brittle-like crack at the connection of a plastic service line to a plastic main [9]. The gas ignited and exploded, destroying the house and burning one person. The investigation determined that vertical loading over the connection generated long-term stress that led to the crack.

A 1973 natural gas explosion and fire in Maryland severely damaged a house, killed three occupants, and injured a fourth [10].

The Safety Board's investigation of a natural gas explosion and fire that resulted in three fatalities in North Carolina in 1975 determined that the gas had accumulated because a concrete drain pipe resting on a plastic service pipe had precipitated two cracks in the plastic pipe [11]. Available documentation suggests that these cracks were brittle-like.

A 1978 natural gas accident in Arizona destroyed one house, extensively damaged two others, partially damaged 11 other homes, and resulted in one fatality and five injuries [12]. Available documentation indicates that the gas line crack that caused the accident was brittle-like.

A 1978 accident in Nebraska involved the same brand of plastic piping as that involved in the Waterloo accident. A crack in a plastic piping fitting resulted in an explosion that injured one person, destroyed one house, and damaged three other houses [13].

The Safety Board determined that inadequate support under the plastic fitting resulted in long-term stress intensification that led to the formation of a circumferential crack in the fitting. Available documentation indicates that the crack was brittle-like.

A December 1981 natural gas explosion and fire in Arizona destroyed an apartment, damaged five other apartments in the same building, damaged nearby buildings, and injured three occupants [14].

The Safety Board's investigation determined that assorted debris, rocks, and chunks of concrete in the excavation backfill generated stress intensification that resulted in a circumferential crack in a plastic pipe at a connection to a plastic fitting. Available documentation indicates that the crack was brittle-like.

A July 1982 natural gas explosion and fire in California destroyed a store and two residences, severely damaged nearby commercial and residential structures, and damaged automobiles [15]. The Safety Board's investigation identified a longitudinal crack in a plastic pipe as the source of the gas leak that led to the explosion. Available documentation indicates that the crack was brittle-like.

A September 1983 natural gas explosion in Minnesota involved the same brand of plastic piping as that involved in the Waterloo and Nebraska accidents [16]. The explosion destroyed one house and damaged several others, and injured five persons. The Safety Board's investigation determined that rock impingement generated stress intensification that resulted in a crack in a plastic pipe. Available documentation indicates that the crack was brittle-like.

One woman was killed and her 9-month-old daughter injured in a December 1983 natural gas explosion and fire in Texas [17]. The Safety Board's investigation determined that the source of the gas leak was a brittle-like crack that had resulted from damage to the plastic pipe during an earlier squeezing operation to control gas flow [18].

A September 1984 natural gas explosion in Arizona resulted in five fatalities, seven injuries, and two destroyed apartments [19]. The Safety Board's investigation determined that a reaction between a segment of plastic pipe and some liquid trapped in the pipe weakened the pipe and led to a brittle-like crack.

During the course of the investigation of the accident at Waterloo, Iowa, the Safety Board learned of several other accidents, not investigated by the Safety Board, that involved cracks in the same brand of plastic piping as that involved in the Waterloo accident. Three of these accidents, which occurred in Illinois (1978 and 1979) and in Iowa (1983), resulted in five injuries and damage to buildings [20]. A 1995 accident in Michigan also involved a crack in this same brand of pipe [21]. Available documentation indicates that the cracks were brittle-like.

Strength Ratings, Ductility, and Material Standards for Plastic Piping

During the 1950s and early 1960s, when plastic piping was beginning to gain acceptance as an alternative to steel piping for the transport of water and gas, no established procedures existed for rating the strength of materials intended for use in plastic pressure piping.

In November 1958, the Thermoplastic Pipe Division of the Society of the Plastics Industry organized a group called the Working Stress Subcommittee [22]. The subcommittee, in January 1963, issued a procedure (hereinafter referred to as the PPI procedure) that specified a uniform protocol for rating the strength of materials used in the manufacture of thermoplastic pipe in the United States. In March 1963, the Thermoplastic Pipe Division adopted its current name, the Plastics Pipe Institute (PPI).

On July 1, 1963, the PPI established a voluntary program of listing the material strengths of plastic piping materials, specifically, those materials designed for water applications. To apply for a PPI listing, applicants sent strength test data to the PPI, often accompanied by the manufacturer's analysis of the data and a proposed material strength rating. The PPI would analyze the data and, if warranted, list the material for the calculated strength. The PPI did not certify or approve the material received or validate the data submitted, nor did it audit or inspect those submitting data [23].

In simplified terms, the PPI procedure, which is performed by the materials manufacturers themselves, involves recording how much time it takes stressed pipe samples to rupture at a standardized temperature of 73 °F. The stresses used in the tests are recorded as "hoop stress," which is tensile stress in the wall of the pipe in a circumferential orientation (hence the term "hoop") due to internal pressure. Although hoop stress is expressed in pounds per square inch, it is a value quite different from the pipe's internal pressure.

The testing process involves subjecting pipe samples to various hoop stress levels, and then recording the time to rupture. For some samples at some pressures, rupture will occur in as little as 10 hours. As hoop stress is reduced, the time-to- failure increases. At some hoop stress level, at least one of the tested specimens will not rupture until at least 10,000 hours (slightly more than 1 year). After the rupture data points (hoop stresses and times-to-failure) for this material have been recorded, the data points are plotted on log-log coordinates as the relationship between hoop stress and time-to-failure. (see Figure 13.3.) A mathematically developed "best-fit" straight line is correlated with the data points to represent the material's resistance to rupturing at various hoop stress levels.

Once the best-fit straight line is calculated to 10,000 hours, it is extrapolated to 100,000 hours (about 11 years). The hoop stress level that coincides with the point at which the line intersects the 100,000-hour time line represents the calculated long-term hydrostatic strength of that particular material.

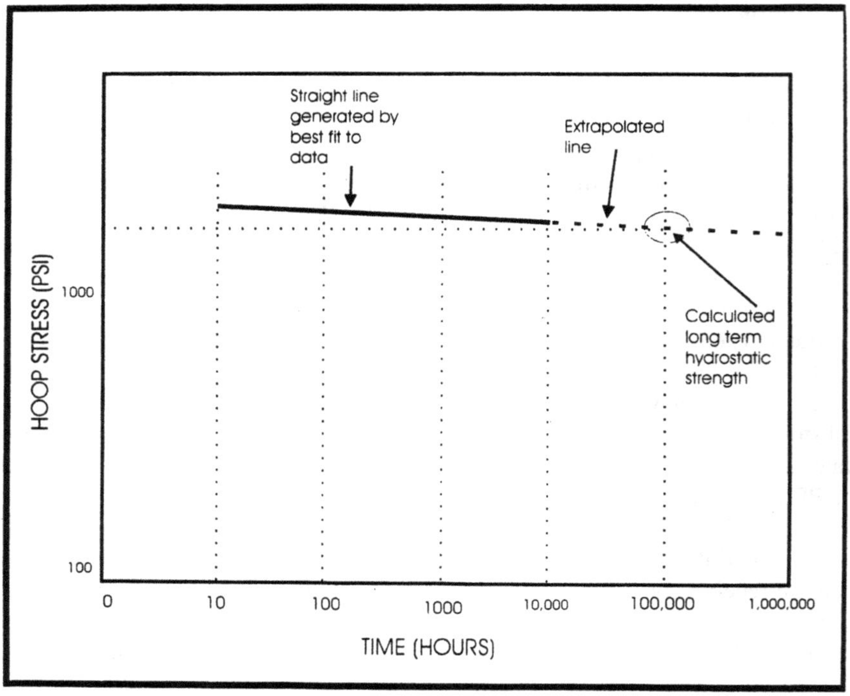

FIGURE 13.3 Stress rupture data plotted as best-fit straight line and extrapolated to determine long-term hydrostatic strength. *(Derived from A.G.A Plastic pipe Manual for Gas Service)*

To simplify the ratings and facilitate standardization, the PPI procedure grouped materials with similar long-term hydrostatic strength ranges into "hydrostatic design basis" categories. For example, those materials having long-term hydrostatic strengths between 1200 and 1520 psi were grouped together and assigned a hydrostatic design basis of 1250 psi. Those materials having long-term hydrostatic strengths between 1530 and 1910 psi were grouped together and assigned a hydrostatic design basis of 1600 psi.

To help ensure the validity of the mathematically derived line, the PPI procedure required the submission of all rupture data points. It further specified the minimum number of data points and minimum number of tested lots. The procedure employed statistical tests to verify the quality of data and quality of fit to the mathematically derived line. These measures excluded materials when the data demonstrated excessive data scatter due to either inadequate quality of data or deviation from straight line behavior through 10,000 hours [24].

The PPI procedure, after some refinement, was issued as an ASTM method in 1969 (ASTM D2837). The PPI adopted a policy document for PPI's listing service in 1968, which remained under PPI jurisdiction [25].

When polyethylene pipe fails during laboratory stress rupture testing at 73 °F, it fails primarily by means of ductile fractures, which are characterized by substantial visible deformation During stress rupture tests, if hoop stress on the test piping is decreased, the time-to-failure increases, and the amount of deformation apparent in the failure decreases [26]. In pipe subjected to prolonged stress rupture testing, slit fractures may begin 26 Mruk, S. A., "The Ductile Failure of Polyethylene Pipe," SPE Journal, Vol. 19, No. 1, January 1963 [27]. Because of the frequent lack of visible deformation associated with them, slit fractures are also referred to as brittle-like fractures. to appear at some point (depending on the specific polyethylene resin material). The PPI procedure did not differentiate between ductile and slit failure types, and, based on most available laboratory test data (at 73°F), assumed that both types of failures would be described by the same extrapolated (straight) line [28].

In 1963-64, the National Sanitation Foundation amended its standard for plastic piping used for potable water service to require that manufacturers furnish evidence of having an appropriate strength rating in accordance with the PPI procedure [29]. Manufacturers then decided to utilize the PPI listing service, having determined that this was the most convenient way to furnish the required evidence.

In 1966, the ASTM issued ASTM D2513, the society's first standard specification covering polyethylene plastic piping for gas service [30]. ASTM D2513 made reference to long-term hydrostatic strength and hydrostatic design stress and included an appendix defining these terms in accordance with the PPI procedure It also required that polyethylene pipe meet certain requirements of ASTM D2239 (a polyethylene pipe specification for water service), which also included references to the PPI procedure. ASTM D2513 did not explicitly require materials to have a PPI listing [31].

Even without an explicit requirement, some manufacturers voluntarily obtained PPI listings for their resin materials intended for gas use, and some others, as noted above, obtained PPI listings for their resins that were intended for water use (but were similar to their resins intended for gas service) as a way of meeting National Sanitation Foundation requirements [32,33] .

In 1967, the United States of America Standards Institute B31.8 code, Gas Transmission and Distribution Piping Systems, for the first time recognized the suitability of plastic piping for gas distribution service and included requirements for the pipings' use [34]. The 1966 issuance of ASTM D2513 and the 1967 inclusion of plastic piping within B31.8 cleared the way for the general use of plastic piping for gas distribution [35]. B31.8 included a design equation (see discussion below), and although the code, like the ASTM standard, did not explicitly require a PPI listing, it did require that material used to manufacture plastic pipe establish its long-term hydrostatic strength in accordance with the PPI procedure.

On August 12, 1968, the Natural Gas Pipeline Safety Act was enacted, requiring the DOT to adopt minimum Federal regulations for gas pipelines. In December 1968, the DOT instituted interim Federal regulations by federalizing the State pipeline safety regulations that were in place at the time. The DOT, having concluded that the majority of the States required compliance with the 1968 version of B31.8, adopted that version of the code for the Federal regulations covering those States not yet having their own natural gas pipeline safety regulations.

Most of these Federal interim standards were replaced in November 1970 by 49 Code of Federal Regulations (CFR) 192; however, the interim provisions concerning the design, installation, construction, initial inspection, and initial testing of new pipelines remained in effect until March 1971. At that time, 49 CFR 192 incorporated the design equation for plastic pipe from B31.8 and also required that plastic piping conform to ASTM D2513 [36].

The 1967 version of B31.8 introduced fixed design factors (subsequently incorporated into 49 CFR 192) as a catch-all mechanism to account for various influences on pipe performance and durability [37]. These influences included external loadings, limitations of and imprecision in the PPI procedure, variations in pipe manufacturing, handling and storage effects, temperature fluctuations, and harsh environments [38]. A design equation was used to determine the allowable gas service pipe pressure rating based on the hydrostatic design basis category, pipe dimensions, and design factor [39]. The design basis for plastic pipe thus "Validating the Hydrostatic Design Basis of PE Piping Materials." The design equation (with the current design factor, 0.32) can be found in 49 CFR 192.121, although 192.121 erroneously references the long-term hydrostatic strength instead of the hydrostatic design basis category. RSPA is used internal pressures as a design criterion but did not directly take into account additional stresses that could be generated by external loadings, despite the fact that field failures in plastic piping systems were frequently associated with external loads but were rarely attributable to internal pressure effects alone [40].

Kulmann and Mruk have reported that no direct basis was established to design for external loads because:

- The industry had no easy means of quantifying external loads and their effects on plastic piping systems [41]
- Many in the industry believed that plastic piping, like steel and copper piping, behaved as a ductile material that would withstand considerable deformation before undergoing damage, thus alleviating and redistributing local stress concentrations that would crack brittle materials such as cast iron. This belief resulted from short-term laboratory tests showing that plastic piping had enormous capacity to deform before rupturing [42]

Because of plastic piping's expected ductile behavior, many manufacturers believed it safe to base their designs on average distributed stress concentrations generated primarily by internal pressure and, within reason, to neglect localized stress concentrations. They believed such stress would be reduced by localized yielding, or deformation. Mruk and Palermo have pointed out that design protocols were predicated on the assumption of such ductile behavior [43].

In contrast, cast iron piping has recognized brittle characteristics. The design basis for cast iron therefore does not assume that localized yielding or deformation will reduce stress intensification. As a result, the design protocol for cast iron includes the quantification and direct input of external loading factors that can generate localized stress intensification [44].

Failures in polyethylene piping that occur under actual service conditions are frequently slit failures; ductile failures are rare [45]. Slit failures in polyethylene, whether occurring during stress rupture testing or under actual service conditions, result from crack initiation and slow crack growth and are similar to brittle cracks in other materials in that they can occur with little or no visible deformation [46].

During the 1960s and 1970s, some experts began to question the validity of the PPI procedure's assumption of a continuing, gradual straight-line decline in strength (Figure 13.3) [47]. By the late 1970s and early 1980s, the plastic piping industry in the United States realized that testing piping materials at elevated temperatures was a way to accelerate failure behavior that would occur much later at lower temperatures (such as 73 °F). Based on data derived from elevated-temperature testing, the industry concluded that the gradual straight-line decline in strength assumed by the PPI procedure was not valid. Instead, two distinct failure zones were indicated for polyethylene piping in stress rupture testing. The first zone is characterized by the gradual straight-line decline in strength accompanied primarily by ductile fractures. The first zone gradually transitions to the second zone, which is characterized by a more rapid decline in strength accompanied by brittle-like fractures only. The time and magnitude of this more rapid decline in strength varies by type and brand of polyethylene. Piping manufacturers have worked to improve their products' resistance to slit-type failures and thus to push this downturn further out in time. The PPI procedure did not account for this downturn, and the difference between the actual falloff shown in Figure 13.4 and the projected straight-line strengths shown in figure 13.3 for listed materials became more pronounced as the lines were extrapolated beyond 100,000 hours.

As manufacturers steadily improved their formulations to delay the onset of the downturn in long-term strength and associated brittle-like behavior, PPI and ASTM industry standards were upgraded to reflect what the major manufacturers were able and willing to accomplish [48]. Accordingly, and because a consensus of manufacturers recognized the relationship between improved elevated-temperature properties and improved longer term pipe performance, the PPI in 1982 recommended that ASTM D2513 specify a minimum acceptable hydrostatic strength at 140°F. In 1984, ASTM D2513 included a statement in its non-mandatory appendix that gas pipe materials should have a specified long-term hydrostatic strength at 140°F. In the 1988 edition, this requirement was moved to the mandatory section of the standard. This strength at 140°F was calculated the same way that the 73°F strength was calculated—data demonstrating a straight line to 10,000 hours was assumed to extrapolate to 100,000 hours without a downturn.

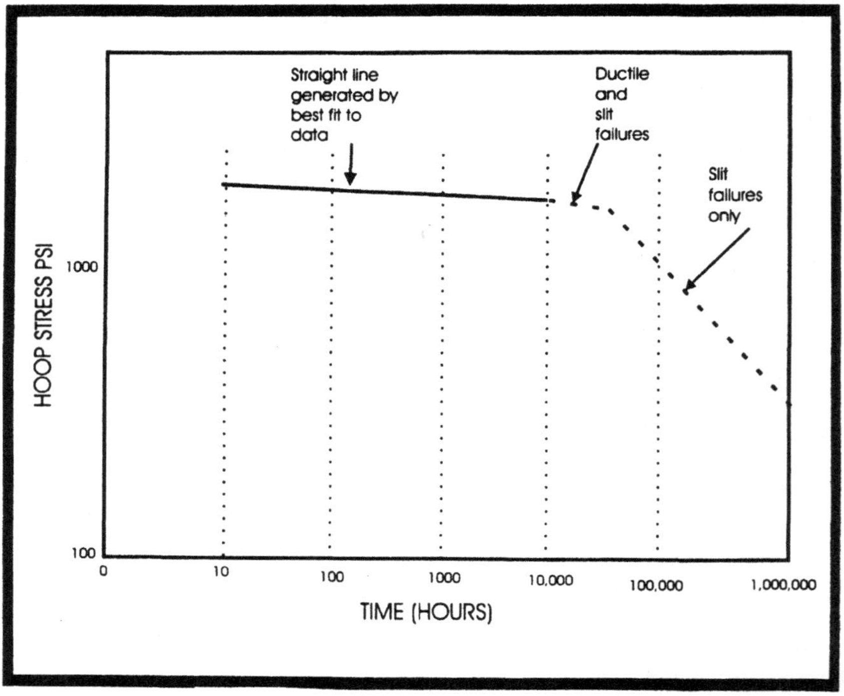

FIGURE 13.4 Stress rupture data plotted as best-fit straight line transitioning to downturn in strength. *(Derived for A.G.A. Plastic Pipe Manual for Gas Service)*

Gradually, more manufacturers obtained PPI listings for their resins intended for gas service, and by the early to mid-1980s, virtually all resins used for gas service had PPI listings. At that time, a consensus of manufacturers supported a change within ASTM D2513 to require PPI listings. In 1985, ASTM D2513 was revised to require that materials for gas service have a PPI listing.

By 1985, manufacturers reached a consensus to exclude materials that deviated from the 73°F extrapolation before 100,000 hours. The PPI adopted this restriction and advised the industry that, effective January 1986, all materials not demonstrating straight-line performance to 100,000 hours would be dropped from its listing.In 1988, ASTM D2837 also included the restriction [49, 50]. The new PPI and ASTM requirements had no effect on pipe installed prior to the effective date of the requirements.

On August 20, 1997, after manufacturers reached a consensus, the PPI issued notice that, effective January 1999, in order for materials to retain their PPI listings for long-term hydrostatic strength at temperatures above 73°F (for example, at 140°F), these materials will have to demonstrate (mathematically, via elevated-temperature testing) that a downturn does not exist prior to 100,000 hours or, alternatively, if a downturn does exist before 100,000 hours, the strength rating will be reduced to reflect the point at which the calculated downturn in strength intercepts 100,000 hours. An ASTM project has been initiated to incorporate this requirement within ASTM D2837. The Safety Board also notes that the PPI has endorsed a proposal to have ASTM D2513 require polyethylene piping to have no downturn in stress rupture testing at 73°F before 50 years, as mathematically determined in elevated-temperature tests.

All available evidence indicates that polyethylene piping's resistance to brittle-like cracking has improved significantly through the years. Several experts in gas distribution plastic piping have told the Safety Board that a majority of the polyethylene piping manufactured in the 1960s and early 1970s had poor resistance to brittle-like cracking, while only a minority of that manufactured by the early 1980s could be so characterized [51]. Several gas system operators have told the Safety Board that they are aware of no instances of brittle-like cracking with their own modern polyethylene piping installations.

Century Pipe Evaluation and History

The Safety Board's investigation of the Waterloo, Iowa, accident determined that the pipe involved in the accident had been manufactured by Amdevco Products Corporation (Amdevco) in Mankato, Minnesota. Amdevco's Mankato plant first began producing plastic pipe in 1970, with plastic piping for gas service as its only piping product. Amdevco made the pipe from Union Carbide's Bakelite DHDA 2077 Tan 3955 (hereinafter referred to as DHDA 2077 Tan) resin material. Century Utility Products, Inc., marketed the pipe to Iowa Public Service Company, and Century's name was marked on the pipe. Century and Amdevco formally merged in 1973 [52]. The combined corporation went out of business in 1979.

Because Amdevco/Century no longer exists, Safety Board investigators could locate no records to indicate the qualification steps Amdevco may have performed before Century marketed its pipe to Iowa Public Service Company. A plastic pipe manufacturer would normally have obtained documentation from its resin supplier indicating that the resin material had a sufficient long-term hydrostatic strength. Code B31.8 required and ASTM D2513 recommended that polyethylene pipe manufacturers perform certain quality control tests on production samples, including twice-per- year sustained pressure tests.

Like many gas operators of that time, Iowa Public Service Company (now MidAmerican Energy Corporation), which had installed the Waterloo piping in 1971, had no formal program for testing or evaluating products. According to MidAmerican Energy, the company accepted representations from a principal of Century, a former DuPont employee, who portrayed himself as being intimately involved with the development and marketing of DuPont's polyethylene piping. MidAmerican Energy has reported that these representations included assertions that Century plastic pipe met industry standards and had the same formulation as DuPont's plastic pipe. In 1970, according to MidAmerican Energy officials, Century offered Iowa Public Service Company attractive commercial terms for its product, with the result that, in 1970, when Amdevco first started to manufacture pipe, Iowa Public Service Company began purchasing all of its plastic pipe from Century [53].

Before the Waterloo accident, a previous accident involving Century pipe had been reported in the Midwest Gas (the operator at the time of the accident) system. That accident occurred in August 1983 in Hudson, Iowa, and resulted in multiple injuries. Midwest Gas, attributing this accident to a rock pressing into the pipe, considered it an isolated incident. During 1992-94, the company had two significant failures with pipe fittings involving brittle-like cracks in Century pipe. Sections of the failed pipe were sent to the two affected pipe fitting manufacturers, and one responded that nothing was wrong with the fitting, suggesting instead that the problem might rest with the piping material.

MidAmerican Energy reported that, as a result of these two failures, Midwest Gas directed inquiries to other utilities operating in the Midwest and, in May 1994, learned of one other accident involving Century pipe. In June 1994, Midwest Gas decided to send samples of Century polyethylene piping to an independent laboratory for test and evaluation. The sample collection was in process at the time of the Waterloo accident. In August 1995, Midwest Gas issued a report, based on the laboratory testing, concluding that the Century samples had poor resistance to slow crack growth.

Subsequent to the accident, Midwest Gas worked to determine if its installations with Century plastic piping had had higher rates of failure than those with piping from other manufacturers. After analyzing the data, Midwest Gas concluded that the piping installations with Century piping had failure rates that were significantly higher than those installations with plastic piping from other manufacturers. Based on this analysis, as well as on other factors—including the severity and consequences of leaks involving Century piping, the laboratory test results, recommendations from two manufacturers of pipe fittings cautioning against use of their fittings with Century pipe because of the pipe's poor resistance to brittle-like cracking, and interviews with field personnel—MidAmerican Energy (the current operator) has replaced all its known Century piping with new piping, completing the replacement program in 1997.

Safety Board investigators found little additional documentation regarding qualification tests of Century plastic pipe by other gas system operators having Century pipe in service. A reference was found to a 1971 Northern States Power Company Testing Department progress report stating that Century pipe complied with ASTM D2513, and that the pipe was acceptable for use with DuPont polyethylene fittings. The actual progress report and records of any tests that may have been performed were not located [54].

Union Carbide DHDA 2077 Tan Resin. The resin used to manufacture the pipe involved in the Waterloo accident was DHDA 2077 Tan. To examine how Union Carbide qualified this material requires some background.

During the late 1960s, several companies manufactured plastic resin and plastic pipe for the gas distribution plastic piping market. At that time, Union Carbide began a process of modifying its DHDA 2077 Black resin (for water distribution) in order to create a DHDA 2077 Tan resin for the gas distribution industry.

Before Union Carbide could market its DHDA 2077 Tan resin material for natural gas service, it needed to generate stress rupture data, in accordance with the PPI procedure, that would support the long-term hydrostatic strength rating it was assigning to the material (a requirement of the interim Federal regulations effective at that time) [55]. The company had three resources to draw upon to support the hydrostatic design basis category:

1. Internal stress rupture data on its DHDA 2077 Tan resin
2. A PPI listing already obtained on its similar black resin
3. Additional internal stress rupture data on its black resin

On June 11, 1968, Union Carbide began stress rupture testing on specimens of pipe made from a pilot-plant batch of its newly developed DHDA 2077 Tan resin. The results of this testing supported Union Carbide's declared hydrostatic design basis category for DHDA 2077 Tan. The number of data points generated by these stress rupture tests for the DHDA 2077 Tan was less than that required by PPI procedure; however, Union Carbide began to market the product for use in gas systems based on these tests and on additional testing performed on the company's black resin material.

Because Union Carbide had not developed the PPI-prescribed number of data points on its DHDA 2077 Tan resin before marketing the product, Safety Board investigators reviewed the data the company developed on its black resin. A review of Union Carbide's laboratory notebooks revealed that a number of adverse data points Union Carbide developed for its black resin were not submitted to the PPI when the company applied for a PPI listing for the black material [56].

Union Carbide first made a commercial version of its DHDA 2077 Tan resin during the spring of 1969, and in April 1970, a first shipment of 80,000 pounds of DHDA 2077 Tan resin was shipped to Amdevco's Mankato plant. The next shipment of the material to Amdevco was not until 1971. Based on Amdevco's June 11, 1970, manufacturing date for the Waterloo pipe, Union Carbide manufactured, sold, and delivered the resin used to make the Waterloo pipe between the spring of 1969 and June 11, 1970, and the resin used to make the pipe involved in the Waterloo accident probably was included in the April 1970 shipment.

Union Carbide began, on December 3, 1970, additional stress rupture tests on its commercial DHDA 2077 Tan resin. These tests generated the results to further support its claimed long-term hydrostatic strength and also provided the number of data points required by the PPI procedure. Additional stress rupture tests on the commercial DHDA 2077 Tan resin beginning on December 28, 1970, and again on January 6, 1972, further supported the material's long-term hydrostatic strength.

During the late 1960s and 1970s, Minnegasco, a gas system operator based in Minneapolis, Minnesota, routinely employed a 1,000-hour sustained pressure test at 100°F detailed in ASTM D2239 and a 1,000-hour sustained pressure test at 73°F detailed in ASTM D2513 to qualify plastic piping for use in its system. Minnegasco went beyond the requirements of ASTM standards by continuing both versions of the testing beyond 1,000 hours until eventual failure occurred. The company used this information to evaluate the relative strengths of different brands of piping.

In 1969-70, Minnegasco began a series of tests on samples from five different suppliers of plastic piping made from DHDA 2077 Tan resin. On March 3, 1972, Minnegasco's laboratory issued an internal report that contained the results of its latest tests on piping made from the resin and referenced earlier tests on several brands of piping (including Amdevco/Century) that were also made from it. Based on this report, Minnegasco rejected for use in its gas system the DHDA 2077 Tan resin. According to the report, the company rejected the material because (1) none of the pipe samples made from this resin could consistently pass the 1,000-hour sustained pressure test at 100°F, and (2) the pipe samples had lower performance in 73°F sustained pressure tests than similar plastic piping materials already in use in the company's gas system.

In 1971, Union Carbide acknowledged to a pipe manufacturer that piping material manufactured by DuPont had a higher pressure rating at 100°F than did its own DHDA 2077 Tan. Union Carbide laboratory notebooks examined by the Safety Board showed test results for the DHDA 2077 Tan material that generally met the 1,000-hour sustained pressure test value at both 100°F and 73°F, although, in the case of the 100°F test, not by a wide margin. The notebooks also showed that the material had an early ductile-to-brittle transition point in stress rupture tests [57].

Information Dissemination Within the Gas Industry. The OPS reports that more than 1,200 gas distribution or master meter system pipeline operators submit reports to the OPS [58]. Additionally, more than 9,000 gas distribution or master meter system pipeline operators are subject to oversight by the States.

As noted earlier, a frequent failure mechanism with polyethylene piping involves crack initiation and slow crack growth. These brittle-like fractures occur at points of stress intensification generated by external loading acting in concert with internal pressure and residual stresses [59].

A 1985 paper analyzed, for linear (straight line) behavior up to 100,000 hours, the stress rupture test performance (by elevated-temperature testing) of six polyethylene piping materials [60]. The results were then correlated with field performance. This paper found that those materials that did not maintain linearity through 100,000 hours had what the author characterized as "known poor" or "questionable" field performance. On the other hand, those materials that maintained linearity through 100,000 hours had what the author characterized as "known good" field performance through their 20-year history logged as of 1985.

By the early to mid-1980s, the industry had developed a method to mathematically relate failure times to temperatures and stresses during stress rupture testing [61]. In the early 1990s, the industry developed "shift functions," another mathematical method to relate failure times to temperatures and stresses [62].

One study pointed out that using mathematical methods to calculate the remaining service life of pipe under the assumption that the pipe would only be exposed to stresses of internal operating pressures would result in unrealistically long service-life predictions [63]. As noted earlier, polyethylene piping systems have failed at points of long-term stress intensification caused by external loading acting in concert with internal pressure and residual stresses; thus, to obtain a realistic prediction of useful service life, stresses from external loadings need to be acknowledged.

Over a number of years, the Gas Research Institute (GRI) sponsored research projects investigating various tests and performance characteristics of polyethylene piping materials. Among these projects was a series of research investigations directed at exploring the fracture mechanics principles behind crack initiation and slow crack growth. These investigations led to the development of slow crack growth tests. The research studies frequently identified the piping and resins studied by codes rather than by specific materials, manufacturers, or dates of manufacture.

In 1984, the GRI published a study that compared and ranked several commercially extruded polyethylene piping materials produced after 1971 [64]. Again, the materials tested were identified by codes. Stress rupture tests were performed using methane and nitrogen as the internal pressure medium and air as the outside environment. Several stress rupture curves showed early transitioning from ductile to brittle failure modes.

The A.G.A.'s Plastic Materials Committee periodically updates the A.G.A Plastic Pipe Manual for Gas Service, which addresses a number of issues covered by this Safety Board special investigation. In 1991, the committee formed a task group to gather and then disseminate to the industry information regarding the performance of older plastic piping systems. The task group disbanded in 1994 without issuing a report.

In 1982 and 1986, DuPont formally notified its customers about brittle-like cracking concerns with the company's pre-1973 pipe. Safety Board investigators could find no record of either Century/Amdevco, Union Carbide, or any other piping or resin manufacturer formally notifying the gas industry of the susceptibility to premature brittle-like failures of their products. Nor does any mechanism exist to ensure that the OPS receives safety-related information from manufacturers.

Regarding Federal actions on this issue, the OPS has not informed the Safety Board of any substantive action it has taken to advise gas system operators of the susceptibility to premature brittle-like failures of any older polyethylene piping [65].

Installation Standards and Practices

The discussion in this section is intended to present a "snapshot" of the regulations and some of the primary standards, practices, and guidance to prevent stress intensification at plastic service connections to steel tapping tees.

Federal Regulations. The OPS establishes, in 49 CFR 192.361, minimum pipeline safety standards for the installation of gas service piping.

> Paragraph 192.361(b) reads as follows:
>
> *Support and backfill.* Each service line must be properly supported on undisturbed or well-compacted soil....
>
> Paragraph 192.361(d) reads:
>
> *Protection against piping strain and external loading.* Each service line must be installed so as to minimize anticipated piping strain and external loading.

Subsequent to the Waterloo accident, personnel from the Iowa Department of Commerce, after discussions with OPS personnel, stated that the Waterloo installation was not in violation of the Federal regulation. They further stated that, while they agree that the installation of protective sleeves at pipeline connections is prudent, a specific requirement to install protective sleeves is beyond the scope of Part 192 and is inconsistent with the regulation's performance orientation [66].

The Transportation Safety Institute (TSI), part of RSPA, conducts training classes for Federal and State pipeline inspectors. TSI instructors advise class participants that many of the performance-oriented regulations within Part 192 can only be found to be violated if the gas system fails in a way that demonstrates that the regulation was not followed. The TSI acknowledges the difficulty of identifying violations under paragraph 192.361(d). A TSI instructor told the Safety Board that, in the case of the failed pipe at Waterloo, an enforcement action faulting the installation would be unlikely to prevail because of the poor brittle-like crack resistance of the failed pipe and the length of time (23 years) between the installation and failure dates.

GPTC Guide for Gas Transmission and Distribution Piping Systems. After the adoption of the Natural Gas Pipeline Safety Act in August 1968, the American Society of Mechanical Engineers, after discussions with the Secretary of Transportation, formed the Gas Piping Standards Committee (later renamed the Gas Piping Technology Committee) to develop and publish "how-to" specifications for complying with Federal gas pipeline safety regulations. The result was the GPTC Guide for Gas Transmission and Distribution Piping Systems (GPTC Guide). The GPTC Guide lists the regulations by section number and provides guidance, as appropriate, for each section of the regulation.

In its investigation of the previously referenced 1971 accident in Texas, the Safety Board determined that protective sleeves were too short to fully protect a series of service connections to a main. The Safety Board noted that a protective sleeve must have the correct inner diameter and length if it is to protect the connection from excessive shear forces. As a result, and in response to a Safety Board safety recommendation, the 1974 and later editions of the GPTC Guide included guidance that "a protective sleeve designed for the specific type of connection should be used to reduce stress concentrations" [67]. No guidance was included as to the importance of a protective sleeve's length, diameter, or placement [68].

The GPTC Guide does not include recommendations to limit bending in plastic piping during the installation of service lines under 49 CFR 192.361. Although the guide references the A.G.A. Plastic Pipe Manual for Gas Service, and this manual does provide recommendations on bending limits, the GPTC Guide does not reference this manual in its guidance material under 49 CFR 192.361.

A.G.A. Plastic Pipe Manual for Gas Service. The most recent edition of the *A.G.A. Plastic Pipe Manual for Gas Service* identifies the connection of plastic pipe to service tees as "a critical junction" needing installation measures "to avoid the potentially high...stresses on the plastic at this point" [69]. The manual recommends proper support and the use of protective sleeves. Although the manual recommends following manufacturers' recommendations, no guidance is included on the importance of a protective sleeve's proper length, diameter, or placement. The manual includes, without elaboration, the following sentence:

> *Installation of the tee outlet at angles up to 45° from the vertical or along the axis of the main as a 'side saddle' or 'swing joint' may be considered to further minimize...stresses.*

The 1994 edition adds that manufacturers' recommended limits on bending at fittings may be more restrictive than for a run of piping alone.

A.G.A. Gas Engineering and Operating Practices (GEOP) Series. The preface to the current Distribution Book D-2 of the GEOP series states that the intent of the books is to offer broad general treatment of their subjects, and that listed references provide additional detailed information.

Figure 13.5 reproduces an illustration from Book D-2. This figure shows a steel tapping tee with a compression coupling joint connected to a plastic service. The illustration shows a protective sleeve and includes a note to extend the protective sleeve to undisturbed or compacted soil or to blocking. But the figure also shows the blocking positioned so that either the edge of the blocking or the edge of the protective sleeve might provide a fixed contact point on the plastic service pipe if the weight of backfill were to cause the pipe to bend down. Additional illustrations within this GEOP series book show this same positioning of the blocking with respect to the plastic pipe.

ASTM. The most recent ASTM standard covering the installation of polyethylene piping was revised in 1994 [70]. This standard addresses the vulnerability of the point-of-service connection to the main.

This standard, advising consultation with manufacturers, recommends taking extra care during bedding and backfilling to provide for firm and uniform support at the point of connection. In addition, the document recommends minimizing bends near tap connections, generally recommending that bends occur no closer than 10 pipe diameters from any fitting and that manufacturers' bend limits be followed. Similar recommendations for avoiding bends close to a fitting can be found in the forward to a water industry standard [71].

This ASTM standard further recommends the use of a protective sleeve if needed to protect against possible differential settlement. Currently, manufacturers that provide protective sleeves have their own criteria for designing sleeve lengths and diameters for their fittings. Some manufacturers' criteria are based on limiting stress to a maximum safe value, while one manufacturer has advised the Safety Board that its sleeve is not designed to limit bending, but only to guard against shear forces at the connection point [72].

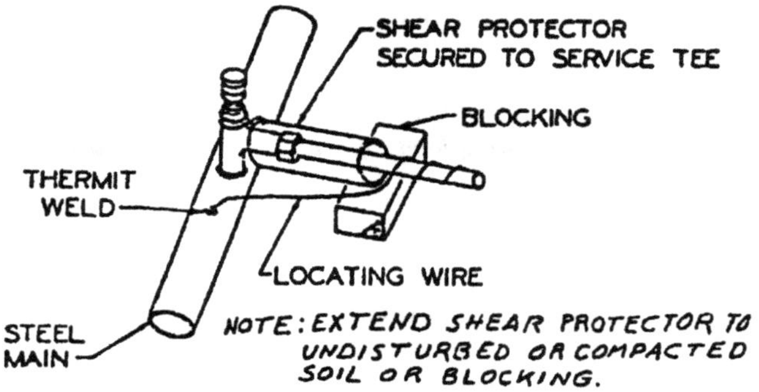

FIGURE 13.5 Reproduction from A.G.A GEOP series illustrating application of protective sleeve. *(Hand-scribed notation from the original)*

Guidance Manual for Operators of Small Natural Gas Systems. *The OPS/RSPA Guidance Manual for Operators of Small Natural Gas Systems* notes that plastic pipe failures have been found at transitions between plastic and metal pipes at mechanical fittings. The manual states the need to firmly compact soil under plastic pipe, advises following manufacturers' instructions for proper coupling procedures, and shows protective sleeves on connections of plastic services to steel tapping tees. The manual indicates that a properly designed protective sleeve should be used. The manual does not caution against bending the piping in proximity to a connection.

Manufacturers' Recommendations. As noted earlier, both the *A.G.A. Plastic Pipe Manual for Gas Service* and ASTM D2774 specifically refer the reader to manufacturers for further guidance on limiting shear and bending forces at plastic service connections made to steel mains via steel tapping tees.

Bending and Shear Forces. Safety Board investigators contacted representatives of the four principal companies that marketed plastic piping for gas service to determine to what extent plastic piping manufacturers were providing recommendations for limiting shear and bending forces at plastic service connections to steel mains via steel tapping tees. The four manufacturers contacted were CSR PolyPipe, Phillips Driscopipe, Plexco, and Uponor Aldyl Company (Uponor).

Three out of four of these manufacturers had published recommendations addressing these issues. These three manufacturers have historically emphasized heat fusion fitting systems instead of field-assembled mechanical fitting systems [73]. Representatives of these manufacturers indicated that mechanical fittings manufacturers should provide installation instructions covering their systems. Accordingly, one of the manufacturers' published literature referred the reader to the manufacturers of mechanical fittings for installation instructions. Nonetheless, these three major polyethylene pipe manufacturers did, in fact, provide recommendations to limit shear and bending forces, and these recommendations can apply to plastic service connections to steel mains via steel tapping tees.

With respect to the specific issue of limiting bends, DuPont, in January 1970, issued recommendations to limit bends for polyethylene pipe. DuPont/Uponor later published bend radius recommendations that differentiated between pipe segments consisting of pipe alone and those with fusion fittings [74]. The recommendations specified much less bending for pipe segments with fusion fittings; however, DuPont/Uponor did not provide bend limits for mechanical fittings. Two of the other major manufacturers (Phillips Driscopipe and Plexco) provide bend limits and differentiate between pipe alone and pipe with fittings, without specifying the type of fittings. None of the manufacturers' literature discusses bending with or against any residual bend remaining in the pipe after it is uncoiled. (See "Pipe Residual Bending" below.)

Of these four major polyethylene gas pipe manufacturers, only CSR PolyPipe had no published recommendations for limiting shear and bending forces at plastic service connections to steel mains via steel tapping tees. Although the company does not manufacture steel tapping tees with compression ends for attachment to plastic services, it does manufacture pipe that will be attached to steel tapping tees via mechanical compression couplings. The company has been supplying polyethylene pipe to the gas industry since the 1980s and is thus relatively new to that business compared to the other three major manufacturers [75]. When CSR PolyPipe entered the market, plastic materials were vastly improved compared to earlier versions with respect to resistance to crack initiation and slow crack growth. For this reason, according to CSR PolyPipe personnel, the company saw less need to publish installation recommendations.

The Safety Board attempted to identify every U.S. steel tee manufacturer that currently manufactures steel tees with a compression end for plastic gas service connections [76]. The Safety Board identified and contacted representatives of Continental Industries (Continental), Dresser Industries, Inc. (Dresser), Inner-Tite Corp. (Inner-Tite), and Mueller Company (Mueller) [77].

Only Continental and Inner-Tite offered protective sleeves to their customers as an option. None of these manufacturers has published installation recommendations to limit shear and bending forces on the plastic pipe that connects to their steel tapping tees. On another issue related to protective sleeves, Safety Board examination of a protective sleeve offered by Continental to its customers revealed that the sleeve that did not have sufficient clearance to allow the application of field wrap (intended to protect the steel tee from corrosion after it is in the ground) to that portion of the steel tee under the sleeve. This observation was confirmed by a Continental representative.

Pipe Residual Bending. The service involved in the Waterloo accident was installed with a bend at the connection point to the main. The plastic service pipe leaving the tee immediately curved horizontally. The pipe was cut out and brought into the laboratory, at which time the bend had a measured horizontal radius of approximately 34 inches. Based on field conditions and photos, MidAmerican Energy estimated the original installed horizontal bend radius to have been about 32 inches. This bend is sharper than that allowed by current industry installation recommendations for modern piping adjacent to fittings.

An issue related to recommended bend radius is residual pipe bending. Plastic pipe often arrives at a job site in banded coils. After the bands are released, the coiled pipe will partially straighten, but some residual bending will remain. The water industry already recognizes that bends in the direction of the residual coil bend should be treated differently than bends against the direction of the bend; however, gas industry field bend radius recommendations do not address residual coil bending [78].

A former Iowa Public Service Company employee stated that Iowa Public Service and marketed the complete assembly. Company, in an effort to reduce stress at connection points, generally attempted to install polyethylene services at an angle to the main to match the residual bend left after uncoiling the pipe. This former employee stated that no set time was specified to allow for complete relaxing of the pipe, but that the pipe would be placed in the ditch, and the crews would weld the tee at what they judged to be the appropriate angle.

MidAmerican Energy Installation Standards. As a result of the Waterloo accident, Safety Board investigators examined some of MidAmerican Energy's construction standards for minimizing shear and bending forces at plastic service connection points to steel mains. Specifically, Safety Board investigators examined MidAmerican Energy's standards pertaining to providing firm support, using protective sleeves, and limiting bends at plastic service connections to steel mains.

According to the company, MidAmerican Energy no longer installed steel tapping tees with mechanical compression ends to connect to plastic service pipe. Instead, it employed steel tapping tees welded at the factory to factory-made steel-to-plastic transition fittings. It then field-fused the plastic ends from the transition fittings to the plastic service pipe. MidAmerican Energy advised the Safety Board that it had no standard calling for firm compacted support under plastic service connection points to steel mains.

MidAmerican Energy designed, constructed, and installed its own protective sleeves for installation on its purchased steel tapping tee/transition fitting assemblies. MidAmerican Energy required its protective sleeves to be a minimum of 12 inches long; however, MidAmerican Energy could provide no design criteria for this length. MidAmerican Energy has reported that the company's unwritten field practice was to install the smallest diameter sleeve that will clear the field wrapped fitting, but MidAmerican Energy had no written requirements or design criteria for the diameter of its protective sleeves. The company's standard showed the sleeve as approximately centered over the steel-to-plastic transition, and no criteria or instructions were provided for the correct positioning of the sleeves.

The Safety Board notes that manufacturers that provide factory-made steel-to-plastic transition fittings will also provide protective sleeves along with the transition fittings and will provide positioning guidance for their use. Effective January 27, 1997, MidAmerican Energy instituted minimum bend radii requirements that differentiated between pipe segments consisting of pipe alone and pipe with fittings.

Gas System Performance Monitoring

This section examines gas system performance monitoring largely in the context of the Waterloo accident. Federal regulations (49 CFR 192.613 and 192.617) require that gas pipeline system operators have procedures in place for monitoring the performance of their gas systems. These procedures must cover surveillance of gas system failures and leakage history, analysis of failures, submission of failed samples for laboratory examination (to determine the causes of failure), and minimizing the possibility of failure recurrences.

Prior to the Waterloo accident, Midwest Gas had two systems for tracking, identifying, and statistically characterizing failures. The first system was the leak data base, which tracked the status of leak reports, documented actions taken, and recorded almost all gas system leaks. The data base received input from two primary sources: leak reports from customers and leak survey results. The data base parameters classified the general type of piping material that leaked (such as "plastic," "cast iron," "bare steel"), and indicated whether the leak occurred in pipe or certain fittings. The parameters did not include manufacturers, manufacturing or installation dates, sizes, or failure conditions commonly found with plastic piping (for example, poor fusions, bending force failures, insufficient soil compaction, rock impingement failures, and lack or improper use of protective sleeves) [79]. The data base indicated that the performance of plastic piping overall was comparable to other piping materials. MidAmerican Energy stated that the parameters chosen for this data base were those required for reporting to the DOT. The company said the parameters were also chosen on the premise that pipe meeting industry specifications would perform similarly.

The second system used by Midwest Gas for tracking failures was the company's material failure report data base, which was intended for use in evaluating the quality and performance histories of products installed in the company's gas system. Input to the data base was by way of a form (or, in some cases, a tag) filled out by field personnel. The form included categories such as the manufacturer, size, and an internal material identification number of the affected pipe or component. It also included areas for a narrative description of the failure. The form did not include dates of manufacture or installation dates or failure conditions commonly found on plastic piping. Field personnel sent the failed item, along with the completed form or tag, to engineering personnel, who examined the item and accompanying information to determine the need for corrections. Midwest Gas personnel then transcribed the narrative description of the failure word-for-word into the data base without attempting to determine and categorize causes of failure. Engineering personnel compiled the available data into periodically issued material summary reports. The company said engineering personnel from time to time sorted available data fields to determine trends.

The material failure report data base included only a portion of the leaks in the Midwest Gas system. For example, if Midwest Gas field personnel corrected a leak by replacing an entire line segment without digging up the leaking component (which the company said was a frequent occurrence with bare steel, cast iron, and certain plastic piping that was difficult to join), the material failure report data base system was not used. Also, field personnel were not required to use the reporting system if they determined that the failed item was related to an operating problem, such as excavation damage, rather than to a material problem. Additionally, the company indicated that the system did not enjoy full participation from field personnel.

When, after the Waterloo accident, Midwest Gas attempted to determine if installations with Century plastic piping had higher rates of failure than those with piping from other manufacturers, it found that its material failure report data base's incomplete coverage of gas leaks made that data base unsuitable for the purpose. The company decided instead to use the leak data base, which the company believed included almost all leaks. But because the leak data base did not list the manufacturers of plastic piping, Midwest Gas took several months to correlate entries in the leak data base with records showing the manufacturers of plastic piping. Midwest Gas, in 1995, concluded that piping installations with Century piping had failure incidence rates that were significantly higher than the balance of its plastic piping system. The company did not correlate entries with the years of installation. Since the Waterloo accident, the current Waterloo gas system operator, MidAmerican Energy, in addition to replacing all its Century pipe, has added parameters such as piping size, installation date, and pressure to the forms used for input into its leak data base. Also since the accident, MidAmerican Energy has added parameters such as installation date, pressure, and component location and position to its form for input into its material failure report data base. The company has also worked to determine if any other plastic piping manufacturers can be linked to piping with unacceptable performance.

The current (1994) edition of the A.G.A. Plastic Pipe Manual for Gas Service recommends the use and provides a sample of a form for recording information on plastic piping failures. The manual recommends collecting this information and then performing a visual examination or, in some cases, a laboratory analysis, to determine the type and cause of failure.

CONCLUSIONS

1. Plastic pipe extruded by Century Utility Products, Inc., and made from Union Carbide's DHDA 2077 Tan resin has poor resistance to brittle-like cracking under stress intensification, and this characteristic contributed to the Waterloo, Iowa, accident.
2. The procedure used in the United States to rate the strength of plastic pipe may have overrated the strength and resistance to brittle-like cracking of much of the plastic pipe manufactured and used for gas service from the 1960s through the early 1980s.
3. Much of the plastic pipe manufactured and used for gas service from the 1960s through the early 1980s may be susceptible to premature brittle-like failures when subjected to stress intensification, and these failures represent a potential public safety hazard.
4. Gas pipeline operators have had insufficient notification that much of the plastic pipe manufactured and used for gas service from the 1960s through the early 1980s may be susceptible to brittle-like cracking and therefore may not have implemented adequate pipeline surveillance and replacement programs for their older piping.

5. Even though the Gas Research Institute has developed a significant amount of data about older plastic piping used for gas service, because the data have been published in codified terms, the information is not sufficiently useful to gas pipeline system operators.
6. Because guidance covering the installation of plastic piping is inadequate for limiting stress intensification at plastic service connections to steel mains, many of these connections may have been installed without adequate protection from shear and bending forces.
7. Because MidAmerican Energy Corporation's gas construction standards do not establish well-defined criteria for supporting plastic pipe connections to steel mains or for designing or installing its protective sleeves at these connections, these standards do not ensure that connections will be adequately protected from stress intensification.
8. Before the Waterloo, Iowa, accident, the systems used by Midwest Gas Company for tracking, identifying, and statistically characterizing plastic piping failures did not permit an effective analysis of system failures and leakage history.
9. Before the Waterloo accident, Midwest Gas Company had had an effective surveillance program that tracked and identified the high leakage rates associated with Century Utility Products, Inc., piping when subjected to stress intensification, the company could have implemented a replacement program for the pipe and may have replaced the failed service connection before the accident.
10. MidAmerican Energy Corporation's current systems for tracking, identifying, and statistically characterizing plastic piping failures do not enable an effective analysis of system failures and leakage history.
11. The use of Continental Industries, Inc., tapping tees with the company's protective sleeves may leave the tapping tees susceptible to corrosion because the sleeves do not provide sufficient clearance for the application of field wrap to the metallic steel tapping tee.

RECOMMENDATIONS

As a result of this special investigation, the National Transportation Safety Board makes the following safety recommendations:

To the Research and Special Programs Administration:

Notify pipeline system operators who have installed polyethylene gas piping extruded by Century Utility Products, Inc., from Union Carbide Corporation

DHDA 2077 Tan resin of the piping's poor brittle-crack resistance. Require these operators to develop a plan to closely monitor the performance of this piping and to identify and replace, in a timely manner, any of the piping that indicates poor performance based on such evaluation factors as installation, operating, and environmental conditions; piping failure characteristics; and leak history. (P-98-1)

Determine the extent of the susceptibility to premature brittle-like cracking of older plastic piping (beyond that piping marketed by Century Utility Products, Inc.) that remains in use for gas service nationwide. Inform gas system operators of the findings and require them to closely monitor the performance of the older plastic piping and to identify and replace, in a timely manner, any of the piping that indicates poor performance based on such evaluation factors as installation, operating, and environmental conditions; piping failure characteristics; and leak history. (P-98-2)

Immediately notify those States and territories with gas pipeline safety programs of the susceptibility to premature brittle-like cracking of much of the plastic piping manufactured from the 1960s through the early 1980s and of the actions that the Research and Special Programs Administration will require of gas system operators to monitor and replace piping that indicates unacceptable performance. (P-98-3)

In cooperation with the manufacturers of products used in the transportation of gases or liquids regulated by the Office of Pipeline Safety, develop a mechanism by which the Office of Pipeline Safety will receive copies of all safety-related notices, bulletins, and other communications regarding any defect, unintended deviation from design specification, or failure to meet expected performance of any piping or piping product that is now in use or that may be expected to be in use for the transport of hazardous materials. (P-98-4)

Revise the *Guidance Manual for Operators of Small Natural Gas Systems* to include more complete guidance for the proper installation of plastic service pipe connections to steel mains. The guidance should address pipe bending limits and should emphasize that a protective sleeve, in order to be effective, must be of the proper length and inner diameter for the particular connection and must be positioned properly. (P-98-5)

To the Gas Research Institute:

Publish the codes used to identify plastic piping products in previous Gas Research Institute studies to make the information contained in these studies more useful to pipeline system operators. (P-98-6)

To the Plastics Pipe Institute:

Advise your members to notify pipeline system operators if any of their piping products, or materials used in the manufacture of piping products, currently in service for natural gas or other hazardous materials indicate poor resistance to brittle-like failure. (P-98-7)

Advise your plastic pipe manufacturing members to develop and publish recommendations for limiting shear and bending forces at plastic service pipe connections to steel mains. (P-98-8) Advise your plastic pipe manufacturing members to revise their pipeline bend radius recommendations as necessary to take into account the effects of residual coil bends in plastic piping. (P-98-9)

To the Gas Piping Technology Committee:

Revise the *Guide for Gas Transmission and Distribution Piping Systems* to include complete guidance for the proper installation of plastic service pipe connections to steel mains. The guidance should emphasize the need to limit pipe bending and should include a discussion of the proper design and positioning of a protective sleeve to limit stress at the connection. (P-98-10)

To the American Society for Testing and Materials:

Revise ASTM D2774 to emphasize that a protective sleeve, in order to be effective, must be of the proper length and inner diameter for the particular connection and must be positioned properly. (P-98-11) Develop and publish standard criteria for the design of protective sleeves to limit stress intensification at plastic pipeline connections. (P-98-12)

To the American Gas Association:

Revise your *Plastic Pipe Manual for Gas Service* and your *Gas Engineering and Operating Practices* series to provide complete and unambiguous guidance for limiting stress at plastic pipe service connections to steel mains. (P-98-13)

To MidAmerican Energy Corporation:

Modify your gas construction standards to require (1) firm compacted support under plastic service connections to steel mains, and (2) the proper design and positioning of protective sleeves at these connections. (P-98-14)

As a basis for the timely replacement of your plastic piping systems that indicate unacceptable performance, review your existing plastic piping surveillance and analysis program and make the changes necessary to ensure that the program is based on sufficiently precise factors such as piping manufacturer, installation date, pipe diameter, geographical location, and conditions and locations of failures. (P-98-15)

To Continental Industries, Inc.:

Provide a means to ensure that the use of your protective sleeves with your tapping tees at plastic pipe connections to steel mains does not compromise corrosion protection for the connection. (P-98-16)

To Continental Industries, Inc. (P-98-17):

To Dresser Industries, Inc. (P-98-18):

To Inner-Tite Corporation (P-98-19):

To Mueller Company (P-98-20):

Develop and publish recommendations and instructions for limiting shear and bending forces at locations where your steel tapping tees are used to connect plastic service pipe to steel mains.

By the National Transportation Safety Board
James E. Hall
Chairman
Robert T. Francis, II
Vice Chairman
John A. Hammerschmidt
Member
John J. Goglia
Member
George W. Black, Jr.
Member
April 23, 1998

REFERENCES

1. See appendix for brief descriptions of the organizations, associations, and agencies referenced in this report.
2. Watts, J., "Plastic Pipe Maintains Lion's Share of Market," Pipeline and Gas Journal, December 1982, p. 19, and National Transportation Safety Board Special Study—-An Analysis of Accident Data from Plastic Pipe Natural Gas Distribution Systems (NTSB/PSS-80/1).
3. The body of the report will make clear the distinction between brittle-like and ductile fractures.
4. National Transportation Safety Board Pipeline Accident Report—San Juan Gas Company, Inc./Enron Corp., Propane Gas Explosion in San Juan, Puerto Rico, on November 21, 1996 (NTSB/PAR-97/01).
5. Investigation No. 97-AI-055, October 31, 1997.
6. For more detailed information, see Pipeline Accident Brief in appendix.
7. Uralil, F. S., et al., The Development of Improved Plastic Piping Materials and Systems for Fuel Gas Distribution—Effects of Loads on the Structural and Fracture Behavior of Polyolefin Gas Piping, Gas Research Institute Topical Report, 1/75 - 6/80, NTIS No. PB82- 180654, GRI Report No. 80/0045, 1981, and Hulbert, L. E., Cassady, M. J., Leis, B. N., Skidmore, A., Field Failure Reference Catalog for Polyethylene Gas Piping, Addendum No. 1, Gas Research Institute Report No. 84/0235.2, 1989, and Brown, N. and Lu, X., "Controlling the Quality of PE Gas Piping Systems by Controlling the Quality of the Resin," Proceedings Thirteenth International Plastic Fuel Gas Pipe Symposium, pp. 327-338, American Gas Association, Gas Research Institute, Battelle Columbus Laboratories, 1993.
8. Stress intensification occurs when stress is higher in one area of a pipe than in those areas adjacent to it. Stress intensification can be generated by external forces or a change in the geometry of the pipe (such as at a connection to a fitting).
9. National Transportation Safety Board Pipeline Accident Report—Lone Star Gas Company, Fort Worth, Texas, October 4, 1971 (NTSB/PAR-72/5).
10. National Transportation Safety Board Pipeline The Safety Board's investigation revealed that a brittle-like crack occurred in a plastic pipe as a result of an occluded particle that created a stress point. Accident Report—Washington Gas Light Company, Bowie, Maryland, June 23, 1973 (NTSB/PAR-74/5).
11. National Transportation Safety Board Pipeline Accident Brief—"Natural Gas Corporation, Kinston, North Carolina, September 29, 1975."
12. National Transportation Safety Board Pipeline Accident Brief—"Arizona Public Service Company, Phoenix, Arizona, June 30, 1978."
13. National Transportation Safety Board Pipeline Accident Brief—"Northwestern Public Service, Grand Island, Nebraska, August 28, 1978."
14. National Transportation Safety Board Pipeline Accident Brief—"Southwest Gas Corporation, Tucson, Arizona, December 3, 1981."
15. National Transportation Safety Board Pipeline Accident Brief—"Pacific Gas and Electric Company, San Andreas, California, July 8, 1982."
16. National Transportation Safety Board Pipeline Accident Brief—"Northern States Power Company, Newport, Minnesota, September 19, 1983."
17. National Transportation Safety Board Pipeline Accident Brief—"Lone Star Gas Company, Terell, Texas, December 9, 1983."

18. Plastic pipe is sometimes squeezed to control the flow of gas. In some cases, squeezing plastic pipe can damage it and make it more susceptible to brittle-like cracking.
19. National Transportation Safety Board Pipeline Accident Report—Arizona Public Service Company Natural Gas Explosion and Fire, Phoenix, Arizona, September 25, 1984 (NTSB/PAR-85/01).
20. Illinois Commerce Commission accident reports dated September 14, 1978, and December 4, 1979. Iowa State Commerce Commission accident report dated August 29, 1983.
21. Research and Special Programs Administration Incident Report—"Gas Distribution System," Report No. 318063, January 8, 1996.
22. This subcommittee was subsequently made into a permanent unit and was renamed the Hydrostatic Stress Board.
23. As a result of Safety Board inquiries to the PPI about its inability to verify the actual data submitted, the institute, in 1997, revised its policy document for its listing service to require a signed statement from applicants that data accompanying applications for a PPI listing are complete, accurate, and reliable.
24. The PPI procedure also had restrictions on the degree of slope of the straight line so that the material's strength would not excessively diminish beyond 100,000 hours.
25. Plastics Pipe Institute, Policies and Procedures for Developing Recommended Hydrostatic Design Stresses for Thermostatic Pipe, PPI-TR3-July 1968.
26. Mruk, S. A., "The Ductile Failure of Polyethylene Pipe," SPE Journal, Vol. 19, No. 1, January 1963.
27. Because of the frequent lack of visible deformation associated with them, slit fractures are also referred to as brittle-like fractures.
28. Kulhman, H. W., Wolter, F., Sowell, S., Smith, R. B., Second Summary Report, The Development of Improved Plastic Pipe for Gas Service, Prepared for the American Gas Association, Battelle Memorial Institute, covering the work from mid-1968 through 1969. Stress rupture tests were performed using methane and nitrogen as the internal pressure medium and air as the outside environment. Some experts have advised the Safety Board that stress rupture testing showing time-to-failure in the slit mode may vary with different pressure media and
29. Now known as NSF International.
30. This standard also included plastic piping materials other than polyethylene.
31. Although adherence to ASTM appendixes is not mandatory, the PPI procedure was the only industry-accepted mechanism to determine long-term hydrostatic strength and hydrostatic design stress.
32. Resins are polymer materials used for the manufacture of plastics.
33. For example, E. I. du Pont de Nemours & Company, Inc., and Union Carbide Corporation.
34. Now known as ASME B31.8.
35. A.G.A. Plastic Pipe Handbook for Gas Service, American Gas Association, Catalog No. X50967, April 1971.
36. RSPA reviews revised editions of ASTM D2513 for acceptability before referencing them in 49 CFR 192.
37. A design factor is similar to a safety factor, except that a design factor attempts to account for other factors not directly included within the design equation that significantly affect the durability of the pipe.
38. Reinhart, F. W., "Whence Cometh the 2.0 Design Factor," Plastics Pipe Institute, undated, and Mruk, S. A.,

39. The design equation (with the current design factor, 0.32) can be found in 49 CFR 192.121, although 192.121 erroneously references the long-term hydrostatic strength instead of the hydrostatic design basis category. RSPA is currently conducting rule-making activities to correct this error.
40. Kulmann, H. W., Wolter, F., Sowell, S., "Investigation of Joint Performance of Plastic Pipe for Gas Service," 1970 Operating Section Proceedings, American Gas Association, pp. D-191 to D-198.
41. Kulmann, Wolter, and Sowell.
42. Mruk, S. A., "Validating the Hydrostatic Design Basis of PE Piping Materials."
43. Mruk, S. and Palermo, E., "The Notched Constant Tensile Load Test: A New Index of the Long Term Ductility of Polyethylene Piping Materials," summary of presentation given in the Technical Information Session hosted by ASTM Committee F17's task group on Project 62-95-02, held in conjunction with ASTM Committee F17's November 1996 meetings, New Orleans, LA.
44. Mruk and Palermo and Hunt, W. J., "The Design of Grey and Ductile Cast Iron Pipe," Cast Iron Pipe News, March/April 1970.
45. Mruk, S. A., "Validating the Hydrostatic Design Basis of PE Piping Materials," and Bragaw, C. G., "Fracture Modes in Medium-Density Polyethylene Gas Piping Systems," Plastics and Rubber: Materials and Applications, pp. 145-148, November 1979.
46. Mruk and Palermo have quantified and discussed the deformation in brittle-like failures in: Mruk, S. and Palermo, E., "The Notched Constant Tensile Load Test: A New Index of the Long Term Ductility of Polyethylene Piping Materials," summary of presentation given in the Technical Information Session hosted by ASTM Committee F17's task group on Project 62-95-02, held in conjunction with ASTM Committee F17's November 1996 meetings, New Orleans, LA, and Mruk, S. A., "Validating the Hydrostatic Design Basis of PE Piping Materials," pp. 202-214, 1985.
47. The 1971 A.G.A. Plastic Handbook for Gas Service noted that the cause and mechanisms of brittle fractures sometimes found with long-term stress rupture testing was not yet well established. Two of the pioneering papers in the United States to suggest a downturn in long-term hydrostatic strength with brittle-like failures or in elevated temperature testing were: Mruk, S. A., "The Ductile Failure of Polyethylene Pipe," SPE Journal, Vol. 19, No. 1, January 1963, and Davis, G. W., "What are Long Term Criteria for Evaluating Plastic Gas Pipe?" Proceedings Third A.G.A. Plastic Pipe Symposium, American Gas Association, pp. 28-35, 1971.
48. Both the PPI and the ASTM work on a consensus principle, meaning that requirements are put into place only when a consensus of voting members is reached. The PPI is a manufacturers' organization. With respect to the ASTM technical committee that generates requirements for plastic piping, the major piping manufacturers participate actively in the committee and are in a position to influence ASTM strength rating requirements.
49. Mruk, S. A., "Validating the Hydrostatic Design Basis of PE Piping Materials."
50. A.G.A. Plastic Pipe Manual for Gas Service, American Gas Association, Catalog No. XR 9401, 1994.
51. A number of these experts considered material to have poor resistance to brittle-like cracking if the material was shown to have a downturn in strength associated with brittle-like fractures in stress rupture testing (at 73 °F) before 100,000 hours.
52. Because of a series of organizational changes and mergers, the name of the owner/operator of the gas system at Waterloo, Iowa, has changed over the years. In 1971, Iowa Public Service Company installed the gas service that ultimately failed. At the time of the accident, the gas system operator was Midwest Gas Company. The current operator is MidAmerican Energy Corporation.

53. Iowa Public Service Company continued to purchase DuPont plastic piping fittings until fittings were available from Century. MidAmerican Energy made technical procurement decisions via a Gas Standards Committee. According to company officials, the company has implemented a process to ensure that it continues to receive quality products once the products have passed an initial qualification process.
54. Northern States Power is based in St. Paul, Minnesota.
55. The company was required to follow the PPI procedure in developing the necessary stress rupture data, but no requirement existed for those data to be submitted to the PPI or for the PPI to assign a listing before the tested material could be marketed.
56. Although the PPI procedure required the submission of all valid data points for statistical analysis, the Union Carbide employee who managed the data indicated that he believed he could discard data that, in his judgment, did not adequately characterize the material's performance. Union Carbide has contended that the non-submitted data may have been invalid because of experimental error, uncompleted tests, or other reasons.
57. The data from the laboratory notebooks suggest that this material's early ductile-to-brittle transition would not have met today's standards.
58. Master meter system refers to a pipeline system that distributes gas to a definable area, such as a mobile home park, a housing project, or an apartment complex, where the master meter operator purchases gas for resale to the ultimate consumer.
59. Kanninen, M. F., O'Donoghue, P. E., Popelar, C. F., Popelar, C. H., Kenner, V. H., Brief Guide for the Use of the Slow Crack Growth Test for Modeling and Predicting the Long-Term Performance of Polyethylene Gas Pipes, Gas Research Institute Report 93/0105, February 1993. Because, after extrusion, the outside of the pipe cools before the inside, residual stresses are usually developed in the wall of the pipe.
60. Mruk, S. A., "Validating the Hydrostatic Design Basis of PE Piping Materials."
61. Bragaw, C. G., "Prediction of Service Life of Polyethylene Gas Piping System," Proceedings Seventh Plastic Fuel Gas Pipe Symposium, pp. 20-24, 1980, and Bragaw, C. G., "Service Rating of Polyethylene Piping Systems by the Rate Process Method," Proceedings Eighth Plastic Fuel Gas Pipe Symposium, pp. 40-47, 1983, and Palermo, E. F., "Rate Process Method as a Practical Approach to a Quality Control Method for Polyethylene Pipe," Proceedings Eighth Plastic Fuel Gas Pipe Symposium, pp. 96-101, 1983, and Mruk, S. A., "Validating the Hydrostatic Design Basis of PE Piping Materials," and Palermo, E. F., "Rate Process Method Concepts Applied to Hydrostatically Rating Polyethylene Pipe," Proceedings Ninth Plastic Fuel Gas Pipe Symposium, pp. 215-240, 1985.
62. Popelar, C. H., "A Comparison of the Rate Process Method and the Bidirectional Shifting Method," Proceedings of the Thirteenth International Plastic Fuel Gas Pipe Symposium, pp. 151-161, and Henrich, R. C., "Shift Functions," 1992 Operating Section Proceedings, American Gas Association.
63. Broutman, L. J., Bartelt, L. A., Duvall, D. E., Edwards, D. B., Nylander, L. R., Stellmack-Yonan, M., Aging of Plastic Pipe Used for Gas Distribution, Final Report, Gas Research Institute report number GRI-88/ 0285, December 1988.
64. Cassady, M. J., Uralil, F. S., Lustiger, A., Hulbert, L. E., Properties of Polyethylene Gas Piping Materials Topical Report (January 1973 - December 1983), GRI Report 84/0169, Gas Research Institute, Chicago, IL, 1984.
65. The Safety Board asked the OPS for information about its actions in regard to older piping, after which, in 1997, the OPS notified State pipeline safety program managers of several issues regarding Century pipe and solicited input on their experiences with this particular piping.

66. Protective sleeves are intended to help shield the pipe at the connection point from bearing loads and shear forces and to limit the maximum pipe bending.
67. Safety Recommendation P-72-64 from National Transportation Safety Board Pipeline Accident Report—Lone Star Gas Company, Fort Worth, Texas, October 4, 1971.
68. The correct positioning of the protective sleeve has a bearing on its effective length.
69. A.G.A. Plastic Pipe Manual for Gas Service, American Gas Association, Catalog No. XR 9401, 1994.
70. ASTM D2774-94, Standard Practice for Underground Installation of Thermoplastic Pressure Piping, American Society for Testing and Materials, 1994.
71. Forward to American Water Works Association Standard C901-96, AWWA Standard for Polyethylene (PE) and Tubing, • In. (13 mm) Through 3 In. (76 mm) for Water Service, effective March 1, 1997.
72. Allman, W. B., "Determination of Stresses and Structural Performance in Polyethylene Gas Pipe and Socket Fittings Due to Internal Pressure and External Soil Loads," 1975 Operating Section Proceedings, American Gas Association, 1975.
73. Heat fusion fittings are used to make piping joints by heating the mating surfaces and pressing them together so that they become essentially one piece.
74. Uponor purchased DuPont's plastic pipe business in 1991.
75. CSR Hydro Conduit Company purchased PolyPipe in 1995. PolyPipe began supplying polyethylene pipe to the gas industry in the 1980s.
76. J. B. Rombach, Inc., which manufactures M. B. Skinner Pipeline products, told the Safety Board that it no longer manufactures or markets its "Punch-It-Tee" line of steel tapping tees. Chicago Fittings Corporation told the Safety Board it no longer manufactures or markets its line of steel tapping tees. The Safety Board therefore made no further inquiry with these companies.
77. Inner-Tite did not manufacture steel tees; it purchased them, affixed its own compression connections,
78. Forward to American Water Works Association Standard C901-96.
79. While sizes of the piping, along with a drawing of the piping assembly, were normally written or drawn on the forms, piping size was not captured in the data base generated by these forms.

APPENDIX A
SPECIFICATIONS FOR MATERIALS AND CONSTRUCTION FOR A NATURAL GAS SYSTEM

GENERAL CONSTRUCTION MATERIALS

The Owner shall supply the Contractor with all materials necessary for the completion of the Work specified herein, except where it is expressly stated that materials shall be furnished by the Contractor.

1.1 Steel Casing Pipe

Steel casing pipe shall be employed when polyethylene pipe is traversing mainline railroad tracks or right of way, interstate highways or by the direction of the Engineer. Under railroads the casing pipe shall be 3 pipe sizes larger than the carrier pipe. At all other locations the casing pipe shall be 2-pipe sizes larger than the carrier pipe.

All steel casing pipe used in the Work shall, at a minimum, conform to API Standard 5L, Grade B specifications. Pipe shall not be primed, coated or taped.

The minimum required wall thicknesses for steel casing pipes are listed in Table A.1.

TABLE A.1 Casing Pipe—Minimum Required Wall Thickness

Nominal Size (Inches)	Wall Thickness (Inches)
30	.406
28	0.377
26	0.344
24	0.312
22	0.281
20	0.281
18	0.250
16	0.219
14	0.188
12	0.188
10	0.188
8 and smaller	0.188

When casing pipe is installed without cathodic protection, the wall thicknesses listed in Table A.1 shall be increased to the nearest standard size, or a minimum of 0.063 inch greater than the thickness shown. For casing pipes less than 12 inches nominal diameter, no wall thickness adjustment is required.

1.2 Casing Vents

All pipe used for casing vents shall be two inch steel conforming to API Standard 5L, Grade B specifications. The minimum wall thickness for the casing vent piping shall conform to Table A.1. All casing vents shall be directed away from streets and highways.

When such pipe is furnished by the Contractor, he shall provide the Engineer with a manufacturer's affidavit of conformance to the above Specifications, and may further be required to furnish mill control check records indicating the results of physical and chemical tests.

1.3 Casing End Seals

End seals for casings shall be modular, mechanical type, consisting of interlocking synthetic rubber links shaped to continuously fill the annular space between the casing and the carrier pipe. The elastomeric element shall be sized as per the manufacturer's recommendations. Where differences in carrier pipe outer diameter and casing pipe inner diameter prohibit the use of modular, mechanical type, link seals, alternate means of sealing the casing ends will be provided for in the Plans.

1.4 Casing Pipe Spacers

Spacers used for the cased installation of the carrier pipe will be provided by the Contractor.

1.4.1 Insulating Type. Insulating casing pipe spacers used for the cased installation of metallic carrier pipe shall be constructed of high density polyethylene and shall be of a type and design approved by the Engineer.

1.4.2 Non-Insulating Type. Non-insulating casing pipe spacers used for the cased installation of non-metallic carrier pipe shall be of a type and design approved by the Engineer.

1.5 Tracer Wire

Tracer wire shall be AWG No. 12, single conductor solid copper with 600 volt insulation designed to meet U.S.E. requirements for buried service. The tracer wire insulated coating shall be yellow. Tracer wire will be provided by the Contractor.

1.6 Warning Tape

Warning tape shall have a minimum 5.0 mil overall thickness. The warning tape, including labeling, shall not contain any dilutants, pigments or other contaminants, and shall resist degradation by elements encountered in the soil. The locating tape shall be color coded yellow for gas and imprinted with the words "Caution—Buried Gas Main Below." Locating tape will be provided by the Contractor.

1.7 Electronic Locating Devices

Electronic locating markers shall be self-leveling ball type markers which emit a signal at a unique frequency to facilitate location.

1.8 Bituminous Paving

All aggregate, mineral filler, bitumen, and prime coat shall be in accordance with the (DOT Specifications), latest edition.

Aggregates shall include stone, gravel, slag and sand

Mineral filler shall include limestone dust, portland cement or other inert material

Bitumen shall include asphalt and tar cement

Prime coat shall include asphalt cutback, tar or asphalt emulsion

1.9 Course Aggregate

Course Aggregate used for road repair and replacement shall consist of crushed stone, crushed slag, or crushed or uncrushed gravel with clean, hard, tough, and durable pieces free from adherent coatings and deleterious amounts of friable, thin, elongated, or laminated pieces; soluble salts; or organic materials, and shall conform to the requirements of the DOT Specifications.

1.10 Riprap and Bedding

Stone for riprap and bedding shall be sound, durable, and free from seams, cracks, and other structural defects. Riprap and bedding shall be crushed stone, minimum Grade B conforming to the requirements of DOT Specifications.

1.11 Backfill

Gravel for porous backfill shall conform to the requirements of the DOT Specifications.

1.12 Select Fill

Material used as foundation for subbase, shoulder surfacing, backfill, or other specific purposes shall conform to the requirements of the DOT Specifications.

1.13 Sand

Sand shall be naturally occurring sand or manufactured stone sand. Natural sand shall consist of grains of hard, sound material, predominantly quartz, occurring in natural deposits. Manufactured sand shall consist of sound crushed particles of minimum DOT Grade F stone, essentially free from flat or elongated pieces, with sharp edges and corners removed. All sand shall be clean and free from foreign matter such as loam, dirt, sticks, roots, leaves, silt, vegetable matter and oil or dyestuffs.

SECTION 2. GENERAL CONSTRUCTION REQUIREMENTS

2.1 Equipment, Tools, Labor and Materials

2.1.1 Equipment, Tools, Labor and Materials To Be Furnished By the Owner. The Owner shall supply the Contractor with all materials necessary for the completion of the Work specified herein, except where it is expressly stated that materials shall be furnished by the Contractor. Material furnished by the Owner will be available to the Contractor at the Contractor's storage facilities.

2.1.2 Equipment, Tools, Labor and Materials To Be Furnished By Contractor. The Contractor shall provide all equipment, tools and labor necessary for the completion of the Work specified herein, including but not limited to: butt fusion, saddle fusion and/or electrofusion equipment; excavation, trenching, boring, and insertion equipment; pipe cutting and welding equipment and supplies; pipeline testing equipment; dewatering equipment; traffic control devices; and any and all applicable safety equipment which may be required.

The Contractor shall supply certain material items necessary for the completion of the Work specified herein, including but not limited to: select fill, sand and gravel; concrete; asphalt; testing equipment and fittings; erosion and sediment control materials; protective rock shields; and field applied pipeline coating materials.

Workmanship, tools, equipment and materials shall be of good quality meeting established industry standards. The Contractor shall furnish satisfactory evidence as to the kind and quality of materials. Only equipment which will not damage the surfacing along any improved surfaces shall be used. When crossing improved surfacing with equipment which will damage it, wood boards, flat pads or other approved methods shall be used to prevent damage to the surface. The Contractor shall furnish a complete list of equipment which he will employ on the job from the commencement of the Work and until the job is accepted.

2.2 Submittals

The Contractor shall be required to submit as-built data and samples, as directed to the Engineer for review and approval in accordance with the Specifications provided herein. All submittals shall be identified as required by the Engineer, and shall be complete with respect to quantities, dimensions, specified performance design criteria, and materials.

2.2.1 As-Built Documents. Upon the completion of each job, the Contractor shall provide the Owner with one complete set of Plans recording the installation of the job. The as-built Plans shall be updated daily during the course of construction. As-built documentation shall, as specified herein, include the following minimum information, as applicable:

1. Any parts of the Work that varies from that indicated in the Construction Plans shall be neatly and clearly marked on the as-built drawings. Where a deviation occurs, the main shall be located relative to the nearest station or at the location of the deviation. Where sizes or types of the materials installed differ from the Construction Plans, the type and size installed shall be clearly noted
2. Where possible, the location of all valves, bends, sleeves, plugged or capped ends, and any other fittings installed shall be measured to the nearest fire hydrant, light pole, sewer manhole or other fixed object. A minimum of two dimensions shall be provided for each item located and shall be labeled on the as-built drawings

2.3 Right-of-Way and Easements

Generally the Work will be carried out within the rights-of-way of state, county and city streets and roadways. The Contractor shall familiarize himself with, and follow all provisions pertinent to construction within such rights-of-way as provided in the latest edition of the DOT Manual. When it is required that the Contractor work on private property, the Contractor shall provide twenty-four (24) hours minimum advance notice to all effected land owners and/or tenants.

The necessary rights-of-way and construction easements for the natural gas system will be provided by the Engineer. The Contractor shall confine his construction operations to the immediate vicinity of the project location, and shall further use due care in placing construction tools, equipment, excavated materials, and pipeline materials and supplies so as to cause the least possible damage to property and the least interference with traffic.

The placing of such tools, equipment, and materials shall be subject to the approval of the Engineer.

The Contractor shall conduct the construction in such a manner as to cause the least inconvenience to the citizens of the area, thereby maintaining good public relations. The Contractor shall not unnecessarily interfere with the use of any public or private improvements, including landscaping; nor shall he unnecessarily damage such improvements. The Contractor shall repair any damage to such improvements to pre-construction condition, or as otherwise directed by the Engineer.

The Contractor shall strive to maintain, at all times during the execution of the Work, continuous ingress and egress to all affected parcels and traveled ways. When ingress and egress to affected parcels must be blocked, due to the direct executing of the Work, 24 hours advance notice must be given to the affected property owner. In no case shall the blocking of ingress and egress be allowed for more than 24 hours consecutively.

2.4 Maintenance of Traffic

The Contractor shall be required to provide maintenance of traffic within the construction area for the duration of the construction period, including during any temporary suspension of Work. Maintenance of traffic shall be performed in accordance with the current additions of the DOT Manuals.

The Contractor may be required to submit a Traffic Maintenance Plan prior to commencing work on a particular portion of the Work. If the Contractor is asked to submit such a plan, work must not commence on the portion of the project covered by the plan until the Traffic Maintenance Plan is approved by the Engineer. The amount of roadway closure shall be limited to the immediate Work area and shall be in accordance with the above mentioned manuals and specifications.

2.5 Pavement Removal and Disposal

Removal of pavement includes cutting of the pavement, breaking of the pavement surface, and excavating the pavement using conventional trenching, hand, and pneumatic equipment. Pavement removal includes removal of all layers of bituminous asphalt and concrete pavement necessary to properly install the pipe and/or appurtenances. Removal of bituminous and concrete pavement shall correspond to ditch widths limited to the nominal diameter of the pipe being installed plus two feet for main line pipe.

Cutting of the pavement for trenches or bellholes shall be performed using appropriate pavement saw(s) and shall be cut-back and squared off in a neat and workmanlike manner. Pavement cutting shall be required in all direct burial applications, as indicated on the construction Plans, as required by permit, or as directed by the Engineer.

Where pavement is cut and replaced, the Contractor shall cut the edges to a straight and even line before repairing the pavement. Non-uniform edges will not be permitted or accepted.

All pavements removed as part of the Work shall be removed from the jobsite and disposed of in accordance with the requirements of Federal, State, County, City, and all applicable environmental regulations.

2.6 Erosion & Sediment Control

The Contractor shall provide a means of protecting and minimizing the effects of erosion and sediment displacement to the construction area and all immediate surrounding areas that may be affected by the construction activity.

Methods most suitable to the site and soil conditions shall be employed to intercept sediment-carrying runoff from the site and remove the maximum practical amount of sediment from the storm runoff. Erosion and sediment control measures, include but are not limited to: silt fences; storm drain inlet protectors; soil stabilization mats; temporary seeding; and permanent seeding shall be installed and maintained as indicated on the Plans, or as otherwise directed by the Engineer, in accordance with the directives of the Department of Environmental Protection and DOT.

2.7 Pipe and Materials Handling

The Contractor shall load, unload, haul, receive, sign for, store, and otherwise be responsible for all materials. All materials shall be handled and placed in a manner which prevents damage and does not interfere with public and private travel.

All pipe handling shall be accomplished using equipment which will not damage the pipe or the pipe coating and/or lining. All pipes shall be lifted, rolled, or otherwise handled so as to not damage the coating and/or lining. All damaged coating and/or lining shall be repaired and acceptance of same shall be contingent upon approval of the Engineer.

Coated steel pipe shall be stacked not more than four layers high on padded skids in a manner which will not damage the coating.

Polyethylene pipe shall be protected from fire, excessive heat, harmful chemicals, and long term exposure to direct sunlight. The Contractor shall exercise due care during handling to prevent gouges, scratches, cuts, kinks, or punctures in the pipe. All defects or damage which could impair the serviceability of the polyethylene pipe, in the opinion of the Engineer, including cuts, gouges or scratches which are deeper than 10 percent of the wall thickness of the pipe shall be removed from the pipe joint or the piping system. When loading, unloading, moving and placing polyethylene pipe, the Contractor shall avoid dropping or dragging the pipe. Chains shall not be used for handling polyethylene pipe. Polyethylene pipe shall be stored in the shade to minimize expansion of the pipe.

The height of polyethylene pipe stacks shall not exceed four feet. Pipe shall not be stored overnight on the job site unless it is stored in an area protected from vandals. Pipe and other materials shall not be placed directly on the ground but rather on wooden pallets or a similar clean, flat surface.

Due care shall be taken during all handling so as not to damage the beveled ends of steel pipe. All ends so damaged shall be repaired by removing the end of the pipe and re-beveling the pipe with a pipe beveling machine.

Fusion operations on polyethylene pipe shall be performed adjacent to the trench and the pipe lifted and lowered into the trench. Where absolutely necessary to fuse polyethylene pipe at another location than adjacent to the trench, the pipe shall be lifted and carried to the trench. Under no circumstances shall any length or portion of the polyethylene pipe be dragged, slid, pushed or pulled, on any surface to the trench.

2.8 Pipe Bending

Pipe bends shall be used, as required, in place of fabricated fittings to change the horizontal and/or vertical alignment of the pipe. All bends in steel pipe shall be approved by the Engineer prior to performing the bending operation.

2.8.1 Bends in Steel Pipe. All bends in steel pipe shall be made by a smooth bending method. They shall be made with a bending shoe, as approved by the Engineer. When bends are required in steel pipe, they shall be made in the pipe section prior to welding said bent section to the rest of the piping.

Bends shall be free of wrinkles, buckles, cracks or other evidence of damage or characteristics which, in the opinion of the Engineer, will reduce the quality of the finished pipeline. Miter bends shall be approved by the Engineer. In no case shall a bend section contain a weld joint. The longitudinal weld of steel pipe shall be at the neutral axis of the bend.

Field bends in steel pipelines that damage the pipe coating shall require the area of damaged coating to be coated with a hot applied wrap, tape, or other approved coating material prior to lowering the pipe.

Bends in steel pipe to be made with fabricated fittings shall be made with standard weight, long radius weld fittings approved by the Engineer.

Bends in steel casing pipe shall not be allowed.

2.8.2 Bends in Polyethylene Pipe. The bending radius for polyethylene pipe shall not be less than the minimum recommended by the manufacturer for the kind, type, grade, wall thickness, and diameter of the particular polyethylene used as listed in Table A2.1.

TABLE A2.1 Minimum Bending Radius of Polyethylene Pipe SDR-11

Nominal Pipe Size (Inches)	Outside Diameter (D) (Inches)	Radius of Curvature R = D(20) (Inches)
1	1.660	33.2
2	2.375	47.5
4	4.500	90.0
6	6.625	132.5
8	8.625	172.5

A prefabricated elbow, bend or tee shall be used if a change in direction cannot be accomplished in accordance with Table A2.1 Care shall be taken to prevent kinking in the polyethylene pipe. If the polyethylene pipe becomes kinked, the kinked section shall be cut out and replaced.

All fittings including butt fused, saddle fused and/or electrofused valves, elbows, tees, couplings, and/or service fittings shall be installed such that they are located on a straight section of pipe, a minimum of three feet from any field bend.

2.9 Pipe Installation

2.9.1 Location of Other Utilities. It shall be the responsibility of the Contractor, to investigate and verify the existence and location of all utilities within the vicinity of the Work.

The Contractor shall comply with all the provisions of the Underground Utility Damage Prevention Act. At least 48 hours prior to starting the Work the Contractor shall verify the existence and location of all underground utilities, structures and associated appurtenances. The Contractor shall notify Dig-Safe to locate all participating underground utilities. The Contractor shall be responsible for identifying all utilities in the Work area which are not participating members of the one-call system. These utility operators shall be provided with a minimum 48 hours notice to have their facilities located prior to starting the Work.

The excavation of test holes may be required to ascertain the existence, location, size, type, and alignment of existing utilities or underground structures. The dimensions of these test holes shall be the minimum required to effectively locate said utilities and underground structures.

In the event that any gas lines, water lines, sewer lines, electric lines, cables, conduit, and/or any other existing utility, either underground or above ground, is damaged by the Contractor during the prosecution of the Work, the owner of the damaged utility shall be notified immediately. If approved and/or requested by the owner of the damaged utility, the Contractor shall immediately make the necessary repairs, to the satisfaction of the utility and the Engineer.

2.9.2 Required Clearance. Regardless of the method of installation, whether by open trench, directional drilling, or underboring, all gas mains and services shall be installed such that a minimum of 12 inches, or as otherwise specified by the Engineer, horizontal and vertical clearance is maintained from all other existing underground utilities and/or structures, thereby permitting proper routine maintenance and protection against damage which may result from proximity to the utilities and/or structures.

2.9.3 Alignment. All gas pipes shall be installed true to the horizontal and vertical alignment indicated on the Plans, or as otherwise directed by the Engineer. The Contractor shall make no deviations to the proposed horizontal and/or vertical alignment of the gas pipes unless otherwise directed to do so by the Engineer.

In such cases where the proposed horizontal and/or vertical pipeline alignment will cause conflict with other utilities and/or structures, or result in less than the specified minimum clearance or cover, the Engineer shall be notified and the pipeline relocated as per his direction.

2.9.4 Required Cover. Regardless of the method of installation; whether by open trench, directional drilling, or underboring, all gas mains shall be installed such that a minimum cover of 36 inches is provided between the top of the pipe or casing pipe and the finished grade.

When the mains cross creeks, land subjected to flooding, or major drainage ditches, a minimum of 48 inches of cover shall be provided.

The Contractor may, upon the approval of the Engineer, install gas pipes with less cover when the specified minimum cover cannot be obtained, provided the pipe is adequately protected from all superimposed loads by means of approved sleeving or shielding.

The Contractor may be required to install the pipe with greater cover than the specified minimum, as directed by the Engineer and/or specific easement or permit requirements.

2.9.5 Direct Burial. The Contractor shall, install all gas pipes and associated facilities by direct burial.

Direct burial of the gas pipes associated facilities shall include, but not be limited to: clearing and grubbing, trench excavation (trenching), rock excavation (as required), trench stabilization (as required), lowering and laying pipe and backfilling, as described herein.

2.9.5.1 Clearing, Grubbing and Tree Removal. The Contractor shall clear all brush and timbers located along the alignment of the proposed pipeline, and properly dispose of such, off-site, in a prompt manner prior to commencing trenching operations.

2.9.5.2 Trenching. Trenching shall include all excavation necessary to prepare the ditch for the pipe to be installed regardless of what means or methods are necessary to produce such ditch. All trench excavation operations shall be performed in accordance with 29 CFR 1926, Subpart P-Excavations.

In cases where the pavement is to be broken, obtain any and all required permits prior to cutting or breaking the pavement. No paved roadways shall be cut without the approval of the Engineer.

Prior to trenching, the Contractor shall verify the existence, location, elevation and orientation of all underground and aboveground facilities within the vicinity of the Work, in accordance with Location of Other Utilities. The Contractor shall exercise care in the vicinity of any and all such obstructions. In the event that any such gas lines, water lines, sewer lines, electric lines, cables, conduits, and/or any other existing utility, either above ground or below ground, is damaged by the Contractor during the prosecution of the Work, the owner of the damaged utility shall be notified immediately. If approved and requested by the owner of the damaged utility, the Contractor shall immediately make the necessary repairs, to the satisfaction of the utility and the Engineer.

The trench shall be excavated to a depth which will provide the minimum required cover. The maximum width of the trench shall be 24 inches plus the nominal pipe diameter, and the minimum width of the trench shall be eight inches plus the nominal pipe diameter.

The trench shall be excavated in a manner which offers smooth, firm and continuous support along the entire length of the pipeline. All sharp objects and debris shall be removed from the trench or the pipe shall be bedded with sand or clean fill to protect the pipe. A minimum of six inches of pipe bedding shall be required in such locations. Where pipe bedding is required, the trench shall be excavated to a depth which will provide the minimum required cover.

Whenever wet or otherwise unsuitable material, which is incapable of properly supporting the pipe, is encountered in the trench bottom, such material shall be over-excavated as directed by the Engineer to a depth necessary to allow for construction of a stable pipe bedding. The over-excavated portion of the trench shall then be backfilled with select fill to proper grade to provide the minimum required cover.

2.9.5.3 Rock Excavation. Rock excavation includes the excavation of rock occurring in mass and ledge formations of such character and structure as to warrant removal by means of hydraulic hammer, specialized rock trenching equipment, and/or explosives.

Rock which has been removed during the excavation process shall not be used for backfilling the trench. The Contractor shall provide sufficient clean backfill to replace any and all reduced volumes of earth resulting from rock excavation. The Contractor shall promptly remove all excavated rock material and properly dispose of such off-site.

2.9.5.3.1 Blasting. All blasting operations required for the purpose of rock excavation, including but not limited to permit acquisition, employee training/certification, explosives handling/storage and charge detonation, shall be performed in accordance with 29 CFR 1926, Subpart U—Blasting and the Use of Explosives.

The Contractor shall be responsible for securing any and all required permits and for providing trained/certified blasting personnel. Prior to blasting, the Contractor shall submit to the Engineer for approval a written blasting procedure which includes addressing the protection of existing subsurface utilities and structures.

2.9.5.3.2 Sand Bedding. Prior to laying a section of pipe in the trench where rock has been excavated, the Contractor shall provide a minimum of six inches of sand bedding. Where sand pipe bedding is required, the trench shall be excavated to a depth which will provide the minimum required cover.

2.9.5.4 Trench Stabilization. Where the depth of the trench and/or the type and condition of the soil requires stabilization, the Contractor shall provide a method of trench stabilization as directed and approved by the Engineer.

All materials and installation methods required for shoring, sheeting, bracing and any other required means of trench stabilization shall conform to any and all requirements of 29 CFR 1926 and applicable appendices.

Trench stabilization system members shall be securely connected together and installed in a manner that prevents sliding, falling, kickouts or other predictable failures of the trench sides. Support systems shall be installed and removed in a manner that protects employees from all forms of trench failure or from being struck by members of the support system.

Cross braces installed above the pipe to support the sheeting shall be removed only after pipe embedment has been completed.

Where trench sheeting is required to be left in place, such sheeting shall be cut-off at a minimum of three feet below finished grade and the cut-off portion removed from the trench. Sheeting left in place shall not be braced against the pipe, but shall be supported in a manner which will eliminate concentrated loads and horizontal thrusts on the pipe.

2.9.5.5 Lowering and Laying Pipe. Belt slings and/or padded calipers which are sized to the particular pipe being laid shall be used to handle the pipe provided such slings or calipers are free of all characteristics which might damage the pipe.

Inspection of the trench shall be made by the Contractor prior to lowering the pipe to ensure that no rocks or other sharp objects which may damage the pipe are located within the trench.

When polyethylene pipe is laid in the trench, sufficient slack in the placed pipe should be provided to allow for the contraction of the placed pipe. When piping is lowered into the trench, care shall be exercised to avoid over stressing or buckling the piping or imposing excessive stress on the joints. Anchors and supports shall be provided as directed and where required for fastening work into place.

Where the Work is suspended, at night or for any other reason, the open ends of the pipe shall be securely plugged or closed to prevent entrance of water and other foreign material.

2.9.5.6 Backfilling. Backfilling operations shall include the furnishing of all labor, materials and equipment necessary for the backfilling and compaction of all trenches, bellholes, and excavations over the entire length of the pipeline, as specified herein and in accordance with the DOT Specifications, latest edition.

Trenches shall not be backfilled until the pipe has proper cover, bedding and smooth, firm and continuous support along the entire length of the pipe. The trench shall be backfilled as soon as possible after the pipe has been properly placed.

Where the trench crosses driveways, roads, streets, or other places used for the travel of vehicles or pedestrians, proper care shall be taken so as not to impede the flow of traffic. All traveled ways, including driveways, walks, streets, or alleys crossed by the trench shall be compacted by mechanical means at +/- 20 percent of optimum moisture content to 95 percent. Where deemed necessary, the Engineer may elect to have density tests performed on the backfilled trench by an independent contractor or consultant.

Unsuitable material encountered during trench excavation shall not be used as backfill. Unsuitable material shall be removed and replaced with select fill, as specified herein.

All backfill material shall be free from all objects which might damage the pipe. Wherever it is deemed necessary by the Engineer, hand labor shall be used in starting the backfill. The backfill placed from the bottom of the ditch to the top of the pipe shall be placed in the trench simultaneously on both sides of the pipe for the full width of the trench in layers not to exceed six inches in depth. The backfill material shall be thoroughly compacted under and on each side of the pipe to provide solid backing against the external surface of the pipe and to remove all voids. The trench may be backfilled from one foot above the pipe to the top of the trench with mechanical equipment provided the machine is operated parallel to the trench, and the material is placed in the trench in layers not to exceed six inches for the full width.

All trenched construction shall be adequately compacted by means of rolling, tamping with mechanical rammers, or hand tamping such that no future settlement of the trench backfill will occur. If vibratory rollers are used for backfill compaction, vibratory motors shall not be activated until at least three feet of backfill has been placed and compacted around the pipe. Flooding shall not be permitted as a means of backfill consolidation. Backfill compaction achieved by means of driving any type of construction equipment and/or vehicles, other than those specifically designed for trench compaction work, across any part of the trench shall not be permitted. The Contractor shall refill and compact backfill areas where settlement occurs.

2.9.6 Directional Drilling. The Contractor may, upon the approval and/or direction of the Engineer, choose or otherwise be directed to utilize directional drilling as an alternative method of installing polyethylene and steel gas pipes.

Prior to commencing directional drilling operations, the Contractor shall be required to provide proof that the personnel performing the drilling operations have a minimum of one year of experience performing directional drilling operations of this type.

The length of each continuous directionally drilled installation shall be limited by the size and type of drilling equipment utilized for the operation.

A minimum of one bellhole per 500 foot interval shall be excavated around the pipe to verify its location, depth and structural integrity. The sending and receiving pits for the directional drilling operation shall not be considered as part of the required number of inspection bellholes. Tracer wire shall be installed along with all directionally drilled polyethylene pipes.

2.9.6.1 Equipment. The directional drilling system/equipment used for pipe installation as specified herein shall be subject to the approval of the Engineer and shall incorporate the following features:

1. The system shall be remotely steerable permitting control of horizontal and vertical alignment within a window of two inches
2. The system shall provide for electronic monitoring of horizontal and vertical alignment. The locating tool shall be calibrated daily to an accuracy of two inches
3. The system shall be capable of turning 90 degrees in a radius of 35 feet
4. The system may utilize an inert and environmentally risk free drilling fluid. No toxic or otherwise hazardous chemical additive shall be added to the drilling fluid. A dry boring system is also acceptable
5. Back reaming bits shall be of a diameter at least two inches larger than the outside diameter of the pipe to be installed
6. Drilling equipment shall be fitted with a permanent alarm system capable of detecting an electric current. The system shall have an audible alarm to warn the operator when the drill head nears electrified cables

2.9.6.2 Procedure. The leading end of the pipe shall be capped prior to insertion through the boring hole or sleeve.

A "weak link," consisting of smaller diameter pipe or tubing, shall be fused to the leading end of the pipe being pulled. The weak link shall be half the diameter of the main or service being installed and shall be a minimum of three feet in length. If the weak link breaks or is otherwise substantially damaged during installation, the drilling operation shall be abandoned and new undamaged piping reinstalled.

The leading six feet of the installed pipe shall be pulled through the receiving pit and inspected. If any abrasions, gouges or lacerations are present which may compromise the integrity of the pipe, the pipe shall be exposed back to the point where the damage originated. All damaged pipes which are determined to be unacceptable shall be removed and replaced.

All fused joints contained within the polyethylene piping to be installed by directional drilling shall be allowed to cool down in accordance with the manufacturer's recommended fusion procedures prior to commencing the pulling operation.

2.9.7 Underboring. The Contractor may choose or otherwise be directed to bore the gas pipe or casing pipe beneath certain traveled ways and/or watercourses.

All underboring methods shall be subject to the approval of the Engineer, and may include: dry boring, boring and jacking, augering, pushing, and piercing.

The boring methods and equipment utilized shall be industry proven and accepted, subjected to the approval of the Engineer. All employees of the Contractor utilized in boring operations shall be trained and experienced with the specific boring method and equipment chosen.

All boring equipment utilized shall be properly sized to install the casing or carrier pipe without removing any excess spoil. The diameter of the auger used in any boring operation shall not, in any case, be greater than four inches larger than the outside diameter of the casing or carrier pipe to be installed.

Boring operations shall be performed in such a manner that settlement, displacement, distortion, or any other damage to the existing ground surface, utilities and or structures will not occur. Where a utility is damaged or severely displaced, the authority having jurisdiction over the utility shall be contacted immediately. The Contractor shall be responsible for promptly repairing or having repaired any such damage, to the affected utility owner's satisfaction.

Boring operations shall, at all times, be conducted in a manner which does not create a hazard or impede the flow of traffic.

Casing or carrier pipe installation shall be performed immediately upon completion of the boring operation. Soil voids which remain around the pipe after installation shall be properly filled with hydraulic cement grout. The grout shall be placed under pressure in a manner approved by the Engineer.

When, in the opinion of the Engineer, a completed bore results in a deficiency which renders the pipe unusable, including but not limited to: insufficient cover; insufficient clearance with existing underground utilities and/or structures; excessive curvature of the pipe; excessive damage to the pipe and/or coating; or failure to stay within the right-of-way, the bore shall be abandoned; the pipe filled with cement grout, plugged and abandoned in place; and a new bore completed. The lengths of all required bores shall be as shown on the Plans or as otherwise directed by the Engineer.

2.9.7.1 Underboring without Casing. Certain boring operations, as indicated on the Plans may not require the pipe to be installed within a casing.

Steel pipes under-bored without a casing pipe shall be bored through an additional distance sufficient to allow inspection for excessive damage to the pipe wall and/or coating. Upon completion of the inspection, the excess pipe shall be properly cut off and removed. Any such bored pipe which, results in excessive damage to the pipe wall and/or coating shall be repaired or replaced, as specified herein or as otherwise directed by the Engineer. All steel pipes under-bored without casing shall be cathodically protected as directed by the Engineer.

Tracer wire shall be installed along with all polyethylene carrier pipes underbored without a casing pipe.

2.9.7.2 Underboring with Casing. Certain boring operations, as indicated on the Plans or as otherwise directed by the Engineer, may require the pipe to be installed within a casing pipe.

2.9.8 Casing Pipe Installation. The Contractor may be required to install the gas pipes within a steel casing pipe, as indicated on the Plans.

The Contractor may choose or otherwise be directed to install the casing pipe by trenching and/or boring. The casing pipe shall be a minimum of two nominal pipe sizes larger than the outside diameter of the carrier pipe, joints or couplings, except for railroad crossings which require three nominal pipe sizes larger.

The Contractor may install a larger diameter casing pipe than is specified or otherwise shown on the Plans. If a larger diameter casing pipe is installed, all minimum cover and clearance requirements, as specified herein, shall be met. The casing pipe shall be installed true to line and grade; sloping to one end with an even bearing throughout its length. The casing pipe installation shall be made so as to allow free and unrestricted movement of the carrier pipe during insertion.

Lengths of steel casing pipe shall be joined by welding the joints completely around the circumference of the pipe. Casing pipe vent(s) shall be installed at the end(s) of the casing pipe as directed by the Engineer. The vents shall be painted above grade with a corrosion resistant primer and paint. The vent opening(s) shall be screened and turned downward. Approved gas warning signs shall be attached to the vent pipe(s) or placed immediately adjacent to the casing vent(s) at each end of the casing pipe. Both ends of all casing pipe installations shall be sealed. Sealed casings shall have a minimum of one two inch vent welded on the casing before the carrier pipe is inserted.

Insulating casing spacers shall be set within three foot of each end of the casing and placed along the carrier pipe at a maximum spacing of 10 feet. Following the insertion of a steel carrier pipe into the casing, the two metallic structures shall be checked to verify electrical isolation. Cathodic protection test stations shall be installed at each casing per the construction Plans. The Engineer shall be contacted to verify and approve all cathodic protection devices and connections prior to backfilling. The Contractor shall correct any shorting to the casing and/or carrier pipe using a method approved by the Engineer.

For a polyethylene piping system, the casing pipe shall be prepared to the extent necessary to remove any sharp edges, projections, or abrasive material which could damage the plastic during and after insertion. Polyethylene pipe shall be inserted into the casing pipe in such a manner so as to protect the polyethylene pipe from damage. The leading end of the polyethylene pipe shall be capped prior to insertion.

2.9.9 Underwater Crossings. Where gas pipes cross watercourses or drainage ditches, the pipe shall be provided with a minimum of 48 inches cover between the top of the pipe and the bottom of the watercourse or ditch. The pipe shall be fitted to the terrain with any necessary bends and/or fittings. Bends shall not exceed the minimum bending radius previously specified.

The Contractor shall perform all pipe installation operations in a manner which will minimize disturbances to the watercourse. The appropriate diversion channels, flumes and/or cofferdams shall be installed and maintained, as indicated on the Plans.

The gas mains shall be installed within casing pipe, as indicated on the Plans. Installation of the casing pipe shall include pipe spacers placed near both ends of the casing pipe and at 10 feet on center (O.C.) throughout the remaining length of the sleeve. The ends of the casing pipe shall not be sealed.

Pre-cast concrete river weights shall be installed along the pipeline, as indicated on the Plans. Rock shields, a minimum of ¼ inch thick, shall be securely installed between the pipe and each river weight. Bolt-on weights shall be securely fitted to the pipe and all bolts firmly tightened. Where river weights are installed on long sections of pipe, care shall be exercised in pipe handling and placing operations to prevent buckling of said pipe. Under no circumstances shall the river weights be used to sink the pipe in a water filled trench.

The watercourse banks shall be backfilled to the original alignment of the bank line, and protected from erosion, as indicated on the Plans and in accordance with the Erosion and Sediment Control Handbook, latest edition, and any other regulatory agencies, as required. The Contractor shall comply with all provisions of the permits.

2.9.10 Installation In Railroad Right-of-Way. Piping passing under the right-of-way of a commercial railroad shall be installed in accordance with Part 5, of the American Railway Engineering Association (AREA) Specifications, or as otherwise directed by the railway company permit requirements. Where necessary, the Engineer will obtain all required permits for installations in railroad rights-of-way. The Contractor shall comply with all provisions of the permits.

2.9.11 Pipe Locating Devices. The Contractor shall install tracer wire with all uncased polyethylene pipe (direct buried, directionally drilled and plowed-in) to facilitate location of the pipe with commercially available electronic pipe locators. Installation of tracer wire and locating tape shall be as included in Table A2.2

TABLE A2.2 Installation of Tracer Wire

Method of Construction	Tracer Wire Location
Direct Bury	6" Min./12" Max. Above Pipe
Directional Drill	Pull Through Drill Hole With Pipe
Plow-in	6" Min./12" Max. Above Pipe
Bored w/out Casing	Pull Through Bore Hole With Pipe
Bored w/Casing	Not Required

2.9.11.1 Electrically Conductive Tracer Wire. The Contractor shall be required to install an electrically conductive tracer wire (tracer wire) as an means of facilitating the location of buried polyethylene pipe.

When polyethylene pipe is installed by directional drilling or boring without a casing pipe, the tracer wire or locating tape shall be attached to the bull-nose in order to facilitate installation. The tracer wire shall be pulled into each locating station with sufficient slack to extend a minimum of 24 inches above finished grade. The tracer wire shall not be cut, but should remain continuous.

In the event that the continuity of the tracer wire is broken during installation, the Contractor shall install a replacement tracer wire by either open trenching or plowing. Tracer wire shall not be mechanically fastened to the pipe. Under no circumstances shall the tracer wire be wrapped around the polyethylene pipe.

Where new tracer wire is connected to existing tracer wire or where separate spools of tracer wire are connected, the tracer wire shall be spliced using an approved split bolt connector or an approved waterproof splicing kit. These connections shall be wrapped using splicing tape and/or plastic electrical tape in order to waterproof the splice.

2.9.11.2 Warning Tape. The contractor shall be required to install warning tape as a means to warn of placement of the gas pipe. The warning tape shall be placed from 6 to12 inches from grade with the text facing up.

2.9.11.3 Locating/Test Stations. Locating/test stations shall be installed at all locations indicated on the Plans. Locating/test station installations shall include valve boxes (top and bottom sections) and a lid. The valve box lid shall be marked "TEST."

Locating/test station installation shall include excavating, setting of the valve box sections(s), coiling the tracer wire into the box, properly setting the valve box lid, backfilling and compacting around the box, and restoration.

2.10 Clean Up

The Contractor shall keep the right-of-way reasonably clear of construction debris during the progress of the Work. Cleanup shall consist of all Work necessary to restore the damaged area to pre-construction condition.

This operation shall include, but not be limited to, the removal of excess excavated materials, equipment, rock and other materials which cannot be placed in the trench backfill. Cleanup shall also consist of the repairing of trenches, disposal of vegetative debris and re-seeding and mulching or sodding as directed by the Engineer.

2.11 Pavement and Concrete Structure Replacement

The Contractor shall replace roadway, driveway and walkway surfaces necessarily removed for the installation of the main and service line piping. It is the intent of these Specifications that the Contractor return all paved surfaces affected by the Work to as near pre-construction condition as possible in conformance with approved methods.

2.11.1 Gravel and Other Surfacing. Gravel roads and dirt roads shall be repaired and replaced to their original condition, or as otherwise directed by the Engineer.

2.11.2 Concrete Structures. Concrete structures, including but not limited to headwalls, curbing and sidewalks damaged during construction, shall be promptly and satisfactorily restored to pre-construction condition, as directed by the Engineer, in accordance with all applicable sections of the DOT Specifications, latest edition.

Curb and gutter or concrete curbs shall be rebuilt to original lines, grade, cross-section and finish. Any curbing that has settled or shifted as a result of the Work shall be replaced.

2.12 Inspection. Prior to installation of the gas system, the Contractor shall inspect all pipe, fittings, valves, and other appurtenances in accordance with all provisions specified herein as well as all applicable manufacturer's standards and specifications. The Contractor shall remove from the Work all materials which do not meet the provisions specified herein, as well as any and all manufacturer's standards and specifications, and replace such with acceptable materials.

The Contractor shall produce evidence that any and all items of the Work have been installed in accordance with the project Plans and Specifications.

SECTION 3. MATERIALS FOR NATURAL GAS SYSTEM INSTALLATION

3.1 Pipe

3.1.1 Polyethylene Gas Pipe. All polyethylene gas pipe shall be PE 2406, medium-density polyethylene pipe as classified in accordance with ASTM D3350, Standard Specifications for Polyethylene Plastics Pipe and Fittings, as a Type II, Grade PE24 material. The polyethylene pipe shall be manufactured and tested in accordance with ASTM specification D2513. Acceptable manufacturers are Phillips Driscopipe (Series 6500) and Plexco Yellowstripe, a division of Chevron Chemical Co. All polyethylene pipes shall be IPS (Iron Pipe Size), unless noted as cts (Copper Tubing Size).

Pipe sections shall be marked as required by ASTM D 2513.

Butt fittings shall conform to ASTM D 3261.

PE 2406 polyethylene pipe properties shall be as listed in Table A3.1

TABLE A3.1 Polyethylene Pipe Properties

Size (Inches)	SDR	Weight (lb./ft.)	Coil/Straight Length (ft.)
8	11.0	8.43	40' (Straight Length)
6	11.0	4.97	40' (Straight Length)
4	11.0	2.30	40' (Straight Length)
2	11.0	0.64	500' Coil
1	11.0	0.31	500' Coil

3.1.2 Steel Gas Pipe. All Steel pipe shall conform to API 5L Grade B, X42, and/or X52 specifications. The pipe shall be seamless or electric resistance welded steel pipe as specified in ASME B31.8. Buttweld fittings shall be in accordance with ASME B16.9. All steel pipe shall be plain end and beveled for welding.

Steel pipe shall be plant coated with Fusion Bonded Epoxy External Line Pipe Coating. Approved products are 3M #206N FBE Coating and NAP-GARD #2500 FBE Coating. Uniform cured film thickness shall be 18 +/- 2 mils. Cathodic protection is required for all buried steel pipe. The minimum steel pipe properties shall be as listed in Table A3.2. The Engineer may require the Contractor to use a higher grade steel pipe, where deemed necessary.

3.2 Pipe Fittings

3.2.1 Polyethylene Pipe Fittings. Polyethylene pipe fittings shall be PE 2406, butt fusion, saddle fusion or electrofusion fittings manufactured by an approved manufacturer and shall be composed of the same material as the pipe. PE fittings shall conform to ASTM D 3350 and ASTM D 2513.

TABLE A3.2 Minimum Steel Pipe Properties

Pipe Size	Grade API 5L	Design Pressure Class 3 Location (psig)	Wall Thickness (inches)	Actual Design Pressure Class 3 Location (psig)	Maximum Operating Pressure (psig) allowed for Hoop Stress to stay below:		Maximum Test Pressure (psig)
					20% SMYS	30% SMYS	
4	B	1440	.188	1462	584	877	1125
4	X42	1440	.156	1456	582	873	1125
4	X52	1440	.125	1444	577	866	1125
6	B	1440	.280	1479	591	887	1125
6	X42	1440	.250	1585	633	951	1125
6	X52	1440	.188	1476	590	885	1125
8	B	1440	.375	1522	608	913	1125
8	X42	1440	.312	1519	607	911	1125
8	X52	1320	.219	1320	528	796	1125
8	X52	1440	.250	1507	602	904	1125
10	B	1440	.500	1628	651	976	1125
10	X42	1440	.438	1711	684	1026	1125
10	X52	1440	.307	1485	594	891	1125

3.2.2 Service Saddles. "Straight outlet" service saddles and branch saddles for polyethylene mains shall be composed of the same material as the pipe. The saddles shall be sized to match the mains being used on and shall be sized for IPS pipe except for one inch and ½ inch connections which shall be CTS (0.090 wall thickness).

3.2.3 Service Tees. Service tapping tees and service saddles for polyethylene mains shall be composed of the same material as the pipe, as specified in 5.1.1 Polyethylene Gas Pipe. Service tees for polyethylene mains shall be "sidewall" fusion fittings, self-tapping with IPS sized outlets, except for ½ inch inch and one inch services which shall have CTS sized outlets. The cap of the service tapping tee shall provide a leak free seal between the cap and the stem of the tee. The base (saddle) of the tapping tee shall match the size of the main that is being tapped.

3.2.4 Transition Fittings. Steel to polyethylene transition fittings shall meet or exceed 49 CFR 192 and ASTM D2513 specifications. The steel portion of the fitting shall be coated with electrostatically applied epoxy and the end shall be beveled for welding and tapered to match the pipe bore. The polyethylene portion of the fittings shall be composed of the same material as the pipe.

3.2.5 Steel Pipe Fittings. Steel fittings one inch and smaller shall conform to ASME B16.11. Steel buttweld fittings two inch and larger, including elbows, tees, reducers, and caps shall be factory made wrought steel buttwelding fittings conforming to the ASME Specification B16.9. All fittings shall be seamless and beveled accordingly to accommodate welding to pipe. All weld elbows shall be triple radiuses to allow for the passage of internal pipe inspection devices.

Steel flanged fittings two inches and larger, including bolts, nuts, and bolt patterns shall be in accordance with ASME B16.5.

3.2.5.1 Flange Gaskets. Gaskets shall be non-asbestos compressed material in accordance with ASME B16.21, 1/16 inch thickness, full face or self-centering flat ring type. The gaskets shall conform to the applicable requirements of ASME B31.8.

3.2.5.2 Flanged Joints. Insulating joint materials shall be provided between flanges pipe systems where shown to isolate galvanic or electrolytic action. Joints for flanged pipe shall consist of full face sandwich-type flange insulating gasket of the dielectric type, with insulating sleeves for flange bolts and insulating washers for flange nuts.

3.2.6 Pipe Coating. All steel pipe shall be plant coated in accordance with specifications for Fusion Bonded Epoxy External Line Pipe Coating. Protective coatings shall be applied mechanically in a factory or field plant especially equipped for the purpose. Valves and fittings that cannot be coated and wrapped mechanically shall have the protective covering applied by hand.

3.2.6.1 Inspection of Pipe Coatings. Any damage to the protective covering during transit and handling shall be repaired before installation. After field coating and wrapping has been applied, the entire pipe shall be inspected by an electric holiday detector with impressed current set at a value in accordance with NACE RP0274 using a full-ring, spring-type coil electrode. The holiday detector shall be equipped with a bell, buzzer, or other type of audible signal which sounds when a holiday is detected. All holidays in the protective covering shall be repaired immediately upon detection. Labor, materials, and equipment necessary for conducting the inspection shall be furnished by the Contractor.

3.2.6.2 Protective Covering for Aboveground Piping Systems—Ferrous Surfaces. Shop primed surfaces shall be touched up with ferrous metal primer same type paint as the shop primer. Shop primer shall be compatible with Keeler & Long P-Series Poly-Silicone Enamel.

Surfaces that have not been shop primed shall be solvent-cleaned in accordance with Steel Structures Painting Council SSPC SP 1. Surfaces that contain loose rust, loose mill scale, and other foreign substances shall be mechanically cleaned by power wire brushing in accordance with SSPC SP 3 or brush-off blast cleaned in accordance with SSPC SP 7 and primed with Keeler & Long No. 3200 Kolor Poxy Primer at 2.5 to 6.0 dry film thickness (DFT) light grey in color. Primed surfaces shall be topcoated with one coat of Keeler & Long P-Series Poly–Silicone Enamel at 1.5 to 2.5 DFT.

3.2.7 Joint Repair. All steel fittings, valves, pipe joints, piping installed below ground that is not plant coated, and holidays in the plant coating shall be wrapped with a tape coating system designed for corrosion protection.

Wrapping shall be performed with tape tested in accordance with ASTM D1737, ASTM G22and ASTM D1000. The thickness of the tape shall be 50 mils. Kendall Polyken™ #1027 Pipeline Primer and Kendall Polyken™ #930-50 Wrapping Tape shall be used. The tape shall be applied with a single, continuous overlap wrapping.

3.2.8 Electrofusion Fittings. Electrofusion fittings shall be manufactured of polyethylene resins compatible with PE 2406, medium-density pipe. The fittings shall be engineered to be used with and meet or exceed the resistance properties of SDR 11, polyethylene pipe.

3.3 Transition Fittings. Gas transition fittings shall be manufactured steel fittings approved for joining steel and polyethylene pipe. Approved transition fittings are those that conform to AGA Manual requirements for transition fittings and shall meet or exceed 49 CFR 192, ASTM D2513 and API 5L GR.B specifications. The steel portion of the fitting shall be coated with electrostatically applied epoxy and the end shall be beveled for welding and tapered to match the pipe bore. The polyethylene portion of the fittings shall be composed of the same material as the polyethylene pipe.

3.4 Valves

All valves to be installed in the natural gas system shall be wrench operated, low maintenance or no maintenance valves as indicated on the Plans. Valves shall be suitable for shutoff or isolation service and shall conform to the following specifications.

3.4.1 Isolation Valves. Valves shall be suitable for shutoff or isolation service and shall conform to API Spec 6D, carbon steel, buttweld ends, ANSI Class 600. Isolation valves shall be ball-type valves.

3.4.2 Polyethylene Valves. All polyethylene valves shall be full opening, ball type and maintenance free, as manufactured by Nordstrom Valve, Inc., or an equal approved by the Engineer. The valves shall be composed of the same material as the pipe, as specified in 5.1.1 Polyethylene Gas Pipe.

3.4.3 Steel Valves. Steel valves 1½ inches and smaller installed underground shall conform to ASME B16.34, carbon steel, socket weld ends, with square wrench operator adapter. Steel valves 1½ inches and smaller installed aboveground shall conform to ASME B16.34, carbon steel, socket weld or threaded ends with handwheel or wrench operator. Steel valves two inches and larger shall conform to API Spec 6D, carbon steel, buttweld ends, with handwheel or wrench operator.

3.4.4 Steel Valve Operators. Valves eight inches and larger shall be provided with worm or spur gear operators, totally enclosed, grease packed, and sealed. The operators shall have Open and Closed stops and position indicators. A locking feature shall be provided where indicated. Wherever the lubricant connections are not conveniently accessible, suitable extensions for the application of lubricant shall be provided. Valves shall be provided with lubricant compatible with gas service.

3.5 Test/Locating Stations and Valve Boxes

Test/locating stations shall be installed to facilitate the location of the mains and to provide a means of monitoring the cathodic protection for the pipe as indicated on the Plans.

Test and valve boxes, extension pieces, collars and covers shall be cast iron as manufactured by Bingham and Taylor or approved equal. Valve box components shall conform to the specifications and the detail as shown on the Plans. Valve box covers shall have the word "GAS" embossed on top. No screw type valve boxes will be allowed. Test/locating station covers shall have the word "TEST" embossed on top.

3.6 Protective Sleeves

Protective sleeves for polyethylene pipe shall be constructed of fiberglass reinforced polyethylene (FRP) and shall be of a type and design approved by the Engineer.

3.7 District Pressure Regulators

Regulators shall have ferrous bodies, shall provide backflow and vacuum protection, and shall be designed to meet the pressure, load and other service conditions. Pressure regulators for main distribution lines, supplied from a source of gas which is at a higher pressure than the maximum allowable operating pressure for the system, shall be equipped with pressure regulating devices of adequate capacity. In addition to the pressure regulating devices, a suitable method shall be provided to prevent overpressuring of the system in accordance with ASME B31.8.

The district regulators will be monitoring regulators installed in series with the primary pressure regulator. The regulators will be Fisher or Mooney pilot-type regulators. Each district regulator station will have site specific Plans and Specifications.

SECTION 4. NATURAL GAS SYSTEM INSTALLATION

4.1 Contractor Qualifications

The Contractor shall use only competent and skilled workmen for the performance of any and all Work on the gas distribution system, as specified herein. The workmen shall not perform any welding or heat fusion operations on any pipe or associated fittings within the system until they have been qualified to perform such operations in accordance with the test requirements specified in Welding Qualifications and Heat Fusion Qualifications. The Contractor shall furnish evidence that the specified testing requirements have been met for each employee prior to their utilization on the Work.

4.2 Welding Qualifications

Testing and certification of welders, whether by destructive or nondestructive inspection methods, shall be in accordance with American Petroleum Institute Standard 1104 (API 1104), "Standard for Welding Pipelines and Related Facilities," which is hereby incorporated by reference and made a part of these Specifications. The Contractor shall provide documentation as evidence that all welders performing welds for the Contractor are certified according to the above requirements.

4.3 Heat Fusion Qualifications

Operators of heat fusion equipment, including: butt fusion, saddle fusion and electrofusion, shall be tested and certified in accordance with the requirements of 49 CFR 192, Subpart F, Paragraph 285 along with any and all additional requirements of the specific pipe and/or fitting manufacturer. In addition to and in accordance with the requirements above, all personnel performing heat fusion operations shall be certified by to join polyethylene pipe approved for use by the following procedures:

> Certification: Each technician making joints in polyethylene pipe must be qualified by the Engineer and the contractor before making joints on polyethylene pipe that will be installed in the gas system.
>
> Testing: Each technician must show proof of satisfactory training and practice in making heat fused joints on polyethylene pipe and fittings. A technician will be tested with the following procedure:

1. Make three joints in two inch polyethylene pipe. The examiner will observe the joining without interference except to prevent damage to the equipment or injury to the personnel. The examiner may waive all or part of these tests if he is familiar with quality of the technicians' work. If it is necessary for the examiner to interfere in the joining procedure, a complete explanation will be made on the record of examination
2. After the joints have cooled, the pipe and joint will be examined visually. Results will be recorded on the examination form
3. The joints will be cut into four longitudinal strips of equal width. The eight surfaces exposed by the cutting will be examined for evidence of voids or discontinuities. Results of this test will be recorded
4. The examiner will comment on each part of the test. If the welder has failed, a complete explanation will be entered with enough detail to aid in retraining and to make clear to the examine why he failed the test and how he can correct the faults causing failure
5. When the examinee has passed the test, his record of training, the completed examination sheet, and the test specimens will be taken to the Engineer. If the Engineer approves the test and specimens they will be so annotated on the examination form and the records maintained for certification of qualification

4.4 Welding

All steel pipe and/or fittings, connections and other fabrications within the gas distribution system shall be welded, unless otherwise specified or directed by the Engineer. All welds shall be performed in accordance with the requirements of API 1104.

4.4.1 Procedure. All welding material and/or equipment shall, at all times, be protected from damage and kept in good working condition. Filler metals and fluxes shall be protected from deterioration and excessive moisture changes. Welding rods and other materials which show signs of deterioration or damage shall be replaced. Welding machines which are in poor repair or are not of sufficient capacity to perform the Work shall be replaced.

Suitable windguards shall be provided to protect the Work during periods of excessive wind. The Contractor shall temporarily suspend all welding operations whenever conditions are not conducive to the performance of good work.

All steel pipe, fittings, connections and fabrications shall be butt welded by either the oxyacetylene or the shielded metal arc welding process using a manual welding technique, unless alternative methods have been submitted to and approved by the Engineer.

All surfaces to be welded shall be properly cleaned and free of material that may be detrimental to the integrity of the completed weld. The ends of pipe and/or fittings at all welded joints shall be properly beveled using an appropriate pipe beveling machine. Each completed weld shall be free of overlaps, undercuts, excessive convexity, scale, oxides, pin holes, non-metallic inclusions, air pockets and all other defects.

Arc burns on the pipe and/or fittings shall be removed by grinding, provided the resulting pipe wall thickness is not less than 90 percent of the required design wall thickness. Arc burns which cannot be repaired by grinding and repair attempts which result in less than 90 percent of the original wall thicknesses shall be cut out. All welds shall be air cooled. Accelerated cooling by any method shall not be permitted.

4.4.2 Inspection. Visual, nondestructive and/or destructive testing procedures shall be implemented to determine the quality of the welds.

The Engineer may, at their discretion, require x-ray or other nondestructive testing of any and all welds prior to the initiation of coating or coating repair procedures. Should any weld prove to be defective for any reason, the Contractor shall assume any and all costs associated with the testing, cutting out and replacement of the weld.

The Contractor shall be required to radiograph all welds within all commercial railroad rights-of-way and all welds within the limits of all bridge crossings. In addition, the Contractor may be required to have certain welds radiographed to verify compliance with API 1104 Standards. Such tests shall be required on all welds identified by the Engineer based upon observation of poor welding techniques, previously identified substandard welds, and/or evidence of leakage during pressure testing.

The Engineer shall make all determinations as to what constitutes an acceptable weld as well as the disposition of all defective welds. These determinations shall be made upon completion of either a visual or a radiograph inspection.

4.5 Heat Fusion

All polyethylene pipe and/or fitting connections and other fabrications within the gas distribution system shall be made by heat fusion, unless otherwise directed by the Engineer. Heat fusion shall include: butt fusion, saddle fusion and electrofusion.

4.5.1 Procedure. All heat fusion jointing procedures shall be performed in accordance with 49 CFR 192 and any and all recommended Specifications and procedures provided by the pipe and/or fitting manufacturer.

Heat fusion equipment shall, at all times, be protected from damage and kept in good working condition. Fusion equipment which shows signs of deterioration or damage shall be replaced. Heat fusion machines which are in poor repair or are not of sufficient capacity to perform the Work shall not be used. Suitable windguards shall be provided to protect the Work during periods of excessive wind or cold weather. When the ambient temperature is below 32°F care must be taken to maintain the proper heater plate temperature. The Contractor shall temporarily suspend all heat fusion operations whenever conditions are not conducive to the performance of good work. All fused joints and other connections shall be air cooled. Accelerated cooling by any method shall not be permitted.

Fusion operations on polyethylene pipe shall be performed adjacent to the trench and the pipe lifted and lowered into the trench. Where absolutely necessary to fuse polyethylene pipe at another location than adjacent to the trench, the pipe shall be lifted and carried to the trench. Under no circumstances shall any length or portion of the polyethylene pipe be dragged, slid, pushed or pulled, on any surface to the trench.

4.5.2 Inspection. Visual, nondestructive and/or destructive testing procedures shall be implemented to determine the quality of the fused joints. The Engineer may, at their discretion, require nondestructive testing and inspection of any or all fused joints prior to the initiation of backfilling or insertion operations.

The Engineer shall make all determinations as to what constitutes an acceptable fused joint as well as the disposition of all defective joints. These determinations shall be made upon completion of a visual inspection. Defective joints shall be removed from the piping system.

4.6 Valves

Valves shall be installed at all locations indicated on the Plans. Valve installations shall include the valve, complete valve box assembly and any required blocking. Prior to installation, all valves shall be fully opened and fully closed a sufficient number of times to ensure that all parts are in proper working order.

All polyethylene valves shall be installed below grade by butt fusion. Butt fusion operations on polyethylene valves shall be in accordance with the section on Heat Fusion. All steel valves shall be installed by welding. Welding operations on steel valves shall be in accordance with the section on Welding.

Valve boxes shall be installed so as not to hinder the operation of the valve.

Valve boxes shall be insulated from the valve by blocking under the valve box with brick, concrete block or suitable masonry material. Similar material shall be used to block under the center of the valve.

Backfill shall be carefully tamped around each valve box to a distance of four feet on all sides of the box, or to the undisturbed trench face if less than four feet, such that the plumbness of the valve box is maintained.

All valves shall be in the open position during pressure testing, and shall remain as such upon completion of the tests.

4.7 Blow-Off Assemblies

Blow-off assemblies shall be installed at all locations indicated on the Plans. Blow-off assembly installations shall include the polyethylene reducer(s), transition fitting, steel piping, ell, cut-off valve, plug, valve box (top and bottom section) and lid as shown in the attached detail drawings. All connections for the steel pipe and fittings shall be threaded coupled connections. The valve box lids shall be marked “gas”. All metallic surfaces (pipe, joints, and components) shall be coated and/or wrapped with protective materials. The protective materials shall be Polyken protective coating tape, Royston paint and tape corrosion resistant kit or an approved equivalent. Practices in using these materials shall be done in the manner recommended by the respective manufacturer of the protective/coating material.

Prior to installation, all valves shall be fully opened and fully closed a sufficient number of times to ensure that all parts are in proper working order. Valve boxes shall be installed so as not to hinder the operation of the cut-off valve.

4.8 Pressure and Leak Testing

Each gas main and service installed shall be pressure and leak tested, in the presence of the Engineer ,as specified in the Test Procedures. When the length of any pipe section exceeds 1,000 feet, the Engineer reserves the right to require the pipe to be tested in sections, the length of which shall be determined by the Engineer. Natural gas shall not be admitted into any gas main or service line prior to the Engineer’s approval and the successful completion of all required pressure tests.

4.9 Purging

Upon the successful completion of the pressure and/or soap test, and under the direction of the Engineer, natural gas will be admitted into the completed system in sufficient quantities such that all air is purged out of the line(s).

All purging operations will be done under the direct supervision of the Engineer. The Contractor shall provide a minimum of 24 hours notice to the Engineer prior to commencing any purging operations.

APPENDIX B

TRANSPORTATION OF NATURAL OR OTHER GAS BY PIPELINE: MINIMUM FEDERAL SAFETY STANDARDS

Source: Department of Transportation

Subpart A–General

Sec.
192.1 Scope of Part.
192.3 Definitions.
192.5 Class locations.
192.7 Incorporation by reference.
192.9 Gathering lines.
192.10 Outer Continental Shelf pipelines.
192.11 Petroleum gas systems.
192.13 General.
192.14 Conversion to service subject to this part.
192.15 Rules of regulatory construction.
192.16 Customer notification.
192.17 [Reserved]

Subpart B–Materials

192.51 Scope.
192.53 General.
192.55 Steel pipe.
192.57 [Reserved]
192.59 Plastic pipe.
192.61 [Reserved]
192.63 Marking of materials.
192.65 Transportation of pipe.

Subpart C–Pipe Design

192.101 Scope.
192.103 General.
192.105 Design formula for steel pipe.
192.107 Yield strength (S) for steel pipe.
192.109 Nominal wall thickness (t) for steel pipe.
192.111 Design factor (F) for steel pipe.
192.113 Longitudinal joint factor (E) for steel pipe.
192.115 Temperature derating factor (T) for steel pipe.
192.117 [Reserved]
192.119 [Reserved]
192.121 Design of plastic pipe.
192.123 Design limitations for plastic pipe.
192.125 Design of copper pipe.

Subpart D–Design of Pipeline Components

192.141 Scope.
192.143 General requirements.
192.144 Qualifying metallic components.
192.145 Valves.
192.147 Flanges and flange accessories.
192.149 Standard fittings.
192.150 Passage of internal inspection devices.
192.151 Tapping.
192.153 Components fabricated by welding.
192.155 Welded branch connections.
192.157 Extruded outlets.
192.159 Flexibility.
192.161 Supports and anchors.
192.163 Compressor stations: Design and construction.
192.165 Compressor stations: Liquid removal.
192.167 Compressor stations: Emergency shutdown.
192.169 Compressor stations: Pressure limiting devices.
192.171 Compressor stations: Additional safety equipment.
192.173 Compressor stations: Ventilation.
192.175 Pipe-type and bottle-type holders.
192.177 Additional provisions for bottle-type holders.

192.179 Transmission line valves.
192.181 Distribution line valves.
192.183 Vaults: Structural design requirements.
192.185 Vaults: Accessibility.
192.187 Vaults: Sealing, venting, and ventilation.
192.189 Vaults: Drainage and waterproofing.
192.191 Design pressure of plastic fittings.
192.193 Valve installation in plastic pipe.
192.195 Protection against accidental overpressuring.
192.197 Control of the pressure of gas delivered from high-pressure distribution systems.
192.199 Requirements for design pressure relief and limiting devices.
192.201 Required capacity of pressure relieving and limiting stations.
192.203 Instrument, control, and sampling pipe and components.

Subpart E–Welding of Steel in Pipelines

192.221 Scope.
192.225 Welding-General.
192.227 Qualification of welders.
192.229 Limitations on welders.
192.231 Protection from weather.
192.233 Miter joints.
192.235 Preparation for welding.
192.241 Inspection and test of welds.
192.243 Nondestructive testing.
192.245 Repair or removal of defects.

Subpart F–Joining of Materials Other Than by Welding

192.271 Scope.
192.273 General.
192.275 Cast iron pipe.
192.277 Ductile iron pipe.
192.279 Copper pipe.
192.281 Plastic pipe.
192.283 Plastic pipe; qualifying joining procedures.
192.285 Plastic pipe; qualifying persons to make joints.
192.287 Plastic pipe; inspection of joints.

Subpart G–General Construction Requirements for Transmission Lines and Mains

192.301 Scope.
192.303 Compliance with specifications or standards.
192.305 Inspection: General.
192.307 Inspection of materials.
192.309 Repair of steel pipe.
192.311 Repair of plastic pipe.
192.313 Bends and elbows.
192.315 Wrinkle bends in steel pipe.
192.317 Protection from hazards.
192.319 Installation of pipe in a ditch.
192.321 Installation of plastic pipe.
192.323 Casing.
192.325 Underground clearance.
192.327 Cover.

Subpart H–Customer Meters, Service Regulators, and Service Lines

192.351 Scope.
192.353 Customer meters and regulators: Location.
192.355 Customer meters and regulators: Protection from damage.

192.357 Customer meters and regulators: Installation.
192.359 Customer meter installations: Operating pressure.
192.361 Service lines: Installation.
192.363 Service lines: Valve requirements.
192.365 Service lines: Location of valves.
192.367 Service lines: General requirements for connections to main piping.
192.369 Service lines: Connections to cast iron or ductile iron mains.
192.371 Service lines: Steel.
192.373 Service lines: Cast iron and ductile iron.
192.375 Service lines: Plastic.
192.377 Service lines: Copper.
192.379 New service lines not in use.
192.381 Service lines: Excess flow valve performance standards.
192.383 Excess flow valve customer notification

Subpart I–Requirements for Corrosion Control

192.451 Scope.
192.452 Applicability to converted pipelines.
192.453 General.
192.455 External corrosion control: Buried or submerged pipelines installed after July 31, 1971.
192.457 External corrosion control: Buried or submerged pipelines installed before August 1, 1971.
192.459 External corrosion control: Examination of buried pipeline when exposed.
192.461 External corrosion control: Protective coating.
192.463 External corrosion control: Cathodic protection.
192.465 External corrosion control: Monitoring
192.467 External corrosion control: Electrical isolation.
192.469 External corrosion control: Test stations.
192.471 External corrosion control: Test leads.
192.473 External corrosion control: Interference currents.
192.475 Internal corrosion control: General.
192.477 Internal corrosion control: Monitoring.
192.479 Atmospheric corrosion control: General.
192.481 Atmospheric corrosion control: Monitoring.
192.483 Remedial measures: General.
192.485 Remedial measures: Transmission lines.
192.487 Remedial measures: Distribution lines other than cast iron or ductile iron lines.
192.489 Remedial measures: Cast iron and ductile iron pipelines.
192.491 Corrosion control records.

Subpart J–Test Requirements

192.501 Scope.
192.503 General requirements.
192.505 Strength test requirements for steel pipeline to operate at a hoop stress of 30 percent or more of SMYS.

192CODE.DOC, DT (Revised) 7/98

192.507 Test requirements for pipelines to operate at a hoop stress less than 30 percent of SMYS and above 100 psig.
192.509 Test requirements for pipelines to operate below 100 psig.
192.511 Test requirements for service lines.
192.513 Test requirements for plastic pipelines.
192.515 Environmental protection and safety requirements.
192.517 Records.

Subpart K–Uprating

192.551 Scope.
192.553 General requirements.
192.555 Uprating to a pressure that will produce a hoop stress of 30 percent or more of SMYS in steel pipelines.
192.557 Uprating: Steel pipelines to a pressure that will produce a hoop stress less than 30 percent of SMYS; plastic, cast iron, and ductile iron pipelines.

Subpart L–Operations

192.601 Scope.
192.603 General provisions.
192.605 Procedural manual for operations, maintenance, and emergencies.
192.607 Initial determination of class location and confirmation or establishment of maximum allowable operating pressure.
192.609 Change in class location: Required study.
192.611 Change in class location: Confirmation or revision of maximum allowable operating pressure.
192.612 Underwater inspection and reburial of pipelines in the Gulf of Mexico and its inlets.
192.613 Continuing surveillance.
192.614 Damage prevention program.
192.615 Emergency plans.
192.616 Public education.
192.617 Investigation of failures.
192.619 Maximum allowable operating pressure: Steel or plastic pipelines.
192.621 Maximum allowable operating pressure: High-pressure distribution systems.
192.623 Maximum and minimum allowable operating pressure: Low-pressure distribution systems.
192.625 Odorization of gas.
192.627 Tapping pipelines under pressure.
192.629 Purging of pipelines.

Subpart M–Maintenance

192.701 Scope.
192.703 General.
192.705 Transmission lines: Patrolling.
192.706 Transmission lines: Leakage surveys.
192.707 Line markers for mains and transmission lines.
192.709 Transmission lines: Record keeping.

192CODE.DOC, DT (Revised) 7/98

Authority: 49 U.S.C 5103, 60102, 60104, 60108, 60109, 60110, 60113, and 60118; and 49 CFR 1.53.

[Amdt. 192-75, 61 FR 18512, Apr. 26, 1996; Amdt. 192-81, 62 FR 61692, Nov. 19, 1997]

Source: 35 FR 13257, Aug. 19, 1970, unless otherwise noted.

Editorial Note: Nomenclature changes to Part 192 appear at 50 FR 45732, Nov. 1, 1985.

Subpart A–General

§192.1 Scope of part.

(a) This part prescribes minimum safety requirements for pipeline facilities and the transportation of gas, including pipeline facilities and the transportation of gas within the limits of the outer continental shelf as that term is defined in the Outer Continental Shelf Lands Act (43 U.S.C. 1331).

(b) This part does not apply to:

(1) Offshore pipelines upstream from the outlet flange of each facility where hydrocarbons are produced or where produced hydrocarbons are first separated, dehydrated, or otherwise processed, whichever facility is farther downstream;

(2) Onshore gathering of gas outside of the following areas:

(i) An area within the limits of any incorporated or unincorporated city, town, or village.

(ii) Any designated residential or commercial area such as a subdivision, business or shopping center, or community development.

(3) Onshore gathering of gas within inlets of the Gulf of Mexico except as provided in §192.612.

(4) Any pipeline system that transports only petroleum gas or petroleum gas/air mixtures to–

(i) Fewer than 10 customers, if no portion of the system is located in a public place; or

(ii) A single customer, if the system is located entirely on the customer's premises (no matter if a portion of the system is located in a public place).

(5) On the Outer Continental Shelf upstream of the point at which operating responsibility transfers from a producing operator to a transporting operator.

[35 FR 13257, Aug. 19, 1970, as amended by Amdt. 192-27, 41 FR 34598, Aug. 16, 1976; Amdt. 192-67, 56 FR 63764, Dec. 5, 1991; Amdt. 192-78, 61 FR 28770, June 6, 1996; Amdt. 192-81, 62 FR 61692, Nov. 19, 1997]

§192.3 Definitions.

As used in this part:

Administrator means the Administrator of the Research and Special Programs Administration or any person to whom authority in the matter concerned has been delegated by the Secretary of Transportation.

Distribution Line means a pipeline other than a gathering or transmission line.

Exposed pipeline means a pipeline where the top of the pipe is protruding above the seabed in water less than 15 feet (4.6 meters) deep, as measured from the mean low water.

Gas means natural gas, flammable gas, or gas which is toxic or corrosive.

Gathering Line means a pipeline that transports gas from a current production facility to a transmission line or main.

Gulf of Mexico and its inlets means the waters from the mean high water mark of the coast of the Gulf of Mexico and its inlets open to the sea (excluding rivers, tidal marshes, lakes, and canals) seaward to include the territorial sea and Outer Continental Shelf to a depth of 15 feet (4.6 meters), as measured from the mean low water.

Hazard to navigation means, for the purpose of this part, a pipeline where the top of the pipe is less than 12 inches (305 millimeters) below the seabed in water less than 15 feet (4.6 meters) deep, as measured from the mean low water.

High pressure distribution system means a distribution system in which the gas pressure in the main is higher than the pressure provided to the customer.

Line section means a continuous run of transmission line between adjacent compressor stations, between a compressor station and storage facilities, between a compressor station and a block valve, or between adjacent block valves.

Listed specification means a specification listed in section I of Appendix B of this part.

Low-pressure distribution system means a distribution system in which the gas pressure in the main is substantially the same as the pressure provided to the customer.

Main means a distribution line that serves as a common source of supply for more than one service line.

Maximum actual operating pressure means the maximum pressure that occurs during normal operations over a period of 1 year.

Maximum allowable operating pressure (MAOP) means the maximum pressure at which a pipeline or segment of a pipeline may be operated under this part.

Municipality means a city, county, or any other political subdivision of a State.

Offshore means beyond the line of ordinary low water along that portion of the coast of the United States that is in direct contact with the open seas and beyond the line marking the seaward limit of inland waters.

Operator means a person who engages in the transportation of gas.

Outer Continental Shelf means all submerged lands lying seaward and outside the area of lands beneath navigable waters as defined in Section 2 of the Submerged Lands Act (43 U.S.C. 1301) and of which the subsoil and seabed appertain to the United States and are subject to its jurisdiction and control.

Person means any individual, firm, joint venture, partnership, corporation, association, State, municipality, cooperative association, or joint stock association, and including any trustee, receiver, assignee, or personal representative thereof.

Petroleum gas means propane, propylene, butane, (normal butane or isobutanes), and butylene (including isomers), or mixtures composed predominantly of these gases, having a vapor pressure not exceeding 208 psi (1434 kPa) gage at 100°F (38°C).

Pipe means any pipe or tubing used in the transportation of gas, including pipe-type holders.

Pipeline means all parts of those physical facilities through which gas moves in transportation, including pipe, valves, and other appurtenance attached to pipe, compressor units, metering stations, regulator stations, delivery stations, holders, and fabricated assemblies.

Pipeline facility means new and existing pipeline, rights-of-way, and any equipment, facility, or building used in the transportation of gas or in the treatment of gas during the course of transportation.

Service Line means a distribution line that transports gas from a common source of supply to (a) a customer meter or the connection to a customer's piping, whichever is farther downstream, or (b) the connection to a customer's piping if there is no customer meter. A customer meter is the meter that measures the transfer of gas from an operator to a consumer.

SMYS means specified minimum yield strength is:

(a) For steel pipe manufactured in accordance with a listed specification, the yield strength specified as a minimum in that specification; or

(b) For steel pipe manufactured in accordance with an unknown or unlisted specification, the yield strength determined in accordance with §192.107(b).

State means each of the several States, the District of Columbia, and the Commonwealth of Puerto Rico.

Transmission Line means a pipeline, other than a gathering line, that:

(a) Transports gas from a gathering line or storage facility to a distribution center, storage facility, or large volume customer that is not downstream from a distribution center;

(b) Operates at a hoop stress of 20 percent or more of SMYS; or

(c) Transports gas within a storage field.

A large volume customer may receive similar volumes of gas as a distribution center, and includes factories, power plants, and institutional users of gas.

Transportation of gas means the gathering, transmission, or distribution of gas by pipeline, or the storage of gas, in or affecting interstate or foreign commerce.

[35 FR 13257, Aug. 19, 1970, as amended by Amdt. 192-13, 38 FR 9084, Apr. 10, 1973; Amdt. 192-27, 41 FR 34598, Aug. 16, 1976; Amdt. 192-58, 53 FR 1633, Jan. 21, 1988; Amdt. 192-67, 56 FR 63764, Dec. 5, 1991; Amdt. 192-72, 59 FR 17281, May 12, 1994; Amdt. 192-78, 61 FR 28770, June 6, 1996; Amdt. 192-81, 62 FR 61692, Nov. 19, 1997; Amdt. 192-85, 63 FR 37500, July 13, 1998]

§192.5 Class locations.

(a) This section classifies pipeline locations for purposes of this part. The following criteria apply to classifications under this section.

(1) A "class location unit" is an onshore area that extends 220 yards (200 meters) on either side of the centerline of any continuous 1-mile (1.6 kilometers) length of pipeline.

(2) Each separate dwelling unit in a multiple dwelling unit building is counted as a

separate building intended for human occupancy.

(b) Except as provided in paragraph (c) of this section, pipeline locations are classified as follows:

(1) A Class 1 location is:

(i) An offshore area; or

(ii) Any class location unit that has 10 or fewer buildings intended for human occupancy.

(2) A Class 2 location is any class location unit that has more than 10 but fewer than 46 buildings intended for human occupancy.

(3) A Class 3 location is:

(i) Any class location unit that has 46 or more buildings intended for human occupancy; or

(ii) An area where the pipeline lies within 100 yards (91 meters) of either a building or a small, well-defined outside area (such as a playground, recreation area, outdoor theater, or other place of public assembly) that is occupied by 20 or more persons on at least 5 days a week for 10 weeks in any 12-month period. (The days and weeks need not be consecutive.)

(4) A Class 4 location is any class location unit where buildings with four or more stories aboveground are prevalent.

(c) The length of Class locations 2, 3, and 4 may be adjusted as follows:

(1) A Class 4 location ends 220 yards (200 meters) from the nearest building with four or more stories aboveground.

(2) When a cluster of buildings intended for human occupancy requires a Class 2 or 3 location, the class location ends 220 yards (200 meters) from the nearest building in the cluster.

[35 FR 13257, Aug. 19, 1970, as amended by Amdt. 192-27, 41 FR 34598, Aug. 16, 1976; Amdt. 192-56, 52 FR 32924, Sept. 1, 1987; Amdt. 192-78, 61 FR 28770, June 6, 1996; Amdt. 192-78B, 61 FR 35139, July 5, 1996; Amdt. 192-85, 63 FR 37500, July 13, 1998]

§192.7 Incorporation by reference.

(a) Any documents or portions thereof incorporated by reference in this part are included in this part as though set out in full. When only a portion of a document is referenced, the remainder is not incorporated in this part.

(b) All incorporated materials are available for inspection in the Research and Special Programs Administration, 400 Seventh Street, SW., Washington, DC, and at the Office of the Federal Register, 800 North Capitol Street, NW., Suite 700, Washington, DC. These materials have been approved for incorporation by reference by the Director of the Federal Register in accordance with 5 U.S.C. 552(a) and 1 CFR part 51. In addition, the incorporated materials are available from the respective organizations listed in Appendix A to this part.

(c) The full titles for the publications incorporated by reference in this part are provided in Appendix A to this part. Numbers in parentheses indicate applicable editions. Earlier editions of documents listed or editions of documents formerly listed in previous editions of Appendix A may be used for materials and components manufactured, designed, or installed in accordance with those earlier editions at the time they were listed. The user must refer to the appropriate previous edition of 49 CFR for a listing of the earlier listed editions or documents.

[35 FR 13257, Aug. 19, 1970, as amended by Amdt. 192-37, 46 FR 10157, Feb. 2, 1981; Amdt. 192-51, 51 FR 15333, Apr. 23, 1986; Amdt. 192-68, 58 FR, 14519, Mar. 18, 1993; Amdt. 192-78, 61 FR 28770, June 6, 1996]

§192.9 Gathering lines.

Except as provided in §§192.1 and 192.150, each operator of a gathering line must comply with the requirements of this part applicable to transmission lines.

[Amdt. 192-72, 59 FR 17281, Apr. 12, 1994]

§192.10 Outer continental shelf pipelines.

Operators of transportation pipelines on the Outer Continental Shelf (as defined in the Outer Continental Shelf Lands Act (43 U.S.C. 1331) must identify on all their respective pipelines the specific points at which operating responsibility transfers to a producing operator. For those instances in which the transfer points are not identifiable by a durable marking, each operator will have until September 15, 1998 to identify the transfer points. If it is not practicable to durably mark a transfer point and the transfer point is located above water, the operator must depict the transfer point on a schematic located near the transfer point. If a transfer point is located subsea, then the operator must identify the transfer point on a schematic which must be maintained at the nearest upstream facility and provided to RSPA upon request. For those cases in which adjoining operators have not agreed on a transfer point by September 15, 1998 the Regional Director and the MMS Regional Supervisor will make a joint determination of the transfer point.

[Amdt. 192-81, 62 FR 61692, Nov. 19, 1997]

§192.11 Petroleum gas systems.

(a) Each plant that supplies petroleum gas by pipeline to a natural gas distribution system must meet the requirements of this part and ANSI/NFPA 58 and 59.

(b) Each pipeline system subject to this part that transports only petroleum gas or petroleum gas/air mixtures must meet the requirements of this part and of ANSI/NFPA 58 and 59.

(c) In the event of a conflict between this part and ANSI/NFPA 58 and 59, ANSI/NFPA 58 and 59 prevail.

[35 FR 13257, Aug. 19, 1970, as amended by Amdt. 192-68, 58 FR 14519, Mar. 18, 1993; Amdt. 192-75, 61 FR 18512, Apr. 26, 1996; Amdt. 192-78, 61 FR 28770, June 6, 1996]

§192.13 General.

(a) No person may operate a segment of pipeline that is readied for service after March 12, 1971, or in the case of an offshore gathering line, after July 31, 1977, unless:

(1) The pipeline has been designed, installed, constructed; initially inspected, and initially tested in accordance with this part; or

(2) The pipeline qualifies for use under this part in accordance with §192.14.

(b) No person may operate a segment of pipeline that is replaced, relocated, or otherwise changed after November 12, 1970, or in the case of an offshore gathering line, after July 31, 1977, unless that replacement, relocation, or change has been made in accordance with this part.

(c) Each operator shall maintain, modify as appropriate, and follow the plans, procedures, and programs that it is required to establish under this part.

[35 FR 13257, Aug. 19, 1970, as amended by Amdt. 192-27, 41 FR 34598, Aug. 16, 1976; Amdt. 192-30, 42 FR 60146, Nov. 25, 1977]

§192.14 Conversion to service subject to this part.

(a) A steel pipeline previously used in service not subject to this part qualifies for use under this part if the operator prepares and follows a written procedure to carry out the following requirements:

(1) The design, construction, operation, and maintenance history of the pipeline must be reviewed and, where sufficient historical records are not available, appropriate tests must be performed to determine if the pipeline is in a satisfactory condition for safe operation.

(2) The pipeline right-of-way, all aboveground segments of the pipeline, and appropriately selected underground segments must be visually inspected for physical defects and operating conditions which reasonably could be expected to impair the strength or tightness of the pipeline.

(3) All known unsafe defects and conditions must be corrected in accordance with this part.

(4) The pipeline must be tested in accordance with Subpart J of this part to substantiate the maximum allowable operating pressure permitted by Subpart L of this part.

(b) Each operator must keep for the life of the pipeline a record of investigations, tests, repairs, replacements, and alterations made under the requirements of paragraph (a) of this section.

[Amdt. 192-30, 42 FR 60146, Nov. 25, 1977]

§192.15 Rules of regulatory construction.

(a) As used in this part:

"Includes" means 'including but not limited to."

"May" means "is permitted to" or "is authorized to."

"May not" means "is not permitted to" or "is not authorized to."

"Shall" is used in the mandatory and imperative sense.

(b) In this part:

(1) Words importing the singular include the plural;

(2) Words importing the plural include the singular; and,

(3) Words importing the masculine gender include the feminine.

§192.16 Customer notification.

(a) This section applies to each operator of a service line who does not maintain the customer's buried piping up to entry of the first building downstream, or, if the customer's buried piping does not enter a building, up to the principal gas utilization equipment or the first fence (or wall) that surrounds that equipment. For the purpose of this section, "customer's buried piping" does not include branch lines that serve yard lanterns, pool heaters, or other types of secondary equipment. Also, "maintain" means monitor for corrosion according to §192.465 if the customer's buried piping is metallic, survey for leaks according to §192.723, and if an unsafe condition is found, shut off the flow of gas, advise the customer of the need to repair the unsafe condition, or repair the unsafe condition.

(b) Each operator shall notify each customer once in writing of the following information:

(1) The operator does not maintain the customer's buried piping.

(2) If the customer's buried piping is not maintained, it may be subject to the potential hazards of corrosion and leakage.

(3) Buried gas piping should be–

(i) Periodically inspected for leaks;

(ii) Periodically inspected for corrosion if the piping is metallic; and

(iii) Repaired if any unsafe condition is discovered.

(4) When excavating near buried gas piping, the piping should be located in advance, and the excavation done by hand.

(5) The operator (if applicable), plumbing contractors, and heating contractors can assist in locating, inspecting, and repairing the customer's buried piping.

(c) Each operator shall notify each customer not later than August 14, 1996, or 90 days after the customer first receives gas at a particular location, whichever is later. However, operators of master meter systems may continuously post a general notice in a prominent location frequented by customers.

(d) Each operator must make the following records available for inspection by the Administrator or a State agency participating under 40 U.S.C. 60105 or 60106;

(1) A copy of the notice currently in use; and

(2) Evidence that notices have been sent to customers within the previous 3 years.

[Amdt. 192-74, 60 FR 41821, Aug. 14, 1995 as amended by 192-74A, 60 FR 63450, Dec. 11, 1995; Amdt. 192-84, 63 FR 7721, Feb. 17, 1998]

§192.17 [Reserved]

[Amdt. 192-1, 35 FR 16405, Oct. 21, 1970 as amended by Amdt. 192-38, 48 FR 37250, July 20, 1981]

Subpart B–Materials

§192.51 Scope.

This subpart prescribes minimum requirements for the selection and qualification of pipe and components for use in pipelines.

§192.53 General.

Materials for pipe and components must be:

(a) Able to maintain the structural integrity of the pipeline under temperature and other environmental conditions that may be anticipated;

(b) Chemically compatible with any gas that they transport and with any other material in the pipeline with which they are in contact; and,

(c) Qualified in accordance with the applicable requirements of this subpart.

§192.55 Steel pipe.

(a) New steel pipe is qualified for use under this part if:

(1) It was manufactured in accordance with a listed specification;

(2) It meets the requirements of–

(i) Section II of Appendix B to this part; or

(ii) If it was manufactured before November 12, 1970, either section II or III of Appendix B to this part; or

(3) It is used in accordance with paragraph (c) or (d) of this section.

(b) Used steel pipe is qualified for use under this part if:

(1) It was manufactured in accordance with a listed specification and it meets the requirements of paragraph II-C of Appendix B to this part;

(2) It meets the requirements of:

(i) Section II of Appendix B to this part; or

(ii) If it was manufactured before November 12, 1970, either section II or III of Appendix B to this part;

(3) It has been used in an existing line of the same or higher pressure and meets the requirements of paragraph II-C of Appendix B to this part; or

(4) It is used in accordance with paragraph (c) of this section.

(c) New or used steel pipe may be used at a pressure resulting in a hoop stress of less than 6,000 psi (41 MPa) where no close coiling or close bending is to be done, if visual examination indicates that the pipe is in good condition and that it is free of split seams and other defects that would cause leakage. If it is to be welded, steel pipe that has not been manufactured to a listed specification must also pass the weldability tests prescribed in paragraph II-B of Appendix B to this part.

(d) Steel pipe that has not been previously used may be used as replacement pipe in a segment of pipeline if it has been manufactured prior to November 12, 1970, in accordance with the same specification as the pipe used in constructing that segment of pipeline.

(e) New steel pipe that has been cold expanded must comply with the mandatory provisions of API Specification 5L.

[35 FR 13257, Aug. 19, 1970, as amended by Amdt. 192-3, 35 FR 17660, Nov. 17, 1970; Amdt. 192-12, 38 FR 4760, Feb. 22, 1973; Amdt. 192-51, 51 FR 15333, Apr. 23, 1986; Amdt. 192-68, 58 FR 14519, Mar. 18, 1993; Amdt. 192-85, 63 FR 37500, July 13, 1998]

§192.57 [Reserved]

[5 FR 13257, Aug. 19, 1970, as amended by Amdt. 192-62, 54 FR 5625, Feb. 6, 1989]

§192.59 Plastic pipe.

(a) New plastic pipe is qualified for use under this part if:

(1) It is manufactured in accordance with a listed specification; and

(2) It is resistant to chemicals with which contact may be anticipated.

(b) Used plastic pipe is qualified for use under this part if:

(1) It was manufactured in accordance with a listed specification;

(2) It is resistant to chemicals with which contact may be anticipated;

(3) It has been used only in natural gas service.

(4) Its dimensions are still within the tolerances of the specification to which it was manufactured; and,

(5) It is free of visible defects.

(c) For the purpose of paragraphs (a)(1) and (b)(1) of this section, where pipe of a diameter included in a listed specification is impractical to use, pipe of a diameter between the sizes included in a listed specification may be used if it:

(1) Meets the strength and design criteria required of pipe included in that listed specification; and

(2) Is manufactured from plastic compounds which meet the criteria for material required of pipe included in that listed specification.

[35 FR 13257, Aug. 19, 1970, as amended by Amdt. 192-19, 40 FR 10472, Mar. 6, 1975; Amdt. 192-58, 53 FR 1633, Jan. 21, 1988]

§192.61 [Reserved]

[35 FR 13257, Aug. 19, 1970, as amended by Amdt. 192-62, 54 FR 5625, Feb. 6, 1989]

§192.63 Marking of materials.

(a) Except as provided in paragraph (d) of this section, each valve, fitting, length of pipe, and other component must be marked–

(1) As prescribed in the specification or standard to which it was manufactured, except that thermoplastic fittings must be marked in accordance with ASTM D 2513; or

(2) To indicate size, material, manufacturer, pressure rating, and temperature rating, and as appropriate, type, grade, and model.

(b) Surfaces of pipe and components that are subject to stress from internal pressure may not be field die stamped.

(c) If any item is marked by die stamping, the die must have blunt or rounded edges that will minimize stress concentrations.

(d) Paragraph (a) of this section does not apply to items manufactured before November 12, 1970, that meet all of the following:

(1) The item is identifiable as to type, manufacturer, and model.

(2) Specifications or standards giving pressure, temperature, and other appropriate criteria for the use of items are readily available.

[35 FR 13257, Aug. 19, 1970, as amended by Amdt. 192-3, 35 FR 17660, Nov. 17, 1970; Amdt. 192-31, 43 FR 13883, Apr. 3, 1978; Amdt. 192-61, 53 FR 36793, Sept. 22, 1988; Amdt. 192-61A, 54 FR 32642, Aug. 9, 1989; Amdt. 192-62, 54 FR 5627, Feb. 6, 1989; Amdt. 192-68, 58 FR 14519, Mar. 18, 1993; Amdt. 192-76, 61 FR 26121, May 25, 1996; Amdt. 192-76A, 61 FR 36825, July 15, 1996]

§192.65 Transportation of pipe.

In a pipeline to be operated at a hoop stress of 20 percent or more of SMYS, an operator may not use pipe having an outer diameter to wall thickness ratio of 70 to 1, or more, that is transported by railroad unless:

(a) The transportation is performed in accordance with API RP 5L1.

(b) In the case of pipe transported before November 12, 1970, the pipe is tested in accordance with subpart J of this part to at least 1.25 times the maximum allowable operating pressure if it is to be installed in a class 1 location and to at least 1.5 times the maximum allowable operating pressure if it is to be installed in a class 2, 3, or 4 location. Notwithstanding any shorter time period permitted under subpart J of this part, the test pressure must be maintained for at least 8 hours.

[Amdt. 192-12, 38 FR 4760, Feb. 22, 1973, as amended by Amdt. 192-17, 40 FR 6346, Feb. 11, 1975; Amdt. 192-68, 58 FR 14519, Mar. 18, 1993]

Subpart C–Pipe Design

§192.101 Scope.

This subpart prescribes the minimum requirements for the design of pipe.

§192.103 General.

Pipe must be designed with sufficient wall thickness, or must be installed with adequate protection, to withstand anticipated external pressures and loads that will be imposed on the pipe after installation.

§192.105 Design formula for steel pipe.

(a) The design pressure for steel pipe is determined in accordance with the following formula:

P = (2 St/D) x F x E x T

P = Design pressure in pounds per square inch (kPa) gage.

S = Yield strength in pounds per square inch (kPa) determined in accordance with §192.107.

D = Nominal outside diameter of the pipe in inches (millimeters).

t = Nominal wall thickness of the pipe in inches (millimeters). If this is unknown, it is determined in accordance with §192.109. Additional wall thickness required for concurrent external loads in accordance with §192.103 may not be included in computing design pressure.

F = Design factor determined in accordance with §192.111.

E = Longitudinal joint factor determined in accordance with §192.113.

T = Temperature derating factor determined in accordance with §192.115.

(b) If steel pipe that has been subjected to cold expansion to meet the SMYS is subsequently heated, other than by welding or stress relieving as a part of welding, the design pressure is limited to 75 percent of the pressure determined under paragraph (a) of this section if the temperature of the pipe exceeds 900°F (482°C) at any time or is held above 600°F (316°C) for more than one hour.

[35 FR 13257, Aug. 19, 1970 as amended by Amdt. 192-47, 49 FR 7569, May. 1, 1984; Amdt. 192-85, 63 FR 37500, July 13, 1998]

§192.107 Yield strength (S) for steel pipe.

(a) For pipe that is manufactured in accordance with a specification listed in section I of Appendix B of this part, the yield strength to be used in the design formula in §192.105 is the SMYS stated in the listed specification, if that value is known.

(b) For pipe that is manufactured in accordance with a specification not listed in section I of Appendix B to this part or whose specification or tensile properties are unknown, the yield strength to be used in the design formula in §192.105 is one of the following:

(1) If the pipe is tensile tested in accordance with section II-D of Appendix B to this part, the lower of the following:

(i) 80 percent of the average yield strength determined by the tensile tests.

(ii) The lowest yield strength determined by the tensile tests.

(2) If the pipe is not tensile tested as provided in paragraph (b)(1) of this section, 24,000 psi (165 MPa).

[35 FR 13257, Aug. 19, 1970, as amended by Amdt. 192-78, 61 FR 28770, June 6, 1996; Amdt. 192-84, 63 FR 7721, Feb. 17, 1998; Amdt. 192-85, 63 FR 37500, July 13, 1998]

§192.109 Nominal wall thickness (t) for steel pipe.

(a) If the nominal wall thickness for steel pipe is not known, it is determined by measuring the thickness of each piece of pipe at quarter points on one end.

(b) However, if the pipe is of uniform grade, size, and thickness and there are more than 10 lengths, only 10 percent of the individual lengths, but not less than 10 lengths, need be measured. The thickness of the lengths that are not measured must be verified by applying a gauge set to the minimum thickness found by the measurement. The nominal wall thickness to be used in the design formula in §192.105 is the next wall thickness found in commercial specifications that is below the average of all the measurements taken. However, the nominal wall thickness used may not be more than 1.14 times the smallest measurement taken on pipe less than 20 inches (508 millimeters) in outside diameter, nor more than 1.11 times the smallest measurement taken on pipe 20 inches (508 millimeters) or more in outside diameter.

[35 FR 13257, Aug. 19, 1970, as amended by Amdt. 192-85, 63 FR 37500, July 13, 1998]

§192.111 Design factor (F) for steel pipe.

(a) Except as otherwise provided in paragraphs (b), (c), and (d) of this section, the design factor to be used in the design formula in §192.105 is determined in accordance with the following table:

Class location	Design factor (F)
1	0.72
2	0.60
3	0.50
4	0.40

(b) A design factor of 0.60 or less must be used in the design formula in §192.105 for steel pipe in Class 1 locations that:

(1) Crosses the right-of-way of an unimproved public road, without a casing;

(2) Crosses without a casing, or makes a parallel encroachment on, the right-of-way of either a hard surfaced road, a highway, a public street, or a railroad;

(3) Is supported by a vehicular, pedestrian, railroad, or pipeline bridge; or

(4) Is used in a fabricated assembly, (including separators, mainline valve assemblies, cross-connections, and river crossing headers) or is used within five pipe

diameters in any direction from the last fitting of a fabricated assembly, other than a transition piece or an elbow used in place of a pipe bend which is not associated with a fabricated assembly.

(c) For Class 2 locations, a design factor of 0.50, or less, must be used in the design formula in §192.105 for uncased steel pipe that crosses the right-of-way of a hard surfaced road, a highway, a public street, or a railroad.

(d) For Class 1 and Class 2 locations, a design factor of 0.50, or less, must be used in the design formula in §192.105 for–

(1) Steel pipe in a compressor station, regulating station, or measuring station, and

(2) Steel pipe, including a pipe riser, on a platform located offshore or in inland navigable waters.

[35 FR 13257, Aug. 19, 1970, as amended by Amdt. 192-27, 41 FR 34598, Aug. 16, 1976]

§192.113 Longitudinal joint factor (E) for steel pipe.

The longitudinal joint factor to be used in the design formula in §192.105 is determined in accordance with the following table:

Specification	Pipe Class	Longitudinal Joint Factor (E)
ASTM A53	Seamless	1.00
	Electric resistance welded	1.00
	Furnace butt welded	0.60
ASTM A106	Seamless	1.00
ASTM A333/A333M	Seamless	1.00
	Electric resistance welded	1.00
ASTM A381	Double submerged arc welded	1.00
ASTM A671	Electric-fusion welded	1.00
ASTM A672	Electric-fusion welded	1.00
ASTM A691	Electric-fusion welded	1.00
API 5L	Seamless	1.00
	Electric resistance welded	1.00
	Electric flash welded	1.00
	Submerged arc welded	1.00
	Furnace butt welded	0.60
Other	Pipe over 4 inches (102 millimeters)	0.80
Other	Pipe 4 inches (102 millimeters) or less	0.60

If the type of longitudinal joint cannot be determined, the joint factor to be used must not exceed that designated for "Other."

[35 FR 13257, Aug. 19, 1970, as amended by Amdt. 192-37, 46 FR 10157, Feb. 2, 1981; Amdt. 192-51, 51 FR 15333, Apr. 23, 1986; Amdt. 192-62 54 FR 5625, Feb. 6, 1989; Amdt. 192-68, 58 FR 14519, Mar. 18, 1993; Amdt. 192-85, 63 FR 37500, July 13, 1998]

§192.115 Temperature derating factor (T) for steel pipe.

The temperature derating factor to be used in the design formula in §192.105 is determined as follows:

Gas Temperature in degrees Fahrenheit (Celsius)	Temperature derating factor (T)
250°F (121°C) or less	1.000
300°F (149°C)	0.967
350°F (177°C)	0.933
400°F (204°C)	0.900
450°F (232°C)	0.867

For intermediate gas temperatures, the derating factor is determined by interpolation.

[35 FR 13257, Aug. 19, 1970, as amended by Amdt. 192-85, 63 FR 37500, July 13, 1998]

§192.117 [Reserved]

[35 FR 13257, Aug. 19, 1970, as amended by Amdt. 192-37, 46 FR 10157, Feb. 2, 1981 and 46 FR 10706, Feb. 4, 1981, effective Mar. 31, 1981; Amdt. 192-62, 54 FR 5625, Feb. 6, 1989]

§192.119 [Reserved]

[35 FR 13257, Aug. 19, 1970, as amended by Amdt. 192-62, 54 FR 5625, Feb. 6, 1989]

§192.121 Design of plastic pipe.

Subject to the limitations of §192.123, the design pressure for plastic pipe is determined in accordance with either of the following formulas:

$$P = 2S\frac{t}{(D-t)}0.32$$

$$P = \frac{2S}{(SDR-1)}0.32$$

where:

P = Design pressure, gage, kPa (psig).

S = For thermoplastic pipe the long-term hydrostatic strength determined in accordance with the listed specification at a temperature equal to 73°F (23°C) 100°F (38°C),120°F (49°C), or 140°F (60°C); for reinforced thermosetting plastic pipe, 11,000 psi (75,842 kPa).

t = Specified wall thickness, mm (in).

D = Specified outside diameter, mm (in).

SDR = Standard dimension ratio, the ratio of the average specified outside diameter to the minimum specified wall thickness, corresponding to a value from a common numbering system that was derived from the American National Standards Institute preferred number series 10.

[Amdt. 192-31, 43 FR 13883, Apr. 3, 1978; 43 FR 43308, Sept. 25, 1978; Amdt. 192-78, 61 FR 28770, June 6, 1996; Amdt. 192-85, 63 FR 37500, July 13, 1998]

§192.123 Design limitations for plastic pipe.

(a) The design pressure may not exceed a gauge pressure of 689 kPa (100 psig) for plastic pipe used in:

(1) Distribution systems; or

(2) Classes 3 and 4 locations.

(b) Plastic pipe may not be used where operating temperatures of the pipe will be:

(1) Below -20°F (-29°C), or -40°F (-40°C) if all pipe and pipeline components whose operating temperature will be below -20°F (-29°C) have a temperature rating by the manufacturer consistent with that operating temperature; or

(2) Above the following applicable temperatures:

(i) For thermoplastic pipe, the temperature at which the long-term hydrostatic strength used in the design formula under §192.121 is determined. However, if the pipe was manufactured before May 18, 1978 and its long-term hydrostatic strength was determined at 73°F (23°C), it may be used at temperatures up to 100°F (38°C).

(ii) For reinforced thermosetting plastic pipe, 150°F (66°C).

(c) The wall thickness for thermoplastic pipe may not be less than 0.062 inches (1.57 millimeters).

(d) The wall thickness for reinforced thermosetting plastic pipe may not be less than that listed in the following table:

Normal size in inches (millimeters)	Minimum wall thickness in inches (millimeters)
2 (51)	0.060 (1.52)
3 (76)	0.060 (1.52)
4 (102)	0.070 (1.78)
6 (152)	0.100 (2.54)

[35 FR 13257, Aug. 19, 1970, as amended by Amdt. 192-31, 43 FR 13883, Apr. 3, 1978; Amdt. 192-78, 61 FR 28770, June 6, 1996; Amdt. 192-85, 63 FR 37500, July 13, 1998]

§192.125 Design of copper pipe.

(a) Copper pipe used in mains must have a minimum wall thickness of 0.065 inches (1.65 millimeters) and must be hard drawn.

(b) Copper pipe used in service lines must have wall thickness not less than that indicated in following table:

Standard size (inch) (millimeters)	Nominal O.D. (inch) (millimeters)	Wall thickness (inch) (millimeters)	
		Nominal	Tolerance
½ (13)	.625 (16)	.040 (1.06)	.0035 (.0889)
5/8 (16)	.750 (19)	.042 (1.07)	.0035 (.0889)
¾ (19)	.875 (22)	.045 (1.14)	.004 (.102)
1 (25)	1.125 (29)	.050 (1.27)	.004 (.102)
1¼ (32)	1.375 (35)	.055 (1.40)	.0045 (.1143)
1½ (38)	1.625 (41)	.060 (1.52)	.0045 (.1143)

(c) Copper pipe used in mains and service lines may not be used at pressures in excess of 100 psi (689 kPa) gage.

(d) Copper pipe that does not have an internal corrosion resistant lining may not be used to carry gas that has an average hydrogen sulfide content of more than 0.3 grains/100 ft^3 (6.9/m^3) under standard conditions. Standard conditions refer to 60°F and 14.7 psia (15.6°C and one atmosphere) of gas.

[35 FR 13257, Aug. 19, 1970, as amended by Amdt. 192-62, 54 FR 5625, Feb. 6, 1989; Amdt. 192-85, 63 FR 37500, July 13, 1998]

Subpart D–Design of Pipeline Components

§192.141 Scope.

This subpart prescribes minimum requirements for the design and installation of pipeline components and facilities. In addition, it prescribes requirements relating to protection against accidental overpressuring.

§192.143 General requirements.

Each component of a pipeline must be able to withstand operating pressures and other anticipated loadings without impairment of its serviceability with unit stresses equivalent to those allowed for comparable material in pipe in the same location and kind of service. However, if design based upon unit stresses is impractical for a particular component, design may be based upon a pressure rating established by the manufacturer by pressure testing that component or a prototype of the component.

[Amdt. 192-48, 49 CFR 19823, May 10, 1984]

§192.144 Qualifying metallic components.

Notwithstanding any requirement of this subpart which incorporates by reference an edition of a document listed in Appendix A of this part, a metallic component manufactured in accordance with any other edition of that document is qualified for use under this part if–

(a) It can be shown through visual inspection of the cleaned component that no defect exists which might impair the strength or tightness of the component; and

(b) The edition of the document under which the component was manufactured has equal or more stringent requirements for the following as an edition of that document currently or previously listed in Appendix A:

(1) Pressure testing;

(2) Materials; and,

(3) Pressure and temperature ratings.

[Amdt. 192-45, 48 FR 30637, July 5, 1983]

§192.145 Valves.

(a) Except for cast iron and plastic valves, each valve must meet the minimum requirements, or equivalent, of API 6D. A valve may not be used under operating conditions that exceed the applicable pressure-temperature ratings contained in those requirements.

(b) Each cast iron and plastic valve must comply with the following:

(1) The valve must have a maximum service pressure rating for temperatures that equal or exceed the maximum service temperature.

(2) The valve must be tested as part of the manufacturing, as follows:

(i) With the valve in the fully open position, the shell must be tested with no leakage to a pressure at least 1.5 times the maximum service rating.

(ii) After the shell test, the seat must be tested to a pressure no less than 1.5 times the maximum service pressure rating. Except for swing check valves, test pressure during the seat test must be applied successively on each side of the closed valve with the opposite side open. No visible leakage is permitted.

(iii) After the last pressure test is completed, the valve must be operated through its full travel to demonstrate freedom from interference.

(c) Each valve must be able to meet the anticipated operating conditions.

(d) No valve having shell components made of ductile iron may be used at pressures exceeding 80 percent of the pressure ratings for comparable steel valves at their listed temperature. However, a valve having shell components made of ductile iron may be used at pressures up to 80 percent of the pressure ratings for comparable steel valves at their listed temperature, if:

(1) The temperature-adjusted service pressure does not exceed 1,000 psi (7 MPa) gage; and

(2) Welding is not used on any ductile iron component in the fabrication of the valve shells or their assembly.

(e) No valve having pressure containing parts made of ductile iron may be used in the gas pipe components of compressor stations.

[35 FR 13257, Aug. 19,1970, as amended by Amdt. 192-3, 35 FR 17660, Nov. 17, 1970; Amdt. 192-22, 41 FR 13590, Mar. 31, 1976; Amdt. 192-37, 46 FR 10159, Feb. 2, 1981; Amdt. 192-62, 54 FR 5625, Feb. 6, 1989; Amdt. 192-85, 63 FR 37500, July 13, 1998]

§192.147 Flanges and flange accessories.

(a) Each flange or flange accessory (other than cast iron) must meet the minimum requirements of ASME/ANSI B16.5, MSS SP-44, or the equivalent.

(b) Each flange assembly must be able to withstand the maximum pressure at which the pipeline is to be operated and to maintain its physical and chemical properties at any temperature to which it is anticipated that it might be subjected in service.

(c) Each flange on a flanged joint in cast iron pipe must conform in dimensions, drilling, face and gasket design to ASME/ANSI B16.1 and be cast integrally with the pipe, valve, or fitting.

[35 FR 13257, Aug. 19, 1970, as amended by Amdt. 192-62, 54 FR 5625, Feb. 6, 1989; Amdt. 192-68, 54 FR 14519, Mar. 18, 1993]

§192.149 Standard fittings.

(a) The minimum metal thickness of threaded fittings may not be less than specified for the pressures and temperatures in the applicable standards referenced in this part, or their equivalent.

(b) Each steel butt-welding fitting must have pressure and temperature ratings based on stresses for pipe of the same or equivalent material. The actual bursting strength of the fitting must at least equal the computed bursting strength of pipe of the designated material and wall thickness, as determined by a prototype that was tested to at least the pressure required for the pipeline to which it is being added.

§192.150 Passage of internal inspection devices.

(a) Except as provided in paragraphs (b) and (c) of this section, each new transmission line and each line section of a transmission line where the line pipe, valve, fitting, or other line component is replaced must be designed and constructed to accommodate the passage of instrumented internal inspection devices.

(b) This section does not apply to:

(1) Manifolds;

(2) Station piping such as at compressor stations, meter stations, or regulator stations;

(3) Piping associated with storage facilities, other than a continuous run of transmission line between a compressor station and storage facilities;

(4) Cross-overs;

(5) Sizes of pipe for which an instrumented internal inspection device is not commercially available;

(6) Transmission lines, operated in conjunction with a distribution system which are installed in Class 4 locations;

(7) Offshore pipelines, other than transmission lines 10 inches (254 millimeters) or greater in nominal diameter, that transport gas to onshore facilities; and,

(8) Other piping that, under §190.9 of this chapter, the Administrator finds in a particular case would be impracticable to design and construct to accommodate the passage of instrumented internal inspection devices.

(c) An operator encountering emergencies, construction time constraints or other unforeseen construction problems need not construct a new or replacement segment of a transmission line to meet paragraph (a) of this section, if the operator determines and documents why an impracticability prohibits compliance with paragraph (a) of this section. Within 30 days after discovering the emergency or construction problem the operator must petition, under §190.9 of this chapter, for approval that design and construction to accommodate passage of instrumented internal inspection devices would be impracticable. If the petition is denied, within 1 year after the date of the notice of the denial, the operator must modify that segment to allow passage of instrumented internal inspection devices.

[Amdt. 192-72, 59 FR 17275, Apr. 12, 1994; Amdt. 192-85, 63 FR 37500, July 13, 1998]

§192.151 Tapping.

(a) Each mechanical fitting used to make a hot tap must be designed for at least the operating pressure of the pipeline.

(b) Where a ductile iron pipe is tapped, the extent of full-thread engagement and the need for the use of outside-sealing service connections, tapping saddles, or other fixtures must be determined by service conditions.

(c) Where a threaded tap is made in cast iron or ductile iron pipe, the diameter of the tapped hole may not be more than 25 percent of the nominal diameter of the pipe unless the pipe is reinforced, except that

(1) Existing taps may be used for replacement service, if they are free of cracks and have good threads; and

(2) A 1¼-inch (32 millimeters) tap may be made in a 4-inch (102 millimeters) cast iron or ductile iron pipe, without reinforcement.

However, in areas where climate, soil, and service conditions may create unusual external stresses on cast iron pipe, unreinforced taps may be used only on 6-inch (152 millimeters) or larger pipe.

[35 FR 13257, Aug. 19, 1970, as amended by Amdt. 192-85, 63 FR 37500, July 13, 1998]

§192.153 Components fabricated by welding.

(a) Except for branch connections and assemblies of standard pipe and fittings joined by circumferential welds, the design pressure of each component fabricated by welding, whose strength cannot be determined, must be established in accordance with paragraph UG-101 of section VIII, Division 1, of the ASME Boiler and Pressure Vessel Code.

(b) Each prefabricated unit that uses plate and longitudinal seams must be designed, constructed, and tested in accordance with section VIII, Division 1, or section VIII, Division 2 of the ASME Boiler and Pressure Vessel Code, except for the following:

(1) Regularly manufactured butt-welding fittings.

(2) Pipe that has been produced and tested under a specification listed in Appendix B to this part.

(3) Partial assemblies such as split rings or collars.

(4) Prefabricated units that the manufacturer certifies have been tested to at least twice the maximum pressure to which they will be subjected under the anticipated operating conditions.

(c) Orange-peel bull plugs and orange-peel swages may not be used on pipelines that are to operate at a hoop stress of 20 percent or more of the SMYS of the pipe.

(d) Except for flat closures designed in accordance with section VIII of the ASME Boiler and Pressure Code, flat closures and fish tails may not be used on pipe that either operates at 100 psi (689 kPa) gage, or more, or is more than 3 inches (76 millimeters) nominal diameter.

192CODE.DOC, DT (Revised) 7/98

[35 FR 13257, Aug. 19, 1970, as amended by Amdt. 192-3, 35 FR 17660, Nov. 17, 1970; Amdt. 192-68, 58 FR 14519, Mar. 18, 1993; Amdt. 192-85, 63 FR 37500, July 13, 1998]

§192.155 Welded branch connections.

Each welded branch connection made to pipe in the form of a single connection, or in a header or manifold as a series of connections, must be designed to ensure that the strength of the pipeline system is not reduced, taking into account the stresses in the remaining pipe wall due to the opening in the pipe or header, the shear stresses produced by the pressure acting on the area of the branch opening, and any external loadings due to thermal movement, weight, and vibration.

§192.157 Extruded outlets.

Each extruded outlet must be suitable for anticipated service conditions and must be at least equal to the design strength of the pipe and other fittings in the pipeline to which it is attached.

§192.159 Flexibility.

Each pipeline must be designed with enough flexibility to prevent thermal expansion or contraction from causing excessive stresses in the pipe or components, excessive bending or unusual loads at joints, or undesirable forces or moments at points of connection to equipment, or at anchorage or guide points.

§192.161 Supports and anchors.

(a) Each pipeline and its associated equipment must have enough anchors or supports to:

(1) Prevent undue strain on connected equipment;

(2) Resist longitudinal forces caused by a bend or offset in the pipe; and,

(3) Prevent or damp out excessive vibration.

(b) Each exposed pipeline must have enough supports or anchors to protect the exposed pipe joints from the maximum end force caused by internal pressure and any additional forces caused by temperature expansion or contraction or by the weight of the pipe and its contents.

(c) Each support or anchor on an exposed pipeline must be made of durable, noncombustible material and must be designed and installed as follows:

(1) Free expansion and contraction of the pipeline between supports or anchors may not be restricted.

(2) Provision must be made for the service conditions involved.

(3) Movement of the pipeline may not cause disengagement of the support equipment.

(d) Each support on an exposed pipeline operated at a stress level of 50 percent or more of SMYS must comply with the following:

(1) A structural support may not be welded directly to the pipe.

(2) The support must be provided by a member that completely encircles the pipe.

(3) If an encircling member is welded to a pipe, the weld must be continuous and cover the entire circumference.

(e) Each underground pipeline that is connected to a relatively unyielding line or other fixed object must have enough flexibility to provide for possible movement, or it must have an anchor that will limit the movement of the pipeline.

(f) Except for offshore pipelines, each underground pipeline that is being connected to new branches must have a firm foundation for both the header and the branch to prevent detrimental lateral and vertical movement.

[35 FR 13257, Aug. 19, 1970, as amended by Amdt. 192-27, 41 FR 34598, Aug. 16, 1976; Amdt. 192-58, 53 FR 1633, Jan. 21, 1988]

§192.163 Compressor stations: Design and construction.

(a) Location of compressor building. Except for a compressor building on a platform located offshore or in inland navigable waters, each main compressor building of a compressor station must be located on property under the control of the operator. It must be far enough away from adjacent property, not under control of the operator, to minimize the possibility of fire being communicated to the compressor building from structures on adjacent property. There must be enough open space around the main compressor building to allow the free movement of fire-fighting equipment.

(b) Building construction. Each building on a compressor station site must be made of noncombustible materials if it contains either–

(1) Pipe more than 2 inches (51 millimeters) in diameter that is carrying gas under pressure; or

(2) Gas handling equipment other than gas utilization equipment used for domestic purposes.

(c) Exits. Each operating floor of a main compressor building must have at least two separated and unobstructed exits located so as to provide a convenient possibility of escape and an unobstructed passage to a place of safety. Each door latch on an exit must be of a type which can be readily opened from the inside without a key. Each swinging door located in an exterior wall must be mounted to swing outward.

(d) Fenced areas. Each fence around a compressor station must have at least two gates located so as to provide a convenient opportunity for escape to a place of safety, or have other facilities affording a similarly convenient exit from the area. Each gate located within 200 feet (61 meters) of any compressor plant building must open outward and, when occupied, must be openable from the inside without a key.

(e) Electrical facilities. Electrical equipment and wiring installed in compressor stations must conform to the National Electrical Code, ANSI/NFPA 70, so far as that code is applicable.

[35 FR 13257, Aug. 19, 1970, as amended by Amdt. 192-27, 41 FR 34598, Aug. 16, 1976; Amdt. 192-37, 46 FR 10157, Feb. 2, 1981; Amdt. 192-68, 58 FR 14519, Mar. 18, 1993; Amdt. 192-85, 63 FR 37500, July 13, 1998]

§192.165 Compressor stations: Liquid removal.

(a) Where entrained vapors in gas may liquefy under the anticipated pressure and temperature conditions, the compressor must be protected against the introduction of those liquids in quantities that could cause damage.

(b) Each liquid separator used to remove entrained liquids at a compressor station must:

(1) Have a manually operable means of removing these liquids.

(2) Where slugs of liquid could be carried into the compressors, have either automatic liquid removal facilities, an automatic compressor shutdown device, or a high liquid level alarm; and,

(3) Be manufactured in accordance with section VIII of the ASME Boiler and Pressure Vessel Code, except that liquid separators constructed of pipe and fittings without internal welding must be fabricated with a design factor of 0.4, or less.

§192.167 Compressor stations: Emergency shutdown.

(a) Except for unattended field compressor stations of 1,000 horsepower (746 kilowatts) or less, each compressor station must have an emergency shutdown system that meets the following:

(1) It must be able to block gas out of the station and blow down the station piping.

(2) It must discharge gas from the blowdown piping at a location where the gas will not create a hazard.

(3) It must provide means for the shutdown of gas compressing equipment, gas fires, and electrical facilities in the vicinity of gas headers and in the compressor building, except, that:

(i) Electrical circuits that supply emergency lighting required to assist station personnel in evacuating the compressor building and the area in the vicinity of the gas headers must remain energized; and

(ii) Electrical circuits needed to protect equipment from damage may remain energized.

(4) It must be operable from at least two locations, each of which is:

(i) Outside the gas area of the station;

(ii) Near the exit gates, if the station is fenced, or near emergency exits, if not fenced; and,

(iii) Not more than 500 feet (153 meters) from the limits of the station.

(b) If a compressor station supplies gas directly to a distribution system with no other adequate source of gas available, the emergency shutdown system must be designed so that it will not function at the wrong time and cause an unintended outage on the distribution system.

(c) On a platform located offshore or in inland navigable waters, the emergency shutdown system must be designed and installed to actuate automatically by each of the following events:

(1) In the case of an unattended compressor station:

(i) When the gas pressure equals the maximum allowable operating pressure plus 15 percent or

(ii) When an uncontrolled fire occurs on the platform; and

(2) In the case of a compressor station in a building:

(i) When an uncontrolled fire occurs in the building; or

(ii) When the concentration of gas in air reaches 50 percent or more of the lower explosive limit in a building which has a source of ignition.

For the purpose of paragraph (c)(2)(ii) of this section, an electrical facility which conforms to Class 1, Group D of the National Electrical Code is not a source of ignition.

[35 FR 13257, Aug. 19, 1970, as amended by Amdt. 192-27, 41 FR 34605, Aug. 16, 1976; Amdt. 192-85, 63 FR 37500, July 13, 1998]

§192.169 Compressor stations: Pressure limiting devices.

(a) Each compressor station must have pressure relief or other suitable protective devices of sufficient capacity and sensitivity to ensure that the maximum allowable operating pressure of the station piping and equipment is not exceeded by more than 10 percent.

(b) Each vent line that exhausts gas from the pressure relief valves of a compressor station must extend to a location where the gas may be discharged without hazard.

§192.171 Compressor stations: Additional safety equipment.

(a) Each compressor station must have adequate fire protection facilities. If fire pumps are a part of these facilities, their operation may not be affected by the emergency shutdown system.

(b) Each compressor station prime mover, other than an electrical induction or synchronous motor, must have an automatic device to shut down the unit before the speed of either the prime mover or the driven unit exceeds a maximum safe speed.

(c) Each compressor unit in a compressor station must have a shutdown or alarm device that operates in the event of inadequate cooling or lubrication of the unit.

(d) Each compressor station gas engine that operates with pressure gas injection must be equipped so that stoppage of the engine automatically shuts off the fuel and vents the engine distribution manifold.

(e) Each muffler for a gas engine in a compressor station must have vent slots or holes in the baffles of each compartment to prevent gas from being trapped in the muffler.

§192.173 Compressor stations: Ventilation.

Each compressor station building must be ventilated to ensure that employees are not endangered by the accumulation of gas in rooms, sumps, attics, pits, or other enclosed places.

§192.175 Pipe-type and bottle-type holders.

(a) Each pipe-type and bottle-type holder must be designed so as to prevent the accumulation of liquids in the holder, in connecting pipe, or in auxiliary equipment, that might cause corrosion or interfere with the safe operation of the holder.

(b) Each pipe-type or bottle-type holder must have minimum clearance from other holders in accordance with the following formula:

C = (D x P x F)/48.33) (C = (3D x P x F)/1,000))

in which:

C = Minimum clearance between pipe containers or bottles in inches (millimeters).

D = Outside diameter of pipe containers or bottles in inches (millimeters).

P = Maximum allowable operating pressure, psi (kPa) gage.

F = Design factor as set forth in §192.111 of this part.

[35 FR 13257, Aug. 19, 1970, as amended by Amdt. 192-85, 63 FR 37500, July 13, 1998]

§192.177 Additional provisions for bottle-type holders.

(a) Each bottle-type holder must be–

(1) Located on a site entirely surrounded by fencing that prevents access by unauthorized persons and with minimum clearance from the fence as follows:

Maximum allowable operating pressure	Minimum clearance feet (meters)
Less than 1,000 psi (7 MPa) gage	25 (7.6)
1,000 psi (7 MPa) gage or more	100 (31)

(2) Designed using the design factors set forth in §192.111; and,

(3) Buried with a minimum cover in accordance with §192.327.

(b) Each bottle-type holder manufactured from steel that is not weldable under field conditions must comply with the following:

(1) A bottle-type holder made from alloy steel must meet the chemical and tensile requirements for the various grades of steel in ASTM A 372/A 372M.

(2) The actual yield-tensile ratio of the steel may not exceed 0.85.

(3) Welding may not be performed on the holder after it has been heat treated or stress relieved, except that copper wires may be attached to the small diameter portion of the

bottle end closure for cathodic protection if a localized thermit welding process is used.

(4) The holder must be given a mill hydrostatic test at a pressure that produces a hoop stress at least equal to 85 percent of the SMYS.

(5) The holder, connection pipe, and components must be leak tested after installation as required by Subpart J of this part.

[35 FR 13257, Aug. 19, 1970 as amended by Amdt 192-58, 53 FR 1635, Jan 21, 1988; Amdt 192-62, 54 FR 5625, Feb. 6, 1989; Amdt 192-68, 58 FR 14519, Mar. 18, 1993; Amdt. 192-85, 63 FR 37500, July 13, 1998]

§192.179 Transmission line valves.

(a) Each transmission line, other than offshore segments, must have sectionalizing block valves spaced as follows, unless in a particular case the Administrator finds that alternative spacing would provide an equivalent level of safety:

(1) Each point on the pipeline in a Class 4 location must be within 2½ miles (4 kilometers) of a valve.

(2) Each point on the pipeline in a Class 3 location must be within 4 miles (6.4 kilometers) of a valve.

(3) Each point on the pipeline in a Class 2 location must be within 7½ miles (12 kilometers) of a valve.

(4) Each point on the pipeline in a Class 1 location must be within 10 miles (16 kilometers) of a valve.

(b) Each sectionalizing block valve on a transmission line, other than offshore segments, must comply with the following:

(1) The valve and the operating device to open or close the valve must be readily accessible and protected from tampering and damage.

(2) The valve must be supported to prevent settling of the valve or movement of the pipe to which it is attached.

(c) Each section of a transmission line, other than offshore segments, between main line valves must have a blowdown valve with enough capacity to allow the transmission line to be blown down as rapidly as practicable. Each blowdown discharge must be located so the gas can be blown to the atmosphere without hazard and, if the transmission line is adjacent to an overhead electric line, so that the gas is directed away from the electrical conductors.

(d) Offshore segments of transmission lines must be equipped with valves or other components to shut off the flow of gas to an offshore platform in an emergency.

[35 FR 13257, Aug. 19, 1970, as amended by Amdt. 192-27, 41 FR 34598, Aug. 16, 1976; Amdt. 192-78, 61 FR 28770, June 6, 1996; Amdt. 192-85, 63 FR 37500, July 13, 1998]

§192.181 Distribution line valves.

(a) Each high-pressure distribution system must have valves spaced so as to reduce the time to shut down a section of main in an emergency. The valve spacing is determined by the operating pressure, the size of the mains, and the local physical conditions.

(b) Each regulator station controlling the flow or pressure of gas in a distribution system must have a valve installed on the inlet piping at a distance from the regulator station sufficient to permit the operation of the valve during an emergency that might preclude access to the station.

(c) Each valve on a main installed for operating or emergency purposes must comply with the following:

(1) The valve must be placed in a readily accessible location so as to facilitate its operation in an emergency.

(2) The operating stem or mechanism must be readily accessible.

(3) If the valve is installed in a buried box or enclosure, the box or enclosure must be installed so as to avoid transmitting external loads to the main.

§192.183 Vaults: Structural design requirements.

(a) Each underground vault or pit for valves, pressure relieving, pressure limiting, or pressure regulating stations, must be able to meet the loads which may be imposed upon it, and to protect installed equipment.

(b) There must be enough working space so that all of the equipment required in the vault or pit can be properly installed, operated, and maintained.

(c) Each pipe entering, or within, a regulator vault or pit must be steel for sizes 10 inches (254 millimeters), and less, except that control and gage piping may be copper. Where pipe extends through the vault or pit structure, provision must be made to prevent the passage of gases or liquids through the opening and to avert strains in the pipe.

[35 FR 13257, Aug. 19, 1970, as amended by Amdt. 192-85, 63 FR 37500, July 13, 1998]

§192.185 Vaults: Accessibility.

Each vault must be located in an accessible location and, so far as practical, away from:

(a) Street intersections or points where traffic is heavy or dense;

(b) Points of minimum elevation, catch basins, or places where the access cover will be in the course of surface waters; and,

(c) Water, electric, steam, or other facilities.

§192.187 Vaults: Sealing, venting, and ventilation.

Each underground vault or closed top pit containing either a pressure regulating or reducing station, or a pressure limiting or relieving station, must be sealed, vented or ventilated, as follows:

(a) When the internal volume exceeds 200 cubic feet (5.7 cubic meters):

(1) The vault or pit must be ventilated with two ducts, each having at least the ventilating effect of a pipe 4 inches (102 millimeters) in diameter;

(2) The ventilation must be enough to minimize the formation of combustible atmosphere in the vault or pit; and,

(3) The ducts must be high enough above grade to disperse any gas-air mixtures that might be discharged.

(b) When the internal volume is more than 75 cubic feet (2.1 cubic meters) but less than 200 cubic feet (5.7 cubic meters):

(1) If the vault or pit is sealed, each opening must have a tight fitting cover without open holes through which an explosive mixture might be ignited, and there must be a means for testing the internal atmosphere before removing the cover;

(2) If the vault or pit is vented, there must be a means of preventing external sources of ignition from reaching the vault atmosphere; or

(3) If the vault or pit is ventilated, paragraph (a) or (c) of this section applies.

(c) If a vault or pit covered by paragraph (b) of this section is ventilated by openings in the covers or gratings and the ratio of the internal volume, in cubic feet, to the effective ventilating area of the cover or grating, in square feet, is less than 20 to 1, no additional ventilation is required.

[35 FR 13257, Aug. 19, 1970, as amended by Amdt. 192-85, 63 FR 37500, July 13, 1998]

§192.189 Vaults: Drainage and waterproofing.

(a) Each vault must be designed so as to minimize the entrance of water.

(b) A vault containing gas piping may not be connected by means of a drain connection to any other underground structure.

(c) Electrical equipment in vaults must conform to the applicable requirements of Class 1, Group D, of the National Electrical Code, ANSI/NFPA 70.

[Amdt. 192-76, 61 FR 26121, May 24, 1996]

§192.191 Design pressure of plastic fittings.

(a) Thermosetting fittings for plastic pipe must conform to ASTM D 2517.

(b) Thermoplastic fittings for plastic pipe must conform to ASTM D 2513.

[35 FR 13257, Aug. 19, 1970, as amended by Amdt. 192-3, 35 FR 17660, Nov. 17, 1970; Amdt. 192-58, 53 FR 1633, Jan. 21, 1988]

§192.193 Valve installation in plastic pipe.

Each valve installed in plastic pipe must be designed so as to protect the plastic material against excessive torsional or shearing loads when the valve or shutoff is operated, and from any other secondary stresses that might be exerted through the valve or its enclosure.

§192.195 Protection against accidental overpressuring.

(a) General requirements. Except as provided in §192.197, each pipeline that is connected to a gas source so that the maximum allowable operating pressure could be exceeded as the result of pressure control failure or of some other type of failure, must have pressure relieving or pressure limiting devices that meet the requirements of §192.199 and §192.201.

(b) Additional requirements for distribution systems. Each distribution system that is supplied from a source of gas that is at a higher pressure than the maximum allowable operating pressure for the system must

(1) Have pressure regulation devices capable of meeting the pressure, load, and other service conditions that will be experienced in normal operation of the system, and that could be activated in the event of failure of some portion of the system; and

(2) Be designed so as to prevent accidental overpressuring.

§192.197 Control of the pressure of gas delivered from high-pressure distribution systems.

(a) If the maximum actual operating pressure of the distribution system is under 60 p.s.i. (414 kPa) gage and a service regulator having the following characteristics is used, no other pressure limiting device is required:

(1) A regulator capable of reducing distribution line pressure to pressures recommended for household appliances.

(2) A single port valve with proper orifice for the maximum gas pressure at the regulator inlet.

(3) A valve seat made of resilient material designed to withstand abrasion of the gas, impurities in gas, cutting by the valve, and to resist permanent deformation when it is pressed against the valve port.

(4) Pipe connections to the regulator not exceeding 2 inches (51 millimeters) in diameter.

(5) A regulator that, under normal operating conditions, is able to regulate the downstream pressure within the necessary limits of accuracy and to limit the build-up of pressure under no-flow conditions to prevent a pressure that would cause the unsafe operation of any connected and properly adjusted gas utilization equipment.

(6) A self-contained service regulator with no external static or control lines.

(b) If the maximum actual operating pressure of the distribution system is 60 p.s.i. (414 kPa) gage or less, and a service regulator that does not have all of the characteristics listed in paragraph (a) of this section is used, or if the gas contains materials that seriously interfere with the operation of service regulators, there must be suitable protective devices to prevent unsafe overpressuring of the customer's appliances if the service regulator fails.

(c) If the maximum actual operating pressure of the distribution system exceeds 60 p.s.i. (414 kPa) gage, one of the following methods must be used to regulate and limit, to the maximum safe value, the pressure of gas delivered to the customer:

(1) A service regulator having the characteristics listed in paragraph (a) of this section, and another regulator located upstream from the service regulator. The upstream regulator may not be set to maintain a pressure higher than 60 p.s.i. (414 kPa) gage. A device must be installed between the upstream regulator and the service regulator to limit the pressure on the inlet of the service regulator to 60 p.s.i. (414 kPa) gage or less in case the upstream regulator fails to function properly. This device may be either a relief valve or an automatic shutoff that shuts, if the pressure on the inlet of the service regulator exceeds the set pressure (60 p.s.i. (414 kPa) gage or less), and remains closed until manually reset.

(2) A service regulator and a monitoring regulator set to limit, to a maximum safe value, the pressure of the gas delivered to the customer.

(3) A service regulator with a relief valve vented to the outside atmosphere, with the relief valve set to open so that the pressure of gas going to the customer does not exceed a maximum safe value. The relief valve may either be built into the service regulator or it may be a separate unit installed downstream from the service regulator. This combination may be used alone only in those cases where the inlet pressure on the service regulator does not exceed the manufacturer's safe working pressure rating of the service regulator, and may not be used where the inlet pressure on

the service regulator exceeds 125 p.s.i. (862 kPa) gage. For higher inlet pressure, the methods in paragraph (c)(1) or (2) of this section must be used.

(4) A service regulator and an automatic shutoff device that closes upon a rise in pressure downstream from the regulator and remains closed until manually reset.

[35 FR 13257, Aug. 19, 1970, as amended by Amdt. 192-3, 35 FR 17660, Nov. 7, 1970; Amdt. 192-85, 63 FR 37500, July 13, 1998]

§192.199 Requirements for design of pressure relief and limiting devices.

Except for rupture discs, each pressure relief or pressure limiting device must:

(a) Be constructed of materials such that the operation of a device will not be impaired by corrosion;

(b) Have valves and valve seats that are designed not to stick in a position that will make the device inoperative;

(c) Be designed and installed so that it can be readily operated to determine if the valve is free, can be tested to determine the pressure at which it will operate, and can be tested for leakage when in the closed position;

(d) Have support made of noncombustible material;

(e) Have discharge stacks, vents, or outlet ports designed to prevent accumulation of water, ice, or snow, located where gas can be discharged into the atmosphere without undue hazard;

(f) Be designed and installed so that the size of the openings, pipe, and fittings located between the system to be protected and the pressure relieving device, and the size of the vent line, are adequate to prevent hammering of the valve and to prevent impairment of relief capacity;

(g) Where installed at a district regulator station to protect a pipeline system from overpressuring, be designed and installed to prevent any single incident such as an explosion in a vault or damage by a vehicle from affecting the operation of both the overpressure protective device and the district regulator; and,

(h) Except for a valve that will isolate the system under protection from its source of pressure, be designed to prevent unauthorized operation of any stop valve that will make the pressure relief valve or pressure limiting device inoperative.

[35 FR 13257, Aug. 19, 1970, as amended by Amdt. 192-3, 35 FR 17660, Nov. 17, 1970]

§192.201 Required capacity of pressure relieving and limiting stations.

(a) Each pressure relief station or pressure limiting station or group of those stations installed to protect a pipeline must have enough capacity, and must be set to operate, to insure the following:

(1) In a low pressure distribution system, the pressure may not cause the unsafe operation of any connected and properly adjusted gas utilization equipment.

(2) In pipelines other than a low pressure distribution system:

(i) If the maximum allowable operating pressure is 60 p.s.i. (414 kPa) gage or more, the pressure may not exceed the maximum allowable operating pressure plus 10 percent or the pressure that produces a hoop stress of 75 percent of SMYS, whichever is lower;

(ii) If the maximum allowable operating pressure is 12 p.s.i. (83 kPa) gage or more, but less than 60 p.s.i. (414 kPa) gage, the pressure may not exceed the maximum allowable operating pressure plus 6 p.s.i. (41 kPa) gage; or

(iii) If the maximum allowable operating pressure is less than 12 p.s.i. (83 kPa) gage, the pressure may not exceed the maximum allowable operating pressure plus 50 percent.

(b) When more than one pressure regulating or compressor station feeds into a pipeline, relief valves or other protective devices must be installed at each station to ensure that the complete failure of the largest capacity regulator or compressor, or any single run of lesser capacity regulators or compressors in that station, will not impose pressures on any part of the pipeline or distribution system in excess of those for which it was designed, or against which it was protected, whichever is lower.

(c) Relief valves or other pressure limiting devices must be installed at or near each regulator station in a low-pressure distribution system, with a capacity to limit the maximum pressure in the main to a pressure that will not exceed the safe operating pressure for any connected and properly adjusted gas utilization equipment.

[35 FR 13257, Aug. 19, 1970, as amended by Amdt. 192-9, 37 FR 20826, Oct. 4, 1972; Amdt. 192-85, 63 FR 37500, July 13, 1998]

§192.203 Instrument, control, and sampling pipe and components.

(a) Applicability. This section applies to the design of instrument, control, and sampling pipe and components. It does not apply to permanently closed systems, such as fluid-filled temperature-responsive devices.

(b) Materials and design. All materials employed for pipe and components must be designed to meet the particular conditions of service and the following:

(1) Each takeoff connection and attaching boss, fitting, or adapter must be made of suitable material, be able to withstand the maximum service pressure and temperature of the pipe or equipment to which it is attached, and be designed to satisfactorily withstand all stresses without failure by fatigue.

(2) Except for takeoff lines that can be isolated from sources of pressure by other valving, a shutoff valve must be installed in each takeoff line as near as practicable to the point of takeoff. Blowdown valves must be installed where necessary.

(3) Brass or copper material may not be used for metal temperatures greater than 400°F (204°C).

(4) Pipe or components that may contain liquids must be protected by heating or other means from damage due to freezing.

(5) Pipe or components in which liquids may accumulate must have drains or drips.

(6) Pipe or components subject to clogging from solids or deposits must have suitable connections for cleaning.

(7) The arrangement of pipe, components, and supports must provide safety under anticipated operating stresses.

(8) Each joint between sections of pipe, and between pipe and valves or fittings, must be made in a manner suitable for the anticipated pressure and temperature condition. Slip type expansion joints may not be used. Expansion must be allowed for by providing flexibility within the system itself.

(9) Each control line must be protected from anticipated causes of damage and must

be designed and installed to prevent damage to any one control line from making both the regulator and the over-pressure protective device inoperative.

[35 FR 13257, Aug. 19, 1970, as amended by Amdt. 192-78, 61 FR 28770, June 6, 1996; Amdt. 192-85, 63 FR 37500, July 13, 1998]

Subpart E–Welding of Steel in Pipelines

§192.221 Scope.

(a) This subpart prescribes minimum requirements for welding steel materials in pipelines.

(b) This subpart does not apply to welding that occurs during the manufacture of steel pipe or steel pipeline components.

§192.223 [Removed]

[35 FR 13257, Aug. 19, 1970, as amended by Amdt. 192-52, 51 FR 20294, June 4, 1986]

§192.225 Welding - General.

(a) Welding must be performed by a qualified welder in accordance with welding procedures qualified to produce welds meeting the requirements of this subpart. The quality of the test welds used to qualify the procedures shall be determined by destructive testing.

(b) Each welding procedure must be recorded in detail, including the results of the qualifying tests. This record must be retained and followed whenever the procedure is used.

[35 FR 13257, Aug. 19, 1970, as amended by Amdt. 192-18, 40 FR 10181, Mar. 5, 1975; Amdt. 192-22, 41 FR 13590, Mar. 31, 1976; Amdt. 192-37, 46 FR 10157, Feb. 2, 1981; Amdt. 192-52, 51 FR 20297, June 4, 1986]

§192.227 Qualification of welders.

(a) Except as provided in paragraph (b) of this section, each welder must be qualified in accordance with section 3 of API Standard 1104 or section IX of the ASME Boiler and Pressure Vessel Code. However, a welder qualified under an earlier edition than listed in Appendix A may weld but may not requalify under that earlier edition.

(b) A welder may qualify to perform welding on pipe to be operated at a pressure that produces a hoop stress of less than 20 percent of SMYS by performing an acceptable test weld, for the process to be used, under the test set forth in section I of Appendix C of this part. Each welder who is to make a welded service line connection to a main must also first perform an acceptable test weld under section II of Appendix C of this part as a requirement of the qualifying test.

[35 FR 13257, Aug. 19, 1970, as amended by Amdt. 192-18, 40 FR 10181, Mar. 5, 1975; Amdt. 192-18A, 40 FR 27222, June 27, 1975; Amdt. 192-22, 41 FR 13590, Mar. 31, 1976; Amdt. 192-37, 46 FR 10157, Feb. 2, 1981; Amdt. 192-43, 47 FR 46850, Oct. 21, 1982; Amdt. 192-52, 51 FR 20294, June 4, 1986; Amdt. 192-75, 61 FR 18512, Apr. 26, 1996; Amdt. 192-78, 61 FR 28770, June 6, 1996]

§192.229 Limitations on welders.

(a) No welder whose qualification is based on nondestructive testing may weld compressor station pipe and components.

(b) No welder may weld with a particular welding process unless, within the preceding 6 calendar months, he has engaged in welding with that process.

(c) A welder qualified under §192.227(a)–

(1) May not weld on pipe to be operated at a pressure that produces a hoop stress of 20 percent or more of SMYS unless within the preceding 6 calendar months the welder has had one weld tested and found acceptable under section 3 or 6 of API Standard 1104, except that a welder qualified under an earlier edition previously listed in Appendix A of this part may weld but may not requalify under that earlier edition; and

(2) May not weld on pipe to be operated at a pressure that produces a hoop stress of less than 20 percent of SMYS unless the welder is tested in accordance with paragraph (c)(1) of this section or requalifies under paragraph (d)(1) or (d)(2) of this section.

(d) A welder qualified under §192.227(b) may not weld unless–

(1) Within the preceding 15 calendar months, but at least once each calendar year, the welder has requalified under §192.227(b); or

(2) Within the preceding 7? calendar months, but at least twice each calendar year, the welder has had–

(i) A production weld cut out, tested, and found acceptable in accordance with the qualifying test; or

(ii) For welders who work only on service lines 2 inches (51 millimeters) or smaller in diameter, two sample welds tested and found acceptable in accordance with the test in section III of Appendix C of this part.

[35 FR 13257, Aug. 19, 1970, as amended by Amdt. 192-18, 40 FR 10181, Mar. 5, 1975; Amdt. 192-18A, 40 FR 27222, June 27, 1975; Amdt. 192-37, 46 FR 10157, Feb. 2, 1981; Amdt. 192-78, 61 FR 28770, June 6, 1996; Amdt. 192-85, 63 FR 37500, July 13, 1998]

§192.231 Protection from weather.

The welding operation must be protected from weather conditions that would impair the quality of the completed weld.

§192.233 Miter joints.

(a) A miter joint on steel pipe to be operated at a pressure that produces a hoop stress of 30 percent or more of SMYS may not deflect the pipe more than 3°.

(b) A miter joint on steel pipe to be operated at a pressure that produces a hoop stress of less than 30 percent, but more than 10 percent of SMYS may not deflect the pipe more than 12½° and must be a distance equal to one pipe diameter or more away from any other miter joint, as measured from the crotch of each joint.

(c) A miter joint on steel pipe to be operated at a pressure that produces a hoop stress of 10 percent or less of SMYS may not deflect the pipe more than 90°.

§192.235 Preparation for welding.

Before beginning any welding, the welding surfaces must be clean and free of any material that may be detrimental to the weld, and the pipe or component must be aligned to provide the most favorable condition for depositing the root bead. This alignment must be preserved while the root bead is being deposited.

§192.237 [Removed]

[35 FR 13257, Aug. 19, 1970, as amended by Amdt. 192-37, 46 FR 10157, Feb. 2, 1981; Amdt. 192-52, 51 FR 20294, June 4, 1986]

§192.239 [Removed]

[35 FR 13257, Aug. 19, 1970, as amended by Amdt. 192-37, 46 FR 10157, Feb. 2, 1981; Amdt. 192-52, 51 FR 20294, June 4, 1986]

§192.241 Inspection and test of welds.

(a) Visual inspection of welding must be conducted to insure that:

(1) The welding is performed in accordance with the welding procedure; and

(2) The weld is acceptable under paragraph (c) of this section.

(b) The welds on a pipeline to be operated at a pressure that produces a hoop stress of 20 percent or more of SMYS must be nondestructively tested in accordance with §192.243, except that welds that are visually inspected and approved by a qualified welding inspector need not be nondestructively tested if:

(1) The pipe has a nominal diameter of less than 6 inches (152 millimeters); or

(2) The pipeline is to be operated at a pressure that produces a hoop stress of less than 40 percent of SMYS and the welds are so limited in number that nondestructive testing is impractical.

(c) The acceptability of a weld that is nondestructively tested or visually inspected is determined according to the standards in section 6 of API Standard 1104. However, if a girth weld is unacceptable under those standards for a reason other than a crack, and if the Appendix to API Standard 1104 applies to the weld, the acceptability of the weld may be further determined under that Appendix.

[35 FR 13257, Aug. 19, 1970, as amended by Amdt. 192-18, 40 FR 10181, Mar. 5, 1975; Amdt. 192-18A, 40 FR 27222, June 27, 1975; Amdt. 192-37, 46 FR 10157, Feb. 2, 1981; Amdt. 192-78, 61 FR 28770, June 6, 1996; Amdt. 192-85, 63 FR 37500, July 13, 1998]

§192.243 Nondestructive testing.

(a) Nondestructive testing of welds must be performed by any process, other than trepanning, that will clearly indicate defects that may affect the integrity of the weld.

(b) Nondestructive testing of welds must be performed:

(1) In accordance with written procedures; and

(2) By persons who have been trained and qualified in the established procedures and with the equipment employed in testing.

(c) Procedures must be established for the proper interpretation of each nondestructive test of a weld to ensure the acceptability of the weld under §192.241(c).

(d) When nondestructive testing is required under §192.241(b), the following percentages of each day's field butt welds, selected at random by the operator, must be nondestructively tested over their entire circumference;

(1) In Class 1 locations, except offshore, at least 10 percent.

(2) In Class 2 locations, at least 15 percent.

(3) In Class 3 and Class 4 locations, at crossings of major or navigable rivers, offshore, and within railroad or public highway rights-of-way, including tunnels, bridges, and overhead road crossings, 100 percent unless impracticable, in which case at least 90 percent. Nondestructive testing must be impracticable for each girth weld not tested.

(4) At pipeline tie-ins, including tie-ins of replacement sections, 100 percent.

(e) Except for a welder whose work is isolated from the principal welding activity, a sample of each welder's work for each day must be nondestructively tested, when nondestructive testing is required under §192.241(b).

(f) When nondestructive testing is required under §192.241(b), each operator must retain, for the life of the pipeline, a record showing by milepost, engineering station, or by geographic feature, the number of girth welds made, the number nondestructively tested, the number rejected, and the disposition of the rejects.

[35 FR 13257, Aug. 19, 1970, as amended by Amdt. 192-27, 41 FR 34598, Aug. 16, 1976; Amdt. 192-50, 50 FR 37191, Sept. 12, 1985; Amdt. 192-78, 61 FR 28770, June 6, 1996]

§192.245 Repair or removal of defects.

(a) Each weld that is unacceptable under §192.241(c) must be removed or repaired. Except for welds on an offshore pipeline being installed from a pipeline vessel, a weld must be removed if it has a crack that is more than 8 percent of the weld length.

(b) Each weld that is repaired must have the defect removed down to sound metal and the segment to be repaired must be preheated if conditions exist which would adversely affect the quality of the weld repair. After repair, the segment of the weld that was repaired must be inspected to ensure its acceptability.

(c) Repair of a crack, or of any defect in a previously repaired area must be in accordance with written weld repair procedures that have been qualified under §192.225. Repair procedures must provide that the minimum mechanical properties specified for the welding procedure used to make the original weld are met upon completion of the final weld repair.

[35 FR 13257, Aug. 19, 1970, as amended by Amdt. 192-27, 41 FR 34598, Aug. 16, 1976; Amdt. 192-46, 48 FR 48669, Oct. 20, 1983]

Subpart F–Joining of Materials Other Than by Welding

§192.271 Scope.

(a) This subpart prescribes minimum requirements for joining materials in pipelines, other than by welding.

(b) This subpart does not apply to joining during the manufacture of pipe or pipeline components.

§192.273 General.

(a) The pipeline must be designed and installed so that each joint will sustain the longitudinal pullout or thrust forces caused by contraction or expansion of the piping or by anticipated external or internal loading.

(b) Each joint must be made in accordance with written procedures that have been proved by test or experience to produce strong gastight joints.

(c) Each joint must be inspected to insure compliance with this subpart.

§192.275 Cast iron pipe.

(a) Each caulked bell and spigot joint in cast iron pipe must be sealed with mechanical leak clamps.

(b) Each mechanical joint in cast iron pipe must have a gasket made of a resilient material as the sealing medium. Each gasket must be suitably confined and retained under compression by a separate gland or follower ring.

(c) Cast iron pipe may not be joined by threaded joints.

(d) Cast iron pipe may not be joined by brazing.

[35 FR 13257, Aug. 19, 1970, as amended by Amdt. 192-62, 54 FR 5628, Feb. 6, 1989]

§192.277 Ductile iron pipe.

(a) Ductile iron pipe may not be joined by threaded joints.

(b) Ductile iron pipe may not be joined by brazing.

[35 FR 13257, Aug. 19, 1970, as amended by Amdt. 192-62, 54 FR 5628, Feb. 6, 1989, effective Mar. 8, 1989]

§192.279 Copper pipe.

Copper pipe may not be threaded except that copper pipe used for joining screw fittings or valves may be threaded if the wall thickness is equivalent to the comparable size of Schedule 40 or heavier wall pipe listed in Table C1 of ASME/ANSI B16.5.

[Amdt. 192-62 , 54 FR 5628, Feb. 6, 1989; Amdt. 192-68, 58 FR 14519, Mar. 18, 1993]

§192.281 Plastic pipe.

(a) General. A plastic pipe joint that is joined by solvent cement, adhesive, or heat fusion may not be disturbed until it has properly set. Plastic pipe may not be joined by a threaded joint or miter joint.

(b) Solvent cement joints. Each solvent cement joint on plastic pipe must comply with the following:

(1) The mating surfaces of the joint must be clean, dry, and free of material which might be detrimental to the joint.

(2) The solvent cement must conform to ASTM Designation: D 2513.

(3) The joint may not be heated to accelerate the setting of the cement.

(c) Heat-fusion joints. Each heat-fusion joint on plastic pipe must comply with the following:

(1) A butt heat-fusion joint must be joined by a device that holds the heater element square to the ends of the piping, compresses

the heated ends together, and holds the pipe in proper alignment while the plastic hardens.

(2) A socket heat-fusion joint must be joined by a device that heats the mating surfaces of the joint uniformly and simultaneously to essentially the same temperature.

(3) An electrofusion joint must be joined utilizing the equipment and techniques of the fittings manufacturer or equipment and techniques shown, by testing joints to the requirements of §192.283(a)(1)(iii), to be at least equivalent to those of the fittings manufacturer.

(4) Heat may not be applied with a torch or other open flame.

(d) Adhesive joints. Each adhesive joint on plastic pipe must comply with the following:

(1) The adhesive must conform to ASTM Designation: D 2517.

(2) The materials and adhesive must be compatible with each other.

(e) Mechanical joints. Each compression type mechanical joint on plastic pipe must comply with the following:

(1) The gasket material in the coupling must be compatible with the plastic.

(2) A rigid internal tubular stiffener, other than a split tubular stiffener, must be used in conjunction with the coupling.

[35 FR 13257, Aug. 19, 1970, as amended by Amdt. 192-34, 44 FR 42968, July 23, 1979; Amdt. 192-58, 53 FR 1635, Jan. 21, 1988; Amdt. 192-61, 53 FR 36793, Sept. 22, 1988; Amdt. 192-68, 58 FR 14519, Mar. 18, 1993; Amdt. 192-78, 61 FR 28770, June 6, 1996]

192CODE.DOC, DT (Revised) 7/98

§192.283 Plastic pipe; qualifying joining procedures.

(a) Heat fusion, solvent cement, and adhesive joints. Before any written procedure established under §192.273(b) is used for making plastic pipe joints by a heat fusion, solvent cement, or adhesive method, the procedure must be qualified by subjecting specimen joints made according to the procedure to the following tests:

(1) The burst test requirements of–

(i) In the case of thermoplastic pipe, paragraph 6.6 (Sustained Pressure Test) or paragraph 6.7 (Minimum Hydrostatic Burst Pressure (Quick Burst)) of ASTM D 2513;

(ii) In the case of thermosetting plastic pipe, paragraph 8.5 (Minimum Hydrostatic Burst Pressure) or paragraph 8.9 (Sustained Static Pressure Test) of ASTM D2517; or

(iii) In the case of electrofusion fittings for polyethylene pipe and tubing, paragraph 9.1 (Minimum Hydraulic Burst Pressure Test), paragraph 9.2 (Sustained Pressure Test), paragraph 9.3 (Tensile Strength Test), or paragraph 9.4 (Joint Integrity Tests) of ASTM Designation F1055.

(2) For procedures intended for lateral pipe connections, subject a specimen joint made from pipe sections joined at right angles according to the procedure to a force on the lateral pipe until failure occurs in the specimen. If failure initiates outside the joint area, the procedure qualifies for use; and,

(3) For procedures intended for nonlateral pipe connections, follow the tensile test requirements of ASTM D 638, except that the test may be conducted at ambient temperature and humidity. If the specimen elongates no less than 25 percent or failure initiates outside the joint area, the procedures qualifies for use.

(b) Mechanical joints. Before any written procedure established under §192.273(b) is used for making mechanical plastic pipe joints that are designed to withstand tensile forces, the procedure must be qualified by subjecting five specimen joints made according to the procedure to the following tensile test:

(1) Use an apparatus for the test as specified in ASTM D 638 (except for conditioning).

(2) The specimen must be of such length that the distance between the grips of the apparatus and the end of the stiffener does not affect the joint strength.

(3) The speed of testing is 0.20 in. (5.0 mm) per minute, plus or minus 25 percent.

(4) Pipe specimens less than 4 inches (102 mm) in diameter are qualified if the pipe yields to an elongation of no less than 25 percent or failure initiates outside the joint area.

(5) Pipe specimens 4 inches (102 mm) and larger in diameter shall be pulled until the pipe is subjected to a tensile stress equal to or greater than the maximum thermal stress that would be produced by a temperature change of 100°F (38°C) or until the pipe is pulled from the fitting. If the pipe pulls from the fitting, the lowest value of the five test results or the manufacturer's rating, whichever is lower must be used in the design calculations for stress.

(6) Each specimen that fails at the grips must be retested using new pipe.

(7) Results obtained pertain only to the specific outside diameter, and material of the pipe tested, except that testing of a heavier wall pipe may be used to qualify pipe of the same material but with a lesser wall thickness.

(c) A copy of each written procedure being used for joining plastic pipe must be available to the persons making and inspecting joints.

(d) Pipe or fittings manufactured before July 1, 1980, may be used in accordance with procedures that the manufacturer certifies will produce a joint as strong as the pipe.

[Amdt. 192-34, 44 FR 42968, July 23, 1979 as amended by Amdt. 192-34A, 45 FR 9931, Feb. 14, 1980; Amdt. 192-34B, 46 FR 39, Jan. 2, 1981; Amdt. 192-34(1), 47 FR 32720, July 29, 1982; Amdt. 192-34(2), 47 FR 49973, Nov. 4, 1982; Amdt. 192-68, 58 FR 14519, Mar. 18, 1993; Amdt. 192-78, 61 FR 28770, June 6, 1996; Amdt. 192-85, 63 FR 37500, July 13, 1998]

§192.285 Plastic pipe; qualifying persons to make joints.

(a) No person may make a plastic pipe joint unless that person has been qualified under the applicable joining procedure by:

(1) Appropriate training or experience in the use of the procedure; and

(2) Making a specimen joint from pipe sections joined according to the procedure that passes the inspection and test set forth in paragraph (b) of this section.

(b) The specimen joint must be:

(1) Visually examined during and after assembly or joining and found to have the same appearance as a joint or photographs of a joint that is acceptable under the procedure; and

(2) In the case of a heat fusion, solvent cement, or adhesive joint;

(i) Tested under any one of the test methods listed under §192.283(a) applicable to the type of joint and material being tested;

(ii) Examined by ultrasonic inspection and found not to contain flaws that would cause failure; or

(iii) Cut into at least three longitudinal straps, each of which is:

(A) Visually examined and found not to contain voids or discontinuities on the cut surfaces of the joint area; and

(B) Deformed by bending, torque, or impact, and if failure occurs, it must not initiate in the joint area.

(c) A person must be requalified under an applicable procedure, if during any 12-month period that person:

(1) Does not make any joints under that procedure; or

(2) Has 3 joints or 3 percent of the joints made, whichever is greater, under that procedure that are found unacceptable by testing under §192.513.

(d) Each operator shall establish a method to determine that each person making joints in plastic pipelines in his system is qualified in accordance with this section.

[Amdt. 192-34, 44 FR 42968, July 23, 1979 as amended by Amdt. 192-34A, 45 FR 9931, Feb. 14, 1980, Amdt. 192-34B, 46 FR 39, Jan. 2, 1981]

§192.287 Plastic pipe; inspection of joints.

No person may carry out the inspection of joints in plastic pipes required by §§192.273(c) and 192.285(b) unless that person has been qualified by appropriate training or experience in evaluating the acceptability of plastic pipe joints made under the applicable joining procedure.

[Amdt. 192-34, 44 FR 42968, July 23, 1979]

Subpart G–General Construction Requirements for Transmission Lines and Mains

§192.301 Scope.

This subpart prescribes minimum requirements for constructing transmission lines and mains.

§192.303 Compliance with specifications or standards.

Each transmission line or main must be constructed in accordance with comprehensive written specifications or standards that are consistent with this part.

§192.305 Inspection: General.

Each transmission line or main must be inspected to ensure that it is constructed in accordance with this part.

§192.307 Inspection of materials.

Each length of pipe and each other component must be visually inspected at the site of installation to ensure that it has not sustained any visually determinable damage that could impair its serviceability.

§192.309 Repair of steel pipe.

(a) Each imperfection or damage that impairs the serviceability of a length of steel pipe must be repaired or removed. If a repair is made by grinding, the remaining wall thickness must a least be equal to either:

(1) The minimum thickness required by the tolerances in the specification to which the pipe was manufactured; or

(2) The nominal wall thickness required for the design pressure of the pipeline.

(b) Each of the following dents must be removed from steel pipe to be operated at a pressure that produces a hoop stress of 20 percent or more, of SMYS:

(1) A dent that contains a stress concentrator such as a scratch, gouge, groove, or arc burn.

(2) A dent that affects the longitudinal weld or a circumferential weld.

(3) In pipe to be operated at a pressure that produces a hoop stress of 40 percent or more of SMYS, a dent that has a depth of:

(i) More than ¼ inch (6.4 millimeters) in pipe 12¾ inches (324 millimeters) or less in outer diameter; or

(ii) More than 2 percent of the nominal pipe diameter in pipe over 12¾ inches (324 millimeters) in outer diameter.

For the purpose of this section a "dent" is a depression that produces a gross disturbance in the curvature of the pipe wall without reducing the pipe-wall thickness. The depth of a dent is measured as the gap between the lowest point of the dent and a prolongation of the original contour of the pipe.

(c) Each arc burn on steel pipe to be operated at a pressure that produces a hoop stress of 40 percent or more, of SMYS must be repaired or removed. If a repair is made by grinding, the arc burn must be completely removed and the remaining wall thickness must be at least equal to either:

(1) The minimum wall thickness required by the tolerances in the specification to which the pipe was manufactured; or

(2) The nominal wall thickness required for the design pressure of the pipeline.

(d) A gouge, groove, arc burn, or dent may not be repaired by insert patching or by pounding out.

(e) Each gouge, groove, arc burn, or dent that is removed from a length of pipe must be removed by cutting out the damaged portion as a cylinder.

[35 FR 13257, Aug. 19, 1970, as amended by Amdt. 192-3, 35 FR 17660, Nov. 17, 1970; Amdt. 192-85, 63 FR 37500, July 13, 1998]

§192.311 Repair of plastic pipe.

Each imperfection or damage that would impair the serviceability of plastic pipe must be repaired by a patching saddle or removed.

§192.313 Bends and elbows.

(a) Each field bend in steel pipe, other than a wrinkle bend made in accordance with §192.315, must comply with the following:

(1) A bend must not impair the serviceability of the pipe.

(2) Each bend must have a smooth contour and be free from buckling, cracks, or any other mechanical damage.

(3) On pipe containing a longitudinal weld, the longitudinal weld must be as near as practicable to the neutral axis of the bend unless:

(i) The bend is made with an internal bending mandrel; or

(ii) The pipe is 12 inches (305 millimeters) or less in outside diameter or has a diameter to wall thickness ratio less than 70.

(b) Each circumferential weld of steel pipe which is located where the stress during bending causes a permanent deformation in the pipe must be nondestructively tested either before or after the bending process.

(c) Wrought-steel welding elbows and transverse segments of these elbows may not be used for changes in direction on steel pipe that is 2 inches (51 millimeters) or more in diameter unless the arc length, as measured along the crotch, is at least 1 inch (25 millimeters).

[Amdt. 192-26, 41 FR 26106, June 24, 1976, as amended by Amdt. 192-29, 42 FR 42865, Aug. 25, 1977; Amdt. 192-29C, 42 FR 60148, Nov. 25, 1977; Amdt. 192-49, 50 FR 13225, Apr. 3, 1985; Amdt. 192-85, 63 FR 37500, July 13, 1998]

§192.315 Wrinkle bends in steel pipe.

(a) A wrinkle bend may not be made on steel pipe to be operated at a pressure that produces a hoop stress of 30 percent or more, of SMYS.

(b) Each wrinkle bend on steel pipe must comply with the following:

(1) The bend must not have any sharp kinks.

(2) When measured along the crotch of the bend, the wrinkles must be a distance of at least one pipe diameter.

(3) On pipe 16 inches (406 millimeters) or larger in diameter, the bend may not have a deflection of more than 1½° for each wrinkle.

(4) On pipe containing a longitudinal weld the longitudinal seam must be as near as practicable to the neutral axis of the bend.

[35 FR 13257, Aug. 19, 1970, as amended by Amdt. 192-85, 63 FR 37500, July 13, 1998]

§192.317 Protection from hazards.

(a) The operator must take all practicable steps to protect each transmission line or main from washouts, floods, unstable soil, landslides, or other hazards that may cause the pipeline to move or to sustain abnormal loads. In addition, the operator must take all practicable steps to protect offshore pipelines from damage by mud slides, water currents, hurricanes, ship anchors, and fishing operations.

(b) Each aboveground transmission line or main, not located offshore or in inland navigable water areas, must be protected from accidental damage by vehicular traffic or other similar causes, either by being placed at a safe distance from the traffic or by installing barricades.

(c) Pipelines, including pipe risers, on each platform located offshore or in inland navigable waters must be protected from accidental damage by vessels.

[Amdt. 192-27, 41 FR 34598, Aug. 16, 1976; Amdt. 192-78, 61 FR 28770, June 6, 1996]

§192.319 Installation of pipe in a ditch.

(a) When installed in a ditch, each transmission line that is to be operated at a pressure producing a hoop stress of 20 percent or more of SMYS must be installed so that the pipe fits the ditch so as to minimize stresses and protect the pipe coating from damage.

(b) When a ditch for a transmission line or main is backfilled, it must be backfilled in a manner that:

(1) Provides firm support under the pipe; and

(2) Prevents damage to the pipe and pipe coating from equipment or from the backfill material.

(c) All offshore pipe in water at least 12 feet (3.7 meters) deep, but not more than 200 feet (61 meters) deep, as measured from the mean low tide, except pipe in the Gulf of Mexico and its inlets under 15 feet (4.6 meters) of water, must be installed so that the top of the pipe is below the natural bottom unless the pipe is supported by stanchions, held in place by anchors or heavy concrete coating, or protected by an equivalent means. Pipe in the Gulf of Mexico and its inlets under 15 feet (4.6 meters) of water must be installed so that the top of the pipe is 36 inches (914 millimeters) below the seabed for normal excavation or 18 inches (457 millimeters) for rock excavation.

[35 FR 13257, Aug. 19, 1970, as amended by Amdt. 192-27, 41 FR 34598, Aug. 16, 1976; Amdt. 192-78, 61 FR 28770, June 6, 1996; Amdt. 192-85, 63 FR 37500, July 13, 1998]

§192.321 Installation of plastic pipe.

(a) Plastic pipe must be installed below ground level unless otherwise permitted by paragraph (g) of this section.

(b) Plastic pipe that is installed in a vault or any other below grade enclosure must be completely encased in gas-tight metal pipe and fittings that are adequately protected from corrosion.

(c) Plastic pipe must be installed so as to minimize shear or tensile stresses.

(d) Thermoplastic pipe that is not encased must have a minimum wall thickness of 0.090 inch (2.29 millimeters), except that pipe with an outside diameter of 0.875 inch (22.3 millimeters) or less may have a minimum wall thickness of 0.062 inch (1.58 millimeters).

(e) Plastic pipe that is not encased must have an electrically conductive wire or other means of locating the pipe while it is underground.

(f) Plastic pipe that is being encased must be inserted into the casing pipe in a manner that will protect the plastic. The leading end of the plastic must be closed before insertion.

(g) Uncased plastic pipe may be temporarily installed aboveground level under the following conditions:

(1) The operator must be able to demonstrate that the cumulative aboveground exposure of the pipe does not exceed the manufacturer's recommended maximum period of exposure or 2 years, whichever is less.

(2) The pipe either is located where damage by external forces is unlikely or is otherwise protected against such damage.

(3) The pipe adequately resists exposure to ultraviolet light and high and low temperatures.

[Amdt. 192-78, 61 FR 28770, June 6, 1996; Amdt. 192-85, 63 FR 37500, July 13, 1998]

§192.323 Casing.

Each casing used on a transmission line or main under a railroad or highway must comply with the following:

(a) The casing must be designed to withstand the superimposed loads.

(b) If there is a possibility of water entering the casing, the ends must be sealed.

(c) If the ends of an unvented casing are sealed and the sealing is strong enough to retain the maximum allowable operating pressure of the pipe, the casing must be designed to hold this pressure at a stress level of not more than 72 percent of SMYS.

(d) If vents are installed on a casing, the vents must be protected from the weather to prevent water from entering the casing.

[35 FR 13257, Aug. 19, 1970, as amended by Amdt. 192-85, 63 FR 37500, July 13, 1998]

§192.325 Underground clearance.

(a) Each transmission line must be installed with at least 12 inches (305 millimeters) of clearance from any other underground structure not associated with the transmission line. If this clearance cannot be attained, the transmission line must be protected from damage that might result from the proximity of the other structure.

(b) Each main must be installed with enough clearance from any other underground structure to allow proper maintenance and to protect against damage that might result from proximity to other structures.

(c) In addition to meeting the requirements of paragraphs (a) or (b) of this section, each plastic transmission line or main must be installed with sufficient clearance, or must be insulated, from any source of heat so as to prevent the heat from impairing the serviceability of the pipe.

(d) Each pipe-type or bottle-type holder must be installed with a minimum clearance from any other holder as prescribed in §192.175(b).

§192.327 Cover.

(a) Except as provided in paragraphs (c), (e), (f), and (g) of this section, each buried transmission line must be installed with a minimum cover as follows:

Location	Normal soil	Consolidated rock
	Inches (Millimeters)	
Class 1 locations	30 (762)	18 (457)
Class 2, 3, and 4 locations	36 (914)	24 (610)
Drainage ditches of public roads and railroad crossings	36 (914)	24 (610)

(b) Except as provided in paragraphs (c) and (d) of this section, each buried main must be installed with at least 24 inches (610 millimeters) of cover.

(c) Where an underground structure prevents the installation of a transmission line or main with the minimum cover, the transmission line or main may be installed with less cover if it is provided with additional protection to withstand anticipated external loads.

(d) A main may be installed with less than 24 inches (610 millimeters) of cover if the law of the State or municipality:

(1) Establishes a minimum cover of less than 24 inches (610 millimeters);

(2) Requires that mains be installed in a common trench with other utility lines; and,

(3) Provides adequately for prevention of damage to the pipe by external forces.

(e) Except as provided in paragraph (c) of this section, all pipe installed in a navigable river, stream, or harbor must be installed with a minimum cover of 48 inches (1219 millimeters) in soil or 24 inches (610 millimeters) in consolidated rock between the top of the pipe and the natural bottom.

(f) All pipe installed offshore, except in the Gulf of Mexico and its inlets, under water not more than 200 feet (60 meters) deep, as measured from the mean low tide, must be installed as follows:

(1) Except as provided in paragraph (c) of this section, pipe under water less than 12 feet (3.66 meters) deep, must be installed with a minimum cover of 36 inches (914 millimeters) in soil or 18 inches (457 millimeters) in consolidated rock between the top of the pipe and the natural bottom.

(2) Pipe under water at least 12 feet (3.66 meters) deep must be installed so that the top of the pipe is below the natural bottom, unless the pipe is supported by stanchions, held in place by anchors or heavy concrete coating, or protected by an equivalent means.

(g) All pipelines installed under water in the Gulf of Mexico and its inlets, as defined in §192.3, must be installed in accordance with §192.612(b)(3).

[35 FR 13257, Aug. 19, 1970, as amended by Amdt. 192-27, 41 FR 34598, Aug. 16, 1976; Amdt. 192-78, 61 FR 28770, June 6, 1996; Amdt. 192-85, 63 FR 37500, July 13, 1998]

Subpart H–Customer Meters, Service Regulators, and Service Lines

§192.351 Scope.

This subpart prescribes minimum requirements for installing customer meters, service regulators, service lines, service line valves, and service line connections to mains.

§192.353 Customer meters and regulators: Location.

(a) Each meter and service regulator whether inside or outside of a building, must be installed in a readily accessible location and be protected from corrosion and other damage. However, the upstream regulator in a series may be buried.

(b) Each service regulator installed within a building must be located as near as practical to the point of service line entrance.

(c) Each meter installed within a building must be located in a ventilated place and not less than 3 feet (914 millimeters) from any source of ignition or any source of heat which might damage the meter.

(d) Where feasible, the upstream regulator in a series must be located outside the building, unless it is located in a separate metering or regulating building.

[35 FR 13257, Aug. 19, 1970, as amended by Amdt. 192-85, 63 FR 37500, July 13, 1998]

§192.355 Customer meters and regulators: Protection from damage.

(a) Protection from vacuum or back pressure. If the customer's equipment might create either a vacuum or a back pressure, a device must be installed to protect the system.

(b) Service regulator vents and relief vents. Service regulator vents and relief vents must terminate outdoors, and the outdoor terminal must:

(1) Be rain and insect resistant;

(2) Be located at a place where gas from the vent can escape freely into the atmosphere and away from any opening into the building; and,

(3) Be protected from damage caused by submergence in areas where flooding may occur.

(c) Pits and vaults. Each pit or vault that houses a customer meter or regulator at a place where vehicular traffic is anticipated, must be able to support that traffic.

[35 FR 13257, Aug. 19, 1970, as amended by Amdt. 192-58, 53 FR 1633, Jan. 21, 1988]

§192.357 Customer meters and regulators: Installation.

(a) Each meter and each regulator must be installed so as to minimize anticipated stresses upon the connecting piping and the meter.

(b) When close all-thread nipples are used, the wall thickness remaining after the threads are cut must meet the minimum wall thickness requirements of this part.

(c) Connections made of lead or other easily damaged material may not be used in the installation of meters or regulators.

(d) Each regulator that might release gas in its operation must be vented to the outside atmosphere.

§192.359 Customer meter installations: Operating pressure.

(a) A meter may not be used at a pressure that is more than 67 percent of the manufacturer's shell test pressure.

(b) Each newly installed meter manufactured after November 12, 1970, must have been tested to a minimum of 10 p.s.i. (69 kPa) gage.

(c) A rebuilt or repaired tinned steel case meter may not be used at a pressure that is more than 50 percent of the pressure used to test the meter after rebuilding or repairing.

[35 FR 13257, Aug. 19, 1970, as amended by Amdt. 192-3, 35 FR 17660, Nov. 17, 1970; Amdt. 192-85, 63 FR 37500, July 13, 1998]

§192.361 Service lines: Installation.

(a) Depth. Each buried service line must be installed with at least 12 inches (305 millimeters) of cover in private property and at least 18 inches (457 millimeters) of cover in streets and roads. However, where an underground structure prevents installation at those depths, the service line must be able to withstand any anticipated external load.

(b) Support and backfill. Each service line must be properly supported on undisturbed or well-compacted soil, and material used for backfill must be free of materials that could damage the pipe or its coating.

(c) Grading for drainage. Where condensate in the gas might cause interruption in the gas supply to the customer, the service line must be graded so as to drain into the main or into drips at the low points in the service line.

(d) Protection against piping strain and external loading. Each service line must be installed so as to minimize anticipated piping strain and external loading.

(e) Installation of service lines into buildings. Each underground service line installed below grade through the outer foundation wall of a building must:

(1) In the case of a metal service line, be protected against corrosion;

(2) In the case of a plastic service line, be protected from shearing action and backfill settlement; and

(3) Be sealed at the foundation wall to prevent leakage into the building.

(f) Installation of service lines under buildings. Where an underground service line is installed under a building:

(1) It must be encased in a gas tight conduit;

(2) The conduit and the service line must, if the service line supplies the building it underlies, extend into a normally usable and accessible part of the building; and,

(3) The space between the conduit and the service line must be sealed to prevent gas leakage into the building and, if the conduit is sealed at both ends, a vent line from the annular space must extend to a point where gas would not be a hazard, and extend above grade, terminating in a rain and insect resistant fitting.

[35 FR 13257, Aug. 19, 1970, as amended by Amdt. 192-75, 61 FR 18512, Apr. 26, 1996; Amdt. 192-85, 63 FR 37500, July 13, 1998]

§192.363 Service lines: Valve requirements.

(a) Each service line must have a service line valve that meets the applicable requirements of Subparts B and D of this part. A valve incorporated in a meter bar, that allows the meter to be bypassed, may not be used as a service line valve.

(b) A soft seat service line valve may not be used if its ability to control the flow of gas could be adversely affected by exposure to anticipated heat.

(c) Each service line valve on a high-pressure service line, installed aboveground or in an area where the blowing of gas would be hazardous, must be designed and constructed to minimize the possibility of the removal of the core of the valve with other than specialized tools.

§192.365 Service lines: Location of valves.

(a) Relation to regulator or meter. Each service line valve must be installed upstream of the regulator or, if there is no regulator, upstream of the meter.

(b) Outside valves. Each service line must have a shutoff valve in a readily accessible location that, if feasible, is outside of the building.

(c) Underground valves. Each underground service line valve must be located in a covered durable curb box or standpipe that allows ready operation of the valve and is supported independently of the service lines.

§192.367 Service lines: General requirements for connections to main piping.

(a) Location. Each service line connection to a main must be located at the top of the main or, if that is not practical, at the side of the main, unless a suitable protective device is installed to minimize the possibility of dust and moisture being carried from the main into the service line.

(b) Compression-type connection to main. Each compression-type service line to main connection must:

(1) Be designed and installed to effectively sustain the longitudinal pullout or thrust forces caused by contraction or expansion of the piping, or by anticipated external or internal loading; and

(2) If gaskets are used in connecting the service line to the main connection fitting, have gaskets that are compatible with the kind of gas in the system.

[Amdt. 192-75, 61 FR 18512, Apr. 26, 1996]

§192.369 Service lines: Connections to cast iron or ductile iron mains.

(a) Each service line connected to a cast iron or ductile iron main must be connected by a mechanical clamp, by drilling and tapping the main, or by another method meeting the requirements of §192.273.

(b) If a threaded tap is being inserted, the requirements of §192.151(b) and (c) must also be met.

§192.371 Service lines: Steel.

Each steel service line to be operated at less than 100 p.s.i. (689 kPa) gage must be constructed of pipe designed for a minimum of 100 p.s.i. (689 kPa) gage.

[35 FR 13257, Aug. 19, 1970, as amended by Amdt. 192-3, 35 FR 17660, Nov. 17, 1970; Amdt. 192-85, 63 FR 37500, July 13, 1998]

192CODE.DOC, DT (Revised) 7/98

§192.373 Service lines: Cast iron and ductile iron.

(a) Cast or ductile iron pipe less than 6 inches (152 millimeters) in diameter may not be installed for service lines.

(b) If cast iron pipe or ductile iron pipe is installed for use as a service line, the part of the service line which extends through the building wall must be of steel pipe.

(c) A cast iron or ductile iron service line may not be installed in unstable soil or under a building.

[35 FR 13257, Aug. 19, 1970, as amended by Amdt. 192-85, 63 FR 37500, July 13, 1998]

§192.375 Service lines: Plastic.

(a) Each plastic service line outside a building must be installed below ground level, except that–

(1) It may be installed in accordance with §192.321(g); and

(2) It may terminate aboveground level and outside the building, if–

(i) The aboveground level part of the plastic service line is protected against deterioration and external damage; and

(ii) The plastic service line is not used to support external loads.

(b) Each plastic service line inside a building must be protected against external damage.

[35 FR 13257, Aug. 19, 1970, as amended by Amdt. 192-78, 61 FR 28770, June 6, 1996]

§192.377 Service lines: Copper

Each copper service line installed within a building must be protected against external damage.

§192.379 New service lines not in use.

Each service line that is not placed in service upon completion of installation must comply with one of the following until the customer is supplied with gas:

(a) The valve that is closed to prevent the flow of gas to the customer must be provided with a locking device or other means designed to prevent the opening of the valve by persons other than those authorized by the operator.

(b) A mechanical device or fitting that will prevent the flow of gas must be installed in the service line or in the meter assembly.

(c) The customer's piping must be physically disconnected from the gas supply and the open pipe ends sealed.

[Amdt. 192-8, 37 FR 20694, Oct. 1972]

§192.381 Service lines: Excess flow valve performance standards.

(a) Excess flow valves to be used on single residence service lines that operate continuously throughout the year at a pressure not less than 10 p.s.i. (69 kPa) gage must be manufactured and tested by the manufacturer according to an industry specification, or the manufacturer's written specification, to ensure that each valve will:

(1) Function properly up to the maximum operating pressure at which the valve is rated;

(2) Function properly at all temperatures reasonably expected in the operating environment of the service line;

(3) At 10 p.s.i. (69 kPa) gage:

(i) Close at, or not more than 50 percent above, the rated closure flow rate specified by the manufacturer; and

(ii) Upon closure, reduce gas flow–

(A) For an excess flow valve designed to allow pressure to equalize across the valve, to no more than 5 percent of the manufacturer's specified closure flow rate, up to a maximum of 20 cubic feet per hour (0.57 cubic meters per hour); or

(B) For an excess flow valve designed to prevent equalization of pressure across the valve, to no more than 0.4 cubic feet per hour (.01 cubic meters per hour); and

(4) Not close when the pressure is less than the manufacturer's minimum specified operating pressure and the flow rate is below the manufacturer's minimum specified closure flow rate.

(b) An excess flow valve must meet the applicable requirements of Subparts B and D of this part.

(c) An operator must mark or otherwise identify the presence of an excess flow valve on the service line.

(d) An operator shall locate an excess flow valve as near as practical to the fitting connecting the service line to its source of gas supply.

(e) An operator should not install an excess flow valve on a service line where the operator has prior experience with contaminants in the gas stream, where these contaminants could be expected to cause the excess flow valve to malfunction or where the excess flow valve would interfere with necessary operation and maintenance activities on the service, such as blowing liquids from the line.

[Amdt. 192-79, 61 FR 31449, June 20, 1996 as amended by Amdt. 192-80, 62 FR 2618, Jan. 17, 1997; Amdt. 192-85, 63 FR 37500, July 13, 1998]

§192.383 Excess flow valve customer notification.

(a) *Definitions.* As used in this section:

Costs associated with installation means the costs directly connected with installing an excess flow valve, for example, costs of parts, labor, inventory and procurement. It does not include maintenance and replacement costs until such costs are incurred.

Replaced service line means a natural gas service line where the fitting that connects the service line to the main is replaced or the piping connected to this fitting is replaced.

Service line customer means the person who pays the gas bill, or where service has not yet been established, the person requesting service.

(b) *Which customers must receive notification.* Notification is required on each newly installed service line or replaced service line that operates continuously throughout the year at a pressure not less than 68.9 m (10 psig) and that serves a single residence. On these lines an operator of a natural gas distribution system must notify the service line customer once in writing.

(c) *What to put in the written notice.*

(1) An explanation for the customer that an excess flow valve meeting the performance standards prescribed under §192.381 is available for the operator to install if the customer bears the costs associated with installation;

(2) An explanation for the customer of the potential safety benefits that may be derived from installing an excess flow valve. The explanation must include that an excess flow valve is designed to shut off the flow of natural gas automatically if the service line breaks;

(3) A description of installation, maintenance, and replacement costs. The notice must explain that if the customer requests the operator to install an EFV, the customer bears all costs associated with installation, and what those costs are. The notice must alert the customer that costs for maintaining and replacing an EFV may later be incurred, and what those costs will be, to the extent known.

(d) *When notification and installation must be made.*

(1) After February 3, 1999 an operator must notify each service line customer set forth in paragraph (b) of this section:

(i) On new service lines when the customer applies for service.

(ii) On replaced service lines when the operator determines the service line will be replaced.

(2) If a service line customer requests installation an operator must install the EFV at a mutually agreeable date.

(e) *What records are required.*

(1) An operator must make the following records available for inspection by the Administrator or a State agency participating under 49 U.S.C. 60105 or 60106:

(i) A copy of the notice currently in use, and

(ii) Evidence that notice has been sent to the service line customers set forth in paragraph (b) of this section, within the previous three years.

(2) [Reserved]

(f) *When notification is not required.*

The notification requirements do not apply if the operator can demonstrate–

(1) That the operator will voluntarily install an excess flow valve or that the state or local jurisdiction requires installation;

(2) That excess flow valves meeting the performance standards of §192.381 are not available to the operator;

(3) That an operator has prior e xperience with contaminants in the gas stream that could interfere with the operation of an excess flow valve, cause loss of service to a residence, or interfere with necessary operation or maintenance activities, such as blowing liquids from the line.

(4) That an emergency or short time notice replacement situation made it impractical for the operator to notify a service line customer before replacing a service line. Examples of these situations would be where an operator has to replace a service line quickly because of–

(i) Third party excavation damage;

(ii) Grade 1 leaks as defined in the Appendix G–192-11 of the Gas Piping Technology Committee guide for gas transmission and distribution systems;

(iii) A short notice service line relocation request.

[Amdt. 192-83, 63 FR 5464, Feb, 3, 1998]

Subpart I–Requirements for Corrosion Control

Source: Amdt. 192-4, 36 FR 12297, June 30, 1971, unless otherwise noted.

§192.451 Scope.

(a) This subpart prescribes minimum requirements for the protection of metallic pipelines from external, internal, and atmospheric corrosion.

[Amdt. 192-4, 36 FR 12297, June 30, 1971, as amended by Amdt. 192-27, 41 FR 34598, Aug. 16, 1976; Amdt. 192-33, 43 FR 39389, Sept. 5, 1978]

§192.452 Applicability to converted pipelines.

Notwithstanding the date the pipeline was installed or any earlier deadlines for compliance, each pipeline which qualifies for use under this part in accordance with §192.14 must meet the requirements of this subpart specifically applicable to pipelines installed before August 1, 1971, and all other applicable requirements within 1 year after the pipeline is readied for service. However, the requirements of this subpart specifically applicable to pipelines installed after July 31, 1971, apply if the pipeline substantially meets those requirements before it is readied for service or it is a segment which is replaced, relocated, or substantially altered.

[Amdt. 192-4, 36 FR 12297, June 30, 1971, as amended by Amdt. 192-30, 42 FR 60146, Nov. 25, 1977]

§192.453 General.

The corrosion control procedures required by §192.605(b)(2), including those for the design, installation, operation, and maintenance of cathodic protection systems, must be carried out by, or under the direction of, a person qualified in pipeline corrosion control methods.

[Amdt. 192-4, 36 FR 12297, June 30, 1971, as amended by Amdt. 192-71, 59 FR 6575, Feb. 11, 1994]

§192.455 External corrosion control: Buried or submerged pipelines installed after July 31, 1971.

(a) Except as provided in paragraphs (b), (c), and (f) of this section, each buried or submerged pipeline installed after July 31, 1971, must be protected against external corrosion, including the following:

(1) It must have an external protective coating meeting the requirements of §192.461.

(2) It must have a cathodic protection system designed to protect the pipeline in accordance with this subpart, installed and placed in operation within 1 year after completion of construction.

(b) An operator need not comply with paragraph (a) of this section, if the operator can demonstrate by tests, investigation, or experience in the area of application, including, as a minimum, soil resistivity measurements and tests for corrosion accelerating bacteria, that a corrosive environment does not exist. However, within 6 months after an installation made pursuant to the preceding sentence, the operator shall conduct tests, including pipe-to-soil potential measurements with respect to either a continuous reference electrode or an electrode using close spacing, not to exceed 20 feet (6 meters), and soil resistivity measurements at potential profile peak locations, to adequately evaluate the potential profile along the entire pipeline. If the tests made indicate that a corrosive condition exists, the pipeline must be cathodically protected in accordance with paragraph (a)(2) of this section.

(c) An operator need not comply with paragraph (a) of this section, if the operator can demonstrate by tests, investigation, or experience that-

(1) For a copper pipeline, a corrosive environment does not exist; or

(2) For a temporary pipeline with an operating period of service not to exceed 5 years beyond installation, corrosion during the 5-year period of service of the pipeline will not be detrimental to public safety.

(d) Notwithstanding the provisions of paragraph (b) or (c) of this section, if a pipeline is externally coated, it must be cathodically protected in accordance with paragraph (a)(2) of this section.

(e) Aluminum may not be installed in a buried or submerged pipeline if that aluminum is exposed to an environment with a natural pH in excess of 8, unless tests or experience indicate its suitability in the particular environment involved.

(f) This section does not apply to electrically isolated, metal alloy fittings in plastic pipelines, if:

(1) For the size fitting to be used, an operator can show by test, investigation, or experience in the area of application that adequate corrosion control is provided by the alloy composition; and

(2) The fitting is designed to prevent leakage caused by localized corrosion pitting.

[Amdt. 192-4, 36 FR 12297, June 30, 1971, as amended by Amdt. 192-28, 42 FR 35654, July 11, 1977; Amdt. 192-39, 47 FR 9842, Mar. 8, 1982; Amdt. 192-78, 61 FR 28770, June 6, 1996; Amdt. 192-85, 63 FR 37500, July 13, 1998]

§192.457 External corrosion control: Buried or submerged pipelines installed before August 1, 1971.

(a) Except for buried piping at compressor, regulator, and measuring stations, each buried or submerged transmission line installed before August 1, 1971, that has an effective external coating must be cathodically protected along the entire area that is effectively coated, in accordance with this subpart. For the purposes of this subpart, a pipeline does not have an effective external coating if its cathodic protection current requirements are substantially the same as if it were bare. The operator shall make tests to determine the cathodic protection current requirements.

(b) Except for cast iron or ductile iron, each of the following buried or submerged pipelines installed before August 1, 1971, must be cathodically protected in accordance with this subpart in areas in which active corrosion is found:

(1) Bare or ineffectively coated transmission lines.

(2) Bare or coated pipes at compressor, regulator, and measuring stations.

(3) Bare or coated distribution lines.

The operator shall determine the areas of active corrosion by electrical survey, or where electrical survey is impractical, by the study of corrosion and leak history records, by leak detection survey, or by other means.

(c) For the purpose of this subpart, active corrosion means continuing corrosion which, unless controlled, could result in a condition that is detrimental to public safety.

[Amdt. 192-4, 36 FR 12297, June 30, 1971, as amended by Amdt. 192-33, 43 FR 39389, Sept. 5, 1978]

§192.459 External corrosion control: Examination of buried pipeline when exposed.

Whenever an operator has knowledge that any portion of a buried pipeline is exposed, the exposed portion must be examined for evidence of external corrosion if the pipe is bare, or if the coating is deteriorated. If external corrosion is found, remedial action must be taken to the extent required by §192.483 and the applicable paragraphs of §§192.485, 192.487, or 192.489.

§192.461 External corrosion control: Protective coating.

(a) Each external protective coating, whether conductive or insulating, applied for the purpose of external corrosion control must–

(1) Be applied on a properly prepared surface;

(2) Have sufficient adhesion to the metal surface to effectively resist underfilm migration of moisture;

(3) Be sufficiently ductile to resist cracking;

(4) Have sufficient strength to resist damage due to handling and soil stress; and,

(5) Have properties compatible with any supplemental cathodic protection.

(b) Each external protective coating which is an electrically insulating type must also have low moisture absorption and high electrical resistance.

(c) Each external protective coating must be inspected just prior to lowering the pipe into the ditch and backfilling, and any damage detrimental to effective corrosion control must be repaired.

(d) Each external protective coating must be protected from damage resulting from adverse ditch conditions or damage from supporting blocks.

(e) If coated pipe is installed by boring, driving, or other similar method, precautions must be taken to minimize damage to the coating during installation.

§192.463 External corrosion control: Cathodic protection.

(a) Each cathodic protection system required by this subpart must provide a level of cathodic protection that complies with one or more of the applicable criteria contained in Appendix D of this part. If none of these criteria is applicable, the cathodic protection system must provide a level of cathodic protection at least equal to that provided by compliance with one or more of these criteria.

(b) If amphoteric metals are included in a buried or submerged pipeline containing a metal or different anodic potential–

(1) The amphoteric metals must be electrically isolated from the remainder of the pipeline and cathodically protected; or

(2) The entire buried or submerged pipeline must be cathodically protected at a cathodic potential that meets the requirements of Appendix D of this part for amphoteric metals.

(c) The amount of cathodic protection must be controlled so as not to damage the protective coating or the pipe.

§192.465 External corrosion control: Monitoring.

(a) Each pipeline that is under cathodic protection must be tested at least once each calendar year, but with intervals not exceeding 15 months, to determine whether the cathodic protection meets the requirements of §192.463. However, if tests at those intervals are impractical for separately protected short sections of mains or transmission lines, not in excess of 100 feet (30 meters), or separately protected service lines, these pipelines may be surveyed on a sampling basis. At least 10 percent of these protected structures, distributed over the entire system must be surveyed each calendar year, with a different 10 percent checked each subsequent year, so that the entire system is tested in each 10-year period.

(b) Each cathodic protection rectifier or other impressed current power source must be inspected six times each calendar year, but with intervals not exceeding 2½ months, to insure that it is operating.

(c) Each reverse current switch, each diode, and each interference bond whose failure would jeopardize structure protection must be electrically checked for proper performance six times each calendar year, but with intervals not exceeding 2½ months. Each other interference bond must be checked at least once each calendar year, but with intervals not exceeding 15 months.

(d) Each operator shall take prompt remedial action to correct any deficiencies indicated by the monitoring.

(e) After the initial evaluation required by paragraphs (b) and (c) of §192.455 and paragraph (b) of §192.457, each operator shall, at intervals not exceeding 3 years, reevaluate its unprotected pipelines and cathodically protect them in accordance with this subpart in areas in which active corrosion is found. The operator shall determine the areas of active corrosion by electrical survey, or where electrical survey is impractical, by the study of corrosion and leak history records, by leak detection survey, or by other means.

[Amdt. 192-4, 36 FR 12297, June 30, 1971, as amended by Amdt. 192-27, 41 FR 34598, Aug. 16, 1976; Amdt. 192-33, 43 FR 39389, Sept. 5, 1978; Amdt. 192-35, 44 FR 75381, Dec. 20, 1979; Amdt. 192-35A, 45 FR 23441, Apr. 7, 1980; Amdt. 192-85, 63 FR 37500, July 13, 1998]

§192.467 External corrosion control: Electrical isolation.

(a) Each buried or submerged pipeline must be electrically isolated from other underground metallic structures, unless the pipeline and the other structures are electrically interconnected and cathodically protected as a single unit.

(b) One or more insulating devices must be installed where electrical isolation of a portion of a pipeline is necessary to facilitate the application of corrosion control.

(c) Except for unprotected copper inserted in a ferrous pipe, each pipeline must be electrically isolated from metallic casings that are a part of the underground system. However, if isolation is not achieved because it is impractical, other measures must be taken to minimize corrosion of the pipeline inside the casing.

(d) Inspection and electrical tests must be made to assure that electrical isolation is adequate.

(e) An insulating device may not be installed in an area where a combustible atmosphere is anticipated unless precautions are taken to prevent arcing.

(f) Where a pipeline is located in close proximity to electrical transmission tower footings, ground cables or counterpoise, or in other areas where fault currents or unusual risk of lightning may be anticipated, it must be provided with protection against damage due to fault currents or lightning, and protective measures must also be taken at insulating devices.

[Amdt. 192-4, 36 FR 12297, June 30, 1971, as amended by Amdt. 192-33, 43 FR 39389, Sept. 5, 1978]

§192.469 External corrosion control: Test stations.

Each pipeline under cathodic protection required by this subpart must have sufficient test stations or other contact points for electrical measurement to determine the adequacy of cathodic protection.

[Amdt. 192-4, 36 FR 12297, June 30, 1971, as amended by Amdt. 192-27, 41 FR 34606, Aug. 16, 1976]

§192.471 External corrosion control: Test leads.

(a) Each test lead wire must be connected to the pipeline so as to remain mechanically secure and electrically conductive.

(b) Each test lead wire must be attached to the pipeline so as to minimize stress concentration on the pipe.

(c) Each bared test lead wire and bared metallic area at point of connection to the pipeline must be coated with an electrical insulating material compatible with the pipe coating and the insulation on the wire.

§192.473 External corrosion control: Interference currents.

(a) Each operator whose pipeline system is subjected to stray currents shall have in effect a continuing program to minimize the detrimental effects of such currents.

(b) Each impressed current type cathodic protection system or galvanic anode system must be designed and installed so as to minimize any adverse effects on existing adjacent underground metallic structures. [Amdt. 192-4, 36 FR 12297, June 30, 1971, as amended by Amdt. 192-33, 43 FR 39389, Sept. 5, 1978]

§192.475 Internal corrosion control: General.

(a) Corrosive gas may not be transported by pipeline, unless the corrosive effect of the gas on the pipeline has been investigated and steps have been taken to minimize internal corrosion.

(b) Whenever any pipe is removed from a pipeline for any reason, the internal surface must be inspected for evidence of corrosion. If internal corrosion is found–

(1) The adjacent pipe must be investigated to determine the extent of internal corrosion:

(2) Replacement must be made to the extent required by the applicable paragraphs of §§192.485, 192.487, or 192,489; and,

(3) Steps must be taken to minimize the internal corrosion.

(c) Gas containing more than 0. 25 grain of hydrogen sulfide per 100 cubic feet (5.8 milligrams/m^3) at standard conditions (4 parts per million) may not be stored in pipe-type or bottle-type holders.

[Amdt. 192-4, 36 FR 12297, June 30, 1971, as amended by Amdt. 192-33, 43 FR 39389, Sept. 5, 1978; Amdt. 192-78, 61 FR 28770, June 6, 1996; Amdt. 192-85, 63 FR 37500, July 13, 1998]

§192.477 Internal corrosion control: Monitoring.

If corrosive gas is being transported, coupons or other suitable means must be used to determine the effectiveness of the steps taken to minimize internal corrosion. Each coupon or other means of monitoring internal corrosion must be checked two times each calendar year, but with interval not exceeding 7½ months.

[Amdt. 192-4, 36 FR 12297, June 30, 1971, as amended by Amdt. 192-33, 43 FR 39389, Sept. 5, 1978]

§192.479 Atmospheric corrosion control; General.

(a) Pipelines installed after July 31, 1971. Each aboveground pipeline or portion of a pipeline installed after July 31, 1971 that is exposed to the atmosphere must be cleaned and either coated or jacketed with a material suitable for the prevention of atmospheric corrosion. An operator need not comply with this paragraph, if the operator can demonstrate by test, investigation, or experience in the area of application, that a corrosive atmosphere does not exist.

(b) Pipelines installed before August 1, 1971. Each operator having an aboveground pipeline or portion of a pipeline installed before August 1, 1971 that is exposed to the atmosphere, shall–

(1) Determine the areas of atmospheric corrosion on the pipeline;

(2) If atmospheric corrosion is found, take remedial measures to the extent required by the applicable paragraphs of §§192.485, 192.487, or 192.489; and,

(3) Clean and either coat or jacket the areas of atmospheric corrosion on the pipeline with a material suitable for the prevention of atmospheric corrosion.

[Amdt. 192-4, 36 FR 12297, June 30, 1971, as amended by Amdt. 192-33, 43 FR 39389, Sept. 5, 1978]

§192.481 Atmospheric corrosion control: Monitoring.

After meeting the requirements of §192.479 (a) and (b), each operator shall, at intervals not exceeding 3 years for onshore pipeline and at least once each calendar year, but with intervals not exceeding 15 months, for offshore pipelines, reevaluate each pipeline that is exposed to the atmosphere and take remedial action whenever necessary to maintain protection against atmospheric corrosion.

[Amdt. 192-4, 36 FR 12297, June 30, 1971, as amended by Amdt. 192-27, 41 FR 34598, Aug. 16, 1976; Amdt. 192-33, 43 FR 39389, Sept. 5, 1978]

§192.483 Remedial measures: General.

(a) Each segment of metallic pipe that replaces pipe removed from a buried or submerged pipeline because of external corrosion must have a properly prepared surface and must be provided with an external protective coating that meets the requirements of §192.461.

(b) Each segment of metallic pipe that replaces pipe removed from a buried or submerged pipeline because of external corrosion must be cathodically protected in accordance with this subpart.

(c) Except for cast iron or ductile iron pipe, each segment of buried or submerged pipe that is required to be repaired because of external corrosion must be cathodically protected in accordance with this subpart.

§192.485 Remedial measures: Transmission lines.

(a) General corrosion. Each segment of transmission line with general corrosion and with a remaining wall thickness less than that required for the maximum allowable operating pressure of the pipeline must be replaced or the operating pressure reduced commensurate with the strength of the pipe based on actual remaining wall thickness. However, if the area of general corrosion is small, the corroded pipe may be repaired. Corrosion pitting so closely grouped as to affect the overall strength of the pipe is considered general corrosion for the purpose of this paragraph.

(b) Localized corrosion pitting. Each segment of transmission line pipe with localized corrosion pitting to a degree where leakage might result must be replaced or repaired, or the operating pressure must be reduced commensurate with the strength of the pipe, based on the actual remaining wall thickness in the pits.

(c) Under paragraphs (a) and (b) of this section, the strength of pipe based on actual remaining wall thickness may be determined by the procedure in ASME/ANSI B31G or the procedure in AGA Pipeline Research Committee Project PR 3-805 (with RSTRENG disk). Both procedures apply to corroded regions that do not penetrate the pipe wall, subject to the limitations prescribed in the procedures.

[Amdt. 192-4, 36 FR 12297, June 30, 1971, as amended by Amdt. 192-33, 43 FR 39389, Sept. 5, 1978; Amdt. 192-78, 61 FR 28770, June 6, 1996]

§192.487 Remedial measures: Distribution lines other than cast iron or ductile iron lines.

(a) General corrosion. Except for cast iron or ductile iron pipe, each segment of generally corroded distribution line pipe with a remaining wall thickness less than that required for the maximum allowable operating pressure of the pipeline, or a remaining wall thickness less than 30 percent of the nominal wall thickness, must be replaced. However, if the area of general corrosion is small, the corroded pipe may be repaired. Corrosion pitting so closely grouped as to affect the overall strength of the pipe is considered general corrosion for the purpose of this paragraph.

(b) Localized corrosion pitting. Except for cast iron or ductile iron pipe, each segment of distribution line pipe with localized corrosion pitting to a degree where leakage might result must be replaced or repaired.

§192.489 Remedial measures: Cast iron and ductile iron pipelines.

(a) General graphitization. Each segment of cast iron or ductile iron pipe on which general graphitization is found to a degree where a fracture or any leakage might result, must be replaced.

(b) Localized graphitization. Each segment of cast iron or ductile iron pipe on which localized graphitization is found to a degree where any leakage might result, must be replaced or repaired, or sealed by internal sealing methods adequate to prevent or arrest any leakage.

§192.491 Corrosion control records.

(a) Each operator shall maintain records or maps to show the location of cathodically protected piping, cathodic protection facilities, galvanic anodes, and neighboring structures bonded to the cathodic protection system. Records or maps showing a stated number of anodes, installed in a stated manner or spacing, need not show specific distances to each buried anode.

(b) Each record or map required by paragraph (a) of this section must be retained for as long as the pipeline remains in service.

(c) Each operator shall maintain a record of each test, survey, or inspection required by this subpart in sufficient detail to demonstrate the adequacy of corrosion control measures or that a corrosive condition does not exist.

These records must be retained for at least 5 years, except that records related to §§192.465(a) and (e) and 192.475(b) must be retained for as long as the pipeline remains in service.

[Amdt. 192-4, 36 FR 12297, June 30, 1971, as amended by Amdt. 192-33, 43 FR 39389, Sept. 5, 1978; Amdt. 192-78, 61 FR 28770, June 6, 1996]

Subpart J–Test Requirements

§192.501 Scope.

This subpart prescribes minimum leak-test and strength-test requirements for pipelines.

§192.503 General requirements.

(a) No person may operate a new segment of pipeline, or return to service a segment of pipeline that has been relocated or replaced, until–

(1) It has been tested in accordance with this subpart and §192.619 to substantiate the maximum allowable operating pressure; and

(2) Each potentially hazardous leak has been located and eliminated.

(b) The test medium must be liquid, air, natural gas, or inert gas that is–

(1) Compatible with the material of which the pipeline is constructed;

(2) Relatively free of sedimentary materials; and,

(3) Except for natural gas, nonflammable.

(c) Except as provided in §192.505(a), if air, natural gas, or inert gas is used as the test medium, the following maximum hoop stress limitations apply:

Class location	Maximum hoop stress allowed as percentage of SMYS	
	Natural gas	Air or inert gas
1	80	80
2	30	75
3	30	50
4	30	40

(d) Each joint used to tie in a test segment of pipeline is excepted from the specific test requirements of this subpart, but each non-welded joint must be leak tested at not less than its operating pressure.

[35 FR 13257, Aug. 19, 1970, as amended by Amdt. 192-58, 53 FR 1633, Jan. 21, 1988; Amdt. 192-60, 53 FR 36028, Sept. 16, 1988; Amdt. 192-60A, 54 FR 5485, Feb. 3, 1989]

§192.505 Strength test requirements for steel pipeline to operate at a hoop stress of 30 percent or more of SMYS.

(a) Except for service lines, each segment of a steel pipeline that is to operate at a hoop stress of 30 percent or more of SMYS must be strength tested in accordance with this section to substantiate the proposed maximum allowable operating pressure. In addition, in a Class 1 or Class 2 location, if there is a building intended for human occupancy within 300 feet (91 meters) of a pipeline, a hydrostatic test must be conducted to a test pressure of at least 125 percent of maximum operating pressure on that segment of the pipeline within 300 feet (91 meters) of such a building, but in no event may the test section be less than 600 feet (183 meters) unless the length of the newly installed or relocated pipe is less than 600 feet (183 meters). However, if the buildings are evacuated while the hoop stress exceeds 50 percent of SMYS, air or inert gas may be used as the test medium.

(b) In a Class 1 or Class 2 location, each compressor station, regulator station, and measuring station, must be tested to at least Class 3 location test requirements.

(c) Except as provided in paragraph (e) of this section, the strength test must be conducted by maintaining the pressure at or above the test pressure for at least 8 hours.

(d) If a component other than pipe is the only item being replaced or added to a pipeline, a strength test after installation is not required, if the manufacturer of the component certifies that–

(1) The component was tested to at least the pressure required for the pipeline to which it is being added; or

(2) The component was manufactured under a quality control system that ensures that each item manufactured is at least equal in strength to a prototype and that the prototype was tested to at least the pressure required for the pipeline to which it is being added.

(e) For fabricated units and short sections of pipe, for which a post installation test is impractical, a preinstallation strength test must be conducted by maintaining the pressure for at least 4 hours.

[35 FR 13257, Aug. 19, 1970, as amended by Amdt. 192-85, 63 FR 37500, July 13, 1998]

§192.507 Test requirements for pipelines to operate at a hoop stress less than 30 percent of SMYS and at or above 100 p.s.i. (689 kPa) gage.

Except for service lines and plastic pipelines, each segment of a pipeline that is to be operated at a hoop stress less than 30 percent of SMYS and at or above 100 p.s.i. (689 kPa) gage must be tested in accordance with the following:

(a) The pipeline operator must use a test procedure that will ensure discovery of all potentially hazardous leaks in the segment being tested.

(b) If, during the test, the segment is to be stressed to 20 percent or more of SMYS and natural gas, inert gas, or air is the test medium–

(1) A leak test must be made at a pressure between 100 p.s.i. (689 kPa) gage and the pressure required to produce a hoop stress of 20 percent of SMYS; or

(2) The line must be walked to check for leaks while the hoop stress is held at approximately 20 percent of SMYS.

(c) The pressure must be maintained at or above the test pressure for at least 1 hour.

[35 FR 13257, Aug. 19, 1970, as amended by Amdt. 192-58, 53 FR 1633, Jan. 21, 1988; Amdt. 192-85, 63 FR 37500, July 13, 1998]

§192.509 Test requirements for pipelines to operate below 100 p.s.i. (689 kPa) gage.

Except for service lines and plastic pipelines, each segment of a pipeline that is to be operated below 100 p.s.i. (689 kPa) gage must be leak tested in accordance with the following:

(a) The test procedure used must ensure discovery of all potentially hazardous leaks in the segment being tested.

(b) Each main that is to be operated at less than 1 p.s.i. (6.9 kPa) gage must be tested to at least 10 p.s.i. (69 kPa) gage and each main to be operated at or above 1 p.s.i. (6.9 kPa) gage must be tested to at least 90 p.s.i. (621 kPa) gage.

[35 FR 13257, Aug. 19, 1970, as amended by Amdt. 192-58, 53 FR 1633, Jan. 21, 1988; Amdt. 192-85, 63 FR 37500, July 13, 1998]

§192.511 Test requirements for service lines.

(a) Each segment of a service line (other than plastic) must be leak tested in accordance with this section before being placed in service. If feasible, the service line connection to the main must be included in the test; if not feasible, it must be given a leakage test at the operating pressure when placed in service.

(b) Each segment of a service line (other than plastic) intended to be operated at a pressure of at least 1 p.s.i. (6.9 kPa) gage but not more than 40 p.s.i. (276 kPa) gage must be given a leak test at a pressure of not less than 50 p.s.i. (345 kPa) gage.

(c) Each segment of a service line (other than plastic) intended to be operated at pressures of more than 40 p.s.i. (276 kPa) gage must be tested to at least 90 p.s.i. (621 kPa) gage, except that each segment of the steel service line stressed to 20 percent or more of SMYS must be tested in accordance with §192.507 of this subpart.

[35 FR 13257, Aug. 19, 1970, as amended by Amdt. 192-75, 61 FR 18512, Apr. 26, 1996; Amdt. 192-85, 63 FR 37500, July 13, 1998]

§192.513 Test requirements for plastic pipelines.

(a) Each segment of a plastic pipeline must be tested in accordance with this section.

(b) The test procedure must insure discovery of all potentially hazardous leaks in the segment being tested.

(c) The test pressure must be at least 150 percent of the maximum operating pressure or 50 p.s.i. (345 kPa) gage, whichever is greater. However, the maximum test pressure may not be more than three times the pressure determined under §192.121, at a temperature not less than the pipe temperature during the test.

(d) During the test, the temperature of thermoplastic material may not be more than 100°F (38°C), or the temperature at which the material's long-term hydrostatic strength has been determined under the listed specification, whichever is greater.

[35 FR 13257, Aug. 19, 1970, as amended by Amdt. 192-77, 61 FR 27789, June 3, 1996; Amdt. 192-77A, 61 FR 45905, Aug. 30, 1996; Amdt. 192-85, 63 FR 37500, July 13, 1998]

§192.515 Environmental protection and safety requirements.

(a) In conducting tests under this s ubpart, each operator shall insure that every reasonable precaution is taken to protect its employees and the general public during the testing. Whenever the hoop stress of the segment of the pipeline being tested will exceed 50 percent of SMYS, the operator shall take all practicable steps to keep persons not working on the testing operation outside of the testing area until the pressure is reduced to or below the proposed maximum allowable operating pressure.

(b) The operator shall insure that the test medium is disposed of in a manner that will minimize damage to the environment.

§192.517 Records.

Each operator shall make, and retain for the useful life of the pipeline, a record of each test performed under §§192.505 and 192.507. The record must contain at least the following information:

(a) The operator's name, the name of the operator's employee responsible for making the test, and the name of any test company used.

(b) Test medium used.

(c) Test pressure.

(d) Test duration.

(e) Pressure recording charts, or other record of pressure readings.

(f) Elevation variations, whenever significant for the particular test.

(g) Leaks and failures noted and their disposition.

Subpart K–Uprating

§192.551 Scope.

This subpart prescribes minimum requirements for increasing maximum allowable operating pressures (uprating) for pipelines.

§192.553 General requirements.

(a) Pressure increases. Whenever the requirements of this subpart require that an increase in operating pressure be made in increments, the pressure must be increased gradually, at a rate that can be controlled, and in accordance with the following:

(1) At the end of each incremental increase, the pressure must be held constant while the entire segment of the pipeline that is affected is checked for leaks.

(2) Each leak detected must be repaired before a further pressure increase is made, except that a leak determined not to be potentially hazardous need not be repaired, if it is monitored during the pressure increase and it does not become potentially hazardous.

(b) Records. Each operator who uprates a segment of pipeline shall retain for the life of the segment a record of each investigation required by this subpart, of all work performed, and of each pressure test conducted, in connection with the uprating.

(c) Written plan. Each operator who uprates a segment of pipeline shall establish a written procedure that will ensure that each applicable requirement of this subpart is complied with.

(d) Limitation on increase in maximum allowable operating pressure. Except as provided in §192.555(c), a new maximum

allowable operating pressure established under this subpart may not exceed the maximum that would be allowed under this part for a new segment of pipeline constructed of the same materials in the same location. However, when uprating a steel pipeline, if any variable necessary to determine the design pressure under the design formula (§ 192.105) is unknown, the MAOP may be increased as provided in §192.619(a)(1).
[Amdt. 192-78, 61 FR 28770, June 6, 1996]

§192.555 Uprating to a pressure that will produce a hoop stress of 30 percent or more of SMYS in steel pipelines.

(a) Unless the requirements of this section have been met, no person may subject any segment of a steel pipeline to an operating pressure that will produce a hoop stress of 30 percent or more of SMYS and that is above the established maximum allowable operating pressure.

(b) Before increasing operating pressure above the previously established maximum allowable operating pressure the operator shall:

(1) Review the design, operating, and maintenance history and previous testing of the segment of pipeline and determine whether the proposed increase is safe and consistent with the requirements of this part; and

(2) Make any repairs, replacements, or alterations in the segment of pipeline that are necessary for safe operation at the increased pressure.

(c) After complying with paragraph (b) of this section, an operator may increase the maximum allowable operating pressure of a segment of pipeline constructed before September 12, 1970, to the highest pressure that is permitted under §192.619, using as test pressure the highest pressure to which the segment of pipeline was previously subjected (either in a strength test or in actual operation).

(d) After complying with paragraph (b) of this section, an operator that does not qualify under paragraph (c) of this section may increase the previously established maximum allowable operating pressure if at least one of the following requirements is met:

(1) The segment of pipeline is successfully tested in accordance with the requirements of this part for a new line of the same material in the same location.

(2) An increased maximum allowable operating pressure may be established for a segment of pipeline in a Class 1 location if the line has not previously been tested, and if:

(i) It is impractical to test it in accordance with the requirements of this part;

(ii) The new maximum operating pressure does not exceed 80 percent of that allowed for a new line of the same design in the same location; and,

(iii) The operator determines that the new maximum allowable operating pressure is consistent with the condition of the segment of pipeline and the design requirements of this part.

(e) Where a segment of pipeline is uprated in accordance with paragraph (c) or (d)(2) of this section, the increase in pressure must be made in increments that are equal to:

(1) 10 percent of the pressure before the uprating; or

(2) 25 percent of the total pressure increase, whichever produces the fewer number of increments.

§192.557 Uprating: Steel pipelines to a pressure that will produce a hoop stress less than 30 percent of SMYS: plastic, cast iron, and ductile iron pipelines.

(a) Unless the requirements of this section have been met, no person may subject:

(1) A segment of steel pipeline to an operating pressure that will produce a hoop stress less than 30 percent of SMYS and that is above the previously established maximum allowable operating pressure; or

(2) A plastic, cast iron, or ductile iron pipeline segment to an operating pressure that is above the previously established maximum allowable operating pressure.

(b) Before increasing operating pressure above the previously established maximum allowable operating pressure, the operator shall:

(1) Review the design, operating, and maintenance history of the segment of pipeline;

(2) Make a leakage survey (if it has been more than 1 year since the last survey) and repair any leaks that are found, except that a leak determined not to be potentially hazardous need not be repaired, if it is monitored during the pressure increase and it does not become potentially hazardous;

(3) Make any repairs, replacements, or alterations in the segment of pipeline that are necessary for safe operation at the increased pressure;

(4) Reinforce or anchor offsets, bends and dead ends in pipe joined by compression couplings or bell and spigot joints to prevent failure of the pipe joint, if the offset, bend, or dead end is exposed in an excavation;

(5) Isolate the segment of pipeline in which the pressure is to be increased from any adjacent segment that will continue to be operated at a lower pressure; and,

(6) If the pressure in mains or service lines, or both, is to be higher than the pressure delivered to the customer, install a service regulator on each service line and test each regulator to determine that it is functioning. Pressure may be increased as necessary to test each regulator, after a regulator has been installed on each pipeline subject to the increased pressure.

(c) After complying with paragraph (b) of this section, the increase in maximum allowable operating pressure must be made in increments that are equal to 10 p.s.i. (69 kPa) gage or 25 percent of the total pressure increase, whichever produces the fewer number of increments. Whenever the requirements of paragraph (b)(6) of this section apply, there must be at least two approximately equal incremental increases.

(d) If records for cast iron or ductile iron pipeline facilities are not complete enough to determine stresses produced by internal pressure, trench loading, rolling loads, beam stresses, and other bending loads, in evaluating the level of safety of the pipeline when operating at the proposed increased pressure, the following procedures must be followed:

(1) In estimating the stresses, if the original laying conditions cannot be ascertained, the operator shall assume that cast iron pipe was supported on blocks with tamped backfill and that ductile iron pipe was laid without blocks with tamped backfill.

(2) Unless the actual maximum cover depth is known, the operator shall measure the actual cover in at least three places where the cover is most likely to be greatest and shall use the greatest cover measured.

(3) Unless the actual nominal wall thickness is known, the operator shall

determine the wall thickness by cutting and measuring coupons from at least three separate pipe lengths. The coupons must be cut from pipe lengths in areas where the cover depth is most likely to be the greatest. The average of all measurements taken must be increased by the allowance indicated in the following table:

Pipe size (inches) (millimeters)	Allowance (inches) (millimeters)		
	Cast iron pipe		Ductile iron pipe
	Pit cast pipe	Centrifugally cast pipe	
3 to 8 (76 to 203)	0.075 (1.91)	0.065 (1.65)	0.065 (1.65)
10 to 12 (254 to 305)	0.08 (2.03)	0.07 (1.78)	0.07 (1.78)
14 to 24 (356 to 610)	0.08 (2.03)	0.08 (2.03)	0.075 (2.03)
30 to 42 (762 to 1067)	0.09 (2.29)	0.09 (2.29)	0.075 (1.91)
48 (1219)	0.09 (2.29)	0.09(2.29)	0.08 (2.03)
54 to 60 (1372 to 1524)	0.09 (2.29)		

(4) For cast iron pipe, unless the pipe manufacturing process is known, the operator shall assume that the pipe is pit cast pipe with a bursting tensile strength of 11,000 p.s.i. (76 MPa) gage and a modulus of rupture of 31,000 p.s.i. (214 MPa) gage.

[35 FR 13257, Aug. 19, 1970, as amended by Amdt. 192-37, 46 FR 10157, Feb. 2, 1981; Amdt. 192-62, 54 FR 5625, Feb. 6, 1989; Amdt. 192-85, 63 FR 37500, July 13, 1998]

Subpart L–Operations

§192.601 Scope.

This subpart prescribes minimum requirements for the operation of pipeline facilities.

§192.603 General provisions.

(a) No person may operate a segment of pipeline unless it is operated in accordance with this subpart.

(b) Each operator shall keep records necessary to administer the procedures established under §192.605.

(c) The Administrator or the State Agency that has submitted a current certification under the pipeline safety laws (49 U.S.C. 60101, *et seq.*) with respect to the pipeline facility governed by an operator's plans and procedures may, after notice and opportunity for hearing as provided in 49 CFR 190.237 or the relevant State procedures, require the operator to amend its plans and procedures as necessary to provide a reasonable level of safety.

[35 FR 13257, Aug. 9, 1970, as amended by 192-66, 56 FR 31087, July 9, 1991; Amdt. 192-71, 59 FR 6575, Feb. 11, 1994; Amdt. 192-75, 61 FR 18512, Apr. 26, 1996]

§192.605 Procedural manual for operations, maintenance, and emergencies.

Each operator shall include the following in its operating and maintenance plan:

(a) General. Each operator shall prepare and follow for each pipeline, a manual of written procedures for conducting operations and maintenance activities and for emergency response. For transmission lines, the manual must also include procedures for handling abnormal operations. This manual must be reviewed and updated by the operator at intervals not exceeding 15 months, but at least once each calendar year. This manual must be prepared before operations of a pipeline system commence. Appropriate parts of the manual must be kept at locations where operations and maintenance activities are conducted.

(b) Maintenance and normal operations. The manual required by paragraph (a) of this section must include procedures for the following, if applicable, to provide safety during maintenance and operations.

(1) Operating, maintaining, and repairing the pipeline in accordance with each of the requirements of this subpart and Subpart M of this part.

(2) Controlling corrosion in accordance with the operations and maintenance requirements of Subpart I of this part.

(3) Making construction records, maps, and operating history available to appropriate operating personnel.

(4) Gathering of data needed for reporting incidents under Part 191 of this chapter in a timely and effective manner.

(5) Starting up and shutting down any part of the pipeline in a manner designed to assure operation within the MAOP limits prescribed by this part, plus the build-up allowed for operation of pressure-limiting and control devices.

(6) Maintaining compressor stations, including provisions for isolating units or sections of pipe and for purging before returning to service.

(7) Starting, operating and shutting down gas compressor units.

(8) Periodically reviewing the work done by operator personnel to determine the effectiveness and adequacy of the procedures used in normal operation and maintenance and modifying the procedure when deficiencies are found.

(9) Taking adequate precautions in excavated trenches to protect personnel from the hazards of unsafe accumulations of vapor or gas, and making available when needed at the excavation, emergency rescue equipment, including a breathing apparatus and, a rescue harness and line.

(10) Systematic and routine testing and inspection of pipe-type or bottle-type holders including–

(i) Provision for detecting external corrosion before the strength of the container has been impaired;

(ii) Periodic sampling and testing of gas in storage to determine the dew point of vapors contained in the stored gas which, if condensed, might cause internal corrosion or interfere with the safe operation of the storage plant; and,

(iii) Periodic inspection and testing of pressure limiting equipment to determine that it is in safe operating condition and has adequate capacity.

(c) Abnormal operation. For transmission lines, the manual required by paragraph (a) of this section must include procedures for the following to provide safety when operating design limits have been exceeded:

(1) Responding to, investigating, and correcting the cause of:

(i) Unintended closure of valves or shutdowns;

(ii) Increase or decrease in pressure or flow rate outside normal operating limits;

(iii) Loss of communications;
(iv) Operation of any safety device; and,
(v) Any other foreseeable malfunction of a component, deviation from normal operation, or personnel error, which may result in a hazard to persons or property.
(2) Checking variations from normal operation after abnormal operation has ended at sufficient critical locations in the system to determine continued integrity and safe operation.
(3) Notifying responsible operator personnel when notice of an abnormal operation is received.
(4) Periodically reviewing the response of operator personnel to determine the effectiveness of the procedures controlling abnormal operation and taking corrective action where deficiencies are found.
(5) The requirements of this paragraph (c) do not apply to natural gas distribution operators that are operating transmission lines in connection with their distribution system.
(d) Safety-related condition reports. The manual required by paragraph (a) of this section must include instructions enabling personnel who perform operation and maintenance activities to recognize conditions that potentially may be safety-related conditions that are subject to the reporting requirements of §191.23 of this subchapter.
(e) Surveillance, emergency response, and accident investigation. The procedures required by §§192.613(a), 192.615, and 192.617 must be included in the manual required by paragraph (a) of this section.

[35 FR 13257, Aug. 19, 1970, as amended by Amdt. 192-59, 53 FR 24942, July 1,1988; Amdt. 192-59C, 53 FR 26560, July 13, 1988; Amdt. 192-71, 59 FR 6579, Feb. 11, 1994; Amdt. 192-71A, 60 FR 14381, Mar. 17, 1995]

§192.607 [Removed and Reserved]

[35 FR 13257, Aug. 10, 1970, as amended by Amdt. 192-5, 36 FR 18194, Sept. 10, 1971; Amdt. 192-78, 61 FR 28770, June 6, 1996]

§192.609 Change in class location: Required study.

Whenever an increase in population density indicates a change in class location for a segment of an existing steel pipeline operating at a hoop stress that is more than 40 percent of SMYS, or indicates that the hoop stress corresponding to the established maximum allowable operating pressure for a segment of existing pipeline is not commensurate with the present class location, the operator shall immediately make a study to determine;
(a) The present class location for the segment involved.
(b) The design, construction, and testing procedures followed in the original construction, and a comparison of these procedures with those required for the present class location by the applicable provisions of this part.
(c) The physical condition of the segment to the extent it can be ascertained from available records;
(d) The operating and maintenance history of the segment;
(e) The maximum actual operating pressure and the corresponding operating hoop stress, taking pressure gradient into account, for the segment of pipeline involved; and,
(f) The actual area affected by the population density increase, and physical barriers or other factors which may limit

further expansion of the more densely populated area.

§192.611 Change in class location: Confirmation or revision of maximum allowable operating pressure.

(a) If the hoop stress corresponding to the established maximum allowable operating pressure of a segment of pipeline is not commensurate with the present class location, and the segment is in satisfactory physical condition, the maximum allowable operating pressure of that segment of pipeline must be confirmed or revised according to one of the following requirements:

(1) If the segment involved has been previously tested in place for a period of not less than 8 hours, the maximum allowable operating pressure is 0.8 times the test pressure in Class 2 locations, 0.667 times the test pressure in Class 3 locations, or 0.555 times the test pressure in Class 4 locations. The corresponding hoop stress may not exceed 72 percent of the SMYS of the pipe in Class 2 locations, 60 percent of SMYS in Class 3 locations, or 50 percent of SMYS in Class 4 locations.

(2) The maximum allowable operating pressure of the segment involved must be reduced so that the corresponding hoop stress is not more than that allowed by this part for new segments of pipelines in the existing class location.

(3) The segment involved must be tested in accordance with the applicable requirements of Subpart J of this part, and its maximum allowable operating pressure must then be established according to the following criteria:

(i) The maximum allowable operating pressure after the requalification test is 0.8 times the test pressure for Class 2 locations, 0.667 times the test pressure for Class 3 locations, and 0.555 times the test pressure for Class 4 locations.

(ii) The corresponding hoop stress may not exceed 72 percent of the SMYS of the pipe in Class 2 locations, 60 percent of SMYS in Class 3 locations, or 50 percent of SMYS in Class 4 locations.

(b) The maximum allowable operating pressure confirmed or revised in accordance with this section, may not exceed the maximum allowable operating pressure established before the confirmation or revision.

(c) Confirmation or revision of the maximum allowable operating pressure of a segment of pipeline in accordance with this section does not preclude the application of §§192.553 and 192.555.

(d) Confirmation or revision of the maximum allowable operating pressure that is required as a result of a study under §192.609 must be completed within 18 months of the change in class location. Pressure reduction under paragraph (a)(1) or (2) of this section within the 18-month period does not preclude establishing a maximum allowable operating pressure under paragraph (a)(3) of this section at a later date.

[35 FR 13257, Aug. 19, 1970, as amended by Amdt. 192-5, 36 FR 18195, Sept. 10, 1971; Amdt. 192-53, 51 FR 34987, Oct. 1, 1986; Amdt. 192-63, 54 FR 24173, June 6, 1989; Amdt. 192-78, 61 FR 28770, June 6, 1996]

§192.612 Underwater inspection and reburial of pipelines in the Gulf of Mexico and its inlets.

(a) Each operator shall, in accordance with this section, conduct an underwater inspection of its pipelines in the Gulf of Mexico and its inlets. The inspection must be conducted after October 3, 1989 and before November 16, 1992.

(b) If, as a result of an inspection under paragraph (a) of this section, or upon notification by any person, an operator discovers that a pipeline it operates is exposed on the seabed or constitutes a hazard to navigation, the operator shall–

(1) Promptly, but not later than 24 hours after discovery, notify the National Response Center, telephone: 1-800-424-8802 of the location, and, if available, the geographic coordinates of that pipeline;

(2) Promptly, but not later than 7 days after discovery, mark the location of the pipeline in accordance with 33 CFR Part 64 at the ends of the pipeline segment and at intervals of not over 500 yards (457 meters) long, except that a pipeline segment less than 200 yards (183 meters) long need only be marked at the center; and,

(3) Within 6 months after discovery, or not later than November 1 of the following year if the 6 month period is later than November 1 of the year the discovery is made, place the pipeline so that the top of the pipe is 36 inches (914 millimeters) below the seabed for normal excavation or 18 inches (457 millimeters) for rock excavation.

[Amdt. 192-67, 56 FR 63764, Dec. 5, 1991 as amended by Amdt. 192-85, 63 FR 37500, July 13, 1998]

§192.613 Continuing surveillance.

(a) Each operator shall have a procedure for continuing surveillance of its facilities to determine and take appropriate action concerning changes in class location, failures, leakage history, corrosion, substantial changes in cathodic protection requirements, and other unusual operating and maintenance conditions.

(b) If a segment of pipeline is determined to be in unsatisfactory condition but no immediate hazard exists, the operator shall initiate a program to recondition or phase out the segment involved, or, if the segment cannot be reconditioned or phased out, reduce the maximum allowable operating pressure in accordance with §192.619(a) and (b).

§192.614 Damage prevention program.

(a) Except as provided in paragraphs (d) and (e) of this section, each operator of a buried pipeline shall carry out, in accordance with this section, a written program to prevent damage to that pipeline from excavation activities. For the purpose of this section, the term "excavation activities" includes excavation, blasting, boring, tunneling, backfilling, the removal of aboveground structures by either explosive or mechanical means, and other earth moving operations.

(b) An operator may comply with any of the requirements of paragraph (c) of this section through participation in a public service program, such as a one-call system, but such participation does not relieve the operator of responsibility for compliance with this section. However, an operator must perform the duties of paragraph (c)(3) of this section through participation in a one-call system, if that one-call system is a qualified one-call system. In areas that are covered by more than one qualified one-call system, an operator need only join one of the qualified one-call systems if there is a central telephone number for excavators to call for excavation activities, or if the one-call systems in those areas communicate with one another. An operator's pipeline system must be covered by a qualified one-call system where there is one in place. For the purpose of this section, a one-call system is considered a "qualified one-call system" if it meets the requirements of section (b)(1) or (b)(2) of this section.

(1) The state has adopted a one-call damage prevention program under §198.37 of this chapter, or

(2) The one-call system:

(i) Is operated in accordance with §198.39 of this chapter;

(ii) Provides a pipeline operator an opportunity similar to a voluntary participant to have a part in management responsibilities; and

(iii) Assesses a participating pipeline operator a fee that is proportionate to the costs of the one-call system's coverage of the operator's pipeline.

(c) The damage prevention program required by paragraph (a) of this section must, at a minimum:

(1) Include the identity, on a current basis, of persons who normally engage in excavation activities in the area in which the pipeline is located.

(2) Provides for notification of the public in the vicinity of the pipeline and actual notification of the persons identified in paragraph (c)(1) of this section of the following as often as needed to make them aware of the damage prevention program:

(i) The program's existence and purpose; and

(ii) How to learn the location of underground pipelines before excavation activities are begun.

(3) Provide a means of receiving and recording notification of planned excavation activities.

(4) If the operator has buried pipelines in the area of excavation activity, provide for actual notification of persons who give notice of their intent to excavate of the type of temporary marking to be provided and how to identify the markings.

(5) Provide for temporary marking of buried pipelines in the area of excavation activity before, as far as practical, the activity begins.

(6) Provide as follows for inspection of pipelines that an operator has reason to believe could be damaged by excavation activities:

(i) The inspection must be done as frequently as necessary during and after the activities to verify the integrity of the pipeline; and

(ii) In the case of blasting, any inspection must include leakage surveys.

(d) A damage prevention program under this section is not required for the following pipelines:

(1) Pipelines located offshore.

(2) Pipelines, other than those located offshore, in Class 1 or 2 locations until September 20, 1995.

(3) Pipelines to which access is physically controlled by the operator.

(e) Pipelines operated by persons other than municipalities (including operators of master meters) whose primary activity does not include the transportation of gas need not comply with the following:

(1) The requirement of paragraph (a) of this section that the damage prevention program be written; and

(2) The requirements of paragraphs (c)(1) and (c)(2) of this section.

[Amdt. 192-40, 47 FR 13818, Apr. 1, 1982; Amdt. 192-57, 52 FR 32798, Aug. 31, 1987; Amdt. 192-73, 60 FR 14646, Mar. 20, 1995; Amdt. 192-78, 61 FR 28770, June 6, 1996; Amdt. 192-82, 62 FR 61695, Nov. 19, 1997; Amdt. 192-84, 63 FR 7721, Feb. 17, 1998; Amdt. 192-84A, 63 FR 38757, July 20, 1998]

§192.615 Emergency plans.

(a) Each operator shall establish written procedures to minimize the hazard resulting from a gas pipeline emergency. At a minimum, the procedures must provide for the following:

(1) Receiving, identifying, and classifying notices of events which require immediate response by the operator.

(2) Establishing and maintaining adequate means of communication with appropriate fire, police, and other public officials.

(3) Prompt and effective response to a notice of each type of emergency, including the following:

(i) Gas detected inside or near a building.

(ii) Fire located near or directly involving a pipeline facility.

(iii) Explosion occurring near or directly involving a pipeline facility.

(iv) Natural disaster.

(4) The availability of personnel, equipment, tools, and materials, as needed at the scene of an emergency.

(5) Actions directed toward protecting people first and then property.

(6) Emergency shutdown and pressure reduction in any section of the operator's pipeline system necessary to minimize hazards to life or property.

(7) Making safe any actual or potential hazard to life or property.

(8) Notifying appropriate fire, police, and other public officials of gas pipeline emergencies and coordinating with them both planned responses and actual responses during an emergency.

(9) Safely restoring any service outage.

(10) Beginning action under §192.617, if applicable, as soon after the end of the emergency as possible.

(b) Each operator shall:

(1) Furnish its supervisors who are responsible for emergency action a copy of that portion of the latest edition of the emergency procedures established under paragraph (a) of this section as necessary for compliance with those procedures.

(2) Train the appropriate operating personnel to assure that they are knowledgeable of the emergency procedures and verify that the training is effective.

(3) Review employee activities to determine whether the procedures were effectively followed in each emergency.

(c) Each operator shall establish and maintain liaison with appropriate fire, police, and other public officials to:

(1) Learn the responsibility and resources of each government organization that may respond to a gas pipeline emergency;

(2) Acquaint the officials with the operator's ability in responding to a gas pipeline emergency;

(3) Identify the types of gas pipeline emergencies of which the operator notifies the officials; and,

(4) Plan how the operator and officials can engage in mutual assistance to minimize hazards to life or property.

[35 FR 13257, Aug. 19, 1970 as amended by Amdt. 192-24, 41 FR 13586, Mar. 31, 1976; Amdt. 192-71, 59 FR 6585, Feb. 11, 1994]

§192.616 Public education

Each operator shall establish a continuing educational program to enable customers, the public, appropriate government organizations, and persons engaged in excavation related activities to recognize a gas pipeline emergency for the purpose of reporting it to the operator or the appropriate public officials. The program and the media used must be as comprehensive as necessary to reach all areas in which the operator transports gas. The program must be conducted in English and in other languages commonly understood by a significant number and concentration of the non-English speaking population in the operator's area.

[Amdt. 192-71, 59 FR 6575, Feb. 11, 1994]

§192.617 Investigation of failures.

Each operator shall establish procedures for analyzing accidents and failures, including the selection of samples of the failed facility or equipment for laboratory examination, where appropriate, for the purpose of determining the causes of the failure and minimizing the possibility of a recurrence.

§192.619 Maximum allowable operating pressure: Steel or plastic pipelines.

(a) Except as provided in paragraph (c) of this section, no person may operate a segment of steel or plastic pipeline at a pressure that exceeds the lowest of the following:

(1) The design pressure of the weakest element in the segment, determined in accordance with Subparts C and D of this part. However, for steel pipe in pipelines being converted under §192.14 or uprated under subpart K of this part, if any variable necessary to determine the design pressure under the design formula (§192.105) is unknown, one of the following pressures is to be used as design pressure:

(i) Eighty percent of the first test pressure that produces yield under section N5.0 of Appendix N of ASME B31.8, reduced by the appropriate factor in paragraph (a)(2)(ii) of this section; or

(ii) If the pipe is 12¾ inches (324 mm) or less in outside diameter and is not tested to yield under this paragraph, 200 p.s.i. (1379 kPa) gage.

(2) The pressure obtained by dividing the pressure to which the segment was tested after construction as follows:

(i) For plastic pipe in all locations, the test pressure is divided by a factor of 1.5.

(ii) For steel pipe operated at 100 p.s.i. (689 kPa) gage or more, the test pressure is divided by a factor determined in accordance with the following table:

	Factors[1], segment		
Class location	Installed before (Nov. 12, 1970)	Installed after (Nov. 11, 1970)	Covered under §192.14
1	1.1	1.1	1.25
2	1.25	1.25	1.25
3	1.4	1.5	1.5
4	1.4	1.5	1.5

[1] For offshore segments installed, uprated or converted after July 31, 1977, that are not located on an offshore platform, the factor is 1.25. For segments installed, uprated or

converted after July 31, 1977, that are located on an offshore platform or on a platform in inland navigable waters, including a pipe riser, the factor is 1.5.

(3) The highest actual operating pressure to which the segment was subjected during the 5 years preceding July 1, 1970 (or in the case of offshore gathering lines, July 1, 1976), unless the segment was tested in accordance with paragraph (a)(2) of this section after July 1, 1965 (or in the case of offshore gathering lines, July 1, 1971), or the segment was uprated in accordance with Subpart K of this part.

(4) The pressure determined by the operator to be the maximum safe pressure after considering the history of the segment, particularly known corrosion and the actual operating pressure.

(b) No person may operate a segment to which paragraph (a)(4) of this section is applicable, unless overpressure protective devices are installed on the segment in a manner that will prevent the maximum allowable operating pressure from being exceeded, in accordance with §192.195.

(c) Notwithstanding the other requirements of this section, an operator may operate a segment of pipeline found to be in satisfactory condition, considering its operating and maintenance history, at the highest actual operating pressure to which the segment was subjected during the 5 years preceding July 1, 1970, or in the case of offshore gathering lines, July 1, 1976, subject to the requirements of §192.611.

[35 FR 13257, Aug. 19, 1970 as amended by Amdt. 192-3, 35 FR 17559, Nov. 17, 1970; Amdt. 192-27, 41 FR 34598, Aug. 16, 1976; Amdt. 192-27A, 41 FR 47252, Oct. 28, 1976; Amdt. 192-30, 42 FR 60146, Nov. 25, 1977; Amdt. 192-78, 61 FR 28770, June 6, 1996; Amdt. 192-85, 63 FR 37500, July 13, 1998]

§192.621 Maximum allowable operating pressure: High-Pressure distribution systems.

(a) No person may operate a segment of a high pressure distribution system at a pressure that exceeds the lowest of the following pressures, as applicable:

(1) The design pressure of the weakest element in the segment, determined in accordance with Subparts C and D of this part.

(2) 60 p.s.i. (414 kPa) gage, for a segment of a distribution system otherwise designated to operate at over 60 p.s.i. (414 kPa) gage, unless the service lines in the segment are equipped with service regulators or other pressure limiting devices in series that meet the requirements of §192.197(c).

(3) 25 p.s.i. (172 kPa) gage in segments of cast iron pipe in which there are unreinforced bell and spigot joints.

(4) The pressure limits to which a joint could be subjected without the possibility of its parting.

(5) The pressure determined by the operator to be the maximum safe pressure after considering the history of the segment, particularly known corrosion and the actual operating pressures.

(b) No person may operate a segment of pipeline to which paragraph (a)(5) of this section applies, unless overpressure protective devices are installed on the segment in a manner that will prevent the maximum allowable operating pressure from being exceeded, in accordance with §192.195.

[35 FR 13257, Aug. 19, 1970 as amended by Amdt. 192-85, 63 FR 37500, July 13, 1998]

§192.623 Maximum and minimum allowable operating pressure; Low-pressure distribution systems.

(a) No person may operate a low-pressure distribution system at a pressure high enough to make unsafe the operation of any connected and properly adjusted low-pressure gas burning equipment.

(b) No person may operate a low pressure distribution system at a pressure lower than the minimum pressure at which the safe and continuing operation of any connected and properly adjusted low-pressure gas burning equipment can be assured.

[Amdt. 192-75, 61 FR 18512, Apr. 26, 1996]

§192.625 Odorization of gas.

(a) A combustible gas in a distribution line must contain a natural odorant or be odorized so that at a concentration in air of one-fifth of the lower explosive limit, the gas is readily detectable by a person with a normal sense of smell.

(b) After December 31, 1976, a combustible gas in a transmission line in a Class 3 or Class 4 location must comply with the requirements of paragraph (a) of this section unless:

(1) At least 50 percent of the length of the line downstream from that location is in a Class 1 or Class 2 location;

(2) The line transports gas to any of the following facilities which received gas without an odorant from that line before May 5, 1975:

(i) An underground storage field;

(ii) A gas processing plant;

(iii) A gas dehydration plant; or

(iv) An industrial plant using gas in a process where the presence of an odorant:

(A) Makes the end product unfit for the purpose for which it is intended;

(B) Reduces the activity of a catalyst; or

(C) Reduces the percentage completion of a chemical reaction;

(3) In the case of a lateral line which transports gas to a distribution center, at least 50 percent of the length of that line is in a Class 1 or Class 2 location; or

(4) The combustible gas is hydrogen intended for use as a feedstock in a manufacturing process.

(c) In the concentrations in which it is used, the odorant in combustible gases must comply with the following:

(1) The odorant may not be deleterious to persons, materials, or pipe.

(2) The products of combustion from the odorant may not be toxic when breathed nor may they be corrosive or harmful to those materials to which the products of combustion will be exposed.

(d) The odorant may not be soluble in water to an extent greater than 2.5 parts to 100 parts by weight.

(e) Equipment for odorization must introduce the odorant without wide variations in the level of odorant.

(f) Each operator shall conduct periodic sampling of combustible gases to assure the proper concentration of odorant in accordance with this section. Operators of master meter systems may comply with this requirement by–

(1) Receiving written verification from their gas source that the gas has the proper concentration of odorant; and

(2) Conducting periodic "sniff" tests at the extremities of the system to confirm that the gas contains odorant.

[35 FR 13257, Aug. 19, 1970 as amended by Amdt. 192-2, 35 FR 17335, Nov. 11, 1970; Amdt. 192-6, 36 FR 25423, Dec. 31, 1971; Amdt. 192-7, 37 FR 17970, Sept. 2, 1972; Amdt. 192-14, 38 FR 14943, June 7, 1973; Amdt. 192-15, 38 FR 35471, Dec. 28, 1973; Amdt. 192-16, 39 FR 45253, Dec. 31, 1974; Amdt. 192-21, 40 FR 20279, May 9, 1975; Amdt. 192-58, 53 FR 1633, Jan. 21, 1988; Amdt. 192-76, 61 FR 26121, May 24, 1996; Amdt. 192-78, 61 FR 28770, June 6, 1996]

§192.627 Tapping pipelines under pressure.

Each tap made on a pipeline under pressure must be performed by a crew qualified to make hot taps.

§192.629 Purging of pipelines.

(a) When a pipeline is being purged of air by use of gas, the gas must be released into one end of the line in a moderately rapid and continuous flow. If gas cannot be supplied in sufficient quantity to prevent the formation of a hazardous mixture of gas and air, a slug of inert gas must be released into the line before the gas.

(b) When a pipeline is being purged of gas by use of air, the air must be released into one end of the line in a moderately rapid and continuous flow. If air cannot be supplied in sufficient quantity to prevent the formation of a hazardous mixture of gas and air, a slug of inert gas must be released into the line before the air.

Subpart M–Maintenance

§192.701 Scope.

This subpart prescribes minimum requirements for maintenance of pipeline facilities.

§192.703 General.

(a) No person may operate a segment of pipeline, unless it is maintained in accordance with this subpart.

(b) Each segment of pipeline that becomes unsafe must be replaced, repaired, or removed from service.

(c) Hazardous leaks must be repaired promptly.

§192.705 Transmission lines: Patrolling.

(a) Each operator shall have a patrol program to observe surface conditions on and adjacent to the transmission line right-of-way for indications of leaks, construction activity, and other factors affecting safety and operation.

(b) The frequency of patrols is determined by the size of the line, the operating pressures, the class location, terrain, weather, and other relevant factors, but intervals between patrols may not be longer than prescribed in the following table:

Class location of line	Maximum interval between patrols	
	At highway and railroad crossings	At all other places
1, 2	7½ months; but at least twice each calendar year.	15 months; but at least once each calendar year.
3	4½ months; but at least four times each calendar year	7½ months; but at least twice each calendar year.
4	4½ months; but at least four times each calendar year.	4½ months; but at least four times each calendar year.

(c) Methods of patrolling include walking, driving, flying or other appropriate means of traversing the right-of-way.

[35 FR 13257, Aug. 19, 1970, as amended by Amdt. 192-21, 40 FR 20283, May 9, 1975; Amdt. 192-43, 47 FR 46850, Oct. 21, 1982; Amdt. 192-78, 61 FR 28770, June 6, 1996]

§192.706 Transmission lines: Leakage surveys.

Leakage surveys of a transmission line must be conducted at intervals not exceeding 15 months, but at least once each calendar year. However, in the case of a transmission line which transports gas in conformity with §192.625 without an odor or odorant, leakage surveys using leak detector equipment must be conducted–

(a) In Class 3 locations, at intervals not exceeding 7½ months, but at least twice each calendar year; and

(b) In Class 4 locations, at intervals not exceeding 4½ months, but at least four times each calendar year.

[Amdt. 192-21, 40 FR 20283, May 9, 1975, as amended by Amdt. 192-43, 47 FR 46850, Oct. 21, 1982; Amdt. 192-71, 59 FR 6575, Feb. 11, 1994]

§192.707 Line markers for mains and transmission lines.

(a) Buried pipelines. Except as provided in paragraph (b) of this section, a line marker must be placed and maintained as close as practical over each buried main and transmission line:

(1) At each crossing of a public road and railroad; and

(2) Wherever necessary to identify the location of the transmission line or main to reduce the possibility of damage or interference.

(b) Exceptions for buried pipelines. Line markers are not required for the following pipelines:

(1) Mains and transmission lines located offshore, or at crossings of or under waterways and other bodies of water.

(2) Mains in Class 3 or Class 4 locations where a damage prevention program is in effect under §192.614.

(3) Transmission lines in Class 3 or 4 locations until March 20, 1996.

(4) Transmission lines in Class 3 or 4 locations where placement of a line marker is impractical.

(c) Pipelines aboveground. Line markers must be placed and maintained along each section of a main and transmission line that is located aboveground in an area accessible to the public.

(d) Marker warning. The following must be written legibly on a background of sharply contrasting color on each line marker:

(1) The word "Warning," "Caution," or "Danger" followed by the words "Gas (or name of gas transported) Pipeline" all of which, except for markers in heavily developed urban areas, must be in letters at least 1 inch (25 millimeters) high with ¼ inch (6.4 millimeters) stroke.

(2) The name of the operator and telephone number (including area code) where the operator can be reached at all times.

[35 FR 13257, Aug. 19, 1970, as amended by Amdt. 192-20A, 41 FR 56808, Dec. 30, 1976; Amdt. 192-20, 40 FR 13505, Mar. 27, 1975; Amdt. 192-27, 41 FR 39752, Sept. 16, 1976; Amdt. 192-40, 47 FR 13818, Apr. 1, 1982; Amdt. 192-44, 48 FR 25206, June 6, 1983; Amdt. 192-73, 60 FR 14646, Mar. 20, 1995; Amdt. 192-85, 63 FR 37500, July 13, 1998]

§192.709 Transmission lines: Record keeping.

Each operator shall maintain the following records for transmission lines for the periods specified:

(a) The date, location, and description of each repair made to pipe (including pipe-to-pipe connections) must be retained for as long as the pipe remains in service.

(b) The date, location, and description of each repair made to parts of the pipeline system other than pipe must be retained for at least 5 years. However, repairs generated by patrols, surveys, inspections, or tests required by subparts L and M of this part must be retained in accordance with paragraph (c) of this section.

(c) A record of each patrol, survey, inspection, and test required by subparts L and M of this part must be retained for at least 5 years or until the next patrol, survey, inspection, or test is completed, whichever is longer.

[Amdt. 192-78, 61 FR 28770, June 6, 1996]

§192.711 Transmission lines: General requirements for repair procedures.

(a) Each operator shall take immediate temporary measures to protect the public whenever:

(1) A leak, imperfection, or damage that impairs its serviceability is found in a segment of steel transmission line operating at or above 40 percent of the SMYS; and

(2) It is not feasible to make a permanent repair at the time of discovery.

As soon as feasible the operator shall make permanent repairs.

(b) Except as provided in §192.717(a)(3), no operator may use a welded patch as a means of repair.

[35 FR 13257, Aug. 19, 1970, as amended by Amdt. 192-27B, 45 FR 3272, Jan. 17, 1980]

§192.713 Transmission lines: Permanent field repair of imperfections and damages.

(a) Except as provided in paragraph (b) of this section, each imperfection or damage that impairs the serviceability of a segment of steel transmission line operating at or above 40 percent of SMYS must be repaired as follows:

(1) If it is feasible to take the segment out of service, the imperfection or damage must

be removed by cutting out a cylindrical piece of pipe and replacing it with pipe of similar or greater design strength.

(2) If it is not feasible to take the segment out of service, a full encirclement welded split sleeve of appropriate design must be applied over the imperfection or damage.

(3) If the segment is not taken out of service, the operating pressure must be reduced to a safe level during the repair operations.

(b) Submerged offshore pipelines and submerged pipelines in inland navigable waters may be repaired by mechanically applying a full encirclement split sleeve of appropriate design over the imperfection or damage.

[35 FR 13257, Aug. 19, 1970, as amended by Amdt. 192-27, 41 FR 34598, Aug. 16, 1976]

§192.715 Transmission lines: Permanent field repair of welds.

Each weld that is unacceptable under §192.241(c) must be repaired as follows:

(a) If it is feasible to take the segment of transmission line out of service, the weld must be repaired in accordance with the applicable requirements of §192.245.

(b) A weld may be repaired in accordance with §192.245 while the segment of transmission line is in service if:

(1) The weld is not leaking:

(2) The pressure in the segment is reduced so that it does not produce a stress that is more than 20 percent of the SMYS of the pipe; and

(3) Grinding of the defective area can be limited so that at least 1/8-inch (3.2 millimeters) thickness in the pipe weld remains.

(c) A defective weld which cannot be repaired in accordance with paragraph (a) or (b) of this section must be repaired by installing a full encirclement welded split sleeve of appropriate design.

[35 FR 13257, Aug. 19, 1970 as amended by Amdt. 192-85, 63 FR 37500, July 13, 1998]

§192.717 Transmission lines: Permanent field repair of leaks.

(a) Except as provided in paragraph (b) of this section, each permanent field repair of a leak on a transmission line must be made as follows:

(1) If feasible, the segment of transmission line must be taken out of service and repaired by cutting out a cylindrical piece of pipe and replacing it with pipe of similar or greater design strength.

(2) If it is not feasible to take the segment of transmission line out of service, repairs must be made by installing a full encirclement welded split sleeve of appropriate design, unless the transmission line:

(i) Is joined by mechanical couplings; and

(ii) Operates at less than 40 percent of SMYS.

(3) If the leak is due to a corrosion pit, the repair may be made by installing a properly designed bolt-on-leak clamp; or, if the leak is due to a corrosion pit and on pipe of not more than 40,000 p.s.i. (276 MPa) gage SMYS, the repair may be made by fillet welding over the pitted area a steel plate patch with rounded corners, of the same or greater thickness than the pipe, and not more than one-half of the diameter of the pipe in size.

(b) Submerged offshore pipelines and submerged pipelines in inland navigable waters may be repaired by mechanically applying a full encirclement split sleeve of appropriate design over the leak.

[35 FR 13257, Aug. 19, 1970, as amended by Amdt. 192-11, 37 FR 21816, Oct. 14, 1972; Amdt. 192-27, 41 FR 34598, Aug. 16, 1976; Amdt. 192-85, 63 FR 37500, July 13, 1998]

§192.719 Transmission lines: Testing of repairs.

(a) Testing of replacement pipe. If a segment of transmission line is repaired by cutting out the damaged portion of the pipe as a cylinder, the replacement pipe must be tested to the pressure required for a new line installed in the same location. This test may be made on the pipe before it is installed.

(b) Testing of repairs made by welding. Each repair made by welding in accordance with §§192.713, 192.715, and 192.717 must be examined in accordance with §192.241.

[35 FR 13257, Aug. 19, 1970, as amended by Amdt. 192-54, 51 FR 41634, Nov. 18, 1986]

§192.721 Distribution systems: Patrolling.

(a) The frequency of patrolling mains must be determined by the severity of the conditions which could cause failure or leakage, and the consequent hazards to public safety.

(b) Mains in places or on structures where anticipated physical movement or external loading could cause failure or leakage must be patrolled–

(1) In business districts, at intervals not exceeding 4? months, but at least four times each calendar year; and

(2) Outside business districts, at intervals not exceeding 7? months, but at least twice each calendar year.

[35 FR 13257, Aug. 19, 1970, as amended by Amdt. 192-43, 47 FR 46850, Oct. 21, 1982; Amdt. 192-78, 61 FR 28770, June 6, 1996]

§192.723 Distribution systems: Leakage surveys.

(a) Each operator of a distribution system shall conduct periodic leakage surveys in accordance with this section.

(b) The type and scope of the leakage control program must be determined by the nature of the operations and the local conditions, but it must meet the following minimum requirements:

(1) A leakage survey with leak detector equipment must be conducted in business districts, including tests of the atmosphere in gas, electric, telephone, sewer, and water system manholes, at cracks in pavement and sidewalks, and at other locations providing an opportunity for finding gas leaks, at intervals not exceeding 15 months, but at least once each calendar year.

(2) Leakage survey with leak detector equipment must be conducted outside business districts as frequently as necessary, but at intervals not exceeding 5 years. However, for cathodically unprotected distribution lines subject to §192.465(e) on which electrical surveys for corrosion are impractical, survey intervals may not exceed 3 years.

[35 FR 13257, Aug. 19, 1970, as amended by Amdt. 192-43, 47 FR 46850, Oct. 21, 1982; Amdt. 192-70, 58 FR 54524, Oct. 22, 1993; Amdt. 192-71, 59 FR 6575, Feb. 11, 1994]

§192.725 Test requirements for reinstating service lines.

(a) Except as provided in paragraph (b) of this section, each disconnected service line must be tested in the same manner as a new service line, before being reinstated.

(b) Each service line temporarily disconnected from the main must be tested from the point of disconnection to the service line valve in the same manner as a new service line, before reconnecting. However, if provisions are made to maintain continuous service, such as by installation of a bypass, any part of the original service line used to maintain continuous service need not be tested.

§192.727 Abandonment or deactivation of facilities.

(a) Each operator shall conduct abandonment or deactivation of pipelines in accordance with the requirements of this section.

(b) Each pipeline abandoned in place must be disconnected from all sources and supplies of gas; purged of gas; in the case of offshore pipelines, filled with water or inert materials; and sealed at the ends. However, the pipeline need not be purged when the volume of gas is so small that there is no potential hazard.

(c) Except for service lines, each inactive pipeline that is not being maintained under this part must be disconnected from all sources and supplies of gas; purged of gas; in the case of offshore pipelines, filled with water or inert materials; and sealed at the ends. However, the pipeline need not be purged when the volume of gas is so small that there is no potential hazard.

(d) Whenever service to a customer is discontinued, one of the following must be complied with:

(1) The valve that is closed to prevent the flow of gas to the customer must be provided with a locking device or other means designed to prevent the opening of the valve by persons other than those authorized by the operator.

(2) A mechanical device or fitting that will prevent the flow of gas must be installed in the service line or in the meter assembly.

(3) The customer's piping must be physically disconnected from the gas supply and the open pipe ends sealed.

(e) If air is used for purging, the operator shall insure that a combustible mixture is not present after purging.

(f) Each abandoned vault must be filled with a suitable compacted material.

[35 FR 13257, Aug. 19, 1970, as amended by Amdt. 192-8, 37 FR 20694, Oct. 3, 1972, Amdt. 192-27, 41 FR 34598, Aug. 16, 1976; Amdt. 192-71, 59 FR 6575, Feb. 11, 1994]

§192.729 [Removed]

[35 FR 13257, Aug. 19, 1970, as amended by Amdt. 192-71, 59 FR 6575, Feb. 11, 1994]

§192.731 Compressor stations: Inspection and testing of relief devices.

(a) Except for rupture discs, each pressure relieving device in a compressor station must be inspected and tested in accordance with §§192.739 and 192.743, and must be operated periodically to determine that it opens at the correct set pressure.

(b) Any defective or inadequate equipment found must be promptly repaired or replaced.

(c) Each remote control shutdown device must be inspected and tested at intervals not exceeding 15 months, but at least once each calendar year, to determine that it functions properly.

[35 FR 13257, Aug. 19, 1970, as amended by Amdt. 192-43, 47 FR 46850, Oct. 21, 1982]

§192.733 [Removed]

[35 FR 13257, Aug. 19, 1970, as amended by Amdt. 192-71, 59 FR 6575, Feb. 11, 1994]

§192.735 Compressor stations: Storage of combustible materials.

(a) Flammable or combustible materials in quantities beyond those required for everyday use, or other than those normally used in compressor buildings, must be stored a safe distance from the compressor building.

(b) Aboveground oil or gasoline storage tanks must be protected in accordance with National Fire Protection Association Standard No. 30.

§192.736 Compressor stations: Gas detection.

(a) Not later than September 16, 1996, each compressor building in a compressor station must have a fixed gas detection and alarm system, unless the building is–

(1) Constructed so that at least 50 percent of its upright side area is permanently open; or

(2) Located in an unattended field compressor station of 1,000 horsepower (746 kilowatts) or less.

(b) Except when shutdown of the system is necessary for maintenance under paragraph (c) of this section, each gas detection and alarm system required by this section must–

(1) Continuously monitor the compressor building for a concentration of gas in air of not more than 25 percent of the lower explosive limit; and

(2) If that concentration of gas is detected, warn persons about to enter the building and persons inside the building of the danger.

(c) Each gas detection and alarm system required by this section must be maintained to function properly. The maintenance must include performance tests.

[Amdt. 192-69, 58 FR 48460, Sept. 16, 1993 as amended by Amdt. 192-85, 63 FR 37500, July 13, 1998]

§192.737 [Removed]

[35 FR 13257, Aug. 19, 1970, as amended by Amdt. 192-71, 59 FR 6575, Feb. 11, 1994]

§192.739 Pressure limiting and regulating stations: Inspection and testing.

Each pressure limiting station, relief device (except rupture discs), and pressure regulating station and its equipment must be subjected at intervals not exceeding 15 months, but at least once each calendar year, to inspections and tests to determine that it is–

(a) In good mechanical condition;

(b) Adequate from the standpoint of capacity and reliability of operation for the service in which it is employed;

(c) Set to function at the correct pressure; and,

(d) Properly installed and protected from dirt, liquids, or other conditions that might prevent proper operation.

[35 FR 13257, Aug. 19, 1970, as amended by Amdt. 192-43, 47 FR 46850, Oct. 21, 1982]

§192.741 Pressure limiting and regulating stations: Telemetering or recording gauges.

(a) Each distribution system supplied by more than one district pressure regulating station must be equipped with telemetering or recording pressure gauges to indicate the gas pressure in the district.

(b) On distribution systems supplied by a single district pressure regulating station, the operator shall determine the necessity of installing telemetering or recording ga uges in the district, taking into consideration the number of customers supplied, the operating pressures, the capacity of the installation, and other operating conditions.

(c) If there are indications of abnormally high- or low-pressure, the regulator and the auxiliary equipment must be inspected and the necessary measures employed to correct any unsatisfactory operating conditions.

§192.743 Pressure limiting and regulating stations: Testing of relief devices.

(a) If feasible, pressure relief devices (except rupture discs) must be tested in place, at intervals not exceeding 15 months, but at least once each calendar year, to determine that they have enough capacity to limit the pressure on the facilities to which they are connected to the desired maximum pressure.

(b) If a test is not feasible, review and calculation of the required capacity of the relieving device at each station must be made at intervals not exceeding 15 months, but at least once each calendar year, and these required capacities compared with the rated or experimentally determined relieving capacity of the device for the operating conditions under which it works. After the initial calculations, subsequent calculations are not required if the review documents that parameters have not changed in a manner which would cause the capacity to be less than required.

(c) If the relieving device is of insufficient capacity, a new or additional device must be installed to provide the additional capacity required.

[35 FR 13257, Aug. 19, 1970, as amended by Amdt. 192-43, 47 FR 46850, Oct. 21, 1982; and Amdt. 192-55, 51 FR 41633. Nov. 18, 1986]

§192.745 Valve maintenance: Transmission lines.

Each transmission line valve that might be required during any emergency must be inspected and partially operated at intervals not exceeding 15-months, but at least once each calendar year.

[35 FR 13257, Aug. 19, 1970, as amended by Amdt. 192-43, 47 FR 46850, Oct. 21, 1982]

§192.747 Valve maintenance: Distribution systems.

Each valve, the use of which may be necessary for the safe operation of a distribution system, must be checked and serviced at intervals not exceeding 15 months, but at least once each calendar year.

[35 FR 13257, Aug. 19, 1970, as amended by Amdt. 192-43, 47 FR 46850, Oct. 21, 1982]

§192.749 Vault maintenance.

(a) Each vault housing pressure regulating and pressure limiting equipment, and having a volumetric internal content of 200 cubic feet (5.66 cubic meters) or more, must be inspected at intervals not exceeding 15 months, but at least once each calendar year, to determine that it is in good physical condition and adequately ventilated.

(b) If gas is found in the vault, the equipment in the vault must be inspected for leaks, and any leaks found must be repaired.

(c) The ventilating equipment must also be inspected to determine that it is functioning properly.

(d) Each vault cover must be inspected to assure that it does not present a hazard to public safety.

[35 FR 13257, Aug. 19, 1970, as amended by Amdt. 192-43, 47 FR 46850, Oct. 21, 1982; Amdt. 192-85, 63 FR 37500, July 13, 1998]

§192.751 Prevention of accidental ignition.

Each operator shall take steps to minimize the danger of accidental ignition of gas in any structure or area where the presence of gas constitutes a hazard of fire or explosion, including the following:

(a) When a hazardous amount of gas is being vented into open air, each potential source of ignition must be removed from the area and a fire extinguisher must be provided.

(b) Gas or electric welding or cutting may not be performed on pipe or on pipe components that contain a combustible mixture of gas and air in the area of work.

(c) Post warning signs, where appropriate.

§192.753 Caulked bell and spigot joints.

(a) Each cast-iron caulked bell and spigot joint that is subject to pressures of 25 p.s.i. (172 kPa) gage or more must be sealed with:

(1) A mechanical leak clamp; or

(2) A material or device which:

(i) Does not reduce the flexibility of the joint;

(ii) Permanently bonds, either chemically or mechanically, or both, with the bell and spigot metal surfaces or adjacent pipe metal surfaces; and,

(iii) Seals and bonds in a manner that meets the strength, environmental, and chemical compatibility requirements of §§192.53(a) and (b) and 192.143.

(b) Each cast iron caulked bell and spigot joint that is subject to pressures of less than 25 p.s.i. (172 kPa) gage and is exposed for any reason, must be sealed by a means other than caulking.

[35 FR 13257, Aug. 19, 1970, as amended by Amdt. 192-43, 47 FR 46850, Oct. 21, 1982; Amdt. 192-85, 63 FR 37500, July 13, 1998]

§192.755 Protecting cast-iron pipelines.

When an operator has knowledge that the support for a segment of a buried cast-iron pipeline is disturbed:

(a) That segment of the pipeline must be protected, as necessary, against damage during the disturbance by:

(1) Vibrations from heavy construction equipment, trains, trucks, buses, or blasting;

(2) Impact forces by vehicles;

(3) Earth movement;

(4) Apparent future excavations near the pipeline; or

(5) Other foreseeable outside forces which may subject that segment of the pipeline to bending stress.

(b) As soon as feasible, appropriate steps must be taken to provide permanent protection for the disturbed segment from damage that might result from external loads, including compliance with applicable requirements of §§192.317(a), 192.319, and 192.361(b) – (d).

[Amdt. 192-23, 41 FR 13589, Mar. 31, 1976]

192CODE.DOC, DT (Revised) 7/98

Appendix A–Incorporated by Reference

I. *List of organizations and addresses.*

A. American Gas Association (AGA), 1515 Wilson Boulevard, Arlington, VA 22209.

B. American National Standards Institute (ANSI), 11 West 42nd Street, New York, NY 10036.

C. American Petroleum Institute (API), 1220 L Street, NW., Washington, DC 20005.

D. The American Society of Mechanical Engineers (ASME), United Engineering Center, 345 East 47th Street, New York, NY 10017.

E. American Society for Testing and Materials (ASTM), 100 Barr Harbor Drive, West Conshohocken, PA 19428.

F. Manufacturers Standardization Society of the Valve and Fittings Industry, Inc. (MSS), 127 Park Street, NW., Vienna, VA 22180.

G. National Fire Protection Association (NFPA), 1 Batterymarch Park, P.O. Box 9101, Quincy, MA 02269-9101.

II. *Documents incorporated by reference. (Numbers in parentheses indicate applicable editions.)*

A. American Gas Association (AGA):

1. AGA Pipeline Research Committee, Project PR-3-805, "A Modified Criterion for Evaluating the Remaining Strength of Corroded Pipe" (December 22, 1989).

B. American Petroleum Institute (API):

1. API Specification 5L "Specification for Line Pipe" (41st edition, 1995).

2. API Recommended Practice 5L1 "Recommended Practice for Railroad Transportation of Line Pipe" (4th edition, 1990).

3. API Specification 6D "Specification for Pipeline Valves (Gate, Plug, Ball, and Check Valves)" (21st edition, 1994).

4. API Standard 1104 "Welding of Pipelines and Related Facilities" (18th edition, 1994).

C. American Society for Testing and Materials (ASTM):

1. ASTM Designation: A 53 "Standard Specification for Pipe, Steel, Black and Hot-Dipped, Zinc-Coated, Welded and Seamless" (A53-96).

2. ASTM Designation: A106 "Standard Specification for Seamless Carbon Steel Pipe for High-Temperature Service" (A106-95).

3. ASTM Designation: A333/A333M "Standard Specification for Seamless and Welded Steel Pipe for Low-Temperature Service" (A333/A333M-94).

4. ASTM Designation: A372/A372M "Standard Specification for Carbon and Alloy Steel Forgings for Thin-Walled Pressure Vessels" (A372/A372M-95).

5. ASTM Designation: A381 "Standard Specification for Metal-Arc-Welded Steel Pipe for Use With High-Pressure Transmission Systems" (A381-93).

6. ASTM Designation: A671 "Standard Specification for Electric-Fusion-Welded Steel Pipe for Atmospheric and Lower Temperatures" (A671-94).

7. ASTM Designation: A672 "Standard Specification for Electric-Fusion-Welded Steel Pipe for High-Pressure Service at Moderate Temperatures" (A672-94).

8. ASTM Designation: A691 "Standard Specification for Carbon and Alloy Steel Pipe, Electric-Fusion-Welded for High-Pressure Service at High Temperatures" (A691-93).

9. ASTM Designation: D638 "Standard Test Method for Tensile Properties of Plastics" (D638-96).

10. ASTM Designation: D2513 "Standard Specification for Thermoplastic Gas Pressure Pipe, Tubing, and Fittings" (D2513-87 edition for §192.63(a)(1), otherwise D2513-96a).

11. ASTM Designation: D2517 "Standard Specification for Reinforced Epoxy Resin Gas Pressure Pipe and Fittings" (D2517-94).

12. ASTM Designation: F1055 "Standard Specification for Electrofusion Type Polyethylene Fittings for Outside Diameter Controlled Polyethylene Pipe and Tubing" (F1055-95).

D. The American Society of Mechanical Engineers (ASME):

1. ASME/ANSI B16.1 "Cast Iron Pipe Flanges and Flanged Fittings" (1989).

2. ASME/ANSI B16.5 "Pipe Flanges and Flanged Fittings" (1988 with October 1988 Errata and ASME/ANSI B16.5a-1992 Addenda).

3. ASME/ANSI B31G "Manual for Determining the Remaining Strength of Corroded Pipelines" (1991).

4. ASME/ANSI B31.8 "Gas Transmission and Distribution Piping Systems" (1995).

5. ASME Boiler and Pressure Vessel Code, Section I "Power Boilers" (1995 edition with 1995 Addenda).

6. ASME Boiler and Pressure Vessel Code, Section VIII, Division 1 "Pressure Vessels" (1995 edition with 1995 Addenda).

7. ASME Boiler and Pressure Vessel Code, Section VIII, Division 2 "Pressure Vessels: Alternative Rules" (1995 edition with 1995 Addenda).

8. ASME Boiler and Pressure Vessel Code, Section IX "Welding and Brazing Qualifications" (1995 edition with 1995 Addenda).

E. Manufacturers Standardization Society of the Valve and Fittings Industry, Inc. (MSS):

1. MSS SP-44-96 "Steel Pipe Line Flanges" (includes 1996 errata)(1996).

2. [Reserved].

F. National Fire Protection Association (NFPA):

1. NFPA 30 'Flammable and Combustible Liquids Code"(1996).

2. ANSI/NFPA 58 'Standard for the Storage and Handling of Liquefied Petroleum Gases'(1995).

3. ANSI/NFPA 59 'Standard for the storage and Handling of Liquefied Petroleum Gases at Utility Gas Plants'(1995).

4. ANSI/NFPA 70 'National Electrical Code"(1996).

[35 FR 13257, Aug. 19, 1970, as amended by Amdt. 192-3, 35 FR 17659, Nov. 17, 1970; Amdt. 192-12, 38 FR 4760, Feb. 22, 1973; Amdt. 192-17, 40 FR 6345, Feb. 11, 1975; Amdt. 192-17C, 40 FR 8188, Feb. 26, 1975; Amdt. 192-18, 40 FR 10181, Mar. 5, 1975; Amdt. 192-19, 40 FR 10471, Mar. 6, 1975; Amdt. 192-22, 41 FR 13589, Mar. 31, 1976; Amdt. 192-32, 43 FR 18553, May 1, 1978; Amdt. 192-34, 44 FR 42968, July 23, 1979; Amdt. 192-37, 46 FR 10157, Feb. 2, 1981; Amdt. 192-41, 47 FR 41381, Sept. 20, 1982; Amdt. 192-42, 47 FR 44263, Oct. 7, 1982; Amdt. 192-51, 51 FR 15333, Apr. 23, 1986; Amdt. 192-61, 53 FR 36793, Sept. 22, 1988; Amdt. 192-62, 54 FR 5625, Feb. 6, 1989; Amdt. 192-64, 54 FR 27881, July 3, 1989; Amdt. 192-65, 54 FR 32344, Aug. 7, 1989; Amdt. 192-68, 58 FR 14519, Mar. 18, 1993; Amdt. 192-76, 61 FR 26121, May 24, 1996; Amdt. 192-78, 61 FR 28770, June 6, 1996; Amdt. 192-78C, 61 FR 41019, Aug. 7, 1996; Amdt. 192-84, 63 FR 7721, Feb. 17, 1998; Amdt. 192-84A, 63 FR 38757, July 20, 1998]

192CODE.DOC, DT (Revised) 7/98

Appendix B–Qualification of Pipe

I. *Listed Pipe Specifications.(Numbers in Parentheses Indicate Applicable Editions.)*

API 5L–Steel pipe (1995).
ASTM A53–Steel pipe (1995a).
ASTM A106–Steel pipe (1994a).
ASTM A333/A333M Steel pipe (1994).
ASTM A381–Steel pipe (1993).
ASTM A671–Steel pipe (1994).
ASTM A672–Steel pipe (1994).
ASTM A691–Steel pipe (1993).
ASTM D2513–Thermoplastic pipe and tubing (1995c). ***Note: ASTM D2513-87 Under Appendix A.II.B(10)***
ASTM D2517–Thermosetting plastic pipe and tubing (1994).

II. *Steel pipe of unknown or unlisted specification.*

A. Bending properties. For pipe 2 inches (51 millimeters) or less in diameter, a length of pipe must be cold bent through at least 90 degrees around a cylindrical mandrel that has a diameter 12 times the diameter of the pipe, without developing cracks at any portion and without opening the longitudinal weld.

For pipe more than 2 inches (51 millimeters) in diameter, the pipe must meet the requirements of the flattening tests set forth in ASTM A53, except that the number of tests must be at least equal to the minimum required in paragraph II-D of this appendix to determine yield strength.

B. Weldability. A girth weld must be made in the pipe by a welder who is qualified under Subpart E of this part. The weld must be made under the most severe conditions under which welding will be allowed in the field and by means of the same procedure that will be used in the field. On pipe more than 4 inches (102 millimeters) in diameter, at least one test weld must be made for each 100 lengths of pipe. On pipe 4 inches (102 millimeters) or less in diameter, at least one test weld must be made for each 400 lengths of pipe. The weld must be tested in accordance with API Standard 1104. If the requirements of API Standard 1104 cannot be met, weldability may be established by making chemical tests for carbon and manganese, and proceeding in accordance with section IX of the ASME Boiler and Pressure Vessel Code. The same number of chemical tests must be made as are required for testing a girth weld.

C. Inspection. The pipe must be clean enough to permit adequate inspection. It must be visually inspected to ensure that it is reasonably round and straight and there are no defects which might impair the strength or tightness of the pipe.

D. Tensile properties. If the tensile properties of the pipe are not known, the minimum yield strength may be taken as 24,000 p.s.i. (165 MPa) or less, or the tensile properties may be established by performing tensile tests as set forth in API Specification 5L. All test specimens shall be selected at random and the following number of tests must be performed:

Number of Tensile Tests–All Sizes	
10 lengths or less	1 set of tests for each length.
11 to 100 lengths	1 set of tests for each 5 lengths, but not less than 10 tests.
Over 100 lengths	1 set of tests for each 10 lengths but not less than 20 tests.

If the yield-tensile ratio, based on the properties determined by those tests, exceeds 0.85, the pipe may be used only as provided in §192.55(c).

III. *Steel pipe manufactured before November 12, 1970, to earlier editions of listed specifications*. Steel pipe manufactured before November 12, 1970, in accordance with a specification of which a later edition is listed in section I of this appendix, is qualified for use under this part if the following requirements are met:

A. Inspection. The pipe must be clean enough to permit adequate inspection. It must be visually inspected to ensure that it is reasonably round and straight and that there are no defects which might impair the strength or tightness of the pipe.

B. Similarity of specification requirements. The edition of the listed specification under which the pipe was manufactured must have substantially the same requirements with respect to the following properties as a later edition of that specification listed in section I of this appendix:

(1) Physical (mechanical) properties of pipe, including yield and tensile strength, elongation, and yield to tensile ratio, and testing requirements to verify those properties.

(2) Chemical properties of pipe and testing requirements to verify those properties.

C. Inspection or test of welded pipe. On pipe with welded seams, one of the following requirements must be met:

(1) The edition of the listed specification to which the pipe was manufactured must have substantially the same requirements with respect to nondestructive inspection of welded seams and the standards for acceptance or rejection and repair as a later edition of the specification listed in section I of this appendix.

(2) The pipe must be tested in accordance with Subpart J of this part to at least 1.25 times the maximum allowable operating pressure if it is to be installed in a class 1 location and to at least 1.5 times the maximum allowable operating pressure if it is to be installed in a class 2, 3, or 4 location. Notwithstanding any shorter time period permitted under Subpart J of this part, the test pressure must be maintained for at least 8 hours.

[35 FR 13257, Aug. 19, 1970; as amended by Amdt. 192-3, 35 FR 17659, Nov. 17, 1970; Amdt. 192-12, 38 FR 4760, Feb. 22, 1973; Amdt. 192-19, 40 FR 10471, Mar. 6, 1975; Amdt. 192-22, 41 FR 13589, Mar. 31, 1976; Amdt. 192-32, 43 FR 18553, May 1, 1978; Amdt. 192-37, 46 FR 10157, Feb. 2, 1981; Amdt. 192-41, 47 FR 41381, Sept. 20, 1982; Amdt. 192-51, 51 FR 15333, Apr. 23, 1986; Amdt. 192-62, 54 FR 5625, Feb. 6, 1989; Amdt. 192-65, 54 FR 32344, Aug. 7, 1989; Amdt. 192-68, 58 FR 14519, Mar. 18, 1993; Amdt. 192-76A, 61 FR 36825, July 15, 1996; Amdt. 192-85, 63 FR 37500, July 13, 1998]

Appendix C–Qualification of Welders for Low Stress Level Pipe

I. *Basic test*. The test is made on pipe 12 inches (305 millimeters) or less in diameter. The test weld must be made with the pipe in a horizontal fixed position so that the test weld includes at least one section of overhead position welding. The beveling, root opening, and other details must conform to the specifications of the procedure under which the welder is being qualified. Upon completion, the test weld is cut into four coupons and subjected to a root bend test. If, as a result of this test, two or more of the four coupons develop a crack in the weld material, or between the weld material and base metal, that is more than 1/8-inch (3.2 millimeters) long in any direction, the weld is unacceptable. Cracks that occur on the corner of the specimen during testing are not considered.

II. *Additional tests for welders of service line connections to mains.* A service line connection fitting is welded to a pipe section with the same diameter as a typical main. The weld is made in the same position as it is made in the field. The weld is unacceptable if it shows a serious undercutting or if it has rolled edges. The weld is tested by attempting to break the fitting off the run pipe. The weld is unacceptable if it breaks and shows incomplete fusion, overlap, or poor penetration at the junction of the fitting and run pipe.

III. *Periodic tests for welders of small service lines.* Two samples of the welder's work, each about 8 inches (203 millimeters) long with the weld located approximately in the center, are cut from steel service line and tested as follows:

(1) One sample is centered in a guided bend testing machine and bent to the contour of the die for a distance of 2 inches (51 millimeters) on each side of the weld. If the sample shows any breaks or cracks after removal from the bending machine, it is unacceptable.

(2) The ends of the second sample are flattened and the entire joint subjected to a tensile strength test. If failure occurs adjacent to or in the weld metal, the weld is unacceptable. If a tensile strength testing machine is not available, this sample must also pass the bending test prescribed in subparagraph (1) of this paragraph.

[35 FR 13257, Aug. 19, 1970 as amended by Amdt. 192-85, 63 FR 37500, July 13, 1998]

192CODE.DOC, DT (Revised) 7/98

Appendix D–Criteria for Cathodic Protection and Determination of Measurements

I. *Criteria for cathodic protection–*

A. Steel, cast iron, and ductile iron structures.

(1) A negative (cathodic) voltage of at least 0.85 volt, with reference to a saturated copper-copper sulfate half cell. Determination of this voltage must be made with the protective current applied, and in accordance with sections II and IV of this appendix.

(2) A negative (cathodic) voltage shift of at least 300 millivolts. Determination of this voltage shift must be made with the protective current applied, and in accordance with sections II and IV of this appendix. This criterion of voltage shift applies to structures not in contact with metals of different anodic potentials.

(3) A minimum negative (cathodic) polarization voltage shift of 100 millivolts. This polarization voltage shift must be determined in accordance with sections III and IV of this appendix.

(4) A voltage at least as negative (cathodic) as that originally established at the beginning of the Tafel segment of the E-log-I curve. This voltage must be measured in accordance with section IV of this appendix.

(5) A net protective current from the electrolyte into the structure surface as measured by an earth current technique applied at predetermined current discharge (anodic) points of the structure.

B. Aluminum structures.

(1) Except as provided in paragraphs (3) and (4) of this paragraph, a minimum negative (cathodic) voltage shift of 150 millivolts, produced by the application of protective current. The voltage shift must be determined

in accordance with sections II and IV of this appendix.

(2) Except as provided in paragraphs (3) and (4) of this paragraph, a minimum negative (cathodic) polarization voltage shift of 100 millivolts. This polarization voltage shift must be determined in accordance with sections III and IV of this appendix.

(3) Notwithstanding the alternative minimum criteria in paragraphs (1) and (2) of this paragraph, aluminum, if cathodically protected at voltages in excess of 1.20 volts as measured with reference to a copper-copper sulfate half cell, in accordance with section IV of this appendix, and compensated for the voltage (IR) drops other than those across the structure-electrolyte boundary may suffer corrosion resulting from the build-up of alkali on the metal surface. A voltage in excess of 1.20 volts may not be used unless previous test results indicate no appreciable corrosion will occur in the particular environment.

(4) Since aluminum may suffer from corrosion under high pH conditions, and since application of cathodic protection tends to increase the pH at the metal surface, careful investigation or testing must be made before applying cathodic protection to stop pitting attack on aluminum structures in environments with a natural pH in excess of 8.

C. Copper structures. A minimum negative (cathodic) polarization voltage shift of 100 millivolts. This polarization voltage shift must be determined in accordance with sections III and IV of this appendix.

D. Metals of different anodic potentials. A negative (cathodic) voltage, measured in accordance with section IV of this appendix, equal to that required for the most anodic metal in the system must be maintained. If amphoteric structures are involved that could be damaged by high alkalinity covered by paragraphs (3) and (4) of paragraph B of this section, they must be electrically isolated with insulating flanges, or the equivalent.

II. *Interpretation of voltage measurement.* Voltage (IR) drops other than those across the structure electrolyte boundary must be considered for valid interpretation of the voltage measurement in paragraphs A(1) and (2) and paragraph B(1) of section I of the appendix.

III. *Determination of polarization voltage shift.* The polarization voltage shift must be determined by interrupting the protective current and measuring the polarization decay. When the current is initially interrupted, an immediate voltage shift occurs. The voltage reading after the immediate shift must be used as the base reading from which to measure polarization decay in paragraphs A(3), B(2), and C of section I of this appendix.

IV. *Reference half cells.*

A. Except as provided in paragraphs B and C of this section, negative (cathodic) voltage must be measured between the structure surface and a saturated copper-copper sulfate half cell contacting the electrolyte.

B. Other standard reference half cells may be substituted for the saturated copper-copper sulfate half cell. Two commonly used reference half cells are listed below along with their voltage equivalent to -0.85 volt as referred to a saturated copper-copper sulfate half cell:

(1) Saturated KCl calomel half cell: -0.78 volt.

(2) Silver-silver chloride half cell used in sea water: -0.80 volt.

C. In addition to the standard reference half cells, an alternate metallic material or structure may be used in place of the saturated copper-copper sulfate half cell if its potential stability is assured and if its voltage equivalent referred to a saturated copper-copper sulfate half cell is established.

[Amdt. 192-4, 36 FR 12297, June 30, 1971]

192CODE.DOC, DT (Revised) 7/98

APPENDIX C
PLASTIC PIPING SYSTEMS COMPONENTS

A Brief Explanation of ASTM Standard Specifications

ASTM D-1784 "Rigid Poly Vinyl Chloride) (PVC) Compounds and Chlorinated Poly (Vinyl Chloride) CPVC Compounds"

This specification covers rigid plastic compounds composed of poly (vinyl chloride), or vinyl copolymers, and the necessary compounding ingredients may consist of lubricants, stabilizers, nonpoly (vinyl chloride) resin modifiers, and pigments essential for processing, property control, and coloring.

ASTM D-2464 "Threaded Poly (Vinyl Chloride) (PVC) Plastic Pipe Fittings, Schedule 80"

This specification covers poly (vinyl chloride) (PVC) threaded Schedule 80 pipe fittings. Included are requirements for materials, workmanship, dimensions and burst pressure.
The tapered pipe threads in this specification refer toANSI B1.20.1 (was B2. 1) and it is this portion to which all George Fischer valves comply.

ASTM F437 (formerly included in D-2464) "Threaded Chlorinated Poly (Vinyl Chloride) (CPVC) Plastic Pipe Fittings, Schedule 80"

ASTM D-2467 "Socket Type Poly (Vinyl Chloride) (PVC) Plastic Pipe Fittings, Schedule 80"

This specification covers poly (vinyl chloride) (PVC) Schedule 80 socket type pipe fittings. Included are requirements for materials, workmanship, dimensions and burst pressure.

ASTM F439 (formerly included in D-2467) "Socket Type Chlorinated Poly (Vinyl Chloride) (CPVC) Plastic Pipe Fittings, Schedule 80"

ANSI B16.5 American National Standard "Pipe Flanges and Flanged Fittings"

This standard covers flanges and flanged fittings in all pressure classes. Bolt hole circle diameters are called out and bolt hole diameters, as well as other dimensions. It is in Class 150 which George Fischer valves comply where applicable.

ASTM D-4101 (formerly D-2146) "Standard Specification for Propylene Plastic Injection and Extrusion Materials"

ASTM D-3222 "Standard Specification for unmodified Poly (Vinylidene Fluoride) (PVDF) Molding Materials"

ASTM D-2657 "Standard Practice for Heat Joining Polyolefin Pipe and Fittings"

Plastic piping system components. *(Courtesy George Fischer Engineering Handbook)*

Plastic Piping Systems Components

The following paragraphs describe key properties and applications for the more popular plastics and fluoroplastics. The reader is encouraged to use this information when selecting plastic components from George Fischer.

Description of Plastics
Within the broad category of plastics as materials, George Fischer offers piping system components constructed of plastics known as fluoroplastics and thermoplastics. The category of fluoroplastics includes those polymers containing one or more atoms of fluorine in their chemical and molecular makeup and structure. The basic group known as fluoroplastics includes PTFE, PVDF, ECTFE, CTFE, ETFE, PFA and FEP. Fluoroplastics as a class offer excellent characteristics, performing well in aggressive chemical elements at temperatures from -328°F (-200°C) to 500°F (260°C). In addition to very good mechanical properties, these materials have low wettability, and in many cases, an exceptionally low coefficient of friction.

The group of plastics known as thermoplastics includes those which are "melt processible," i.e., may be welded or injection molded using standard manufacturing techniques. PTFE and UHMW PE are two popular plastics which have well defined melt points but are not thermoplastics since they are not melt processible. These plastics are compression moldable into stock shapes for machining into piping system components.

The Period Table of the Elements and How it Relates to Plastics
In illustrating the differences in plastics, it is best to highlight the distinctions between members of the classes of polymers including fluoroplastics, thermoplastics, and polyolefins (a class including PP, PE and UHMW PE), using a comparison of their basic chemical structures. Many of the physical property and chemical resistance differences stem directly from the type and arrangement of atoms in the polymer chains.

The periodic table arranges the elements in horizontal rows, called periods, according to their electron structure and atomic weights. The table arranges the basic elements of nature not only by atomic structure but by chemical nature as well. Chemists have placed the elements into classes which have similar properties, i.e. the elements and their compounds which exhibit similar chemical behavior. These classes are the alkali metals, alkaline earth metals, transition metals, rare earth series, actinide series, other metals, nonmetals, and noble (inert) gases.

One group in the periodic table of particular importance to fluoroplastics, within the nonmetals category is known as the halogens. These elements include fluorine, chlorine, bromine, and iodine. These elements are the most electronegative elements in the period table, meaning they have the strongest attraction to gain an electron (from another element) and become a stable structure. Fluorine is the most electronegative halogen of them all. Having this property, fluorine bonds strongly with carbon and hydrogen atoms but not well with fluorine itself. The carbon-fluorine bond, the predominant bond in PVDF and PTFE and which gives it such important properties, is among the strongest known organic compounds. The fluorine acts like a protective sheath for other bonds of lesser strength within the main chain of the polymer. The carbon-hydrogen bond, of which plastics such as PP and PE are composed, is considerably weaker. The carbon-chlorine bond, a key bond in PVC, is weaker yet.

Between parallel polymer chains, similar bond forces between such adjacent elements become established and greatly determine the physical properties and strength of the plastic

Plastic piping system components. *(Courtesy George Fischer Engineering Handbook)*

itself. The intermolecular bond forces are known as Van der Waals forces.

Although Van der Waals forces between adjacent atoms along polymer chains are far smaller than the bond strengths between atoms within a discrete molecule itself, the forces are additive along the length of the polymer chain and subsequently dictate physical characteristics of the plastic such as tensile strength, melt point, impact strength, heat deflection temperature, creep resistance, and more.

The General Bond Strength of Plastics
A good indication of the comparative strength of Van de Waals forces between polymer chains is data which shows relative bond strengths between atoms of polyatomic molecules. Interpretation of the data in the table below substantiates the expected advantage of having a high degree of fluorine in a plastic (i.e. PVDF). Fluorine shows the highest affinity for hydrogen (the most electropositive element in the Periodic Table) and a strong attraction for carbon as well.

While it is extremely important as to the specific elements contained in a plastic's structure, the actual arrangement of elements in the molecule, the symmetry of the structure, and the degree of branching of the polymer chains are highly important as well. The plastics that contain carbon-hydrogen bonds (i.e. PP and PE) and carbon-chlorine bonds (i.e. PVC, ECTFE, CTFE) are noticeably different in the important property of chemical resistance from a fully fluorinated plastic such as PTFE.

Within the following paragraphs describing individual plastics, the simplified molecular structures may be examined to see how structurally alike many plastics are. However, some seemingly minor structure result in considerable differences in melt points, use temperatures, and physical strengths.

General Bond Strengths for Polyatomic Molecules (Bond Dissociation Energy)

Bond Type	Typical Plastic	Bond Energy (kcal/mole)	Molecular Separation (Angstroms)
F - F	PTFE	37.7	1.42
Cl - Cl	PVC	57.9	1.99
Br - Br	-	46.1	2.28
F - H	PTFE, PVDF	134.6	0.96
Cl - H	PVC, ECTFE	103.1	1.27
Br - H	-	87.5	1.41
O - H	-	110.6	0.96
H - H	PP, PE	104.2	0.74
C - C	All	83.0	1.54
C - F	PTFE	107.0	1.32
C - Cl	PVC	78.5	1.76
C - H	PP, PE	98.8	1.07
C - O	-	84.0	1.43

Plastic piping system components. *(Courtesy George Fischer Engineering Handbook)*

I A	II A	III B	IV B	V B	VI B	VII B	VIII	VIII	VIII	I B	II B	III A	IV A	V A	VI A	VII A	0
1 **H** 1.00797																1 **H** 1.00797	2 **He** 4.0026
3 **Li** 6.939	4 **Be** 9.0122											5 **B** 10.811	6 **C** 12.01115	7 **N** 14.0067	8 **O** 15.9994	9 **F** 18.9984	10 **Ne** 20.183
11 **Na** 22.9898	12 **Mg** 24.312											13 **Al** 26.9815	14 **Si** 28.086	15 **P** 30.9738	16 **S** 32.064	17 **Cl** 35.453	18 **Ar** 39.948
19 **K** 39.102	20 **Ca** 40.08	21 **Sc** 44.956	22 **Ti** 47.90	23 **V** 50.942	24 **Cr** 51.996	25 **Mn** 54.938	26 **Fe** 55.847	27 **Co** 58.933	28 **Ni** 58.71	29 **Cu** 63.54	30 **Zn** 65.37	31 **Ga** 69.72	32 **Ge** 72.59	33 **As** 74.922	34 **Se** 78.96	35 **Br** 79.909	36 **Kr** 83.80
37 **Rb** 85.47	38 **Sr** 87.62	39 **Y** 88.905	40 **Zr** 91.22	41 **Nb** 92.906	42 **Mo** 95.94	43 **Tc** (98)	44 **Ru** 101.07	45 **Rh** 102.905	46 **Pd** 106.4	47 **Ag** 107.870	48 **Cd** 112.40	49 **In** 114.82	50 **Sn** 118.69	51 **Sb** 121.75	52 **Te** 127.60	53 **I** 126.904	54 **Xe** 131.30
55 **Cs** 132.905	56 **Ba** 137.34	57 **La** 138.91	72 **Hf** 178.49	73 **Ta** 180.948	74 **W** 183.85	75 **Re** 186.2	76 **Os** 190.2	77 **Ir** 192.2	78 **Pt** 195.09	79 **Au** 196.967	80 **Hg** 200.59	81 **Tl** 204.37	82 **Pb** 207.19	83 **Bi** 208.980	84 **Po** (210)	85 **At** (210)	86 **Rn** (222)
87 **Fr** (223)	88 **Ra** (226)	89 **Ac** (227)															
				58 **Ce** 140.12	59 **Pr** 140.907	60 **Nd** 144.24	61 **Pm** (147)	62 **Sm** 150.35	63 **Eu** 151.96	64 **Gd** 157.25	65 **Tb** 158.924	66 **Dy** 162.50	67 **Ho** 164.930	68 **Er** 167.26	69 **Tm** 168.934	70 **Yb** 173.04	71 **Lu** 174.97
				90 **Th** 232.038	91 **Pa** (231)	92 **U** 238.04	93 **Np** (237)	94 **Pu** (242)	95 **Am** (243)	96 **Cm** (247)	97 **Bk** (247)	98 **Cf** (249)	99 **Es** (254)	100 **Fm** (253)	101 **Md** (256)	102 **No** (254)	103 **Lw** (257)

Plastic piping system components. *(Courtesy George Fischer Engineering Handbook)*

Abrasion Resistance of Plastics

Plastics as a general class offer sizeable performance differences compared to metals with respect to abrasion resistance, coefficient of friction, wettability, and self lubricating qualities. PTFE, with the lowest surface energy of any common solid, has exceptional antistick properties. Additionally, PTFE has an excellent low coefficient of friction, especially at highloads where it can be less than 0.01 . However, PTFE is not optimal with respect to wear resistance, and PTFE may not be ideal for moving parts where friction is prevalent. Plastics such as UHMW PE, PVDF, CTFE, and Nylon have such excellent wear resistance that they prove to be 5 to 10 times better in this regard than 304 Stainless Steel. One widely known test method is the Taber Abrasion Test, wherein an abrasive wheel is cycled over the face of a plate of the material being tested, and the resultant weight loss of the plate is measured after 1000 cycles of the ring.

Taber Abrasion Tester
(Barasion Ring CS-10, Load 1 kg)

Nylon 6-10	5mg / 1000 cycles
UHMW PE	5
PVDF	5 - 10
PVC (rigid)	12 - 20
PP	15 - 20
CPVC	20
CTFE	13
PS	40 - 50
Steel (304 SS)	50
ABS	60 - 80
PTFE	500 - 1000

Deflection Temperature
(Heat Distortion Temperature)

The heat deflection temperature test is an ideal method for comparing the high temperature capabilities of commercial polymers. It is strongly linked to the softening point of the material, and gives a measure of the rigidity of the polymer under a load as well as temperature. Unexpectedly, the fluoroplastics with the highest melt points of the group have the lowest deflection temperatures. The heat distortion test basically is one in which a bar of the plastic in question is heated uniformly in a closed chamber while a load of 66 psi or 264 psi is placed at the center of the horizontal bar. When a slight deflection of 0.25mm at the center is noted, as the temperature is raised, it is noted and called its HDT. HDT is a test to show how easily, or not so easily as the case may be, the Van der Waals forces of intermolecular bonding can be overcome in a plastic. In a practical sense, HDT indicates how much mass (weight) of which the object must be constructed to maintain the desired form stability and strength rating.

ASTM D648

	66 psi	264 psi	Melt Point
PTFE	250°F	132°F	620°F
PVC	135°F	140°F	<285°F
LDPE	-	104°F	221°F
UHMW PE	155°F	110°F	265°F
PP	225°F	120°F	330°F
PFA	164°F	118°F	590°F
FEP	158°F	124°F	554•F
PVDF	298°F	235°F	352°F
ECTFE	240°F	170°F	464°F
CTFE	258°F	167°F	424°F
ETFE	220°F	165°F	518°F

Plastic piping system components. *(Courtesy George Fischer Engineering Handbook)*

The tensile strength of a material is calculated by dividing the maximum load applied to a material prior to its breaking, divided by the original cross-sectional area of the piece. Thus, as opposed to toughness which is a measure of the energy required to break a material, tensile strength is a measure of the stress required to deform the material prior to breakage. Stress is defined as the force applied over the area on which it operates. Tensile strength alone must not be used to determine the ability of a plastic to resist deformation and maintain form stability, and other mechanical characteristics such as elasticity, ductility, creep resistance, hardness, and toughness should be used to determine the optimal plastic. If the chemical structure of plastics is analyzed, it may be seen that those plastics which have atom types with the highest likelihood of intermolecular bonding, indeed do show to be the highest in tensile strength. In many applications, the lower tensile strength of a plastic is offset by adding extra thickness and mass to the part being designed.

Tensile Strength of Plastics @ 73°F (25°C) @ Break

PVDF	8000 psi
ETFE	6500 psi
CTFE	4500 - 6000 psi
PFA	4000 - 4300 psi
ECTFE	7000 psi
PTFE	2500 - 6000 psi
FEP	2700 - 3100 psi
PVC	6000 - 7500 psi
PE	1200 - 4550 psi
PP	4500 - 6000 psi
UHMW PE	5600 psi

The driving force for gases and vapors penetrating and/or diffusing through plastics is basically the difference in concentrations between the environments inside and outside of the plastic barrier. The mathematic equations known as Fick's First and Second Laws of Diffusion enable certain parameters to be identified for individual plastics. These are known as diffusion coefficients, solubility coefficients, and permeability coefficients. Measurements must be made to finally determine equations which may be used to calculate the thicknesses required for protective barriers, etc. Whereas permeation is a major consideration in dealing with thin films, thick sections of 1/8 inch (3mm) or more generally eliminate any permeation situation unless one is dealing with media detection levels of 1 ppm or lower.

Comparative Resistance pf Plastics to Permeation* of Gases

	Oxygen	Hydrogen	Nitrogen	Helium	Carbon Dioxide
SYGEF	2	10	4	25	8
PP	25	180	10	200	100
HPDE	30 - 40	22	18	20	200
PVC	3	10	1	16	16

*Permeability through sheet thickness of 1 mm film in cm^3 / m^2 x24h x 1bar

Comparative Resistance to Gas Permeability for Fluoropolymers at 73°F through a 100 um Film

Gas	Units	Fluoropolymer Type			
		PVDF	PTFE	FEP	PFA
Water Vapor	g / m^2 x day x bar	2	5	1	8
Air	cm^3 / m^2 x day x bar	27	2000	600	1150
Oxygen	cm^3 / m^2 x day x bar	20	1500	2900	-
Nitrogen	cm^3 / m^2 x day x bar	30	500	1200	-
Helium	cm^3 / m^2 x day x bar	600	35000	18000	17000
Carbon Dioxide	cm^3 / m^2 x day x bar	100	15000	4700	7000

Plastic piping system components. *(Courtesy George Fischer Engineering Handbook)*

Description of Individual Plastics

PTFE (Polytetrafluorethylene)

PTFE is made in larger amounts than all other fluoroplastics combined. It is resistant to practically every known chemical or solvent in combination with the highest useful temperature limit of commercially available plastics. Fabricated PTFE has a melt point of 620°F (327°C) and a useful temperature range from -436°F (-260°C) to 500°F (260°C). PTFE for plastic applications has a very high molecular weight which results in a melt viscosity about 1 million times higher than is acceptable for conventional thermoplastics. As a result, the usual processing techniques like injection molding are not possible. PTFE resin is pressed into shapes under high pressure at room temperature and then heated to 700°F (371°C) to complete the molding (sintering process) and adjust the crystalline content. The high melt viscosity provides form stability and this sintering operation is almost always carried out with unsupported article sin air ovens. Ram extrusion of granular resin is used to make rods, tubes, and other continuous shapes. Impact strength is high and PTFE has an exceptionally low coefficient of friction. Since its commercial development in 1948, PTFE has served as the standard against which other plastics, particularly fluoroplastics, are compared.

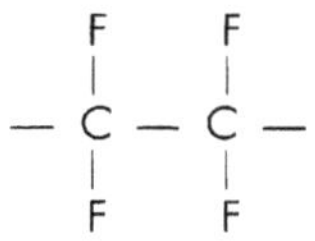

PVC (Polyvinyl Chloride)

PVC is the largest volume of the vinyl family of plastics. Overall it has excellent basic properties, may be easily processed and welded, and is exceptionally economical in cost. Homopolymer grades of PVC comprise over 80% of all PVC used, and contain 56.8% chlorine by weight. Chlorinated PVC, in which the chlorine content is increased from 56.8% to about 67%, increases the heat deflection temperature at 264 psi from 155°F (68°C) to 218°F (103°C). PVC is a thermally sensitive thermoplastic to which compounding ingredients must be added to allow it to be processible. Heat stabilizers are required, in addition to lubricants, fillers, plasticizers, impact modifiers, pigments, and processing aids. Since PVC is amorphous and not crystalline in form as most other thermoplastics are, it has a difficult melt point to detect. The glass transition temperature where major softening occurs is 167°F (75°C) for PVC and 230°F (110°C) for CPVC. The structures for PVC and CPVC are as shown:

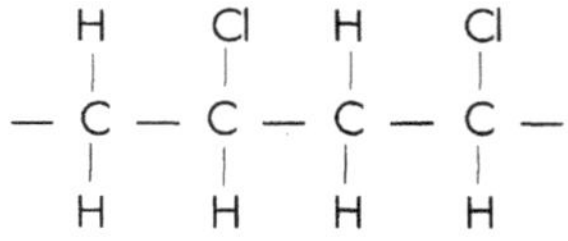

PVC

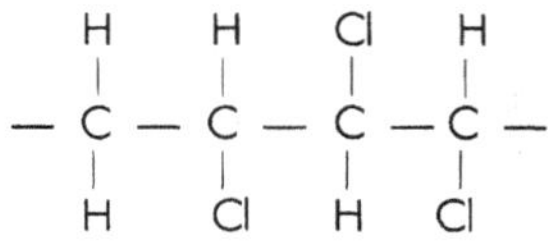

CPVC

Plastic piping system components. *(Courtesy George Fischer Engineering Handbook)*

PE (low density polyethylene)

Polyethylene is a thermoplastic material which varies from type to type according to the particular molecular structure of each type, i.e. its crystallinity, molecular weight, and molecular weight distribution. These variations are made possible through changes in polymerization conditions used in the manufacture of PE. The terms low, medium, and high density PE usually refer to the ASTM designations based on density of the unmodified PE. For example, low density PE has a specific gravity of 0.91 to 0.925 g/cc whereas medium density and high density versions have ranges of 0.926 to 0.940 and 0.941 to 0.959 g/cc respectively. The densities, being related to basic molecular structure, are indicators of the end product's properties and processing characteristics. Low density PE has increased toughness, stress cracking resistance, clarity, flexibility, and elongation. It also has reduced creep and mold shrinkage. Low density PE has a melt point of 221°F (105°C) if it is the common branched version. Another type, known as linear low density PE (LDPE) has a melt point of 252°F (122°C). PE of higher density has better permeation barrier properties, hardness, abrasion resistance, chemical resistance, and surface gloss. With all various PE types, the basic molecular unit is,

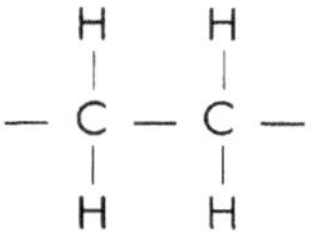

It is important to note that photo or light oxidation will occur when natural PE is exposed to UV radiation, usually from the sun. With the exception of carbon black, other UV stabilizers do just a partial job of screening UV light.

UHMW PE

Ultra-high molecular weight polyethylene is set apart from other polyethylenes by its molecular weight, and its extremely long linear molecular chain. As defined by ASTM, UHMW PE must have a molecular weight of at least 3.1 million. As a comparison, the molecular weight of pipe grade PE is 500,000. The long linear chains of UHMW PE provide great impact strength, abrasion resistance (sliding), toughness, and freedom stresscracking, in addition to the typical PE characteristics of chemical inertness, lubricity, cyclical fatigue resistance, and low coefficient of friction. Because of its extremely high molecular weight, UHMW resin cannot be processed using conventional processing techniques for plastics. As UHMW is heated above its melt point of 265°F (129°C), it will not change much in shape but simply changes from opaque to clear. Compression molding of sheets and ram extrusion of profiles and pipe are the standard manufacturing techniques. Parts may be made into their final form by machining, sintering, or forging. Since UHMW is very nearly chemically inert, it has FDA and USDA sanction for use in direct food contact, as well as complying with National Bureau of Standards requirements for handling of water and medical supplies. One important consideration to keep in mind is the attack of UHMW by UV light, seriously deteriorating impact strength within a relatively short period of time.

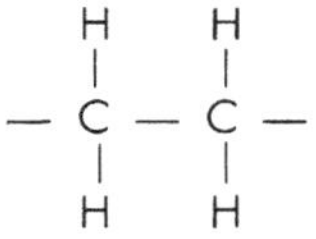

Plastic piping system components. *(Courtesy George Fischer Engineering Handbook)*

PP (polypropylene)

Polypropylene is a crystalline polymer with a melting point of 330°F (165°C). It is the lightest of the most common thermoplastics with a specific gravity of 0.91 if unmodified. The key properties in addition to this are its high heat resistance, stiffness, and chemical resistance. It may be usable for low stress structural applications up to 275°F (135°C) but for piping applications has an upper limit of 212°F (100°C). Although excellent in chemical resistance with respect to handling caustics, solvents, acids, and other organic chemicals, it is not recommended for use with oxidizing type acids, detergents, low-boiling hydrocarbons, alcohols, and some chlorinated organic materials. Unpigmented, natural polypropylene is degraded by UV light unless it is shielded, pigmented, or otherwise stabilized. The heat deflection temperature for PP ranges from 195°F (91°C) to 240°F (116°C) which is higher than other common plastics. Within the structural nature of PP is a distinction between isotactic PP (which composes 97% of PP and wherein the polymer units are highly ordered as the chemical structure is drawn to look), and atactic PP which is a viscous liquid type PP found with the PP polymer matrix. Also, in some cases, PP is actually a combination of PE and PP which is accomplished during a second stage of the polymerization. This is called copolymer PP, and gives the plastic much less of a brittle characteristic than PP which is entirely homopolymer. Copolymer withstands impact forces down to -20°F (-29°C) whereas homopolymer is very brittle below -40°F (-40°C).

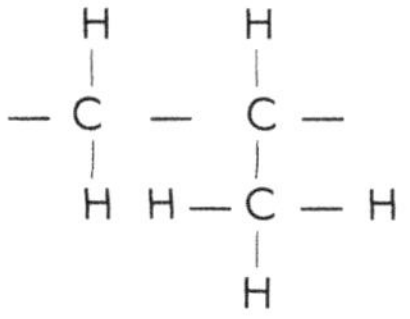

ECTFE (ethylene chlorotrifluoroethylene)

ECTFE is a melt processible thermoplastic with a melt point of 473°F (245°C). It contains 6.5% hydrogen and 24.6% chlorine. It has useful applications in the temperature range from -105°F (-76°C) to 302°F (150°C). Overall it has good physical properties and excellent resistance to creep. To obtain good extrusion characteristics it must be compounded with a small amount of processing aid. Otherwise it is weldable using hot inert gas (nitrogen). Chemically it has excellent resistance to solvents, caustics, chlorides, and most corrosive chemicals. ECTFE is sold under the Halar trademark by Ausimont.

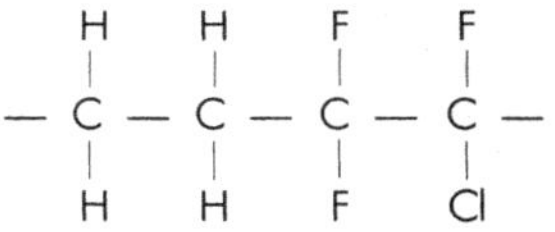

ETFE (ethylene tetrafluoroethylene)

ETFE has a high melt point of 518°F (270°C) and retains good mechanical properties from cyrogenic temperature levels to an upper limit of 350°F (177°C). It has a continuous working temperature limit of 300°F (149°C). ETFE is a tough plastic with good tensile strength, high impact resistance, and is more like the conventional engineering plastics than softer fluoropolymers. It lacks the full chemical resistance properties of PTFE since it is predominantly composed of alternating units of ethylene and tetrafluorethylene. Next to PVDF, ETFE has the highest tensile strength of polymers containing fluorine. However, it lacks good elongation properties at elevated temperatures. Tefzel is a familiar trademark of DuPont for ETFE.

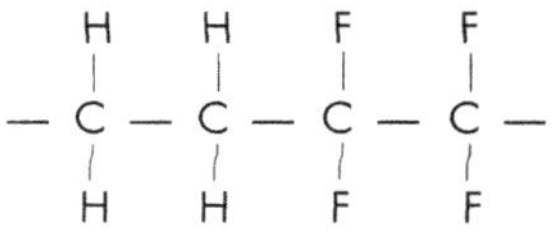

Plastic piping system components. *(Courtesy George Fischer Engineering Handbook)*

PVDF (polyvinylidene fluoride)

PVDF is a thermoplastic fluoropolymer with a melt point of 352°F (178°C) and a wide service range from -40°F (-40°C) to 284° F (140° C). It has a very linear chemical structure, and similar to PTFE with the exception of not being fully fluorinated, i.e. having 3% hydrogen by weight. Its drawbacks in the area of chemical resistance include unsuitability with strong alkalis, fuming acids, polar solvents, amines, ketones, and esters. It has a high tensile strength as well as a high heat deflection temperature. It is readily weldable, offers high purity qualities, and is resistant to permeation of gases. The trademark Solef is used by Solvay and Kynar by Atochem for PVDF resin.

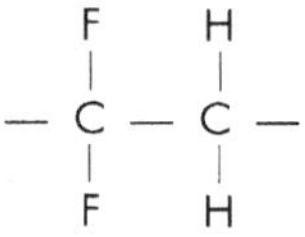

CTFE (chlorotrifluoroethylene)

CTFE is a crystalline plastic related structurally to PTFE with the exception of the addition of a chlorine atom in the basic chemical structure. CTFE has a melt point of 424° F (218°C) and is melt processible, but processing is difficult because of its high melt viscosity and tendency to degrade too readily with a resultant loss of mechanical properties. CTFE has 30.5% chlorine by weight and although this chlorine in its structure allows it to become weldable, there are subsequent limitations to its use. It has high compressive strength, low creep, and good elastic memory. It is non-wetting and experiences no moisture absorption. In its noncrystalline form CTFE is relatively resilient and flexible, although by heat treatment between 300°F (149°C) and 380° F (193°C) it can be crystallized and rendered considerably harder and less flexible. Molded articles of CTFE are therefore very sensitive to conditions of the molding cycle, and under some conditions impact strength is relatively low. The trademark Kel-F is used by 3M Corporation for the resin.

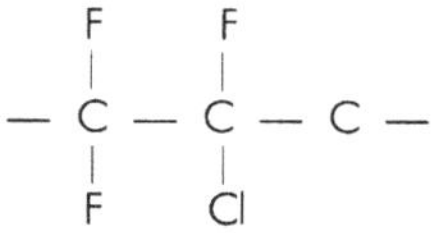

Plastic piping system components. *(Courtesy George Fischer Engineering Handbook)*

FEP (fluorinated ethylene propylene)

FEP is a fully fluorinated thermoplastic which was introduced in 1960. It was chiefly designed to provide melt processibility properties to the unique and high desirable properties of PTFE. Although FEP is thermally superior to other plastics, it still sacrificed some temperature resistance compared to PTFE. This is primarily evidenced by its lower tensile strength and low heat deflection temperature. FEP has some degree of branching, but consists predominantly of linear chinas and has a crystalline melting point of 554°F (290°C). However, FEP exhibits changes in physical strength after prolonged exposure above 400°F (204°C) which largely account for the lower temperature rating for this product compared to a fully fluorinated polymer such as PTFE. FEP is a relatively soft plastic with lower tensile strength, wear resistance, and creep resistance than other plastics. It is insensitive to notched impact forces and has excellent permeation resistance to most liquids except some chlorinated hydrocarbons.

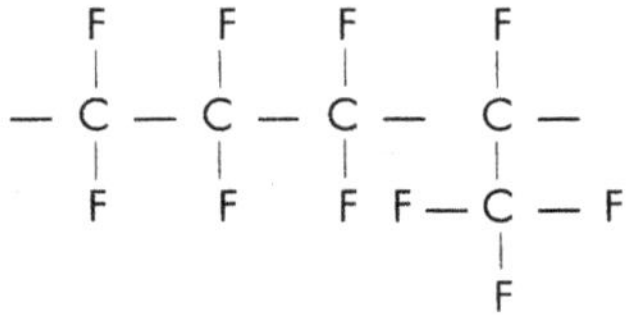

PFA (perfluoroalkoxy resin)

In 1972 DuPont produced another fully fluorinated polymer known as PFA, one which is also melt processible with better melt flow and molding properties than FEP. Although PFA has somewhat better physical and mechanical properties than FEP above 300°F (149°C) and can be used up to 500°F (260°C), it lacks the physical strength of PTFE at elevated temperatures and must be reinforced or designed with thickness to compensate for its softness. For instance, although PFA has reasonable tensile strength at 68°F (20°C), its heat deflection temperature is the lowest of all fluoroplastics. PFA melts at 590°F (310°C) and a service temperature of 500°F (260°C) is claimed due to the fact that pyrolysis studies show that weight losses at that temperature are essentially negligible. Although PFA matches the hardness and impact strength of PTFE, it sustains on 1/4 of the life of PTFE in flexibility tests. PTFE is generally advantageous over PFA in the areas of deformation under load (creep resistance), Taber abrasion resistance, low temperature Izod impact strength, heat distortion temperature, continuous service temperature, fatigue resistance (cycles to failure), and elongation.

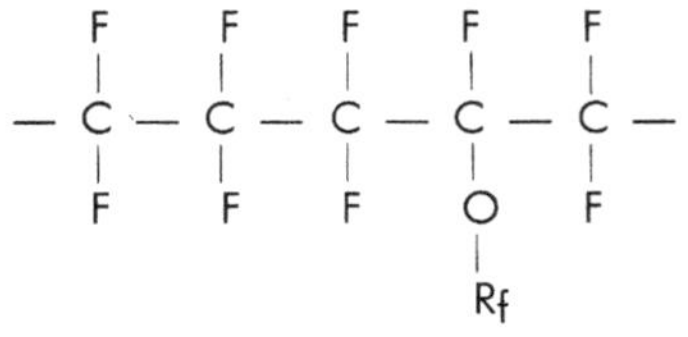

Plastic piping system components. *(Courtesy George Fischer Engineering Handbook)*

Physical Properties

Introduction
Today, plastics materials are used extensively in pipeline construction.

These are frequently unsuitable for plastics piping, and should therefore be avoided.

It is essential, and has in certain countries become a regulation, that only persons familiar with the material may be employed in plastics pipeline construction.

This does not mean that plastics present the engineer with greater mechanical and technical problems.

On the contrary.

Sometimes, however, plastics piping is designed and installed on the basis of techniques traditionally associated with piping in metal, even though it is an entirely different material.

Substantial advantages result from simplified handling methods, while prefabrication and shorter assembly times reduce the cost of installation and maintenance. Technical requirements can in many cases be better fulfilled, with a corresponding increase in the efficiency of the installation.

The correct choice of material, based on the physical properties, is therefore of the utmost importance.

Summary:
A successful installation necessitates careful planning and execution, which takes into account the characteristics of the material.

General Properties
The plastics materials in use today, with their various physical properties, meet most of the requirements of pipeline construction. The suitability of a material should be tested before it is chosen for a particular application.

Apart from the special characteristics of each type of material, there are some which are common to all plastics.

Light weight	easy handling and transport
Non-corrosive	no protective coating necessary
Smooth surfaces	hydraulic advantages, little or no build-up of scale or abrasion
Flexibility	'give' in plastics pipe is an aid to installation
Insulating properties	no electrolytic corrosion, good soundproofing
Poor thermal conductivity	minimum heat loss through uninsulated piping
Chemical resistance	good resistance against most normal corrosive actions of acids and alkalis
Simple joining procedure	reduced installation costs
Quality control and standardization	operational safety dimensional accuracy, interchangeable components

Plastic piping system components. *(Courtesy George Fischer Engineering Handbook)*

Abbreviations Used for Plastics

*ABS	= Acrylonitrile-butadiene-styrene
CA	= Cellulose acetate
CAB	= Cellulose acetate butyrate
CAP	= Cellulose acetate propionate
CF	= Cresol - formaldehyde
CMC	= Carboxymethyl cellulose
CN	= Cellulose nitrate
*CPVC	= Chlorinated polyvinyl chloride
*CR	= Chloroprene rubber (Neoprene®)
CS	= Casein
*CSM	= Chlorine sulphonyl polyethylene (Hypalon®)
EC	= Ethyl cellulose
ECTFE	= Ethylene chlorotrifluoroethylene (Halar® **)
*EP	= Epoxide, epoxy
*EPDM	= Ethylene propylene rubber
FEP	= Perfluorethylenepropylene
*FPM	= Fluorine rubber (Viton®)
*HDPE	= High-density polyethylene
HP	= Laminated paper
*IIR	= Isobutene isoprene (butyl) rubber
MF	= Melamine formaldehyde
*NBR	= Nitrile (butadiene) rubber
*NR	= Natural rubber
PFA	= Perfluoroalkoxy resin
*PA	= Polyamide
*PB	= Polybutylene
*PC	= Polycarbonate
*PCTFE	= Polychlorotrifluoroethylene
PDAP	= Polydiallyl phthalate
PEC	= Chlorinated polyethylene
PETP	= Polyethylene terephthalate
*PF	= Phenol-formaldehyde
PIB	= Polyisobutylene
PMMA	= Polymethyl methacrylate
*POM	= Polyoxymethylene (Kematal® *)
*PP	= Polypropylene
PS	= Polystryne
*PTFE	= Polytetrafluoroethylene (Teflon®)
PUR	= Polyurethane
PVAC	= Polyvinyl acetate
PVAL	= Polyvinyl alcohol
PVB	= Polyvinyl butyral
*PVC	= Polyvinyl chloride
PVCA	= Polyvinyl chloride acetate
PVDF	= Polyvinylidene fluoride
PVF	= Polyvinyl fluoride
PVFM	= Polyvinyl formal
PVK	= Polyvinyl carbazol
SAN	= Styrene-acrylonitrile
SB	= Styrene-butadiene
SI	= Silicone
UF	= Urea-formaldehyde
UP	= Unsaturated polyester

* used in pipeline construction ®Du Pont's registered trade name ®** Ausimont's registered trade name ®* I.C.I.'s registered trade name

Plastic piping system components. *(Courtesy George Fischer Engineering Handbook)*

Physical Data and Aids to Identification

The values below are generic values and not to be used for calculation purposes. Please contact the material product managers at the factory for current detailed values.

Properties	PVC	CPVC	*PP (copolymer)	HDPE	PVDF
Delivery of the pipes	in straight lenghts	in straight lenghts	in straight lenghts	in straight lengths or coiled	in straight lenghts
Surface feel	smooth	smooth	waxy	waxy	smooth
Appearance (water pipes)	dark grey	grey-beige	pale grey	black	opaque natural
Sound produced when dropped	high clatter	high clatter	high clatter	medium clatter	high clatter
Combustibility and appearance of the flame	carbonizes in flame; extinguishes away from flame	carbonizes in flame; extinguishes away from flame	bright flame; drops continue to burn while falling	bright flame; drops continue to burn while falling	carbonizes in flame; extinguishes away from flame
Odor of smoke after flame is extinguished	pungent (like hydro-chloric acid)	pungent (like hydro-chloric acid)	like resin	like candles	pungent
Nail test (Impression made with fingernail)	impression not possible	impression not possible	very slight impression possible	impression possible	impression not possible
Floats in water	no	no	yes	yes	no
Notch sensitivity	yes	yes	slight	no	slight
Weather resistance	stabilized good	stabilized good	stabilized good	stabilized good	unstabilized excellent
Method of permanent joining	solvent cement	solvent cement	fusion	fusion	fusion
Suitable for mechanical joining	yes	yes	yes	yes	yes
Stress crack sensitivity with regard to joining for safe media, e.g. water	none	none	slight	some	none
Linear expansion in/in°F (10^{-5})	4.44	3.89	8.33	11.1	6.72
Thermal conductivity BTU/Ft·HR·°F	0.094	0.094	0.128	0.269	0.073
Specific heat BTU/lb°F	0.23	0.23	0.4	0.42	0.23
Specific weight g/ml	1.38	1.5	0.905	0.955	1.78
Tensile strength at 20°C (psi)	8085	8085	4700	3530	8450
Modulus of elasticity at 20°C (psi) (in tension)	441,000	441,000	220,000	118,000	300,000

* PP listed is a copolymer. Refer to the Beta Polypropylene Section or contact the factory for specifications on ß PP-H (homopolymer) as supplied for our pigmented material.

Plastic piping system components. *(Courtesy George Fischer Engineering Handbook)*

APPENDIX D
PLASTIC PIPE FITTINGS

P/L 02: PVC Schedule 80 (Large Diameter) Fittings
Material: PVC Type I Gray (Cell Classification 12454-B)
Meets ASTM D-2467 (Socket) (up to 12") and ASTM D-2464 (Threaded) (up to 6")

Non-Stocked Fabricated Fittings
Not all of the Fabricated Fittings are carried in our warehouse. Please check with our Customer Service Department for availability. All orders for non-stocked Fabricated Fittings are non-cancellable and non-returnable. Freight is paid by the customer (F.O.B. Plant).

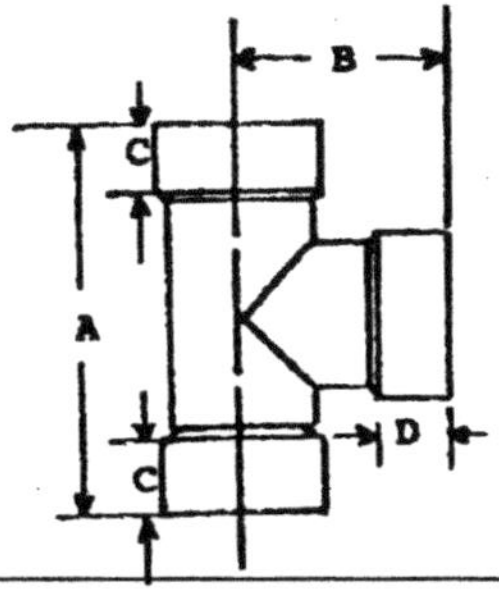

Tee (S x S x S)

Inch Size	Part Number		mp	lbs. each	a	b	c	d
10*	801-100		1	30	22 1/2	11 1/4	5 1/2	5 1/2
12*	801-120		1	35	26 3/4	13 3/8	6 1/2	6 1/2
14	801-140H		1	40	33	16 1/2	7	7
16	801-160H		1	45	39	19 1/2	9	9

*Molded Fittings

Large diameter PVC pressure fittings. *(Courtesy George Fischer Engineering Handbook)*

Reducing Tee (S x S x S)

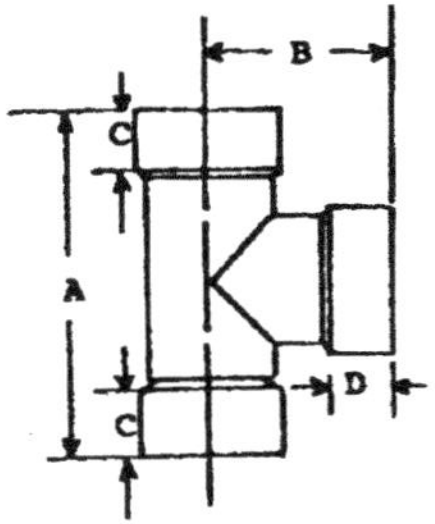

Inch Size	Part Number		mp	lbs. each	a	b	c	d
10 x 2	801-619H		1	30	18 1/2	9	6	2
10 x 3	801-621H		1	30	19 1/2	9 1/2	6	2 1/2
10 x 4	801-623H		1	30	20 1/2	10 1/8	6	3 1/4
10 x 6	801-626H		1	30	22 1/2	10 5/8	6	3 1/2
10 x 8	801-628H		1	30	24 1/2	11 7/8	6	4 3/4
12 x 2	801-659H		1	35	20	10	7	2
12 x 3	801-661H		1	35	21	10	7	2 1/2
12 x 4	801-663H		1	35	22 1/2	11 1/8	7	3 1/4
12 x 6	801-666H		1	35	24 1/2	11 5/8	7	3 1/2
12 x 8	801-668H		1	35	27	12 7/8	7	4 3/4
12 x 10	801-670H		1	35	29 1/2	14 3/8	7	6
14 x 2	801-692H		1	40	20	11	7	2
14 x 3	801-694H		1	40	21	11	7	2 1/2
14 x 4	801-696H		1	40	22 1/2	11 3/4	7	3 1/4
14 x 6	801-698H		1	40	24 1/2	12 1/4	7	3 1/2
14 x 8	801-700H		1	40	27	13 1/2	7	4 3/4
14 x 10	801-702H		1	40	29	15	7	6
14 x 12	801-704H		1	40	31	17	7	7
16 x 2	801-726H		1	45	22	12	9	2
16 x 3	801-728H		1	45	23	12	9	2 1/2
16 x 4	801-730H		1	45	24 1/2	12 3/4	9	3 1/4
16 x 6	801-732H		1	45	28 1/2	13 1/4	9	3 1/2
16 x 8	801-734H		1	45	30	14 1/2	9	4 3/4
16 x 10	801-736H		1	45	32	16	9	6
16 x 12	801-738H		1	45	34	18	9	7
16 x 14	801-740H		1	45	36	18	9	7

Large diameter PVC pressure fittings. *(Courtesy George Fischer Engineering Handbook)*

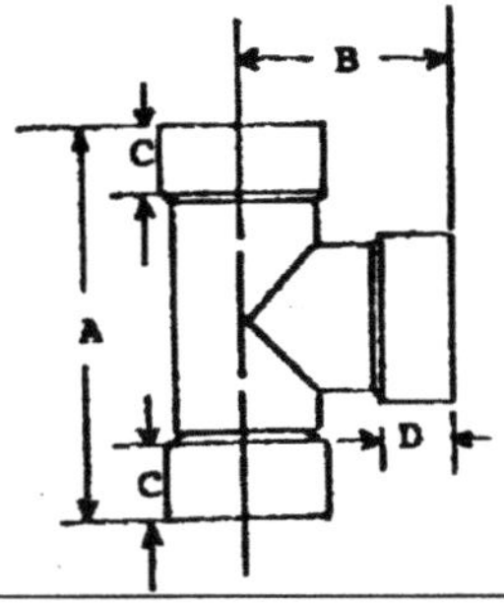

Reducing Tee (S x S x FT)

Inch Size	Part Number		mp	lbs. each	a	b	c	d
10 x 2	802-619H		1	30	18 1/2	9	6	2
10 x 3	802-621H		1	30	19 1/2	9 1/2	6	2
10 x 4	802-623H		1	30	20 1/2	10 1/2	6	2 1/2
12 x 2	802-659H		1	35	20	10	7	2
12 x 3	802-661H		1	35	21	10	7	2
12 x 4	802-663H		1	35	22 1/2	11	7	2 1/2

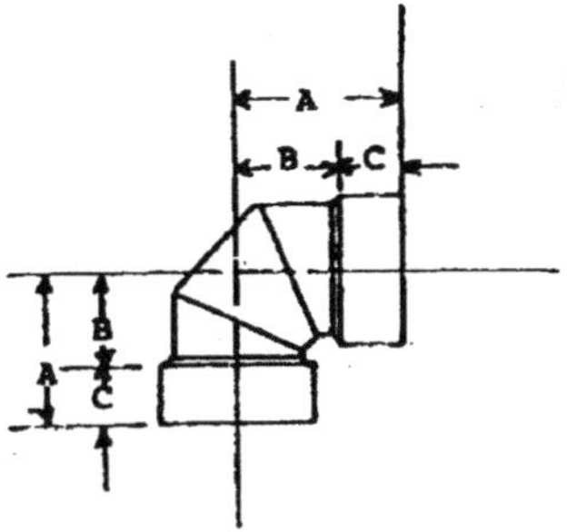

90° Ell (S x S)

Inch Size	Part Number		mp	lbs. each	a	b	c
10*	806-100		1	45	11 1/4	5 3/4	5 1/2
12*	806-120		1	50	13 3/8	6 7/8	6 1/2
14	806-140H		1	55	21	15	6
16	806-160H		1	60	24	16	8

Large diameter PVC pressure fittings. *(Courtesy George Fischer Engineering Handbook)*

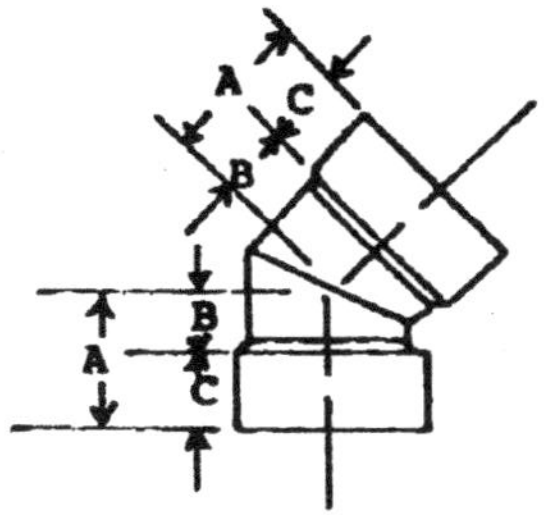

11 $^1/_4$° Ell (S x S)

Inch Size	Part Number		mp	lbs. each	a	b	c	
10	811-100H		1	20	8	2	6	
12	811-120H		1	25	9	2	7	
14	811-140H		1	30	10	3	7	
16	811-160H		1	35	11	3	9	

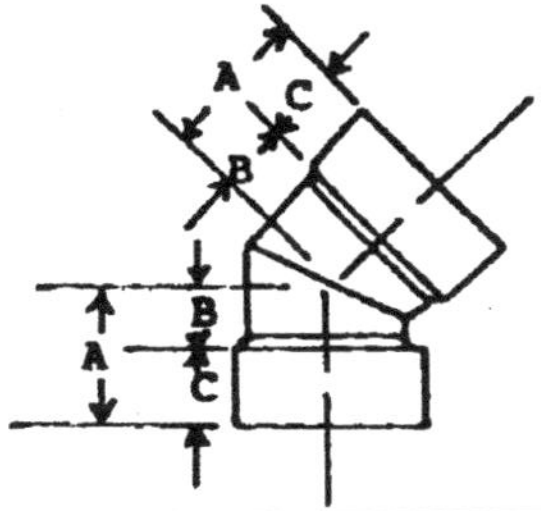

30° Ell (S x S)

Inch Size	Part Number		mp	lbs. each	a	b	c	
10	815-100H		1	20	8	2	6	
12	815-120H		1	25	10	3	7	
14	815-140H		1	30	10	3	7	
16	815-160H		1	35	12	3	9	

Large diameter PVC pressure fittings. *(Courtesy George Fischer Engineering Handbook)*

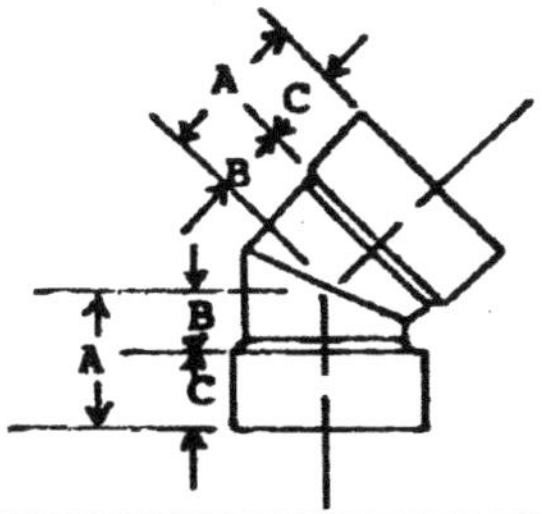

22 1/2° Ell (S x S)

Inch Size	Part Number		mp	lbs. each	a	b	c
10	816-100H		1	20	8 1/2	2 1/2	6
12	816-120H		1	25	10	3	7
14	816-140H		1	30	10 1/2	3 1/2	7
16	816-160H		1	35	11	2	9

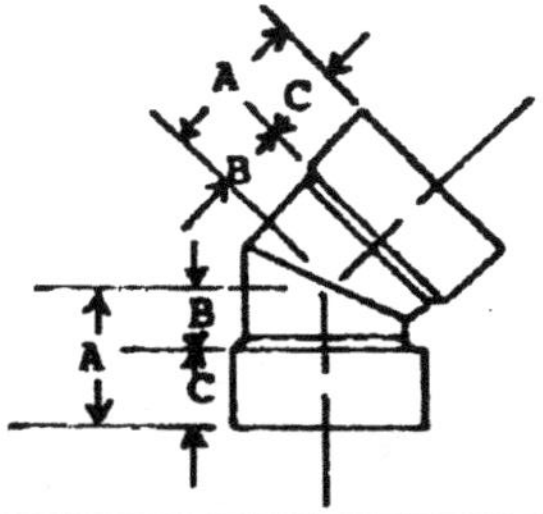

45° Ell (S x S)

Inch Size	Part Number		mp	lbs. each	a	b	c
10*	817-100		1	20	8 1/8	2 5/8	5 1/2
12*	817-120		1	25	9 5/8	3 1/8	6 1/2
14	817-140H		1	30	13 1/2	5 1/2	7
16	817-160H		1	35	16 3/4	8 3/4	9

Large diameter PVC pressure fittings. *(Courtesy George Fischer Engineering Handbook)*

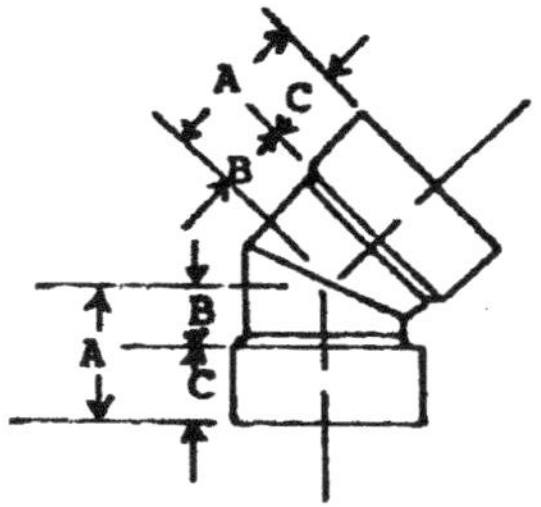

15° Ell (S x S)

Inch Size	Part Number		mp	lbs. each	a	b	c	
10	818-100H		1	20	8	2	6	
12	818-120H		1	25	9	2	7	
14	818-140H		1	30	10	3	7	
16	818-160H		1	35	11	3	9	

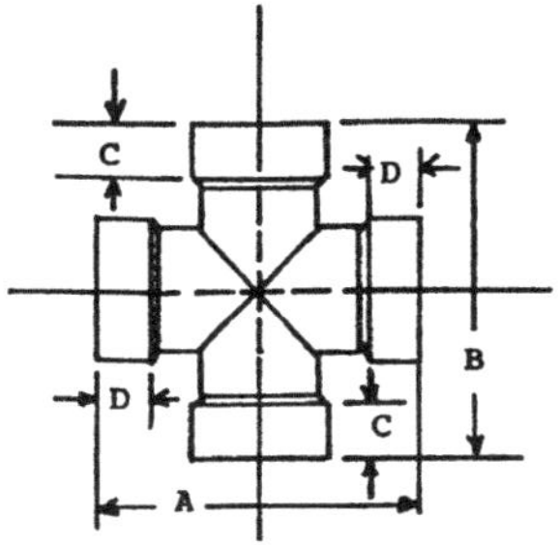

Cross (S x S x S x S)

Inch Size	Part Number		mp	lbs. each	a	b	c	d
10	820-100H		1	25	27 1/2	28	6	6
12	820-120H		1	30	32	32	7	7
14	820-140H		1	35	33	33	7	7
16	820-160H		1	40	39	39	9	9

Large diameter PVC pressure fittings. *(Courtesy George Fischer Engineering Handbook)*

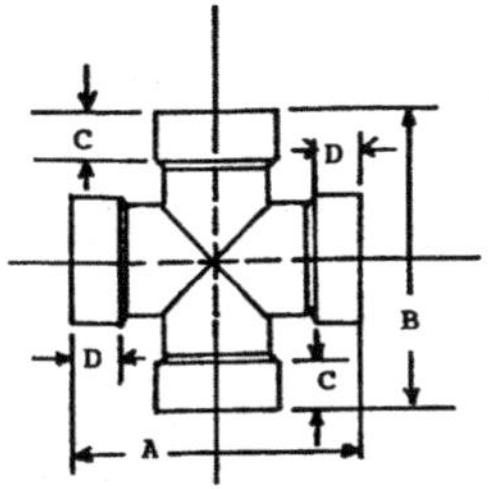

Reducing Cross (S x S x S x S)

Inch Size	Part Number		mp	lbs. each	a	b	c	d
10 x 2	820-619H		1	25	18	18 1/2	6	2
10 x 3	820-621H		1	25	19	19 1/2	6	2 1/2
10 x 4	820-623H		1	25	20 1/4	20 1/2	6	3 1/4
10 x 6	820-626H		1	25	21 1/4	22 1/2	6	3 1/2
10 x 8	820-628H		1	25	24	24 1/2	6	3 5/8
12 x 2	820-659H		1	30	20	20	7	2
12 x 3	820-661H		1	30	20	21	7	2 1/2
12 x 4	820-663H		1	30	22 1/4	22 1/2	7	3 1/4
12 x 6	820-666H		1	30	23 1/4	24 1/2	7	3 1/2
12 x 8	820-668H		1	30	25 3/4	27	7	3 5/8
12 x 10	820-670H		1	30	28 3/4	29 1/2	7	6
14 x 2	820-692H		1	35	22	20	7	2
14 x 3	820-694H		1	35	23 1/2	22 1/2	7	3 1/4
14 x 4	820-696H		1	35	24 1/2	24 1/2	7	3 1/2
14 x 6	820-698H		1	35	27	27	7	3 5/8
14 x 8	820-700H		1	35	30	29	7	6
14 x 10	820-702H		1	35	34	31	7	6
14 x 12	820-704H		1	35	34	31	7	7
16 x 2	820-726H		1	40	24	22	9	2
16 x 3	820-728H		1	40	24	23	9	2 1/2
16 x 4	820-730H		1	40	25 1/2	26 1/2	9	3 1/4
16 x 6	820-732H		1	40	26 1/2	28 1/2	9	3 1/2
16 x 8	820-734H		1	40	29	31	9	3 5/8
16 x 10	820-736H		1	40	32	33	9	6
16 x 12	820-738H		1	40	36	35	9	7
16 x 14	820-740H		1	40	36	37	9	7

Large diameter PVC pressure fittings. *(Courtesy George Fischer Engineering Handbook)*

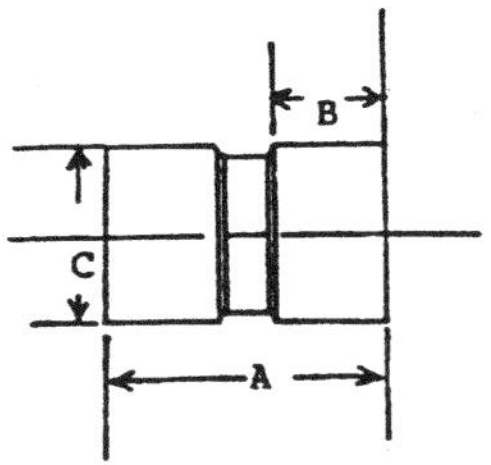

Coupling (S x S)

Inch Size	Part Number		mp	lbs. each	a	b	c
10	829-100H		1	15	14	6	12
12	829-120H		1	20	16	7	14
14	829-140H		1	25	16	7	15 1/2
16	829-160H		1	30	19 1/2	9	17 3/4

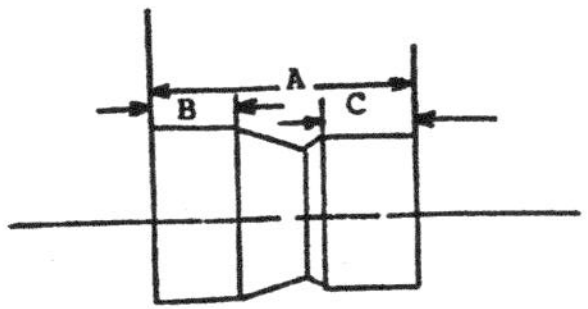

Reducer Coupling (S x S)

Inch Size	Part Number		mp	lbs. each	a	b	c
10 x 4	829-623H		1	15	24	6	3 1/4
10 x 6	829-626H		1	15	20	6	3 1/2
10 x 8	829-628H		1	15	12 3/4	6	3 5/8
12 x 8	829-668H		1	20	21	7	3 5/8
12 x 10	829-670H		1	20	14 1/2	7	6
14 x 10	829-702H		1	25	22 1/2	7	6
14 x 12	829-704H		1	25	15 1/2	7	7
16 x 12	829-738H		1	30	25	9	7
16 x 14	829-740H		1	30	17	9	7

Large diameter PVC pressure fittings. *(Courtesy George Fischer Engineering Handbook)*

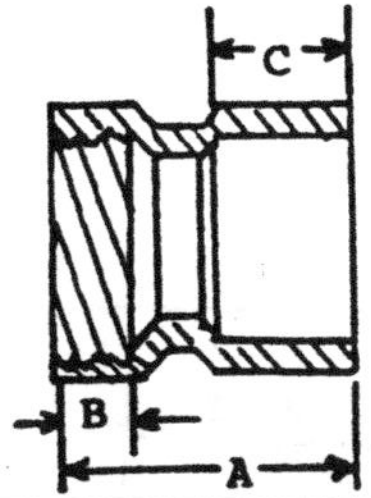

Female Adapter (S x FT)

Inch Size	Part Number		mp	lbs. each	a	b	c	
10	835-100H		1	20	10 3/8	2	6	
12	835-120H		1	25	12	1 3/4	6 1/2	
14	835-140H		1	30	12	2 3/4	6	

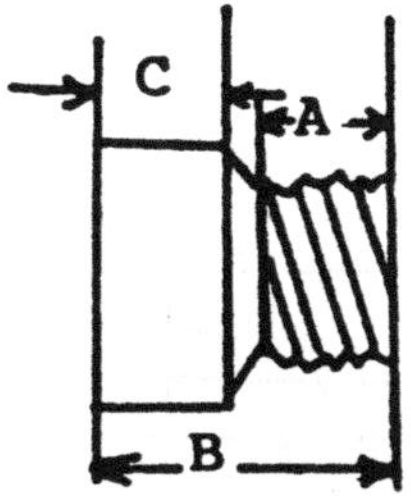

Male Adapter (S x MT)

Inch Size	Part Number		mp	lbs. each	a	b	c	
10	836-100H		1	20	2	10	6	
12	836-120H		1	25	2	10 1/2	6 1/2	
14	836-140H		1	30	2 1/2	12	6	

Large diameter PVC pressure fittings. *(Courtesy George Fischer Engineering Handbook)*

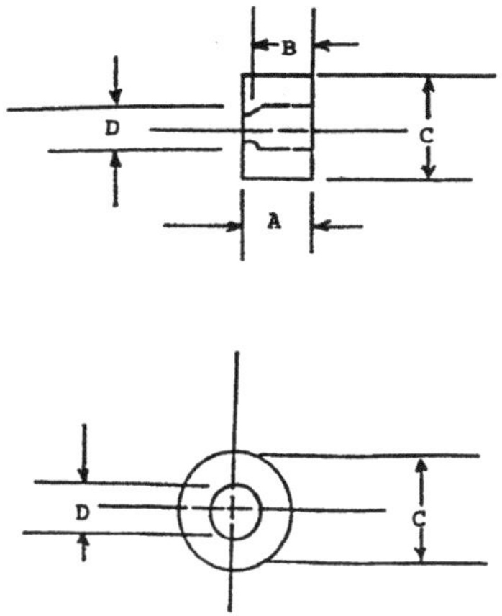

Reducer Bushing (flush) (Spg x S)

Inch Size	Part Number		mp	lbs. each	a	b	c	d
10 x 4	837-623H		1	15	4 1/2	2 1/4	10 3/4	4 1/2
10 x 6	837-626H		1	15	4 1/2	3 1/4	10 3/4	6 5/8
10 x 8	837-628H		1	15	4 1/2	3 7/8	10 3/4	8 5/8
12 x 6	837-666H		1	20	5 3/8	3 1/4	12 3/4	6 5/8
12 x 8	837-668H		1	20	5 3/8	3 7/8	12 3/4	8 5/8
12 x 10	837-670H		1	20	5 3/8	4 7/8	12 3/4	10 3/4
14 x 4	837-696H		1	25	6 1/2	2 1/4	14	4 1/2
14 x 6	837-698H		1	25	6 1/2	3 1/4	14	6 5/8
14 x 8	837-700H		1	25	6 1/2	3 7/8	14	8 5/8
14 x 10	837-702H		1	25	6 1/2	4 7/8	14	10 3/4
14 x 12	837-704H		1	25	6 1/2	5 3/8	14	12 3/4
16 x 4	837-730H		1	30	7	2 1/4	16	4 1/2
16 x 6	837-732H		1	30	7	3 1/4	16	6 5/8
16 x 8	837-734H		1	30	7	3 7/8	16	8 5/8
16 x 10	837-736H		1	30	7	4 7/8	16	10 3/4
16 x 12	837-738H		1	30	7	5 3/8	16	12 3/4
16 x 14	837-740H		1	30	7	6	16	14

Large diameter PVC pressure fittings. *(Courtesy George Fischer Engineering Handbook)*

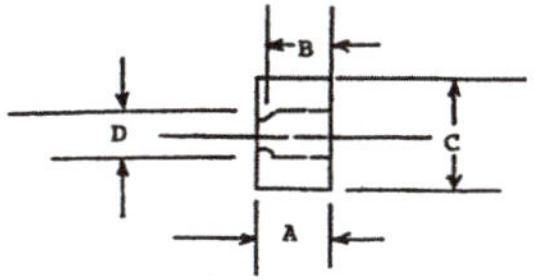

Reducer Bushing (flush) (Spg x FT)

Inch Size	Part Number		mp	lbs. each	a	b	c	d
10 x 4	838-623H		1	15	4 1/2	2	10 3/4	4 1/2
10 x 6	838-626H		1	15	4 1/2	2	10 3/4	6 5/8
10 x 8	838-628H		1	15	4 1/2	2	10 3/4	8 5/8
12 x 4	838-663H		1	20	5 3/8	2	12 3/4	4 1/2
12 x 6	838-666H		1	20	5 3/8	2	12 3/4	6 5/8
12 x 8	838-668H		1	20	5 3/8	2	12 3/4	8 5/8
12 x 10	838-670H		1	20	5 3/8	2	12 3/4	10 3/4

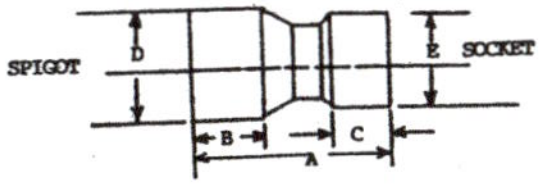

Reducer Bushing Extended Style (Spg x S)

Inch Size	Part Number		mp	lbs. each	a	b	c	d	e
10 x 6	840-626H		1	15	13	6	3 1/2	10 3/4	6 5/8
10 x 8	840-628H		1	15	11 7/8	6	3 5/8	10 3/4	8 5/8
12 x 6	840-666H		1	20	13	6 1/2	3 1/2	12 3/4	6 5/8
12 x 8	840-668H		1	20	14 1/4	6 1/2	3 5/8	12 3/4	8 5/8
12 x 10	840-670H		1	20	12	6 1/2	6	12 3/4	10 3/4
14 x 8	840-700H		1	25	15	6 1/2	3 5/8	14	8 5/8
14 x 10	840-702H		1	25	13 1/2	6 1/2	6	14	10 3/4
14 x 12	840-704H		1	25	13 1/2	6 1/2	7	14	12 3/4
16 x 10	840-736H		1	30	36	9	6	16	10 3/4
16 x 12	840-738H		1	30	25	9	7	16	12 3/4
16 x 14	840-740H		1	30	17	9	7	16	14

Large diameter PVC pressure fittings. *(Courtesy George Fischer Engineering Handbook)*

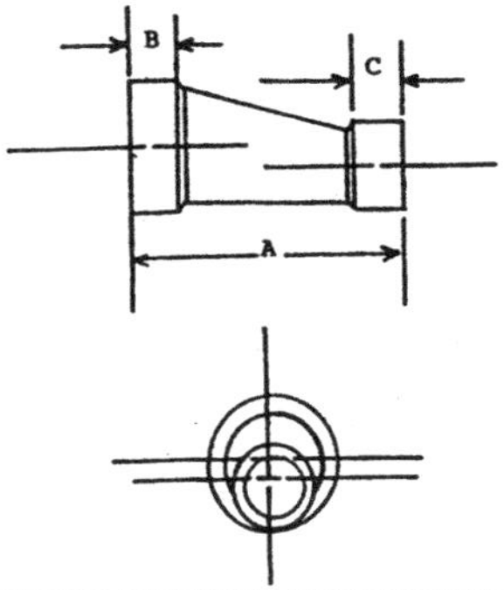

Eccentric Reducers (S x S)

Inch Size	Part Number		mp	lbs. each	a	b	c	
10 x 4	841-623H		1	15	20 1/2	6	3 1/4	
10 x 6	841-626H		1	15	21	6	3 1/2	
10 x 8	841-628H		1	15	18	6	3 5/8	
12 x 4	841-663H		1	20	24	7	3 1/4	
12 x 6	841-666H		1	20	24	7	3 1/2	
12 x 8	841-668H		1	20	20	7	3 5/8	
12 x 10	841-670H		1	20	23	7	6	

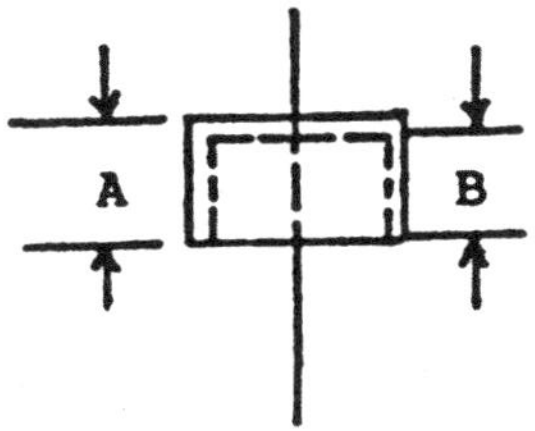

Caps (S)

Inch Size	Part Number		mp	lbs. each	a	b		
10	847-100H		1	20	6 3/4	6		
12	847-120H		1	25	8	7		
14	847-140H		1	30	9	7		
16	847-160H		1	35	9 3/4	9		

Large diameter PVC pressure fittings. *(Courtesy George Fischer Engineering Handbook)*

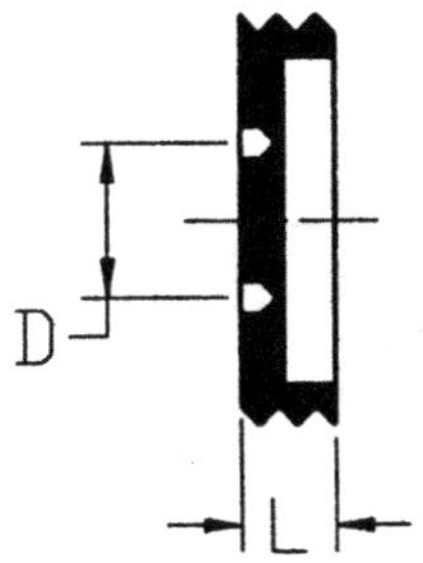

Plug (MT) (1″ thick)

Inch Size	Part Number		mp	lbs. each	d	l		
6	850-060H		1	10	4	1		
8	850-080H		1	15	4	1		
10	850-100H		1	20	4	1		
12	850-120H		1	25	4	1		

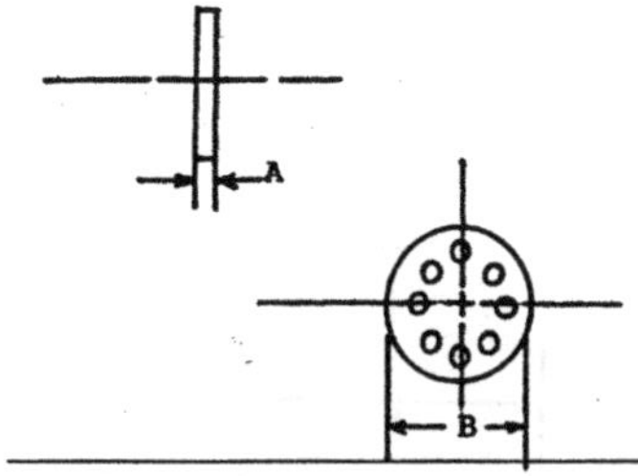

Blind Flange (1″ thick)

Inch Size	Part Number		mp	lbs. each	a	b		
10	853-100H		1	20	1	16		
12	853-120H		1	25	1	19		
14	853-140H		1	30	1	21		
16	853-160H		1	35	1	23 1/2		

Large diameter PVC pressure fittings. *(Courtesy George Fischer Engineering Handbook)*

E—Bolt Circle
F—No. Holes
G—Dia. Holes

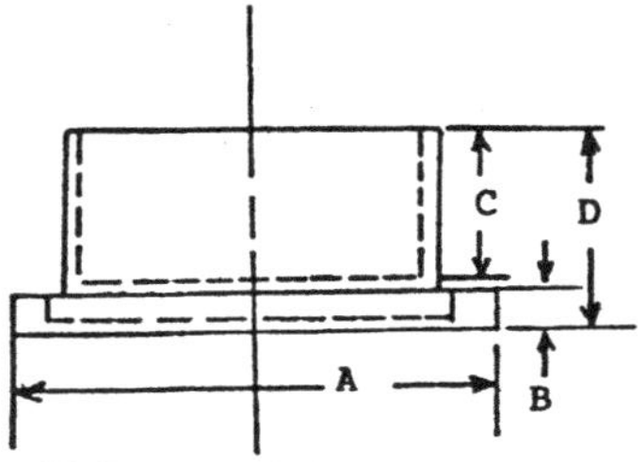

Van Stone Flange (S)

Inch Size	Part Number		mp	lbs. each	a	b	c	d	e	f	g
14*	854-140H		1	30	21	1 3/8	7	7 3/4	7 3/4	12	1 1/8
16*	854-160H		1	35	23 1/2	1 1/2	9	10 3/4	10 3/4	16	1 1/8

*(epoxy coated steel rings)

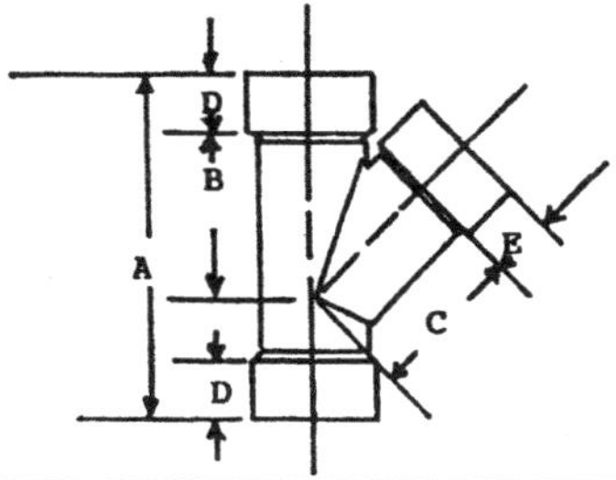

Tee 45° (Wye) (S x S x S)

Inch Size	Part Number		mp	lbs. each	a	b	c	d	e
10	875-100H		1	30	34	16	16	6	6
12	875-120H		1	35	39	19	19	6 1/2	6 1/2
14	875-140H		1	40	41	22	22	7	7
16	875-160H		1	45	47 1/2	25 1/2	25 1/2	9	9

Large diameter PVC pressure fittings. *(Courtesy George Fischer Engineering Handbook)*

Tee 45° (Wye) Reducing (S x S x S)

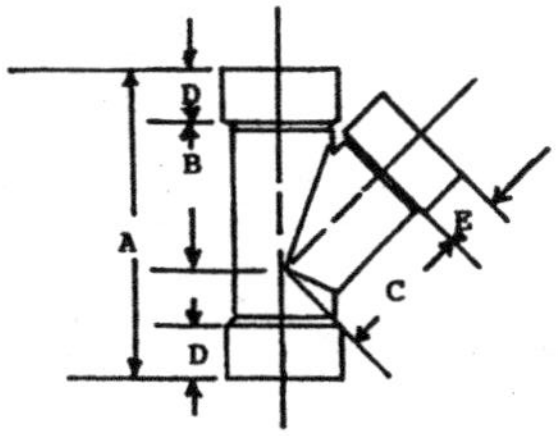

Inch Size	Part Number	sp	lbs. each	a	b	c	d	e
10 x 2	875-619H	1	30	21 1/2	12	12	6	2
10 x 2 1/2	875-620H	1	30	22	12	13	6	2 1/2
10 x 3	875-621H	1	30	22	12 1/2	13	6	2 1/2
10 x 4	875-623H	1	30	24	13	13 3/4	6	3 1/4
10 x 6	875-626H	1	30	26	13 1/2	14	6	4
10 x 8	875-628H	1	30	29	15 1/4	15 3/4	6	4 3/4
12 x 2	875-659H	1	35	24	12 1/2	12 1/2	7	2
12 x 3	875-661H	1	35	24 1/2	13	13	7	2 1/2
12 x 4	875-663H	1	35	26 1/2	13 1/4	13 1/2	7	3 1/4
12 x 6	875-666H	1	35	29	14 1/4	16 1/6	7	4
12 x 8	875-668H	1	35	32	16	16 1/2	7	4 3/4
12 x 10	875-670H	1	35	38	18	15	7	6
14 x 4	875-696H	1	40	25	12	13	7	3 1/4
14 x 6	875-698H	1	40	28	14	15 1/2	7	4
14 x 8	875-700H	1	40	31 1/2	16	16 1/2	7	4 3/4
14 x 10	875-702H	1	40	34	17 1/2	17 1/2	7	6
14 x 12	875-704H	1	40	38 1/2	19 1/2	19 1/2	7	7
16 x 4	875-730H	1	45	31	17	17	9	3 1/4
16 x 6	875-732H	1	45	34	20	20	9	4
16 x 8	875-734H	1	45	37	20 1/2	20 1/2	9	4 3/4
16 x 10	875-736H	1	45	40	21 1/2	21 1/2	9	6
16 x 12	875-738H	1	45	43	21 1/2	21 1/2	9	7
16 x 14	875-740H	1	45	45	24	24	9	7

Large diameter PVC pressure fittings. *(Courtesy George Fischer Engineering Handbook)*

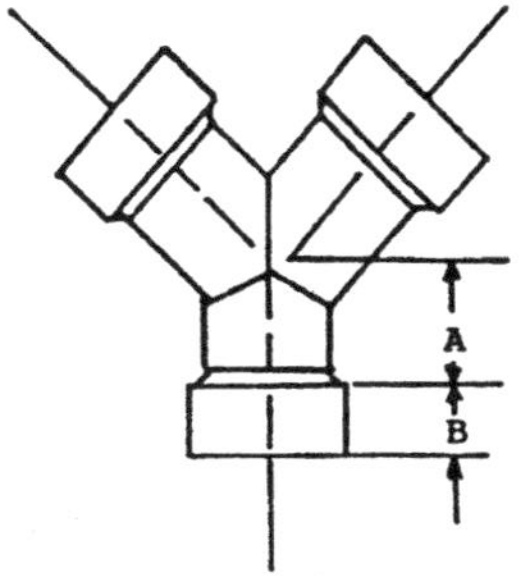

True Wye (S x S x S)

Inch Size	Part Number		sp	lbs. each	a	b		
10	876-100H		1	30	4	6		
12	876-120H		1	35	5	6 1/2		
14	876-140H		1	40	5 3/8	7		
16	876-160H		1	45	6	9		

Large diameter PVC pressure fittings. *(Courtesy George Fischer Engineering Handbook)*

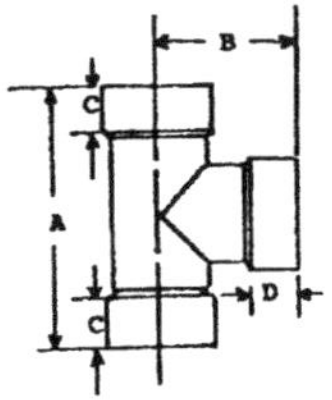

Tee (S x S x S)

Inch Size	Part Number		mp	lbs. each	a	b	c	d
10*	9801-100		1	30	22 1/2	11 1/4	5 1/2	5 1/2
12*	9801-120		1	35	26 3/4	13 3/8	6 1/2	6 1/2

*Molded Fittings

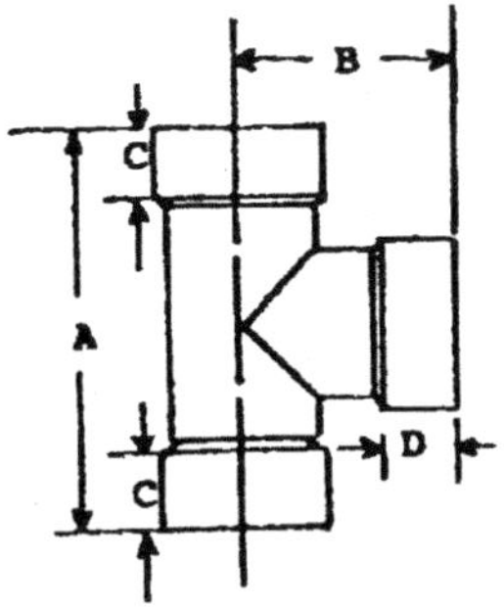

Reducing Tee (S x S x S)

Inch Size	Part Number		mp	lbs. each	a	b	c	d
10 x 2	9801-619H		1	30	18 1/2	9	6	2
10 x 3	9801-621H		1	30	19 1/2	9 1/2	6	2 1/2
10 x 4	9801-623H		1	30	20 1/2	10 1/8	6	3 1/4
10 x 6	9801-626H		1	30	22 1/2	10 5/8	6	3 1/2
10 x 8	9801-628H		1	30	24 1/2	11 7/8	6	4 3/4
12 x 2	9801-659H		1	35	20	10	7	2
12 x 3	9801-661H		1	35	21	10	7	2 1/2
12 x 4	9801-663H		1	35	22 1/2	11 1/8	7	3 1/4
12 x 6	9801-666H		1	35	24 1/2	11 5/8	7	3 1/2
12 x 8	9801-668H		1	35	27	12 7/8	7	4 3/4
12 x 10	9801-670H		1	35	29 1/2	14 3/8	7	6

Large diameter CPVC pressure fittings. (*Courtesy George Fischer Engineering Handbook)*

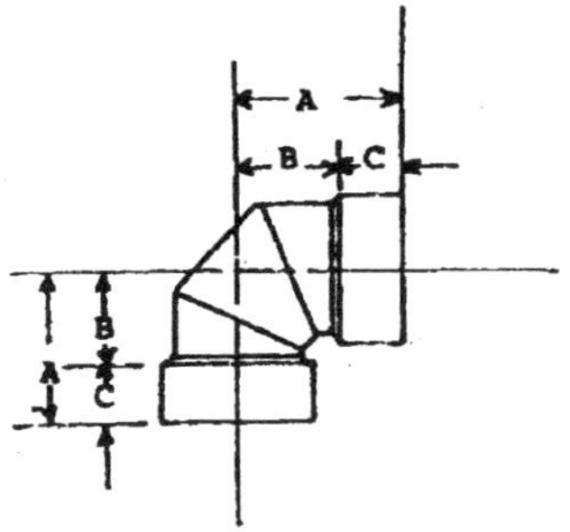

90° Ell (S x S)

Inch Size	Part Number		mp	lbs. each	a	b	c	
10*	9806-100		1	45	11 1/4	5 3/4	5 1/2	
12*	9806-120		1	50	13 3/8	6 7/8	6 1/2	

*Molded Fittings

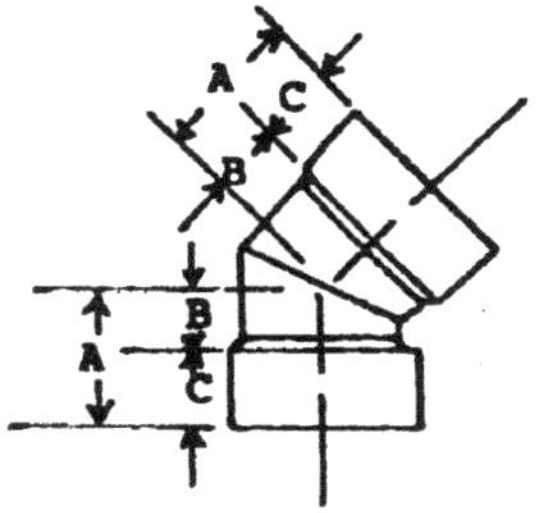

11 1/4° Ell (S x S)

Inch Size	Part Number		mp	lbs. each	a	b	c	
10	9811-100H		1	20	8	2	6	
12	9811-120H		1	25	9	2	7	

Large diameter CPVC pressure fittings. (*Courtesy George Fischer Engineering Handbook*)

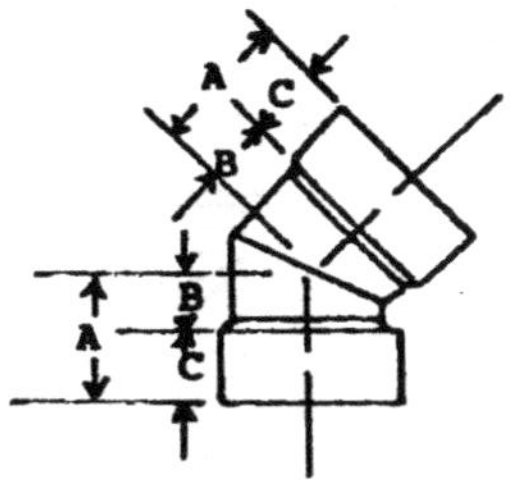

30° Ell (S x S)

Inch Size	Part Number		mp	lbs. each	a	b	c	
10	9815-100H		1	20	8	2	6	
12	9815-120H		1	25	10	3	7	

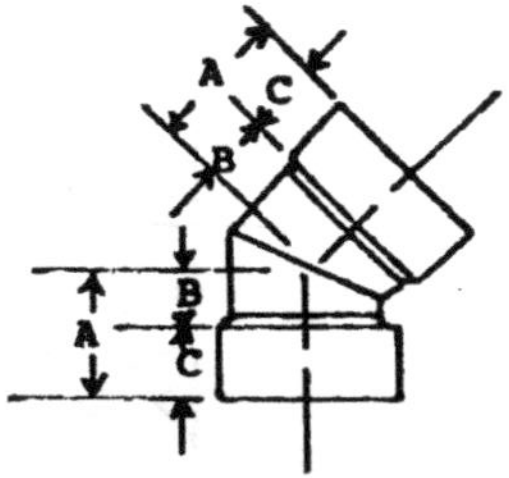

22 1/2° Ell (S x S)

Inch Size	Part Number		mp	lbs. each	a	b	c	
10	9816-100H		1	20	8 1/2	2 1/2	6	
12	9816-120H		1	25	10	3	7	

Large diameter CPVC pressure fittings. (*Courtesy George Fischer Engineering Handbook*)

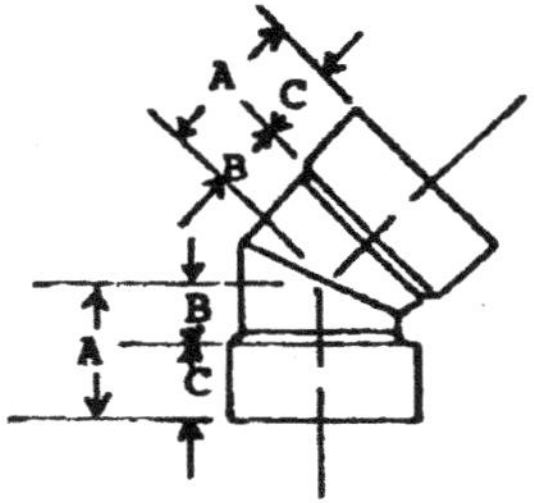

45° Ell (S x S)

Inch Size	Part Number		mp	lbs. each	a	b	c	
10*	9817-100		1	20	8 1/8	2 5/8	5 1/2	
12*	9817-120		1	25	9 5/8	3 1/8	6 1/2	

*Molded Fittings

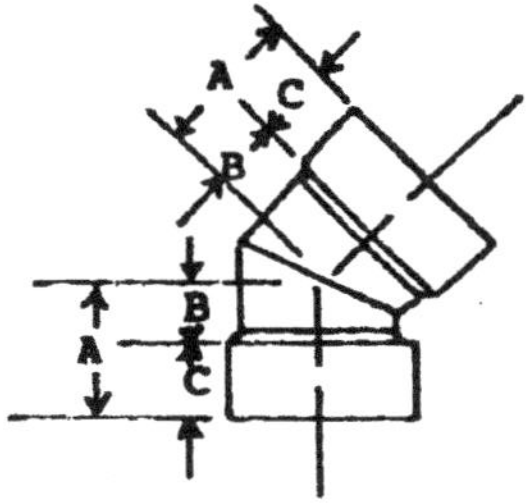

15° Ell (S x S)

Inch Size	Part Number		mp	lbs. each	a	b	c	
10	9818-100H		1	20	8	2	6	
12	9818-120H		1	25	9	2	7	

Large diameter CPVC pressure fittings. (*Courtesy George Fischer Engineering Handbook*)

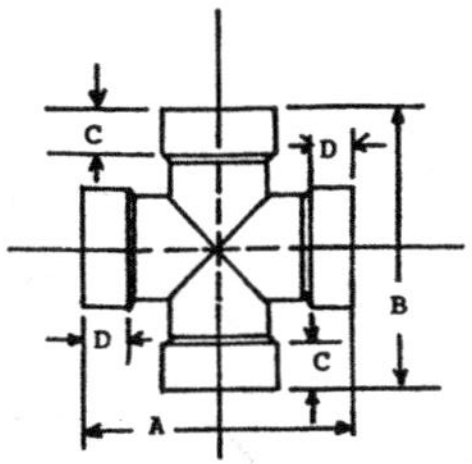

Cross (S x S x S x S)

Inch Size	Part Number		mp	lbs. each	a	b	c	d
10	9820-100H		1	25	27 1/2	28	6	6
12	9820-120H		1	30	32	32	6 1/2	6 1/2

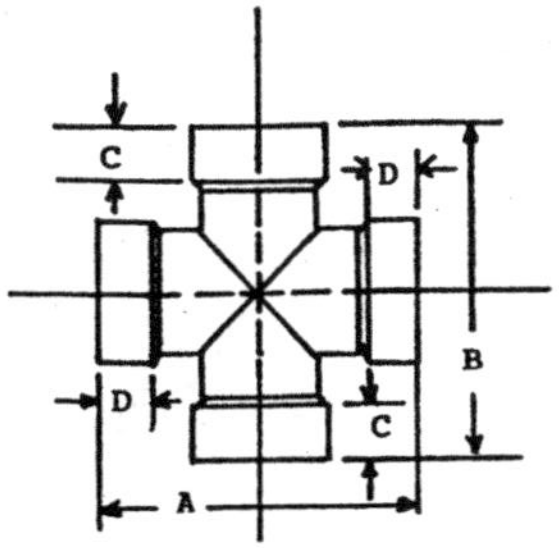

Reducing Cross (S x S x S x S)

Inch Size	Part Number		mp	lbs. each	a	b	c	d
10 x 2	9820-619H		1	25	18	18 1/2	6	2
10 x 3	9820-621H		1	25	19	19 1/2	6	2 1/2
10 x 4	9820-623H		1	25	20 1/4	20 1/2	6	3 1/4
10 x 6	9820-626H		1	25	21 1/4	22 1/2	6	3 1/2
10 x 8	9820-628H		1	25	24	24 1/2	6	3 5/8
12 x 2	9820-659H		1	30	20	20	7	2
12 x 3	9820-661H		1	30	20	21	7	2 1/2
12 x 4	9820-663H		1	30	22 1/4	22 1/2	7	3 1/4
12 x 6	9820-666H		1	30	23 1/4	24 1/2	7	3 1/2
12 x 8	9820-668H		1	30	25 3/4	27	7	3 5/8
12 x 10	9820-670H		1	30	28 3/4	29 1/2	7	6

Large diameter CPVC pressure fittings. (*Courtesy George Fischer Engineering Handbook*)

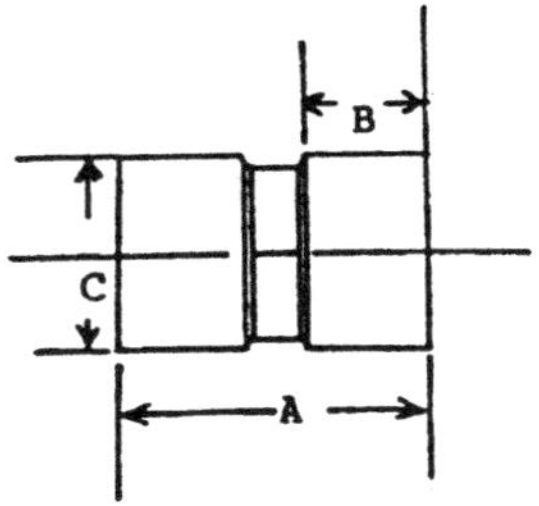

Coupling (S x S)

Inch Size	Part Number		mp	lbs. each	a	b	c	
10	9829-100H		1	15	14	6	12	
12	9829-120H		1	20	16	6 1/2	14	

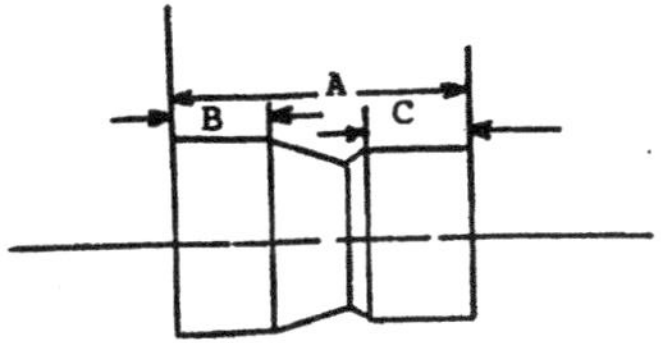

Reducer Coupling (S x S)

Inch Size	Part Number		mp	lbs. each	a	b	c	
10 x 4	9829-623H		1	15	24	5 1/2	3 1/4	
10 x 6	9829-626H		1	15	20	5 1/2	3 1/2	
10 x 8	9829-628H		1	15	12 3/4	6	3 5/8	
12 x 8	9829-668H		1	20	21	6 1/2	3 5/8	
12 x 10	9829-670H		1	20	14 1/2	5 1/2	6	

Large diameter CPVC pressure fittings. (*Courtesy George Fischer Engineering Handbook*)

Female Adapter (S x FT)

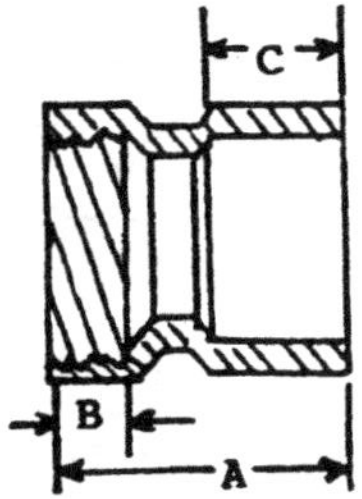

Inch Size	Part Number		mp	lbs. each	a	b	c	
6	9835-060H		1	10	8	2	3 1/2	
8	9835-080H		1	15	8 3/4	1 7/8	3 5/8	
10	9835-100H		1	20	10 3/8	2	6	
12	9835-120H		1	25	12	1 3/4	6 1/2	

Male Adapter (S x MT)

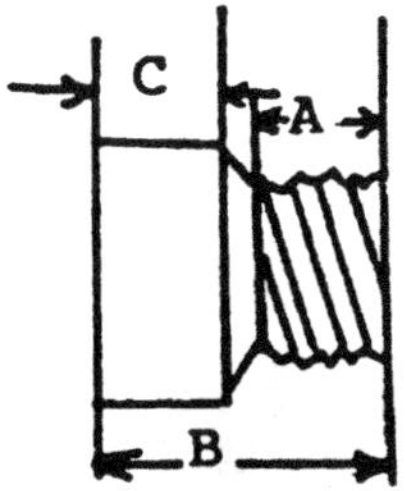

Inch Size	Part Number		mp	lbs. each	a	b	c	
6	9836-060H		1	15	1 3/4	7 1/2	4 1/4	
8	9836-080H		1	20	2	8 1/2	4 3/4	
10	9836-100H		1	25	2	10	6	
12	9836-120H		1	30	2	10 1/2	6 1/2	

Large diameter CPVC pressure fittings. (*Courtesy George Fischer Engineering Handbook*)

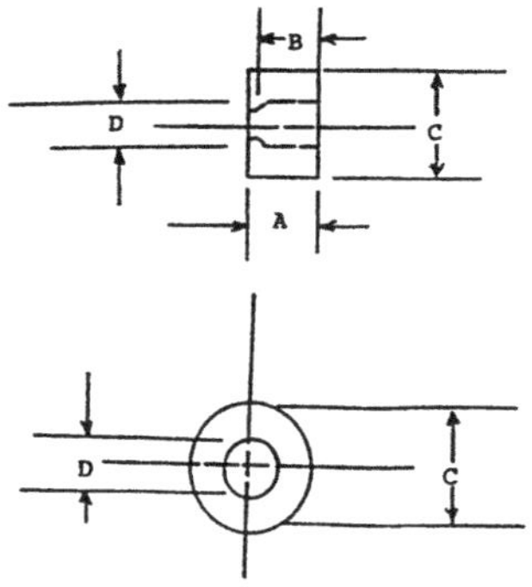

Reducer Bushing (flush) (Spg x S)

Inch Size	Part Number		mp	lbs. each	a	b	c	d
10 x 4	9837-623H		1	15	4 1/2	2 1/4	10 3/4	4 1/2
10 x 6	9837-626H		1	15	4 1/2	3 1/4	10 3/4	6 1/2
10 x 8	9837-628H		1	15	4 1/2	3 7/8	10 3/4	8 1/2
12 x 6	9837-666H		1	20	5 3/8	3 1/4	12 3/4	6 1/2
12 x 8	9837-668H		1	20	5 3/8	3 7/8	12 3/4	8 5/8
12 x 10	9837-670H		1	20	5 3/8	4 7/8	12 3/4	10 3/4

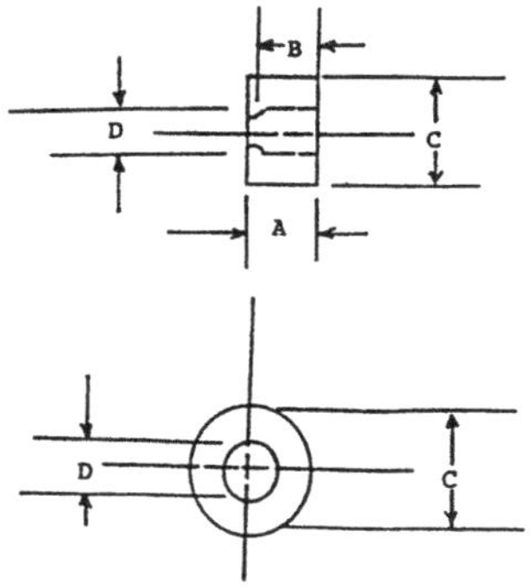

Reducer Bushing (flush) (Spg x FT)

Inch Size	Part Number		mp	lbs. each	a	b	c	d
10 x 6	9838-626H		1	15	4 1/2	2	10 3/4	6 1/2
10 x 8	9838-628H		1	15	4 1/2	2	10 3/4	8 1/2
12 x 8	9838-668H		1	20	5 3/8	2	12 3/4	8 5/8
12 x 10	9838-670H		1	20	5 3/8	2	12 3/4	10 3/4

Large diameter CPVC pressure fittings. (*Courtesy George Fischer Engineering Handbook*)

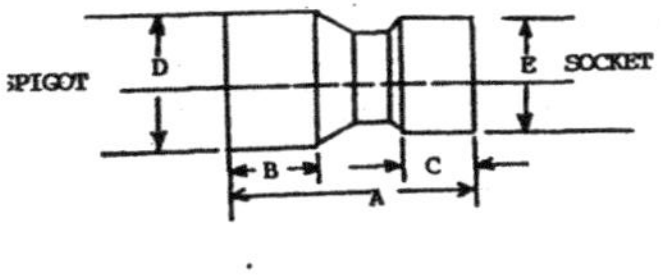

Reducer Bushing Extended Style (Spg x S)

Inch Size	Part Number		mp	lbs. each	a	b	c	d	e
10 x 6	9840-626H		1	15	13	6	3 1/2	10 3/4	6 5/8
10 x 8	9840-628H		1	15	11 7/8	6	3 5/8	10 3/4	8 5/8
12 x 8	9840-668H		1	20	14 1/4	6 1/2	3 5/8	12 3/4	8 5/8
12 x 10	9840-670H		1	20	12	6 1/2	6	12 3/4	10 3/4

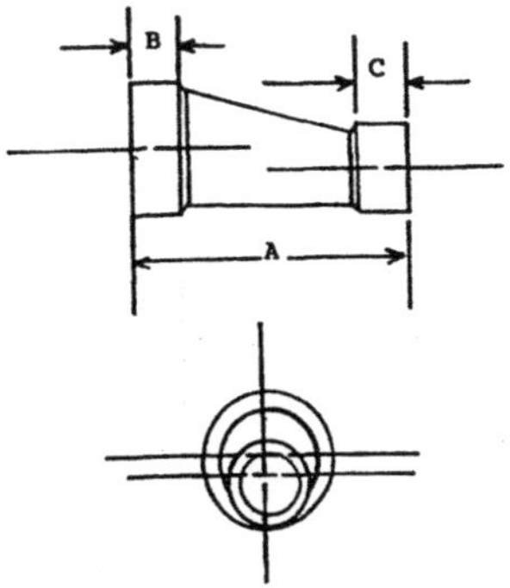

Eccentric Reducers (S x S)

Inch Size	Part Number		mp	lbs. each	a	b	c	
10 x 4	9841-623H		1	15	20 1/2	6	3 1/4	
10 x 6	9841-626H		1	15	21	6	3 1/2	
10 x 8	9841-628H		1	15	18	6	3 5/8	
12 x 6	9841-666H		1	20	24	6 1/2	3 1/2	
12 x 8	9841-668H		1	20	20	6 1/2	3 5/8	
12 x 10	9841-670H		1	20	23	6 1/2	6	

Large diameter CPVC pressure fittings. (*Courtesy George Fischer Engineering Handbook*)

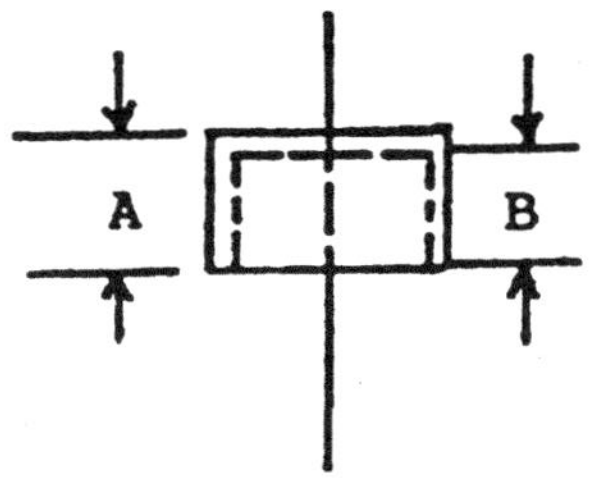

Caps (S)

Inch Size	Part Number		mp	lbs. each	a	b		
10	9847-100H		1	20	6 3/4	6		
12	9847-120H		1	25	8	7		

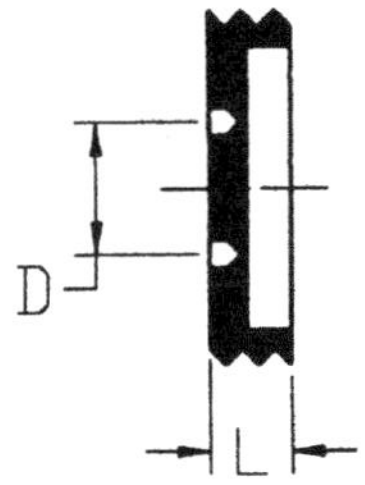

Plug (MT) (1″ thick)

Inch Size	Part Number		mp	lbs. each	d	l		
6	9850-060H		1	10	4	1		
8	9850-080H		1	15	4	1		
10	9850-100H		1	20	4	1		
12	9850-120H		1	25	4	1		

Large diameter CPVC pressure fittings. (*Courtesy George Fischer Engineering Handbook*)

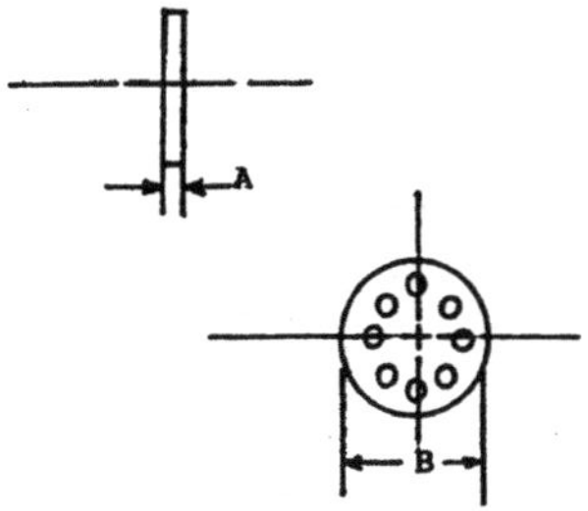

Blind Flange (1" thick)

Inch Size	Part Number		mp	lbs. each	a	b		
10	9853-100H		1	20	1	16		
12	9853-120H		1	25	1	19		

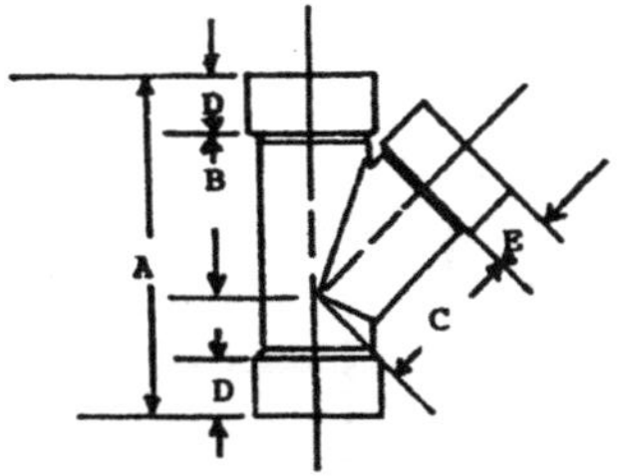

Tee 45° (Wye) (S x S x S)

Inch Size	Part Number		mp	lbs. each	a	b	c	d	e
10	9875-100H		1	30	34	16	16	6	6
12	9875-120H		1	35	39	19	19	6 1/2	6 1/2

Large diameter CPVC pressure fittings. (*Courtesy George Fischer Engineering Handbook*)

Tee 45° (Wye) Reducing (S x S x S)

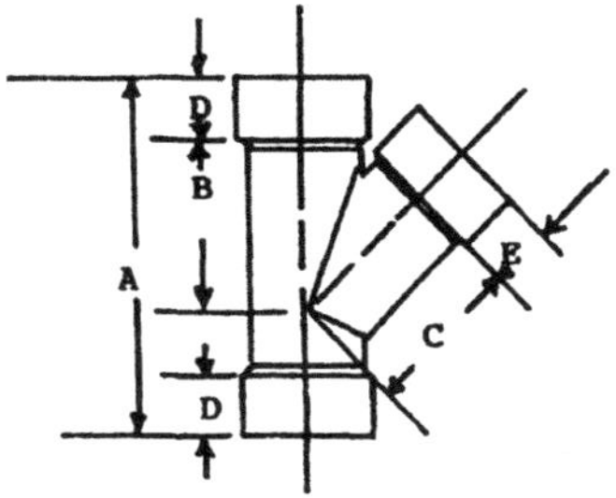

Inch Size	Part Number		mp	lbs. each	a	b	c	d	e
10 x 2	9875-619H		1	30	21 1/2	12	12	6	2
10 x 3	9875-621H		1	30	22	12	13	6	2 1/2
10 x 4	9875-623H		1	30	24	13	13 3/4	6	3 1/4
10 x 6	9875-626H		1	30	26	13 1/2	14	6	4
10 x 8	9875-628H		1	30	29	15 1/4	15 3/4	6	4 3/4
12 x 2	9875-659H		1	35	24	12 1/2	12 1/2	7	2
12 x 3	9875-661H		1	35	24 1/2	13	13	7	2 1/2
12 x 4	9875-663H		1	35	26 1/2	13 1/4	13 1/2	7	3 1/4
12 x 6	9875-666H		1	35	29	14 1/4	16 1/6	7	4
12 x 8	9875-668H		1	35	32	16	16 1/2	7	4 3/4
12 x 10	9875-670H		1	35	38	18	15	7	6

Large diameter CPVC pressure fittings. (*Courtesy George Fischer Engineering Handbook*)

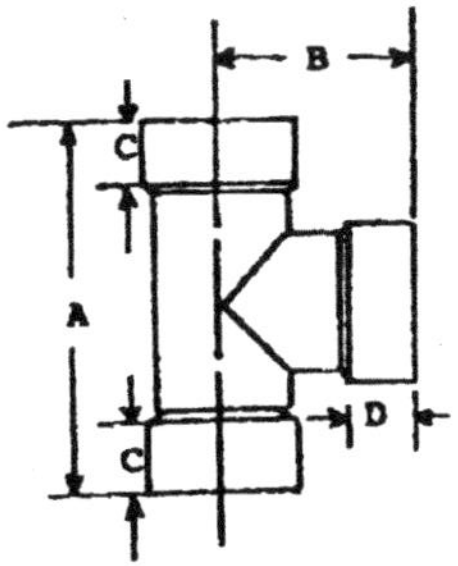

Tee (S x S x S)

Inch Size	Part Number		sp	lbs. each	a	b	c	d
10*	401-100		1	25	22 1/2	11 1/4	5 1/2	5 1/2
12*	401-120		1	30	26 3/4	13 3/8	6 1/2	6 1/2
14	401-140H		1	35	30 1/2	15	6 1/2	6 1/2
16	401-160H		40	32 1/2	15	6 1/2	6 1/2	

*Molded Fittings

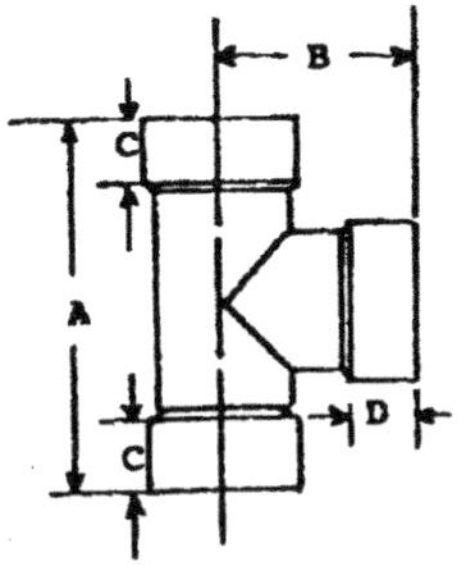

Reducing Tee (S x S x FT)

Inch Size	Part Number		sp	lbs. each	a	b	c	d
10 x 2	402-619H		1	25	16 1/2	7 1/2	5	2
10 x 3	402-621H		1	25	16 1/2	7 1/2	5	2
10 x 4	402-623H		1	25	16 1/2	9	5	2
12 x 2	402-659H		1	30	16 1/2	8 1/2	5 1/2	2
12 x 3	402-661H		1	30	18	8 1/2	5 1/2	2
12 x 4	402-663H		1	30	18 3/4	10	5 1/2	2

Large diameter schedule 40 PVC fittings. *(Courtesy George Fischer Engineering Handbook)*

Reducing Tee (S x S x S)

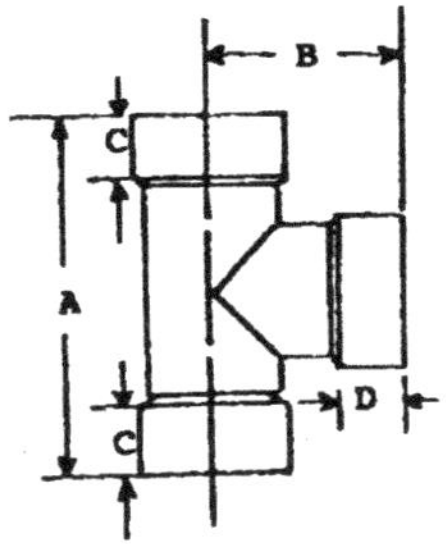

Inch Size	Part Number		sp	lbs. each	a	b	c	d
10 x 2	401-619H		1	25	16 1/2	7 1/2	5	2
10 x 3	401-621H		1	25	16 1/2	7 1/2	5	2 1/2
10 x 4	401-623H		1	25	16 1/2	9	5	2 1/2
10 x 6	401-626H		1	25	18 3/4	11 1/4	5	3 3/4
10 x 8	401-628H		1	25	21	10 1/2	5	3 1/2
12 x 2	401-659H		1	30	16 1/2	8 1/2	5 1/2	2
12 x 3	401-661H		1	30	18	8 1/2	5 1/2	2 1/2
12 x 4	401-663H		1	30	18 3/4	10	5 1/2	2 1/2
12 x 6	401-666H		1	30	19 3/4	11	5 1/2	3
12 x 8	401-668H		1	30	22 1/2	11	5 1/2	3 5/8
12 x 10	401-670H		1	30	25	12 7/8	5 1/2	4 1/2
14 x 2	401-692H		1	35	18	9 1/2	6 1/2	2
14 x 3	401-694H		1	35	20 1/2	9 1/2	6 1/2	2 1/2
14 x 4	401-696H		1	35	20 1/2	10	6 1/2	2 1/2
14 x 6	401-698H		1	35	22	11	6 1/2	3
14 x 8	401-700H		1	35	24 1/2	12	6 1/2	3 5/8
14 x 10	401-702H		1	35	26 1/2	13	6 1/2	4 1/2
14 x 12	401-704H		1	35	28 1/2	14	6 1/2	5 1/2
16 x 2	401-726H		1	40	18 1/2	10	6 1/2	2
16 x 3	401-728H		1	40	20	10	6 1/2	2 1/2
16 x 4	401-730H		1	40	20	10	6 1/2	2 1/2
16 x 6	401-732H		1	40	22	11	6 1/2	3
16 x 8	401-734H		1	40	24 1/2	12	6 1/2	3 5/8
16 x 10	401-736H		1	40	26 1/2	13	6 1/2	4 1/2
16 x 12	401-738H		1	40	28 1/2	14	6 1/2	5 1/2
16 x 14	401-740H		1	40	30 1/2	15	6 1/2	6 1/2

Large diameter schedule 40 PVC fittings. *(Courtesy George Fischer Engineering Handbook)*

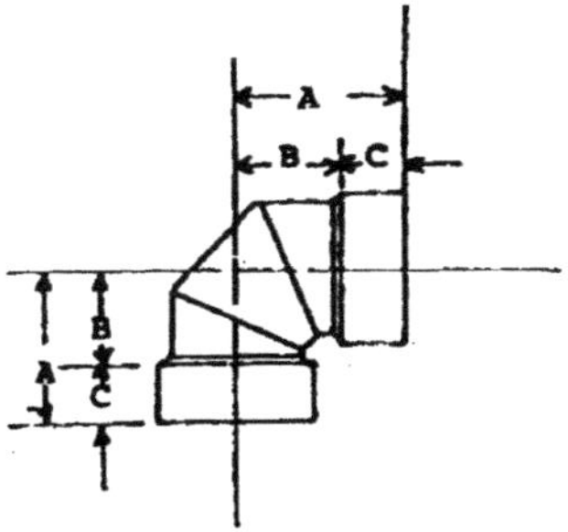

90° Ell (S x S)

Inch Size	Part Number		sp	lbs. each	a	b	c	
10*	406-100		1	25	11 1/4	5 3/4	5 1/2	
12*	406-120		1	30	13 3/8	6 7/8	6 1/2	
14	406-140H		1	35	21 1/4	14 3/4	6 1/2	
16	406-160H		1	40	21 1/2	15	6 1/2	

*Molded Fittings

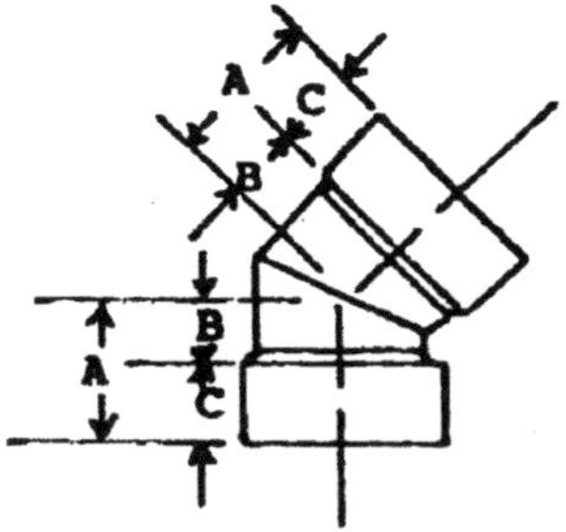

30° Ell (S x S)

Inch Size	Part Number		sp	lbs. each	a	b	c	
10	415-100H		1	15	8	3	5	
12	415-120H		1	20	8 3/4	3 1/4	5 1/2	
14	415-140H		1	25	10 1/4	3 3/4	6 1/2	
16	415-160H		1	30	10 3/4	4 1/4	6 1/2	

Large diameter schedule 40 PVC fittings. *(Courtesy George Fischer Engineering Handbook)*

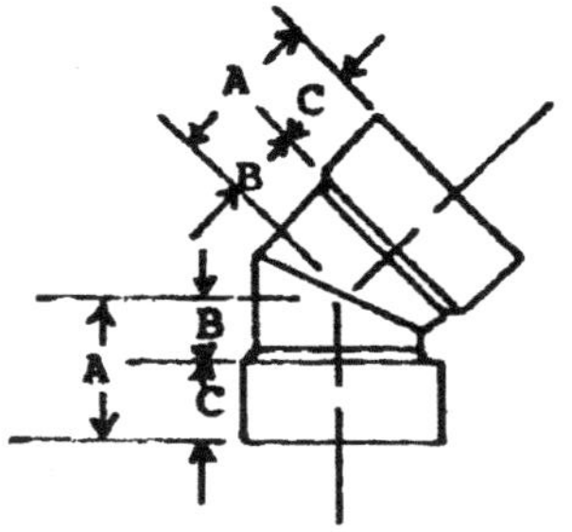

22 1/2° Ell (S x S)

Inch Size	Part Number		sp	lbs. each	a	b	c	
10	416-100H		1	15	8	3	5	
12	416-120H		1	20	9 1/2	4	5 1/2	
14	416-140H		1	25	10	3 1/2	6 1/2	
16	416-160H		1	30	10 1/4	3 3/4	6 1/2	

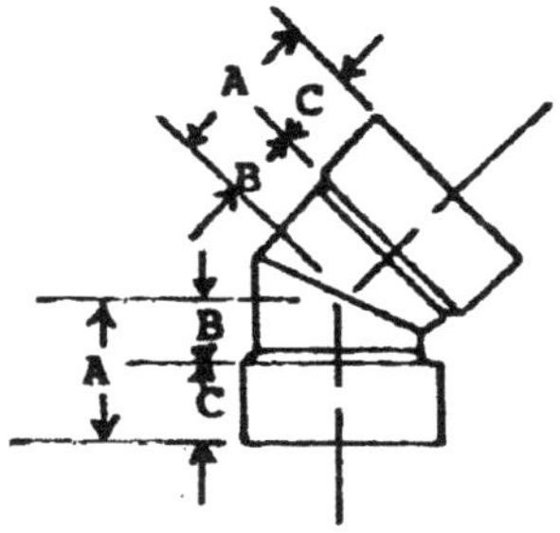

45° Ell (S x S)

Inch Size	Part Number		sp	lbs. each	a	b	c	
10*	417-100		1	15	8 1/8	2 5/8	5 1/2	
12*	417-120		1	20	9 5/8	3 1/8	6 1/2	
14	417-140H		1	25	12	6	5 1/4	
16	417-160H		1	30	14	6	5 1/2	

Large diameter schedule 40 PVC fittings. *(Courtesy George Fischer Engineering Handbook)*

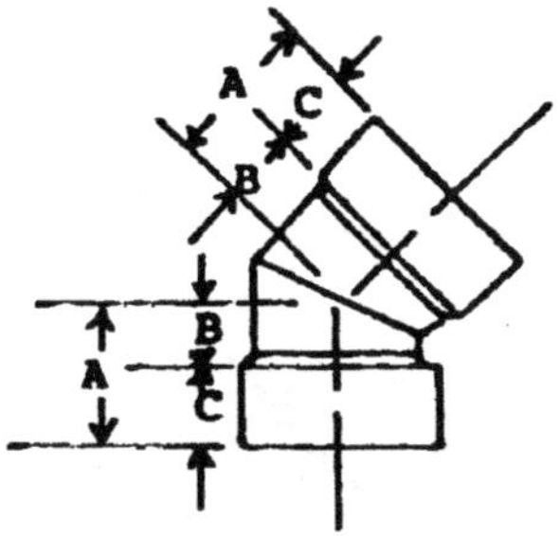

15° Ell (S x S)

Inch Size	Part Number		sp	lbs. each	a	b	c	
10	418-100H		1	15	8	3	5	
12	418-120H		1	20	9 1/2	4	5 1/2	
14	418-140H		1	25	10	3 1/2	6 1/2	
16	418-160H		1	30	10 1/4	3 3/4	6 1/2	

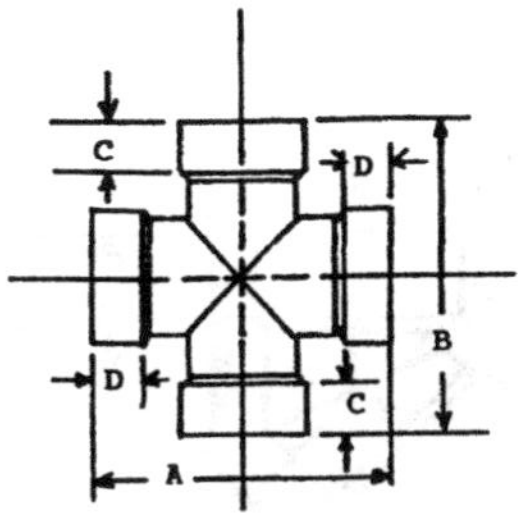

Cross (S x S x S x S)

Inch Size	Part Number		sp	lbs. each	a	b	c	d
10	420-100H		1	25	24	24	4 1/2	4 1/2
12	420-120H		1	30	27 1/2	27 1/2	5	5
14	420-140H		1	35	15	30	6 1/2	6 1/2
16	420-160H		1	40	16	32 1/2	6 1/2	6 1/2

Large diameter schedule 40 PVC fittings. *(Courtesy George Fischer Engineering Handbook)*

Reducing Cross (S x S x S x S)

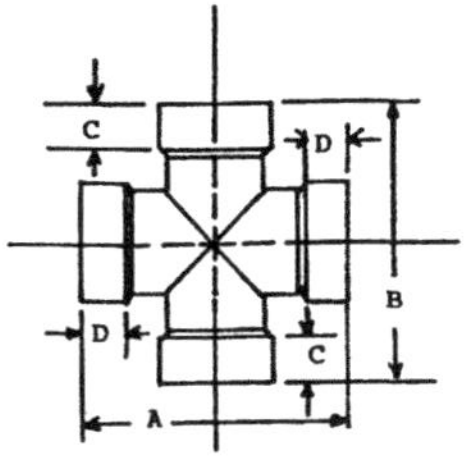

Inch Size	Part Number		sp	lbs. each	a	b	c	d
10 x 2	420-619H		1	25	15	16 1/2	5	2
10 x 3	420-621H		1	25	15	16 1/2	5	2 1/2
10 x 4	420-623H		1	25	18	16 1/2	5	2 1/2
10 x 6	420-626H		1	25	22 1/2	18 3/4	5	3
10 x 8	420-628H		1	25	21	21	5	3 5/8
12 x 2	420-659H		1	30	17	16 1/2	5 1/2	2
12 x 3	420-661H		1	30	17	18	5 1/2	2 1/2
12 x 4	420-663H		1	30	20	18 3/4	5 1/2	2 1/2
12 x 6	420-666H		1	30	21	19 3/4	5 1/2	3
12 x 8	420-668H		1	30	22	22 1/2	5 1/2	3 5/8
12 x 10	420-670H		1	30	27 1/2	25 1/2	5 1/2	4 1/2
14 x 2	420-692H		1	35	19	18	6 1/2	2
14 x 3	420-694H		1	35	10	20 1/2	6 1/2	2 1/2
14 x 4	420-696H		1	35	11	22	6 1/2	3
14 x 6	420-698H		1	35	24	24 1/2	6 1/2	3 5/8
14 x 8	420-700H		1	35	26	26 1/2	6 1/2	4 1/2
14 x 10	420-702H		1	35	28	28 1/2	6 1/2	5
14 x 12	420-704H		1	35	28	28 1/2	6 1/2	5 1/2
16 x 2	420-726H		1	40	20	18 1/2	6 1/2	2
16 x 3	420-728H		1	40	20	20	6 1/2	2 1/2
16 x 4	420-730H		1	40	20	20	6 1/2	2 1/2
16 x 6	420-732H		1	40	22	22	6 1/2	3
16 x 8	420-734H		1	40	24	24 1/2	6 1/2	3 5/8
16 x 10	420-736H		1	40	26	26 1/2	6 1/2	4 1/2
16 x 12	420-738H		1	40	28	28 1/2	6 1/2	5 1/2
16 x 14	420-740H		1	40	30	30 1/2	6 1/2	6 1/2

Large diameter schedule 40 PVC fittings. *(Courtesy George Fischer Engineering Handbook)*

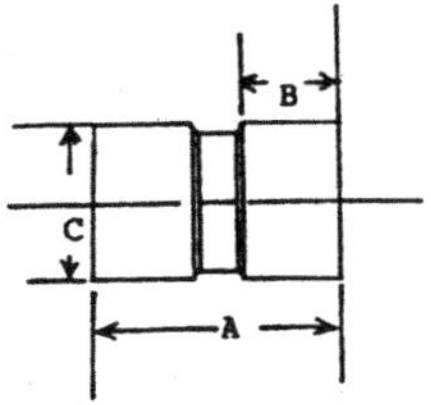

Coupling (S x S)

Inch Size	Part Number		sp	lbs. each	a	b	c	
10	429-100H		1	15	7 1/2	5	11 1/2	
12	429-120H		1	20	7 1/2	5 1/2	13 9/16	
14	429-140H		1	25	7 1/2	6 1/2	14 7/8	
16	429-160H		1	30	7 1/2	6 1/2	17	

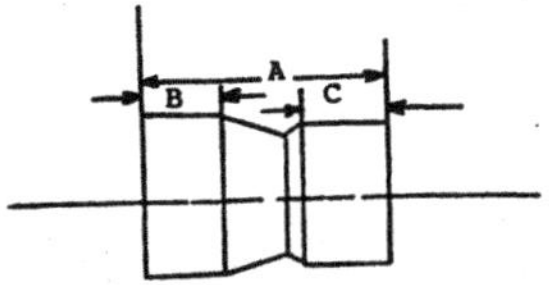

Reducer Coupling (S x S)

Inch Size	Part Number		sp	lbs. each	a	b	c	
10 x 4	429-623H		1	15	16	5	3 1/2	
10 x 6	429-626H		1	15	16	5	3 1/2	
10 x 8	429-628H		1	15	10	5	3 5/8	
12 x 8	429-668H		1	20	17	5 1/2	3 5/8	
12 x 10	429-670H		1	20	12	5 1/2	5	
14 x 10	429-702H		1	25	18	6 1/2	5	
14 x 12	429-704H		1	25	10 1/2	6 1/2	5 1/2	
16 x 12	429-738H		1	30	20 3/4	6 1/2	5 1/2	
16 x 14	429-740H		1	30	13	6 1/2	6 1/2	

Large diameter schedule 40 PVC fittings. *(Courtesy George Fischer Engineering Handbook)*

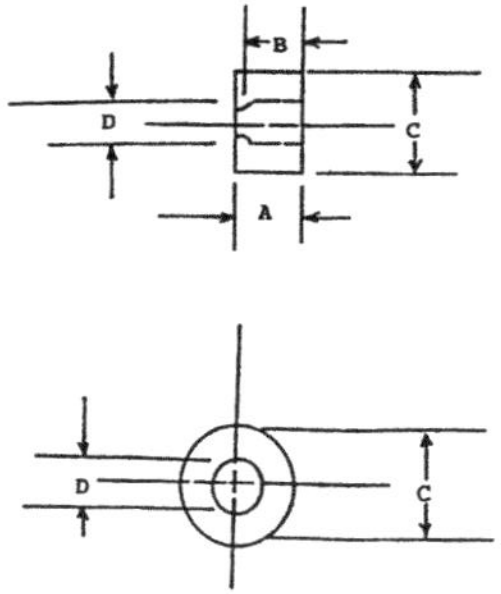

Reducer Bushing (flush) (Spg x S)

Inch Size	Part Number		sp	lbs. each	a	b	c	d
10 x 4	437-623H		1	15	4	2 1/4	10 3/4	4 1/2
10 x 6	437-626H		1	15	4	3	10 3/4	6 5/8
10 x 8	437-628H		1	15	4 3/4	4 1/2	10 3/4	8 1/2
12 x 6	437-666H		1	20	4 1/2	3	12 3/4	6 5/8
12 x 8	437-668H		1	20	4 1/2	4 1/2	12 3/4	8 5/8
12 x 10	437-670H		1	20	5 1/4	5	12 3/4	10 3/4
14 x 8	437-700H		1	25	5	4 1/2	14	8 5/8
14 x 10	437-702H		1	25	5	4	14	10 3/4
14 x 12	437-704H		1	25	6 1/2	5 1/2	14	12 3/4
16 x 10	437-736H		1	30	5 1/2	4	16	10 3/4
16 x 12	437-738H		1	30	5 1/2	5	16	12 3/4
16 x 14	437-740H		1	30	6 1/2	6	16	14

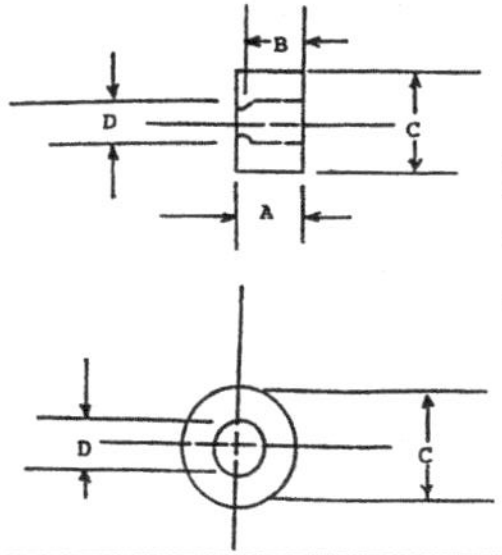

Reducer Bushing (flush) (Spg x FT)

Inch Size	Part Number		sp	lbs. each	a	b	c	d
10 x 6	438-626H		1	15	4	2	10 3/4	6 5/8
10 x 8	438-628H		1	15	4	2	10 3/4	8 1/2
12 x 8	438-668H		1	20	4 1/2	2	12 3/4	8 5/8
12 x 10	438-670H		1	20	4 1/2	2	12 3/4	10 3/4

Large diameter schedule 40 PVC fittings. *(Courtesy George Fischer Engineering Handbook)*

Reducer Bushing Extended Style (Spg x S)

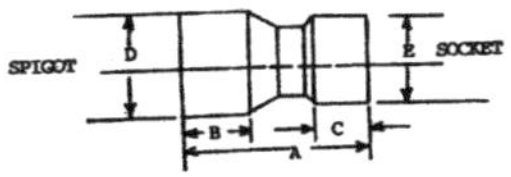

Inch Size	Part Number		sp	lbs. each	a	b	c	d	e
10 x 6	440-626H		1	15	16	5	4	10 3/4	6 5/8
10 x 8	440-628H		1	15	9	5	3 5/8	10 3/4	8 1/2
12 x 6	440-666H		1	20	19 1/2	5 1/2	3	12 3/4	6 5/8
12 x 8	440-668H		1	20	13	5 1/2	3 5/8	12 3/4	8 1/2
12 x 10	440-670H		1	20	10	5 1/2	5	12 3/4	10 3/4
14 x 8	440-700H		1	25	13	6 1/2	3 5/8	14	8 5/8
14 x 10	440-702H		1	25	10	6 1/2	5	14	10 3/4
14 x 12	440-704H		1	25	6 1/2	6 1/2	5 1/2	14	12 3/4
16 x 10	440-736H		1	30	19 1/2	6 1/2	5	16	10 3/4
16 x 12	440-738H		1	30	14 1/2	6 1/2	5 1/2	16	12 3/4
16 x 14	440-740H		1	30	13 1/2	6 1/2	6 1/2	16	14

Eccentric Reducers (S x S)

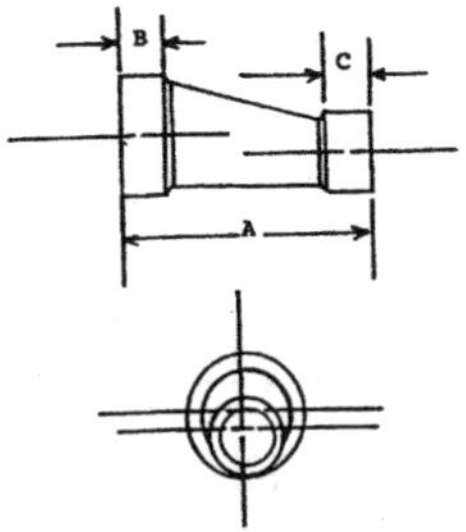

Inch Size	Part Number		sp	lbs. each	a	b	c
10 x 4	441-623H		1	15	21 1/4	5	2 1/2
10 x 6	441-626H		1	15	20	5	3
10 x 8	441-628H		1	15	15 3/4	5	3 5/8
12 x 4	441-663H		1	20	21 1/2	5 1/2	2 1/2
12 x 6	441-666H		1	20	19	5 1/2	3
12 x 8	441-668H		1	20	20	5 1/2	3 5/8
12 x 10	441-670H		1	20	21 1/2	5 1/2	5

Large diameter schedule 40 PVC fittings. *(Courtesy George Fischer Engineering Handbook)*

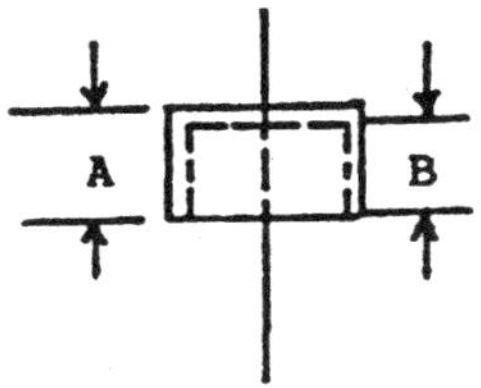

Caps (S)

Inch Size	Part Number		sp	lbs. each	a	b		
10	447-100H		1	15	5	4 1/2		
12	447-120H		1	20	5 1/2	5		
14	447-140H		1	25	7	6 1/2		
16	447-160H		1	30	7	6 1/2		

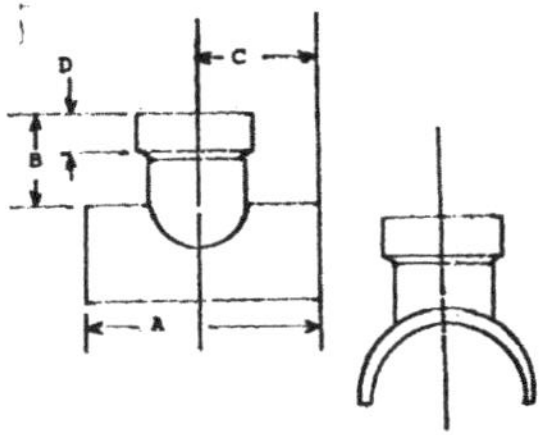

Saddle IPS O.D. (S)

Inch Size	Part Number		sp	lbs. each	a	b	c	d
10 x 4	463-623H		1	15	10 1/2	5 1/2	5 1/4	2 1/2

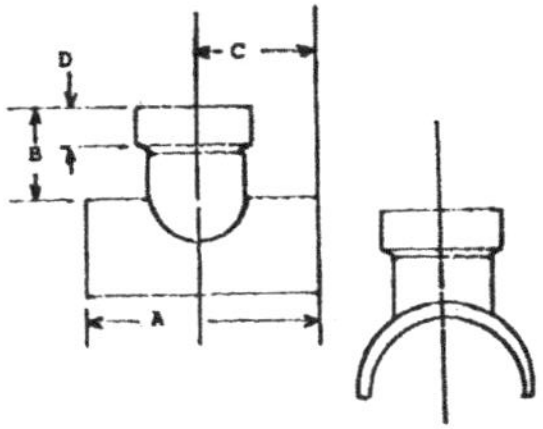

Saddle IPS O.D. (FT)

Inch Size	Part Number		sp	lbs. each	a	b	c	d
10 x 4	464-623H		1	15	10 1/2	5 1/2	5 1/4	2

Large diameter schedule 40 PVC fittings. *(Courtesy George Fischer Engineering Handbook)*

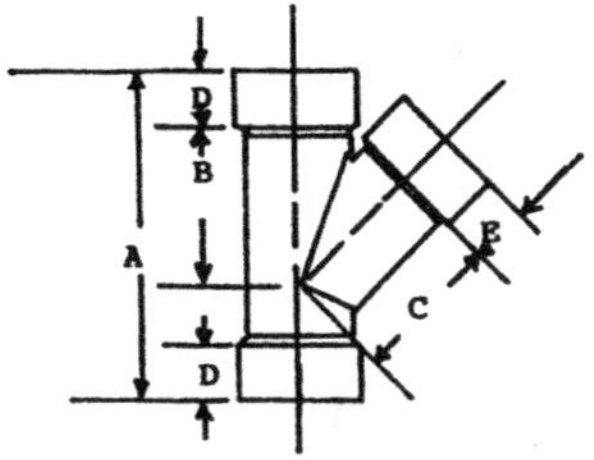

Tee 45° (Wye) (S x S x S)

Inch Size	Part Number		sp	lbs. each	a	b	c	d	e
10	475-100H		1	25	30 1/2	16	16	5	5
12	475-120H		1	30	33 1/2	17 1/2	17 1/2	5 1/2	5 1/2
14	475-140H		1	35	37 1/2	20 3/4	20 3/4	6 1/2	6 1/2
16	475-160H		1	40	39 1/4	23 1/2	22 1/2	6 1/2	6 1/2

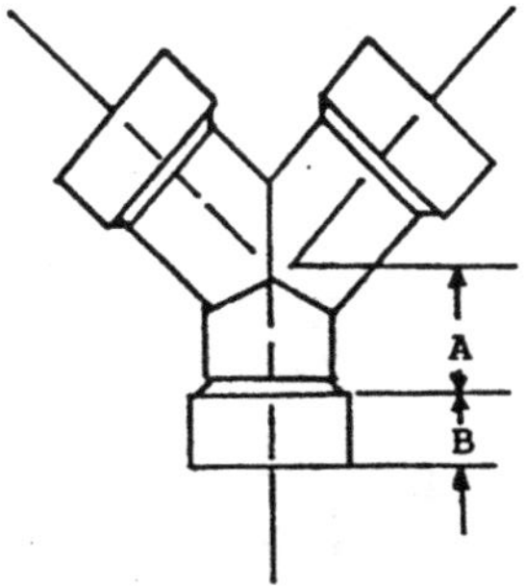

True Wye (S x S x S)

Inch Size	Part Number		sp	lbs. each	a	b		
10	476-100H		1	25	4 1/4	4 1/2		
12	476-120H		1	30	5	5		
14	476-140H		1	35	5	6 1/2		
16	476-160H		1	40	7 1/2	6 1/2		

Large diameter schedule 40 PVC fittings. *(Courtesy George Fischer Engineering Handbook)*

Tee 45° (Wye) Reducing (S x S x S)

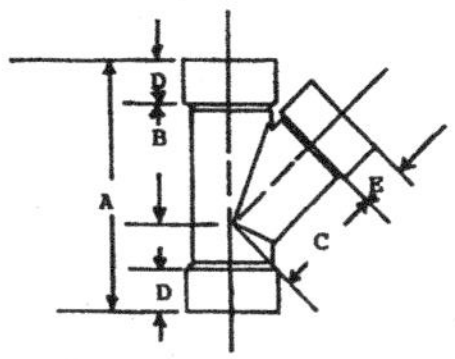

Inch Size	Part Number		sp	lbs. each	a	b	c	d	e
10 x 2	475-619H		1	25	18	8	10	5 1/2	2
10 x 3	475-621H		1	25	18 1/2	8	11	5 1/2	2 1/2
10 x 4	475-623H		1	25	19 1/2	9 1/2	11 1/2	5 1/2	2 1/2
10 x 6	475-626H		1	25	22 1/4	11	12 1/2	5 1/2	3
10 x 8	475-628H		1	25	24 3/4	11 1/2	13 1/2	5 1/2	3 5/8
12 x 2	475-659H		1	30	19	13	11	5 1/2	2
12 x 3	475-661H		1	30	20	13	11 1/2	5 1/2	2 1/2
12 x 4	475-663H		1	30	20 1/2	14	12 3/4	5 1/2	2 1/2
12 x 6	475-666H		1	30	24 1/2	14 1/4	14	5 1/2	3
12 x 8	475-668H		1	30	26 1/2	14 3/4	14 3/4	5 1/2	3 5/8
12 x 10	475-670H		1	30	31	18	16 1/4	5 1/2	5
14 x 2	475-692H		1	35	21 1/2	11	12	6 1/2	2
14 x 3	475-694H		1	35	22	11	13	6 1/2	2 1/2
14 x 4	475-696H		1	35	22 1/2	11 1/2	13 3/4	6 1/2	2 1/2
14 x 6	475-698H		1	35	26	13 1/2	14 3/4	6 1/2	3
14 x 8	475-700H		1	35	28 1/2	14 1/2	15 3/4	6 1/2	3 5/8
14 x 10	475-702H		1	35	32 1/2	16	17 1/4	6 1/2	5
14 x 12	475-704H		1	35	35 1/2	18	18	6 1/2	5 1/2
16 x 2	475-726H		1	40	21	11 1/2	13	6 1/2	2
16 x 3	475-728H		1	40	22	11 1/2	14	6 1/2	2 1/2
16 x 4	475-730H		1	40	22 1/2	12 1/2	14 1/2	6 1/2	2 1/2
16 x 6	475-732H		1	40	26	14 1/2	16 1/4	6 1/2	3
16 x 8	475-734H		1	40	28 1/2	15 1/2	17 1/4	6 1/2	3 5/8
16 x 10	475-736H		1	40	32 1/2	18	18 1/2	6 1/2	4 1/2
16 x 12	475-738H		1	40	35 1/2	18 1/2	19 1/2	6 1/2	5 1/2
16 x 14	475-740H		1	40	37 1/2	19 1/2	21 3/8	6 1/2	6 1/2

Large diameter schedule 40 PVC fittings. *(Courtesy George Fischer Engineering Handbook)*

APPENDIX E
GLOSSARY

Glossary

Abbrasion Resistance: Ability to withstand the effects of repeated wearing, rubbing, scraping, etc.

ABS: Acrylonitrile-butadiene-styrene

Acceptance Test: An investigation performed on an individual lot of a previously qualified product, by, or under the observation of, the purchaser to establish conformity with a purchase agreement.

Acetal Plastics: Plastics based on resins having a predominance of acetal linkages in the main chain.

Acids: One of a class of substances compounded of hydrogen and one or more other elements, capable of uniting with a base to form a salt, and in aqueous solution, turning blue litmus paper red.

Acrylate Resins: A class of thermoplastic resins produced by polymerization of acrylic and acid derivatives.

Acrylonitrile-Butadiene-Sytrene (ABS) Pipe and Fitting Plastics: Plastics containing polymers and/or blends of polymers, in which the minimum butadiene content is 6 percent, the minimum acrylonitrile content is 15 percent, the minimum styrene and/or substituted styrene content is 15 percent, and the maximum content of all other monomers is not more than 5 percent, and lubricants, stabilizers and colorants.

Adhesive: A substance capable of holding materials together by surface attachment.

Adhesive, solvent: An adhesive having a volatile organic liquid as a vehicle.

Aging: The effect of time on plastics exposed indoors at ordinary conditions of temperature and relatively clean air.

Alkalies: Compounds capable of neutralizing acids and usually characterized by an acrid taste. Can be mild like baking soda or highly caustic like lye.

Aliphatic: Derived from or related to fats and other derivatives of the parrafin hydrocarbons, including unsaturated compounds of the ethylene and acetylene series.

Alkyd Resins: A class of resins produced by condensation of a polybasic acid or anhydride and a polyhydric alcohol.

Allyl Resins: A class of resins produced from an ester or other derivative of allyl alcohol by polymerization.

Alternate Product: A product whose use is restricted to incidences where the approved components will not satisfy the needs of the system.

Anneal: To prevent the formation of or remove stresses in plastic parts by controlled cooling from a suitable temperature.

Antioxidant: A compounding ingredient added to a plastic composition to retard possible degradation from contact with oxygen (air), particularly in processing at or exposures to high temperatures.

Approved Product: A product that has been designated for selection, installation and use within an Intel UPW system.

Aromatic: A large class of cyclic organic compounds derived from, or characterized by the presence of the benezene ring and its homologs.

Artificial Weathering: The exposure of plastics to cyclic laboratory conditions involving changes in temperature, relative humidity, and ultraviolet radiant energy, with or without direct water spray, in an attempt to produce changes in the material similar to those observed after long-term continuous outdoor exposure.

ASTM: American Society for Testing Materials

BCF: Bead and Crevice Free, or a welding technique offered within the range of SYGEF® HP products.

Bell End: The enlarged portion of a pipe that resembles the socket portion of a fitting and that is inteded to be used to make a joint by inserting a pieced of

(Courtesy George Fischer Engineering Handbook)

pipe into it. Joining may be accomplished by solvent cements, adhesives, or mechanical techniques.

Beam Loading: The application of a load to a pipe between two pints of support, usually expressed in pounds and the distance between the centers of the supports.

Blister: Undesirable rounded elevation of the surface of a plastic, whose boundaries may be either more or less sharply defined, somewhat resembling in shape or blister on the human skin. A blister may burst and become flattened.

Bond: To attach by means of an adhesive.

Burned: Showing evidence of thermal decomposition through some discoloration, distortion, or destruction of the surface of the plastic.

Burst Strength: The internal pressure required to break a pipe or fitting. This pressure will vary with the rate of build-up of the pressure and the time during which the pressure is held.

Butylene Plastics: Plastics based on resins made by the polymerization of butene or copolymerization of butene with one or more unsaturated compounds, the butene being in in greatest amount by weight.

CA: Cellulose acetate

CAB: Cellulose acetate butyrate

CAP: Cellulose acetate propionate

Calendering: A process by which a heated rubber plastic product is squeezed between heavy rollers into a thin sheet or film. The film may be frictioned into the interstices of cloth, or it may be coated onto cloth or paper.

Cast Resin: A resinous product prepared by pouring liquid resins into a mold and heat-treating the mass to harden it.

Catalysis: The acceleration (or retardation) of the speed of a chemical reaction by the presence of a comparatively small amount of a foreign substance called a catalyst.

Cellulose: Inert substance, chemically a carbohydrate, which is the chief component of the solid structure of plants, wood, cotton, linen, etc.

Cellulose Acetate: A class of resins made from a cellulose base, either cotton linters or purified wood pulp, by the action of acetic anhydride and acetic acid.

Cement: A dispersion of "solution" of unvulcanized rubber or a plastic in a volatile solvent. This meaning is peculiar to the plastics and rubber industries and may or may not be an adhesive composition.

Centipoise: A subunit of the poise, P, the unit of viscosity in the cgs system not preferred in the International units standard. $1P=0.1\ N\ s/m^2 = 0.1 Pa\ s$, (or 0.1 pascal-second) $1cP=0.001\ N\ s/m^2 = 0.001\ Pa\ s$ or m Pa s

Certificate of Analysis: A certification from the manufacturer of a product that it has been tested against a specified standard and meets the requirements.

Certificate of Compliance: A certificate from the manufacturer of a product that it meets the requirements based on previous testing and manufacturing process.

Chemical Resistance: (1) The effect of specific chemicals on the properties of plastic piping with respect to concentration, temperature and time of exposure. (2) The ability of a specific plastic pipe to render service for a useful period in the transport of a specific chemical at a specified concentration and temperature.

Class 1000 (example): The level of cleanliness as defined in Federal Standard 209E equivalent to allowing no more than 1000 particles of a size greater than or equal to 0.5 micron per cubic foot of air. The smaller the class number the cleaner the area. A class 100 area would only be allowed 100 of such particles per cubic foot of air.

(Courtesy George Fischer Engineering Handbook)

Clean Room: A particle controlled area which has filtered air being supplied to maintain a specified level of cleanliness or class.

cm: centimeter, or the basic length unit of the International system

CMC: Carboxymethyl cellulose

CN: Cellulose nitrate

Coalescence: The union or fusing together of fluid globules or particles to form larger drops or a continuous mass.

Cold Flow: Change in dimensions or shape of some materials when subjected to external weight or pressure at room temperature.

Compound: A combination of ingredients before being processed or made into a finished product. Sometimes used as a synonym for material, formulation.

Condensation: A chemical reaction in which two or more molecules combine, usually with the separation of water or some other substance.

Copolymer: The product of simultaneous polymerization of two or more polymerizeable chemicals, commonly known as monomers.

CP: Cellulose propionate

CPVC: Chlorinated polyvinyl chloride

CR: Chloroprene rubber (Neoprene®)

Crazing: Fine cracks at or under the surface of a plastic.

Creep: The unit elongation of a particular dimension under load for a specific time following the initial elastic elongation caused by load application. It is expressed usually in inches per inch per unit of time.

CS: Casein

CSM: Chlorine sulphonyl polyethylene (Hypalon®)

Cure: To change the properties of a polymeric system into a final, more stable, usable condition by the use of heat, radiation, or reaction with chemical additives.

Deflection Temperature: The temperature at which a specimen will deflect a given distance at a given load under prescribed conditions of test.

Degradation: A deleterious change in the chemical structure of a plastic.

Delamination: The separation of the layers of material in a laminate.

Deterioration: A permanent change in the physical properties of a plastic evidenced by impairment of these properties.

Dielectric Constant: Specific inductive capacity. The dielectric constant of a material is the ratio of the capacitance of a condenser having that material as dielectric to the capacity of the same condenser having a vacuum as dielectric.

Dielectric Strength: This is the force required to drive an electric current through a definite thickness of the material; the voltage required to break down a specified thickness of insulation.

Diffusion: The migration or wandering of the particles or molecules of a body of fluid matter away from the main body through a medium or into another medium.

Dimension Ratio: The diameter of a pipe divided by the wall thickness. Each pipe can have two dimension ratios depending on whether the outside or inside diameter is used. In practice, the outside diameter is used if the standards requirement and manufacturing control are based on this diameter. The inside diameter is used when this measurement is the controlling one.

Dimensional Stability: Ability of a plastic part to maintain its original proportions under conditions of use.

Dry-Blend: A free-flowing dry compound prepared without fluxing or addition of solvent.

Durometer: Trade name of the Shore Instrument Company for the instrument that measures hardness. The rubber or

(Courtesy George Fischer Engineering Handbook)

plastics durometer determines the "hardness" of rubber or plastics by measuring the depth of penetration (without puncturing) of blunt needles compressed on the surface for a short period of time.

EC: Ethyl cellulose

ECTFE: Ethylene chlorotrifluoroethylene (Halar®)

Elastic Limit: The load at which a material will no longer return to its original form when the load is released.

Elastomer: The name applied to substances having rubberlike properties.

Electrical Properties: Primarily the resistance of a plastic to the passage of electricity, e.g. dielectric strength.

Elevated Temperature Testing: Tests on plastic pipe above 23°C (73°F).

Elongation: The capacity to take deformation before failure in tension and is expressed as a percentage of the original length.

Emulsion: A dispersion of one liquid in another — possibly only when they are mutually insoluble.

Environmental Stress Cracking: Cracks that develop when the material is subjected to stress in the presence of specific chemicals.

EP: Epoxide, epoxy

EPDM: Ethylene propylene rubber

EPSS: Electropolished stainless steel

Ester: A compound formed by the elimination of waste during the reaction between an alcohol and an acid; many esters are liquids. They are frequently used as plasticizers in rubber and plastic compounds.

Ethyl Cellulose: A thermoplastic material prepared by the ethylation of cellulose by diethyl sulfate halides and alkali.

Ethylene Plastics: Plastics based on resins made by the polymerization of ethylene or copolymerization of ethylene with one or more unsaturated compounds, the ethylene being in greatest amount by weight.

Extrusion: Method of processing plastic in a continuous or extended form by forcing heat-softened plastic through an opening shaped like the cross-section of the finished project.

Extender: A material added to a plastic composition to reduce its cost.

Fabricate: Method of forming a plastic into a finished article by machining, drawing and similar operations.

Failure, adhesive: Rupture of an adhesive bond, such that the place of separation appears to be at the adhesive-adherence interface.

Federal Standard 209E: A US federal standard which specifies clean rooms and clean room production.

FEP: Fluorinated ethylene propylene

Fiber Stress: The unit stress, usually in pounds per square inch (psi), in a piece of material that is subjected to external load.

Filler: A material added to a plastic composition to impart certain qualities in the finished article.

Flexural Strength: The outer fiber stress which must be attained in order to produce a given deformation under a beam load.

Fluorocarbon: See Fluoropolymer

Fluoropolymer: Also known as fluorocarbon. Any of a number of organic compounds analogous to hydrocarbons, in which the hydrogen atoms have been replaced by fluorine. However, in the context of this documentation they are limited to the plastic materials used to create components for piping systems.

Formulation: A combination of ingredients before being processed or made into a finished product. Sometimes used as a synonym for material, compound.

FPM: Fluorine rubber (Viton®)*

(Courtesy George Fischer Engineering Handbook)

ft: foot, or the basic length unit of the English system

Fungi Resistance: The ability of plastic pipe to withstand fungi growth and/or their metabolic products under normal conditions of service or laboratory tests simulating such conditions.

Fuse: To join two plastic parts by softening the material by heat or solvents.

g: gram, or the basic mass (absolute weight) unit of the International system

Generic: Common names for types of plastic materials. They may be either chemical terms or coined names. They contrast with trademarks which are the property of one company.

gf: gram-force, or the conventional gravitational force units used in technology and engineering in the cgs system but not the International system

GPM: U.S. gallons per minute

Hardness: A comparative gauge of resistance to identation, not of surface hardness or abrasion resistance.

HDPE: High-density polyethylene

Heat Joining: Making a pipe joint by heating the edges of the parts to be joined so that they fuse and become essentially one piece with or without the use of additional material.

Heat Resistance: The ability to withstand the effects of exposure to high temperature, Care must be exercised in defining precisely what is meant when this term is used. Descriptions pertaining to heat resistance properties include: boilable, washable, cigaretteproof, sterilizable, etc.

HEPA: High Efficiency Particulate Air, or one method of filtration used to achieve clean room conditions.

Hoop Stress: The circumferential stress imposed on a cylindrical wall by internal pressure loading.

HP^1: (primary meaning) High purity, or the manufacturing and use of a product in its purest state.

HP^2: (secondary meaning) Laminated paper

Hydrostatic Design Stress: The estimated maximum tensile stress in the wall of the pipe in the circumferential orientation due to internal hydrostatic pressure that can be applied continuously with a high degree of certainty that failure of the pipe will not occur.

Hydrostatic Strength (quick): The hoop stress calculated by means of the ISO equation at which the pipe breaks due to an internal pressure build-up, usually within 60 to 90 seconds.

Hz: hertz, or 1 cycle per second

IIR: Isobutene isoprene (butyl) rubber

Impact, Izod: A specific type of impact test made with a pendulum type machine. The specimens are molded or extruded with a machined notch in the center.

Impact Strength: Resistance or mechanical energy absorbed by a plastic part to such shocks as dropping and hard blows.

Impact, Tup: A falling weight (tup) impact test developed specifically for pipe and fittings. There are several variables that can be selected.

Impermeability: Permitting no passage into or through a material.

Impurity: Any unwanted or potentially damaging material or substance which could be present in an ultrapure water component or system.

Injection Molding: A method of repetitively producing similar objects by means of forcing molten plastic into a cavity known as a mold. The inside surface of the component is formed by the core.

(Courtesy George Fischer Engineering Handbook)

IR: Infrared, or a welding technique offered within the range of SYGEF® HP products.

ISO: International Organization for Standardization. The scope of ISO 9001 is to specify the quality system requirements for use where a supplier's capability to design and supply conforming product must be demonstrated. These requirements are aimed at achieving customer satisfaction by preventing nonconformity at all stages from design through to servicing. ISO 9001 Quality Systems are a model for quality assurance in design, development, production, installation and servicing whereas ISO 9002 focuses on production, installation and servicing.

ISO Equation: An equation showing the interrelations between stress, pressure and dimensions of pipe, namely

$$S = \frac{P(ID + t)}{2t} \quad \text{or} \quad \frac{P(OD - t)}{2t}$$

where:

S = stress
P = pressure
ID = average inside diameter
OD = average outside diameter
t = minimum wall thickness

J: The joule of newton-meter is the International unit of energy or work of thermal, electrical, mechanical or chemical origin. A force of 1 newton over a distance of 1 meter produces 1 newton meter or 1 joule of energy.
1J=1Nm

Ketones: Compounds containing the carbonyl group (CO) to which is attached two alkyl groups. Ketones, such as methyl ethyl ketone, are commonly used as solvents for resins and plastics.

kg: kilogram, or the basic mass (absolute weight) unit of the International system

kgf: kilogram-force, or the conventional gravitational force units used in technology and engineering in the mks system but not the International system. The kilogram force is equal to the "kilo-pond" of German technology which uses the symbol "kp".

kp: see kgf

KYNAR® A trade name for the raw polyvinylidene fluoride material produced by Elf Atochem.

lb.: pound, or the basic mass (absolute weight) unit of the English system

lbf: pound-force, or the conventional gravitational force units used in technology and engineering in the English system

Leach Out: The test method of filling a container or component to be tested with ultrapure water, waiting a specified period of time and analyzing the impurity content.

Light Stability: Ability of a plastic to retain its original color and physical properties upon exposure to sun or artificial light.

Light Transmission: The amount of light that a plastic will pass.

Longitudinal Stress: The stress imposed on the long axis of any shape. It can be either a compressive or tensile stress.

Long-Term Hydrostatic Strength: The estimated tensile stress in the wall of the pipe in the circumferential orientation (hoop stress) that when applied continuously will cause failure of the pipe at 100,000 hours (11.43 years). These strengths are usually obtained by extrapolation of log-log regression equations or plots.

LPM: Liters per minute

Lubricant: A substance used to decrease the friction between solid faces and sometimes used to improve processing characteristics of plastic compositions.

m: meter, or the basic length unit of the International system

Melt Flow Index: A measure of the flowing ability of a plastic material at a

(Courtesy George Fischer Engineering Handbook)

specified temperature and pressure.

Melt Point: The temperature at which a material melts.

MF: Melamine formaldehyde

Modulus: The load in pounds per square inch or kilos per square centimeter of initial cross-sectional area necessary to produce a stated percentage-elongation which is used in the physical testing of plastics.

Moisture Resistance: Ability to resist absorption of water.

Molding, Compression: A method of forming objects from plastics by placing the material in a confining mold cavity and applying pressure and usually heat.

Molding, Injection: A method of forming plastics from granular or powdered plastics by the fusing of plastic in a chamber with heat and pressure and then forcing part of the mass into a cooler chamber where it solidifies.

Note: This method is commonly used to form objects from thermoplastics.

Monomer: A relatively simple chemical which can react to form a polymer.

N: Newton, or the basic force unit of the mks metric system used throughout the Standard International System. The newton is the analog of the poundal and the dyne, the more "scientific" force units of the English and cgs systems which produce unit acceleration of unit mass. The force of 1 newton accelerates 1 kilogram at 1 meter/second^2.
$1N = 1kg \times 1\ m/s^2$.

N/m^2 (or Pa) - the pascal (Pa) is the International unit of force/area (stress or pressure) expressed equivalently as the newton per square meter. The choice was to avoid the use of the pascal as neither particulary useful or informative, and stress is expressed as N/cm^2, a valid submultiple of N/m^2,
although it is not a "preferred" form in the International units standard.
$1Pa = 1N/m^2 1N/cm^2 = 10kN/m^2$ (where kN = kilonewtons = 1,000 newtons)

NBR: Nitrile (butadiene) rubber

Nm: see J (or joule)

Non-Flammable: Will not support combustion.

Nonrigid Plastic: A plastic which has a stiffness or apparent modulus of elasticity of not over 10,000 psi at 23°C which determined in accordance with the Standard Method of Test for Stiffness in Flexure of Plastics.

Non-Toxic: Non-poisonous.

NR: Natural rubber

Nylon Plastics: Plastics based on resins composed principally of a long-chain synthetic polymeric amide which has recurring amide groups as an integral part of the main polymer chain.

Olefin Plastics: Plastics based on resins made by the polymerization of olefins or copolymerization of olefins with other unsaturated compounds, the olefins being in greatest amount by weight. Polyethylene, polypropylene and polybutylene are the most common olefin plastics encountered in pipe.

Orange-Peel: Uneven surface somewhat resembling an orange peel.

Organic Chemical: Originally applied to chemicals derived from living organisms, as distinguished from "inorganic" chemicals found in minerals and inanimate substances; modern chemists define organic chemicals more exactly as those which contain the element carbon.

Outdoor Exposure: Plastic pipe placed in service or stored so that it is not protected from the elements of normal weather conditions, i.e., the sun's rays, rain, air and wind. Exposure to industrial and waste gases, chemicals, engine exhausts, etc. are not considered normal "outdoor exposure."

P: see cP

PA: Polyamide

Pa: see N/m^2

(Courtesy George Fischer Engineering Handbook)

PB: Polybutylene

PC: Polycarbonate

PCTFE: Polychlorotrifluoroethylene

PDAP: Polydiallyl phthalate

PEC: Chlorinated polyethylene

Permanence: The property of a plastic which describes its resistance to appreciable changes in characteristics with time and environment.

PETP: Polyethylene terephthalate

PF: Phenol-formaldehyde

PFA: Perfluoroalkoxy resin

PIB: Polyisobutylene

Phenolic Resins: Resins made by reaction of a phenolic compound or tar acid with an aldehyde; more commonly applied to thermosetting resins made from pure phenol.

Plasticity: A property of plastics and resins which allows the material to be deformed continuously and permanently without rupture upon the application of a force that exceeds the yield value of the material.

Plasticizer: A liquid or solid incorporated in natural and synthetic resins and related substances to develop such properties as resiliency, elasticity and flexibility.

Plastics Conduit: Plastic pipe or tubing used as an enclosure for electrical wiring.

Plastics Pipe: A hollow cylinder of a plastic material in which the wall thicknesses are usually small when compared to the diameter and in which the inside and outside walls are essentially concentric.

Platics Tubing: A particular size of plastics pipe in which the outside diameter is essentially the same as that of copper tubing.

PMMA: Polymethyl methacrylate

Polybutylene: A polymer prepared by the polymerization of butene-1 as the sole monomer.

Polybutylene Plastics: Plastics based on polymers made with butene-1 as essentially the sole monomer.

Polyethylenes: A class of resins formed by polymerizing ethylene, a gas obtained from petroleum hydrocarbons.

Polymer: A product resulting from a chemical change involving the successive addition of a large number of relatively small molecules (monomer) to form the polymer, and whose molecular weight is usually a multiple of that of the original substance.

Polymerization: Chemical change resulting in the formation of a new compound whose molecular weight is usually a multiple of that of the original substance.

Polyolefin: A polymer prepared by the polymerization of an olefin(s) as the sole monomer(s).

Polyolefin Plastics: Plastics based on polymers made with an olefin(s) as essentially the sole monomer(s).

Polypropylene: A polymer prepared by the polymerization of propylene as the sole monomer.

Polypropylene Plastics: Plastics based on polymers made with propylene as the sole monomer.

Polystyrene: A plastic based on a resin made by polymerization of styrene as the sole monomer.

Note: Pollystyrene may contain minor proportions of lubricants, stabilizers, fillers, pigments and dyes.

Polyvinyl Chloride: Polymerized vinyl chloride, a synthetic resin, which when plasticized or softened with other chemicals has some rubber-like properties. It is derived from acetylene and

(Courtesy George Fischer Engineering Handbook)

anhydrous hydrochloric acid.

Polyvinyl Chloride Plastics: Plastics made by combining polyvinyl chloride with colorants, fillers, plasticizers, stabilizers, lubricants, other polymers and other compounding ingredients. Not all of these modifiers are used in pipe compounds.

POM: Polyoxymethylene (Kematal®)

Porosity: Presence of numerous visible voids.

Power Factor: The ratio of the power in watts delivered in an alternating current circuit (real power) to the voltampere input (apparent power). The power factor of an insulation indicates the amount of the power input which is consumed as a result of the impressed voltage forcing a small leakage current through the material.

PP: Polypropylene

Pressure: When expressed with reference to pipe the force per unit area exerted by the medium in the pipe.

Pressure Rating: The estimated maximum pressure that the medium in the pipe can exert continuously with a high degree of certainty that failure of the pipe will not occur.

Propylene Plastics: Plastics based on resins made by the polymerization of propylene or copolymerization of propylene with one or more other unsaturated compounds, the propylene being in greatest amount by weight.

PS: Polystyrene

PTFE: Polytetrafluoroethylene (Teflon®)

PUR: Polyurethane

PVAC: Polyvinyl acetate

PVAL: Polyvinyl alcohol

PVB: Polyvinyl butyral

PVC: Polyvinyl chloride

PVCA: Polyvinyl chloride acetate

PVDC: Polyvinylidene chloride

PVDF: Polyvinylidene fluoride, or the fluoropolymer (plastic) material itself used in either extruding, injection molding or machining of SYGEF® HP products.

PVF: Polyvinyl fluoride

PVFM: Polyvinyl formal

PVK: Polyvinyl carbazol

Qualification Test: An investigation, independent of a procurement action, performed on a product to determine whether or not the product conforms to all requirements of the applicable specification.

Note: The examination is usually conducted by the agency responsible for the specification, the purchaser, or by a facility approved by the purchaser, at the request of the supplier seeking inclusion of his product on a qualified products list.

Quick Burst: The internal pressure required to burst a pipe or fitting due to an internal pressure build-up, usually within 60 to 70 seconds.

Resilience: Usually regarded as another name for elasticity. While both terms are fundamentally related, there is a distinction in meaning. Elasticity is a general term used to describe the property of recovering original shape after a deformation. Resilience refers more to the energy of recovery; that is, a body may be elastic but not highly relient.

Resin: An organic substance, generally synthetic, which is used as a base material for the manufacture of some plastics.

Reworked Material: A plastic material that has been reprocessed, after having been previously processed by molding, extrusions, etc., in a fabricator's plant.

Rigid Plastic: A plastic which has a stiffness or apparent modulus of elasticity greater than 100,000 psi at 23°C

(Courtesy George Fischer Engineering Handbook)

when determined in accordance with the Standard Method of Test for Stiffness in Flexure of Plastics.

Rubber: A material that is capable of recovering from large deformations quickly and forcibly.

Sample: A small part or portion that is capable of recovering from large deformations quickly and forcibly.

SAN: Styrene-acrylonitrile

Saran Plastics: Plastics based on resins made by the polymerization of vinylidene chloride or copolymerization of vinylidene chloride with other unsaturated compounds, the vinylidene chloride being in greatest amount of weight.

SB: Styrene-butadiene

Schedule: A pipe size system (outside diameters and wall thicknesses) originated by the iron pipe industry.

Self-Extinguishing: The ability of a plastic to resist burning when the source of heat or flame that ignited it is removed.

SEMATECH: Semiconductor Manufacturing Technology Corporation, or a consortium of member semiconductor companies joined together to exchange technology for the advancement of semiconductor production. SEMATECH is located in Austin, Texas.

Service Factor: A factor which is used to reduce a strength value to obtain an engineering design stress. The factor may vary depending on the service conditions, the hazard, the length of service desired and the properties of the pipe.

Set: To convert an adhesive into a fixed or hardened state by chemical or physical action, such as condensation, polymerization, oxidation, vulcanization, gelatin, hydration or evaporation of volatile constituents.

SI: Silicone

Simulated Weathering: The exposure of plastics to cyclic laboratory conditions of high and low temperatures, high and low relative humidities and ultraviolet radiant energy in an attempt to produce changes in their properties similar to those observed on long-term continuous exposure outdoors. The laboratory exposure conditions are usually intensified beyond those encountered in actual outdoor exposure in an attempt to achieve an accelerated effect.

Simulated Aging: The exposure of plastics to cyclic laboratory conditions of high and low temperatures, and high and low relative humidities in an attempt to produce changes in their properties similar to those observed on long-time continuous exposure to conditions of temperature and relative humidity commonly encountered indoors or to obtain an acceleration of the effects of ordinary indoor exposure. The laboratory exposure conditions are usually intensified beyond those actually encountered in an attempt to achieve an accelerated effect.

Softening Range: The range of temperature in which a plastic changes from a rigid to a soft nature.

Note: Actual values will depend on the method of test. Sometimes referred to as softening point.

SOLEF®: A trade name for the raw polyvinylidene fluoride material produced by SOLVAY.

Solvent: The medium within which a substance is dissolved; most commonly applied to liquids used to bring particular solids into solution, e.g., acetone is a solvent for PVC.

Solvent Cement: In the plastic piping field, a solvent adhesive that contains a solvent that dissolves or softens the surfaces being bonded so that the bonded assembly becomes essentially one piece of the same type of plastic.

Solvent Cementing: Making a pipe joint with a solvent cement.

Specific Gravity: Ratio of the mass of a body to the mass of an equal body of volume of water at 4°C, or some other specified temperature.

(Courtesy George Fischer Engineering Handbook)

Specific Heat: Ratio of the thermal capacity of a substance to that of water at 15°C.

Specimen: An individual piece or portion of a sample used to make a specific test. Specific tests usually require specimens of specific shape and dimensions.

Stabilizer: A chemical substance which is frequently added to plastic compounds to inhibit undesirable changes in the material, such as discoloration due to heat or light.

Standard Dimension Ratio: A selected series of numbers in which the dimension ratios are constants for all sizes of pipe for each standard dimension, ratio and which are the USASI Preferred Number Series 10 modified by +1 or -1. If the outside diameter (OD) is used the modifier is +1. If the inside diameter (ID) is used the modifier is -1.

Standard Thermoplastic Pipe Materials Designation Code: A means for easily identifying a thermoplastic pipe material by means of three elements. The first element is the abbreviation for the chemical type of the plastic in accordance with ASTM D-1600. The second is the type and grade (based on properties in accordance with the ASTM materials specification): in the case of ASTM specifications which have no types and grades or those in the cell structure system, two digit numbers are assigned by the PI that are used in place of the larger numbers. The third is the recommended hydrostatic design stress (RHDS) for water at 23°C (73°F) in pounds per square inch divided by 100 and with decimals dropped, e.g., PVC 1120 indicates that the plastic in polyvinyl chloride, Type I, Grade 1 according to ASTM D-1748 with a RHDS of 2000 psi for water at 73°F. PE 3306 indicates that the plastic is polyethylene. Type III Grade 3 according to ASTM D-1248 with a RHDS of 630 psi for water at 73°F. PP 1208 is polypropylene. Class I-19509 in accordance with ASTM D-2146 with a RHDS of 800 psi for water at 73°F; the designation of PP 12 for polypropylene Class I-19509 will be covered in the ASTM and Product Standards for polypropylene pipe when they are issued.

Stiffness Factor: A physical property of plastic pipe that indicates the degree of flexibility of the pipe when subjected to external loads.

Strain: The ratio of the amount of deformation to the length being deformed caused by the application of a load on a piece of material.

Strength: The mechanical properties of a plastic, such as a load or weight-carrying ability, and ability to withstand sharp blows. Strength properties include tensile, flexural and tear strength, toughness, flexibility, etc.

Stress: When expressed with reference to pipe the force per unit area in the wall of the pipe in the circumferential orientation due to internal hydrostatic pressure.

Stress-Crack: External or internal cracks in a plastic caused by tensile stresses less than that of its short-term mechanical strength.

Note: The development of such cracks is frequently accelerated by the environment to which the plastic is exposed. The stresses which cause cracking may be present internally or externally or may be combinations of these stresses. The appearance of a network of fine cracks is called crazing.

Stress Relaxation: The decrease of stress with respect to time in a piece of plastic that is subject to an external load.

Styrene Plastics: Plastics based on resins made by the polymerization of styrene or copolymerization of styrene with other unsaturated compounds, the styrene being in greatest amount by weight.

Styrene-Rubber (SR) Pipe and Fittings Plastics: Plastics containing at least 50 percent styrene plastics combined with rubbers and other compounding materials, but not more than 15 percent acrylonitrile.

Styrene-Rubber Plastics: Compositions based on rubbers and styrene plastics, the styrene plastics being in greatest amount by weight.

(Courtesy George Fischer Engineering Handbook)

Sustained Pressure Test: A constant internal pressure test for 1000 hours.

SYGEF® HP: A Georg Fischer trade name designating **"SYstem GEorg Fischer - High Purity"**, or a group of components manufactured under strict cleanliness manufacturing conditions and provided to the high purity industry. SYGEF® HP pipe is made by SYMALIT exclusively for Georg Fischer.

Tear Strength: Resistance of a material to tearing (strength).

Tensile Strength: The capacity of a material to resist a force tending to stretch it. Ordinarily the term is used to denote the force required to stretch a material to rupture, and is known variously as "breaking load," "breaking stress," "ultimate tensile strength," and sometimes erroneously as "breaking strain." In plastics testing, it is the load in pounds per square inch or kilos per square centimeter of original cross-sectional area, supported at the moment of rupture by a piece of test sample on being elongated.

Thermal Conductivity: Capacity of a plastic material to conduct heat.

Thermal Expansion: The increase in length of a dimension under the influence of a change in temperature.

Thermoforming: Forming with the aid of heat.

Thermoplastic Materials: Materials which soften when heated to normal processing temperatures without the occurrence of appreciable chemical change, but are quickly hardened by cooling. Unlike the thermosetting materials they can be reheated to soften, and recooled to "set," almost indefinitely; they may be formed and reformed many times by heat and pressure.

Thermoset: A plastic which, when cured by application of heat or chemical means, changes into a substantially infusible and insoluble product.

Thermosetting: Plastic materials which undergo a chemical change and harden permanently when heated in processing. Further heating will not soften these materials.

Translucent: Permitting the passage of light, but diffusing it so that objects beyond cannot be clearly distinguished.

UF: Urea-formaldehyde

UP: Unsaturated polyester

UPWR: Ultrapure water supply

UPWS: Ultrapure water return

Vinyl Chloride Plastics: Plastics based on resins made by the polymerization of vinyl chloride or copolymerization of vinyl chloride with minor amounts (not over 50 percent) of other unsaturated compounds.

Vinyl Plastics: Plastics based on resins made from vinyl monomers, except those specifically covered by other classifications, such as acrylic and styrene plastics. Typical vinyl plastics are polyvinyl chloride, polyvinyl acetate, polyvinyl alcohol, and polyvinyl butyral, and copolymers of vinyl monomers with unsaturated compounds.

Virgin Material: A plastic material in the form of pellets, granules, powder, floc or liquid that has not been subjected to use or processing other than that required for its original manufacture.

Viscosity: Internal friction of a liquid because of its resistance to shear, agitation or flow.

Volatile: Property of liquids to pass away by evaporation.

Volume Resistivity: The electrical resistance of a 1-centimeter cube of the material expressed in ohm-centimeters.

W: the watt, or 1 joule per second. This is the International unit of power. 1W=1J/s=1Nm/s

Water Absorption: The percentage by weight of water absorbed by a sample immersed in water. Dependent upon area exposed.

(Courtesy George Fischer Engineering Handbook)

Water Vapor Transmission: The penetration of a plastic by moisture in the air.

Weather Resistance: Ability of a plastic to retain its original physical properties and appearance upon prolonged exposure to outdoor weather.

Weld-or-Knitline: A mark on a molded plastic formed by the union of two or more streams of plastic flowing together.

Welding: The joining of two or more pieces at adjoining or nearby areas either with or without the addition of plastic from another source.

Yield Point: The point at which a material will continue to elongate at no substantial increase in load during a short test period.

Yield Stress: The force which must be applied to a plastic to initiate flow.

(Courtesy George Fischer Engineering Handbook)

INDEX

www.ingramcontent.com/pod-product-compliance
Lightning Source LLC
Chambersburg PA
CBHW060419220326
41598CB00021BA/2225
9780071634007